HANDBOOK OF DEFEASIBLE REASONING AND UNCERTAINTY MANAGEMENT SYSTEMS

VOLUME 7

HANDBOOK OF DEFEASIBLE REASONING AND UNCERTAINTY MANAGEMENT SYSTEMS

EDITORS:

DOV M. GABBAY
King's College, London, U.K.

PHILIPPE SMETS
IRIDIA - Université Libre de Bruxelles, Belgium

Volume 1: Quantified Representation of Uncertainty and Imprecision
Edited by P. Smets
Volume 2: Reasoning with Actual and Potential Contradictions
Edited by P. Besnard and A. Hunter
Volume 3: Belief Change
Edited by D. Dubois and H. Prade
Volume 4: Abductive Reasoning and Learning
Edited by Dov M. Gabbay and Rudolf Kruse
Volume 5: Algorithms for Uncertainty and Defeasible Reasoning
Edited by Jürg Kohlas and Serafín Moral
Volume 6: Dynamics and Management of Reasoning Processes
Edited by J.-J. Ch. Meyer and J. Treur
Volume 7: Agent-based Defeasible Control in Dynamic Environments
Edited by J.-J. Ch. Meyer and J. Treur

HANDBOOK OF DEFEASIBLE REASONING AND UNCERTAINTY MANAGEMENT SYSTEMS

VOLUME 7

AGENT-BASED DEFEASIBLE CONTROL IN DYNAMIC ENVIRONMENTS

Volume Editors:

J.-J. CH. MEYER
Utrecht University, The Netherlands

and

J. TREUR
Free University, Amsterdam, The Netherlands

KLUWER ACADEMIC PUBLISHERS
DORDRECHT / BOSTON / LONDON

A C.I.P. Catalogue record for this book is available from the Library of Congress.

ISBN 978-90-481-6109-6

Published by Kluwer Academic Publishers,
P.O. Box 17, 3300 AA Dordrecht, The Netherlands.

Sold and distributed in North, Central and South America
by Kluwer Academic Publishers,
101 Philip Drive, Norwell, MA 02061, U.S.A.

In all other countries, sold and distributed
by Kluwer Academic Publishers,
P.O. Box 322, 3300 AH Dordrecht, The Netherlands.

Printed on acid-free paper

CONTENTS

PREFACE

This volume, the 7th volume in the DRUMS Handbook series, is part of the aftermath of the successful ESPRIT project DRUMS (Defeasible Reasoning and Uncertainty Management Systems) which took place in two stages from 1989–1996. In the second stage (1993 - 1996) a work package was introduced devoted to the topics Reasoning and Dynamics, covering both the topics of "Dynamics of Reasoning", where reasoning is viewed as a process, and "Reasoning about Dynamics", which must be understood as pertaining to how both designers of and agents within dynamic systems may reason about these systems.

The present volume presents work done in this context extended with some work done by outstanding researchers outside the project on related issues. While the previous volume in this series had its focus on the dynamics of reasoning processes, the present volume is more focused on "reasoning about dynamics', viz. how (human and artificial) agents reason about (systems in) dynamic environments in order to control them. In particular we consider modelling frameworks and generic agent models for modelling these dynamic systems and formal approaches to these systems such as logics for agents and formal means to reason about agent-based and compositional systems, and action & change more in general.

We take this opportunity to mention that we have very pleasant recollections of the project, with its lively workshops and other meetings, with the many sites and researchers involved, both within and outside our own work package.

We thank everyone involved, in particular the authors of the papers of this volume, and the series editors Dov Gabbay and Philippe Smets for their encouragement and patience. Philippe has also been the overall project leader with a good taste for the quality of both science and life (including food). The meetings he organised were always accompanied by excellent wining and dining, thus helping the project members to get into the right DRUMS spirit. Last, but by no means least, we thank Jane Spurr for the splendid job she did (again), helping us to get this volume together, and in particular for the meticulous translation of Word files into Latex, ably helped by Anna Maros.

John-Jules Meyer
Jan Treur

PART I

INTRODUCTION AND BASIC CONCEPTS

JOHN-JULES MEYER AND JAN TREUR

INTRODUCTION

PART I — GENERAL

One of the recognized problems in AI is the gap between applications and formal foundations. This book (as the previous one in the DRUMS Handbook series) does not present the final solution to this problem, but at least does an attempt to reduce the gap by bringing together state-of-the-art material from both sides and to clarify their mutual relation. In this book the main theme is agents and dynamics: dynamics of reasoning processes (as we have seen in the previous volume in this series), but also dynamics of the external world. Agents often reason about both types of dynamics.

Agents are (hardware or software) entities that act on the basis of a "mental state". They possess both informational and motivational attitudes, which means that while performing their actions they are guided by their knowledge and beliefs as well as their desires, intentions and goals (often referred to as 'BDI notions'), and, moreover, they are able to modify their knowledge, intentions, etc. in the process of acting as well. Clearly the description of agent behaviour involves reasoning about the dynamics of acting, and if agents are supposed to be reflective, they should also themselves be able to do so. Furthermore, since the actions of agents may — apart from actions that change the external world directly - also include reasoning (for example, performing some belief-revising action or an action comprising of reasoning by default), it may be clear that in the context of agent systems the dynamics of reasoning (as a special kind of mental action) and reasoning about dynamics go hand in hand.

PART II — MODELLING

The different phenomena to be modelled in real-world applications show a very wide variety. To model them, models covering quite different aspects are needed. One approach is to build a large collection of models for each phenomenon separately. Another approach is to define one "grand universal model" that covers as many of the phenomena as possible. Both solutions have serious drawbacks. The first solution would end up in a large and ad hoc collection of non-related models. The second solution would impose very high requirements on the universal model to be developed; any proposed model would have strong limitations. Besides, a very complex model would result from which in a given application only a small part is relevant.

The solution that has been chosen in practice, in a sense combines the two options pointed out. Indeed generic models for different phenomena

J.J.Ch. Meyer and J. Treur (eds.),
Handbook of Defeasible Reasoning and Uncertainty Management Systems, Vol. 7, 3–8.

have been developed, however, not in an *ad hoc*, non-related manner, but in a form that is based on one unifying modelling framework. In such a unifying modelling framework, different models can be compared and combined if needed. In this part two of such proposed modelling frameworks are presented. Furthermore, also a generic model for reasoning tasks and agents is presented.

MODELLING FRAMEWORKS AND GENERIC AGENT MODELS

First in this part two modelling frameworks are presented, which both employ modularisation and control techniques. The first one is the underlying modelling framework of the compositional development method for multi-agent systems DESIRE, based on the notions of component and compositionality. The second one is based on the architecture MILORD II for knowledge-based and multi-agent systems, which combines modularisation techniques with both implicit and explicit control mechanisms, and with approximate reasoning facilities based on multi-valued logics. Both modelling frameworks support reflective reasoning.

The third paper shows how the language Concurrent METATEM, which is based on an executable fragment of temporal logic, can be used as a coordination language, that is, as a high-level mechanism to control the coordination of processes that themselves are written in other ('subject') languages (and may deal with the internals of agents, like BDI-like notions).

The fourth paper shows a generic agent model based on the weak notion of agency: an autonomous agent able to observe, communicate and act in the world, and doing so, can behave in a reactive, pro-active or social manner. An agent based on this model is able to reason about dynamics of the world, but also about the dynamics of its own reasoning and communication. This generic agent model has been applied in a large number of applications, such as information brokering, negotiation for load balancing of electricity use, and distributed Call Center support.

PART III — FORMAL ANALYSIS

In this part of the book we will consider various formal means regarding the topics agents and dynamics. By formal means we mean formal logics / calculi as well as formal specification languages together with their formal semantics. We will see how in particular temporal and dynamic logic / semantics are useful means to perform this formal analysis.

Formal foundations of models and modelling techniques are of importance for different reasons. First, by defining semantics of a model or modelling technique in a formal manner, a precise and unambiguous meaning of the syntactical constructs is obtained, which may help designers. This requires that designers are familiar with the formal techniques used to define such formal semantics. Unfortunately, this requirement is often not fulfilled for developers in practice, and there is no reason to expect that this will change in the short term. However, more realistically, those who develop a modelling technique often have more knowledge of formal methods. Therefore they can benefit a lot from knowledge of formal foundations during development of their modelling technique, and use that also as a basis to informally or semi-formally describe the semantics for others (users of the modelling technique) with a less formal background.

Secondly, formal foundations are especially important to obtain the possibility of verification of a design or verification of requirements. Verification is usually a rather technical and tedious matter, only feasible for specialists ("verification engineers"). They need to know about the formal foundations, including formal semantics and proof systems.

The problem of how to cover a large variety of phenomena, for example, as discussed for modelling, also occurs in the formal analysis of such phenomena or models. Many contributions in the literature address different phenomena in an ad hoc and unrelated manner. For example, different theories have been proposed for diagnosis from first principles, nonmonotonic reasoning, BDI-agents, and many other types of agents. Relations between these foundational approaches are almost non-existent, and one 'grand unifying theory' is not within sight at all. Moreover, if such a theory would be constructed, many believe that it would have such a high complexity, that it is hard to use, whereas in a given application only a small part would be relevant.

As far as the relation to applications is concerned, within these foundational approaches especially the dynamic aspects of the agent's internal (reasoning) processes are often not, or at least not fully covered. For example, theories of diagnosis from first principles address diagnosis from a static perspective, but do not cover process-oriented questions such as, on which hypothesis should the process be focused, and which observations should be done at which moment. Similarly, most contributions in the area of logics for nonmonotonic reasoning only offer a formal description of the possible conclusions of such a reasoning process, and not of the dynamics of the reasoning process itself. Moreover, although some degree of dynamics is captured by them, BDI-logics lack a clear relationship to agent architectures in applications of BDI-agents. In such applications, for example, the rather complex issue of revision of beliefs, desires and/or intentions has to be addressed; e.g., when to revise only an intention and not the underlying desire, when to revise both, when to revise a desire and still keep a (committed) intention which was based on the desire?

At this point something can be learned from the way in which the problem of unification of a variety of different models was solved for modelling: the different models are specified within one unifying modelling framework.

For the question to obtain formal semantics, this implies two different types of semantics. The first type is the semantics as used for the modelling framework as a whole. The second type is the semantics used for a specific model. So specific models can share the semantics they inherit from a modelling technique, but differ (and be incomparable) for their more specific semantics. Semantics of a modelling framework abstract from the inherent meaning of concepts specific for a particular model at hand, but still can address how all concepts in a model behave dynamically, given the specification of the model. Inherent meaning of the concepts specific for a given model, can be found, for example, for the BDI-agent model (or for non-monotonic reasoning, as we have seen in the previous DRUMS volume). In this part both types of semantics are addressed.

First, in Part IIIA semantics for modelling frameworks are presented, and their use in verification. Next, in Part IIIB semantics of an agent model is presented.. In Part IIIC a number of approaches are collected for reasoning about the dynamics of the world.

PART IIIA — FORMAL ANALYSIS: MODELLING FRAMEWORKS

We now first look at the analysis of systems, and in particular that of agent systems and compositional systems, from a more general perspective. Of course, the aspects of dynamics and reasoning are omnipresent here, the focus covers both on reasoning about dynamics and the dynamics of reasoning. Most papers in this subpart contain both. In this part semantics for modelling frameworks are presented, and their use in verification. In the first paper it is shown how semantics for multi-agent systems can be defined based on principles of locality and compatibility. For each process a local set of behaviours is defined, based on temporal models. To obtain behaviour of the whole system, local behaviours are selected based on mutual compatibility due to interactions. In the second paper the use of Descriptive Dynamic Logic to describe semantics of a modelling framework for multi-language reflective architectures for reasoning tasks is shown. In the third paper it is shown how temporal epistemic logic can be used to formalize behavioural properties of models specified in a compositional modelling framework, and how verification proofs for properties of such a model can be specified.

PART IIIB — FORMAL ANALYSIS: LOGICS FOR AGENTS

Several papers are devoted to the logical description of intelligent agents, in particular focusing on the various attitudes that are commonly ascribed to them [2]. such as handling knowledge from various sources (observation, communication), pursuing goals and making commitments. The basis of the approach in the papers below is dynamic logic. This logic is also employed to describe / specify agents that are characterized in a multi-level, multi-language logical architecture, providing (strong) semantics of an agent model. In the first paper the basic framework is presented, which is based on modal logic, and a blend of dynamic and epistemic logic in particular. This logical framework, called KARO, enables one to analyse the Knowledge, Abilities, Results, Opportunities of agents, and is intended as a basis for reasoning about the knowledge and the results of actions of agents. In the second paper this basic framework is extended so that one can adequately analyse the agent's informational attitudes such as coping with belief revision on the basis of observations, communication and some simple form of default reasoning. The third paper extends the KARO framework to capture the agent's motivational attitudes so as to get to a full-scale BDI-like logic but now grounded in dynamic logic (and thus the performance of action!). The fourth paper, finally, extends this even further by adding a deontic layer (expressing obligations and permissions) in order to incorporate social attitudes in multi-agent systems, and speech acts in particular, resulting in a very expressive logical framework.

PART IIIC — FORMAL ANALYSIS: REASONING ABOUT DYNAMICS

In this part we will concentrate on the formal analysis of reasoning about dynamics, viz. the formal analysis of (reasoning about) actions that take place in a system (possibly initiated by an agent) and the changes that these cause to happen in the world. Here we touch upon a well-known area in knowledge representation and commonsense reasoning with problems like the infamous *frame problem* [1]. The problem here is to express the effects (and, perhaps more importantly, the *non*-effects) due to the performance of actions (of agents) in an "economic" way. In general, when describing the effects of actions it is assumed that *by default* things (or perhaps more precisely, fluents) do not change (their truth value) by the execution of an action, unless stated so explicitly. In this section we see approaches that employ Dijkstra's weakest precondition/strongest postcondition formalism and the dynamic logic formalism to describe the effect of actions, combined with some way of expressing that fluents are subject to some form of "epistemic inertia": unless it is *known* to be otherwise (i.e. "normally" or "by default') things remain the same. For the latter we see the use of default logic and also a mechanism for specifying in the language of actions what things are known to be changed and which things are likely to stay the same. Furthermore a detailed analysis is made of the different dynamic properties that are required to guarantee that an agent's actions in a dynamic world take place in a co-ordinated manner. Finally an application concerning reasoning about dynamics is discussed, viz. one for agents negotiating for balancing the load of (sometimes strongly fluctuating) electricity use over time.

J.-J. Ch. Meyer
Utrecht University, The Netherlands.

J. Treur
Vrije Universiteit, Amsterdam, The Netherlands.

BIBLIOGRAPHY

[1] E. Sandewall and Y. Shoham, Nonmonotonic Temporal Reasoning, in: *Handbook of Logic in Artificial Intelligence and Logic Programming* Vol.4 (Epistemic and Temporal Reasoning) (D.M. Gabbay, C.J. Hogger & J.A. Robinson, eds.), Oxford University Press, Oxford, 1994.

[2] M.J. Wooldridge & N.R. Jennings (eds.), *Intelligent Agents*, Springer, Berlin, 1995.

J.-J. CH. MEYER AND J. TREUR

BASIC CONCEPTS

In this chapter we give a short treatment of two logical formalisms that are used in this book. The first is *dynamic logic*, a logic to reason about *actions*. The second is a formalism to reason about *time*, and is in fact a framework to reason about the dynamics of a system by considering the evolution of states of that system.

1 DYNAMIC LOGIC

1.1 *Introduction*

A basic formalism that plays an important role in this book is that of *dynamic logic*. Dynamic logic is a modal logic especially designed to reason about actions. Historically it dates back to work by Vaughan Pratt [1976], Bob Moore [1985], and David Harel [1979; 1984], and it has been used for reasoning about programs, thus providing a formalism for program verification and specification [Kozen and Tiuryn, 1990; Cousot, 1990].

In this section we will treat the basic idea behind an elementary form of (propositional) dynamic logic,[1] which should be a sufficient introduction for the reader to recognize and understand the more elaborate forms of dynamic logic as they appear in later chapters.

1.2 *Language*

For the purpose of this basic treatment we introduce the following logical language $\mathcal{L}_{DL}$. Assume a set $\mathcal{P}$ of propositional atoms, and a set $\mathcal{A}$ of atomic actions.

DEFINITION 1. The logical language $\mathcal{L}_{DL}$ and action language $\mathcal{L}_{ACT}$ are given as the least sets closed under the clauses:

- $\mathcal{P} \subseteq \mathcal{L}_{DL}$
- $\mathcal{A} \subseteq \mathcal{L}_{ACT}$
- $\varphi, \psi \in \mathcal{L}_{DL}$ implies $\neg\varphi, \varphi \wedge \psi, \varphi \vee \psi, \varphi \rightarrow \psi, \varphi \leftrightarrow \psi \in \mathcal{L}_{DL}$
- $\varphi \in \mathcal{L}_{DL}, \alpha \in \mathcal{L}_{ACT}$ implies $[\alpha]\varphi, \langle\alpha\rangle\varphi \in \mathcal{L}_{DL}$
- $\varphi \in \mathcal{L}_{DL}$ implies $\varphi? \in \mathcal{L}_{ACT}$
- $\alpha, \beta \in \mathcal{L}_{ACT}$ implies $\alpha; \beta, \alpha + \beta, \alpha^* \in \mathcal{L}_{ACT}$

[1] Propositional Dynamic Logic or PDL is due to Fischer & Ladner [1979]

J.J.Ch. Meyer and J. Treur (eds.),
Handbook of Defeasible Reasoning and Uncertainty Management Systems, Vol. 7, 9–16.

Here we see that the logical language $\mathcal{L}_{DL}$ is an extension of propositional logic with modal operators of the form $[\alpha]$ (and duals $\langle\alpha\rangle$) where α is an element from the action language $\mathcal{L}_{ACT}$. This action language is here taken to be a very basic programming language, viz. that of the regular expressions over 'action alphabet' A. The ';' operator denotes sequential composition ('followed by'), '+' means (nondeterministic) choice, and '*' denotes arbitrary finite repetition. (This choice of operators is 'classic' for dynamic logic (cf. [Harel, 1984; Kozen and Tiuryn, 1990; Stirling, 1992]); in the chapters of this book also different operators corresponding with the perhaps more familiar (deterministic) if and while programming constructs are employed, but the basic ideas remain the same.) Finally the *action* expression $\varphi?$ stands for a test whether the *logical* expression φ holds in the current state. This is typically used in an expression involving a choice operator to guide control of this choice, as in e.g. the expression $p?; a + \neg p?; b$, where $p \in \mathcal{P}$ and $a, b \in \mathcal{A}$.

1.3 Models

The language $\mathcal{L}_{DL}$ is given an interpretation on the basis of Kripke models, as is usual for a modal language. We will use tt and ff for the truth values.

Formally a Kripke model for language $\mathcal{L}_{DL}$ is a structure of the following form:

DEFINITION 2. A Kripke model for $\mathcal{L}_{DL}$ is a structure $\mathcal{M}$ of the form $\langle S, \pi, r\rangle$, where

1. S is a non-empty set (the set of *states*);
2. $\pi : S \to (\mathcal{P} \to \{tt, ff\})$ is a truth assignment function to the atoms per state;
3. $r : \mathcal{L}_{ACT} \to 2^{S\times S}$ are state transition relations per action, satisfying the following properties:
   - $r(\varphi?) = \{< s, s > \mid \mathcal{M}, s \models \varphi\}$, where $\mathcal{M}, s \models \varphi$ is defined below;
   - $r(\alpha; \beta) = r(\alpha) \circ r(\beta)$, where $\circ$ stands for the relational composition;
   - $r(\alpha + \beta) = r(\alpha) \cup r(\beta)$;
   - $r(\alpha^*) = r(\alpha)^\star$, where $\star$ stands for the reflexive, transitive closure operator on relations

Truth of a formula $\varphi \in \mathcal{L}_{DL}$ in a state $s \in S$ in a model $\mathcal{M} = \langle S, \pi, r\rangle$ (written $(\mathcal{M}, s) \models \varphi$), is defined by the following.

DEFINITION 3. Let $\mathcal{M} = \langle S, \pi, r\rangle$ be a given model and $s \in S$. Then:

1. $\mathcal{M}, s \models p$ iff $\pi(s)(p) = tt$, for $p \in \mathcal{P}$;
2. $\mathcal{M}, s \models \neg\varphi$ iff not $\mathcal{M}, s \models \varphi$;

3. $\mathcal{M}, s \models \varphi \wedge \psi$ iff $\mathcal{M}, s \models \varphi$ and $\mathcal{M}, s \models \psi$;
4. $\mathcal{M}, s \models \varphi \vee \psi$ iff $\mathcal{M}, s \models \varphi$ or $\mathcal{M}, s \models \psi$;
5. $\mathcal{M}, s \models \varphi \to \psi$ iff $\mathcal{M}, s \models \varphi$ implies $\mathcal{M}, s \models \psi$;
6. $\mathcal{M}, s \models \varphi \leftrightarrow \psi$ iff $\mathcal{M}, s \models \varphi$ bi-implies $\mathcal{M}, s \models \psi$;
7. $\mathcal{M}, s \models [\alpha]\varphi$ iff $\mathcal{M}, s' \models \varphi$ for all s' with $r(\alpha)(s, s')$;
8. $\mathcal{M}, s \models \langle\alpha\rangle\varphi$ iff $\mathcal{M}, s' \models \varphi$ for some s' with $r(\alpha)(s, s')$;

A formula φ is *valid in a model* $\mathcal{M} = \langle S, \pi, r\rangle$, denoted $\mathcal{M} \models \varphi$, if $\mathcal{M}, s \models \varphi$ for every $s \in S$. A formula φ is *valid* with respect to a set MOD of models, denoted $MOD \models \varphi$, if $\mathcal{M} \models \varphi$ for every model $\mathcal{M} \in MOD$. If MOD is the set of *all* Kripke models of the above form, we generally write $\models \varphi$ instead of $MOD \models \varphi$.

1.4 Logic

The simple propositional dynamic logic introduced above can be finitely axiomatized by the following system

1. any axiomatisation of propositional logic
2. $[\alpha](\varphi \to \psi) \to ([\alpha]\varphi \to [\alpha]\psi)$;
3. $[\varphi?]\psi \leftrightarrow (\varphi \to \psi)$
4. $[\alpha; \beta]\varphi \leftrightarrow [\alpha][\beta]\varphi$
5. $[\alpha + \beta]\varphi \leftrightarrow [\alpha]\varphi \wedge [\beta]\varphi$
6. $[\alpha^*]\varphi \to \varphi$;
7. $[\alpha^*]\varphi \to [\alpha][\alpha^*]\varphi$;
8. $[\alpha^*](\varphi \to [\alpha]\varphi) \to (\varphi \to [\alpha^*]\varphi)$;
9. $[\alpha]\varphi \leftrightarrow \neg\langle\alpha\rangle\neg\varphi$

and rules modus ponens (MP) and

$$\frac{\varphi}{[\alpha]\varphi}$$

We note that taking different choices for the basic operators in the action language, and especially further extensions to these, has a large influence on the axiomatisation. This is the reason that in some of the papers of this volume a quite different axiomatic system is employed. However, as the above system is rather standard in the literature (e.g. [Stirling, 1992]), we have included it here.

2 TEMPORALIZED LOGIC

2.1 Introduction

To formalize dynamics, often a paradigm involving states and evolution of states over time is adopted.

The notion of state of an agent often is interpreted as 'information state': a representation of the (local) information the agent has at a given moment of time. Depending on the logic to be used for the static aspects, information states can be formalised in different manners; for example by sets of formulae, by classical (propositional) two-valued models, by (partial) three-valued models, or by Kripke structures. In particular, if the dynamics of the agent's reasoning process is addressed, a notion of state is required that represents the incomplete information the agent has; also an information ordering between states may be useful, for example, to indicate the agent's growth of information during a reasoning process, or lack thereof. For example, classical two-valued models are less useful then.

Also evolution of states over time can be formalized in different manners. Examples of approaches are transition systems and Petri nets. An often used approach, also in this book, to describe evolution over time is temporal logic. The more standard forms of temporal logic are based on a notion of state represented by two-valued models. As argued above, in particular for the dynamics of reasoning processes, information states are required that represent incomplete information. Therefore temporal logics are required that can be built on top of another logic in which incomplete information is represented in the semantics, e.g., on top of a logic based on three-valued models or Kripke structures. In other words, a construction is required that can be used to temporalize a given logic. Our approach is in line with the approach introduced in [Finger and Gabbay, 1992]. For the given logic, classical propositional logic can be taken, but also, for example, a partial or epistemic logic.

We start defining the flows of time we use, next we define temporalized models and finally we define temporal formulae and their interpretation.

DEFINITION 4 (Flow of time). A *flow of time* $(T, <)$ is a pair consisting of a nonempty set T of time points, and an irreflexive, antisymmetric and transitive (time ordering) relation $<$ on $T \times T$. A flow of time is called *linear* if $\ll = <^+$, the transitive closure of $<$, is a total ordering.

As we want to be able to describe temporal changes in any domain, we will just assume we have an object-level language, $\mathcal{L}_o$, whose formulae describe the domain. The domain states based on this language will be supposed to form a class $\mathcal{M}_o$ of object models. An object-level satisfaction relation $\models_o \subseteq \mathcal{M}_o \times \mathcal{L}_o$ indicates which formulae are true in a model. Thus for $\mathbf{M} \in \mathcal{M}_o$ and $\varphi \in \mathcal{L}_o, \mathbf{M} \models_o \varphi$ means that φ is true in $\mathbf{M}$. We could take, for example, a propositional language with classical propositional models. We could also take the same language but with three-valued models under the Strong Kleene semantics. Or we could take a modal language with modal Kripke models. Thus the choice of language and models can be varied at will. From now on we will assume a fixed object-level language, model class and satisfaction relation.

2.2 *Temporal models*

DEFINITION 5. Let $(T, <)$ be a flow of time. A *temporalized model* $\mathbb{M}$ based on flow of time $(T, <)$ is a triple $(M, T, <)$, where M is a mapping $M : T \rightarrow \mathcal{M}_o$.

So at any point in time we have an object-level model describing what is true in the domain at that time. We will sometimes refer to M as a temporal model based on $(T, <)$. If φ is an object-level formula, and t is a time point in T, and $M_t \models_o \varphi$, then we say that in this model M *at time point* t the formula φ is true.

2.3 *Language*

We will now define the temporal language $\mathcal{L}_T$ in terms of the object-level language using temporal operators to describe truth of object-level and temporal formulae over time. For specific classes of temporal models different operators are used. We consider branching time vs linear time, and discrete vs non-discrete. In the discrete case by $<_0$ the successor relation is denoted, i.e., no other time points exist in between two time points related by $<_0$. Also, we do not want any interaction between object-level formulae and temporal formulae. Therefore the object-level formulae are 'shielded' by an operator C, which informally can be read as 'is true in the current information state'. The temporal operator F has as its meaning 'true at some time point in the future (in some future information state)' (where for the branching time case the quantifier expresses whether this should hold in at least one branch, or in all branches). The operator G means 'true for all future time points', P means 'true for some time point in the past', and H 'true for all time points in the past'. For the discrete case, X means 'true in the next point in time', and Y 'true at the previous point in time'.

DEFINITION 6. The temporal language $\mathcal{L}_T$ is defined to be the least set such that:

- $\varphi \in \mathcal{L}_o \Rightarrow C\varphi \in \mathcal{L}_T$
- $\varphi, \psi \in \mathcal{L}_T \Rightarrow \neg\varphi, \varphi \wedge \psi, \varphi \vee \psi, \varphi \rightarrow \psi \in \mathcal{L}_T$
- $\varphi \in \mathcal{L}_T \rightarrow O\varphi \in \mathcal{L}_T$

where for

–linear time models	$O \in \{F, G, P, H\}$
–discrete linear time models	$O \in \{F, G, X, Y, P, H\}$
–branching time models	$O \in \{\exists F, \forall F, \exists G, \forall G, P, H\}$
–discrete branching time models	$O \in \{\exists F, \forall F, \exists G, \forall G, \exists X, \forall X, Y, P, H\}$

The temporal language is similar to a modal propositional language where the atomic propositions consist of the C operator applied to an object-level formula. Furthermore we introduce the following abbreviations:

- $\varphi \vee \psi \equiv_{\text{def}} \neg(\neg\varphi \wedge \neg\psi)$
- $\varphi \rightarrow \psi \equiv_{\text{def}} \neg\varphi \vee \psi$
- $\top \equiv_{\text{def}} C\alpha \vee \neg C\alpha$ (for an $\alpha \in \mathcal{L}_0$)
- $\bot \equiv_{\text{def}} \neg\top$
- $\forall F\varphi \equiv_{\text{def}} \neg\exists G(\neg\varphi)$
- $\forall G\varphi \equiv_{\text{def}} \neg\exists F(\neg\varphi)$
- $\forall X\varphi \equiv_{\text{def}} \neg\exists X(\neg\varphi)$
- $H\varphi \equiv_{\text{def}} \neg P(\neg\varphi)$

2.4 *Semantics*

In the following, for a temporal model M based on $(T,<)$, $t \in T$, and $\alpha \in \mathcal{L}_T$, $(M,t) \models \alpha$ means that α is true in M at time point t.

DEFINITION 7. Let a temporal model M based on $(T,<)$, and a time point $t \in T$ be given, then inductively define:

- for $\alpha \in \mathcal{L}_o$:

$$(M,t) \models C\alpha \quad \Leftrightarrow \quad M_t \models_o \alpha$$

- for $\varphi, \psi \in \mathcal{L}_T$:

$$\begin{array}{lll} a) & (M,t) \models \neg\varphi & \Leftrightarrow \text{ it is not the case that } (M,t) \models \varphi \\ b) & (M,t) \models \varphi \wedge \psi & \Leftrightarrow (M,t) \models \varphi \text{ and } (M,t) \models \psi \end{array}$$

- for $\varphi \in \mathcal{L}_T$:

For the branching time case:

$a)$	$(M,t) \models \exists F\varphi$	$\Leftrightarrow$	$\exists s \in T[t < s \;\&\; (M,s) \models \varphi]$
$b)$	$(M,t) \models \exists G\varphi$	$\Leftrightarrow$	there exists a branch including t such that for all s in that branch $[t < s \Rightarrow (M,s) \models \varphi]$
$c)$	$(M,t) \models P\varphi$	$\Leftrightarrow$	$\exists s \in T[s < t \;\&\; (M,s) \models \varphi]$
$d)$	in the discrete case		
	$(M,t) \models \exists X\varphi$	$\Leftrightarrow$	$\exists s \in T[t <_0 s \;\&\; (M,s) \models \varphi]$
	$(M,t) \models Y\varphi$	$\Leftrightarrow$	$0 < t \;\&\; \exists s \in T[s <_0 t \;\&\; (M,s) \models \varphi]$

For the linear time case:

$a)$	$(M,t) \models F\varphi$	$\Leftrightarrow$	$\exists s \in T[t < s \;\&\; (M,s) \models \varphi]$
$b)$	$(M,t) \models G\varphi$	$\Leftrightarrow$	for all $s[t < s \Rightarrow (M,s) \models \varphi]$
$c)$	$(M,t) \models P\varphi$	$\Leftrightarrow$	$\exists s \in T[s < t \;\&\; (M,s) \models \varphi]$
$d)$	in the discrete case		
	$(M,t) \models X\varphi$	$\Leftrightarrow$	$\forall s \in T[t <_0 s \;\&\; (M,s) \models \varphi]$
	$(M,t) \models Y\varphi$	$\Leftrightarrow$	$0 < t \;\&\; \exists s \in T[s <_0 t \;\&\; (M,s) \models \varphi]$

For a temporal model M, by $M \models \varphi$ we mean $(M,t) \models \varphi$ for all $t \in T$ and by $M \models K$ we mean $M \models \varphi$ for all $\varphi \in K$, where K is a set of temporal formulae.

The notion of temporalizing a given logic as described above has been successfully used in a variety of cases to describe dynamic phenomena related to agents and their reasoning processes. Examples as can be found in later chapters of this book, cover the control of reasoning in a meta-level architecture, reasoning with and about dynamic assumptions, and default reasoning.

J.-J. Ch. Meyer
Utrecht University, The Netherlands.

J. Treur
Vrije Universiteit, Amsterdam, The Netherlands.

BIBLIOGRAPHY

[Cousot, 1990] P. Cousot, Methods and Logics for Proving Programs, in: J. van Leeuwen (ed.), *Handbook of Theoretical Computer Science, Vol. B: Formal Models and Semantics*, Elsevier, Amsterdam, 1990, pp. 841–993.

[Finger and Gabbay, 1992] M. Finger & D. Gabbay, Adding a Temporal Dimension to a Logic System, *J. of Logic, Language and Computation* 1, 1992, pp. 203–233.

[Fischer and Ladner, 1979] M. Fischer & R. Ladner, Propositional Dynamic Logic of Regular Programs, *J. Comput. System Sci.* 18, 1979, pp. 194–211.

[Harel, 1984] D. Harel, Dynamic Logic, in: D. Gabbay & F. Guenthner (eds.), *Handbook of Philosophical Logic, Vol. II*, Reidel, Dordrecht/Boston, 1984, pp. 497–604.

[Harel, 1979] D. Harel, *First-Order Dynamic Logic*, Lectures Notes in Computer Science 68, Springer, Berlin, 1979.

[Kozen and Tiuryn, 1990] D. Kozen & J. Tiuryn, Logics of Programs, in: J. van Leeuwen (ed.), *Handbook of Theoretical Computer Science, Vol. B: Formal Models and Semantics*, Elsevier, Amsterdam, 1990, pp.789–840.

[Moore, 1985] R.C. Moore, A Formal Theory of Knowledge and Action, in: J.R. Hobbs & R.C. Moore (eds.), *Formal Theories of the Commonsense World*, Ablex, Norwood NJ, 1985, pp. 319–358.

[Pratt, 1976] V. Pratt, Semantical Considerations on Floyd-Hoare Logic, in Proc. 17th IEEE Symp. on Foundations of Computer Science, 1976, pp. 109–121.

[Stirling, 1992] C. Stirling, Modal and Temporal Logics, in: S. Abramsky, D.M. Gabbay & T.S.E. Maibaum (eds.), *Handbook of Logic in Computer Science, Vol. II*, Carendon Press, Oxford, 1992, pp. 477–563.

PART II

MODELLING FRAMEWORKS AND GENERIC AGENT MODELS

FRANCES M.T. BRAZIER, CATHOLIJN M. JONKER AND JAN TREUR

COMPOSITIONAL DESIGN OF MULTI-AGENT SYSTEMS: MODELLING DYNAMICS AND CONTROL

1 INTRODUCTION

The compositional multi-agent design method DESIRE (DEsign and Specification of Interacting REasoning components) supports the design of autonomous interacting agents. Both the *intra-agent functionality* (i.e., the expertise required to perform the tasks for which an agent is responsible in terms of the knowledge, and reasoning and acting capabilities) and the *inter-agent functionality* (i.e., the expertise required to perform and guide co-ordination, co-operation and other forms of social interaction in terms of knowledge, and reasoning and acting capablities) are explicitly modelled. DESIRE views both the individual agents and the overall system as compositional structures—hence all functionality is designed in terms of interacting, compositionally structured components. Complex distributed processes are the result of tasks performed by agents in interaction with their environment.

1.1 The design process

The design of a multi-agent system is an iterative process, which aims at the identification of the parties involved (i.e., human agents, system agents, external worlds), and the processes involved, in addition to the types of knowledge needed. Conceptual descriptions of specific processes and knowledge are often first attained. Further explication of these conceptual design descriptions results in detailed design descriptions, most often in iteration with conceptual design. During the design of these models, partial prototype implementations may be used to analyse or verify the resulting behaviour. On the basis of examination of these partial prototypes, new designs and prototypes are generated and examined, and so on and so forth. This approach to *evolutionary development* of systems, is characteristic to the development of multi-agent systems in DESIRE.

During a multi-agent system design process, DESIRE distinguishes the following descriptions (see Figure 1):

- problem description
- conceptual design
- detailed design

J.J.Ch. Meyer and J. Treur (eds.),
Handbook of Defeasible Reasoning and Uncertainty Management Systems, Vol. 7, 19–63.

- operational design
- design rationale

The *problem description* includes the *requirements* imposed on the design. The rationale specifies the choices made during design at each of the levels, and assumptions with respect to its use.

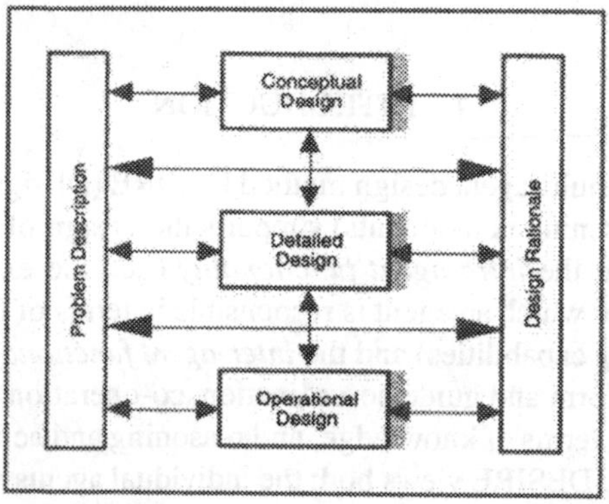

Figure 1. Problem description, levels of design and design rationale

The relationship between the levels of design (conceptual, detailed, operational) is well-defined and structure-preserving. The *conceptual design* includes conceptual models for each individual agent, the external world and the interaction between agents, and between agents and the external world. The *detailed design* of a system, based on the conceptual design, specifies all aspects of a system's knowledge and behaviour. A detailed design is an adequate basis for *operational design*. Prototype implementations, are automatically generated from the detailed design.

There is no fixed sequence of design: depending on the specific situation, different types of knowledge are available at different points during system design. The end result, the final multi-agent system design, is specified by the system designer at the level of detailed design. In addition, important assumptions and design decisions are specified in the design rationale. Alternative design options together with argumentation are included. On the basis of verification during the design process, properties of models can be documented with the related assumptions. The assumptions define the limiting conditions under which the model will exhibit specific behaviour.

1.2 *Compositionality of processes and knowledge*

Compositionality is a general principle that refers to the use of components to structure a design. Within the DESIRE method components are often complex compositional structures in which a number of other, more specific components are grouped. During design different levels of process abstraction are identified. Processes at each of these levels (except the lowest level) are modelled as (process) *components* composed of components at the adjacent lower level.

Processes within a multi-agent system may be viewed as being the result of interaction between more specific processes. A complete multi-agent system may, for example, be seen to be one single component responsible for the performance of the overall process. Within this one single component a number of agent components and an external world can be distinguished, each responsible for a more specific process. Each agent component may, in turn, have a number of internal components responsible for more specific parts of this process. These components may themselves be composed, again entailing interaction between other more specific processes.

The *ontology* used to express the knowledge needed to reason about a specific domain may also be seen as a single (knowledge) component. This *knowledge structure* can often be combined from a number of more specific knowledge structures which, in turn, may again be composed of other even more specific knowledge structures.

As shown in Figure 2 *compositionality of processes* and *compositionality of knowledge* are two separate, orthogonal dimensions. The compositional knowledge structures are referenced by compositional process structures, when needed.

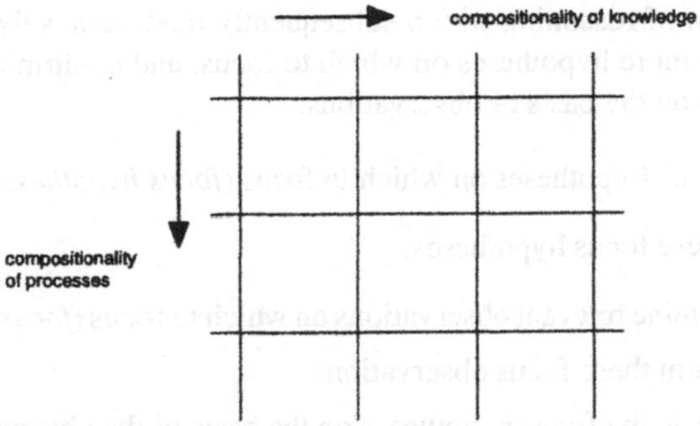

Figure 2. Compositionality of processes and compositionality of knowledge

Compositionality is a means to acquire *information and process hiding* within a model: by defining processes and knowledge at different levels of abstraction, unnecessary detail can be hidden. Compositionality also makes it possible to *integrate different* types of components in one agent. Components and groups of components can be easily included in new designs, supporting *reuse* of components at all levels of design.

2 PROBLEM DESCRIPTION

Which techniques are used to acquire a *problem description* is not pre-defined. Techniques vary in their applicability, depending on, for example, the situation, the task, the type of knowledge on which the system developer wishes to focus. Acquisition of requirements to be imposed on the system as part of the problem description is crucial. These requirements are part of the initial problem definition, but may also evolve during the development of a system. To illustrate the concepts introduced in this chapter an example is used of an information gathering and analysis task: diagnosis of fridge problems.

To make decisions, often agents have to analyse the situation they are in by gathering information and drawing conclusions from the information found. Gathering (the right type of) information requires effort. To use the agent's resources economically, the information gathering process should concentrate on gathering only information that is relevant for the conclusions the agent wants to determine. To achieve this, the agent has to perform strategic reasoning to direct the information gathering process. A situation in which such a reasoning process is required can occur when an environment shows unexpected behaviour, and the cause of this deviant behaviour has to be analysed (this process is sometimes called *diagnosis*). In this section an example of a strategic reasoning process to guide information gathering is discussed. The conclusions in which the agent is interested are called *hypotheses*. Information gathering is assumed to take place by *observation* (for information gathering based on, for example, communication, the pattern is similar).

Viewed from a global perspective, the agent performs a coherent and well-structured pattern of reasoning which subsequently (and iteratively) involves determining one or more hypotheses on which to focus, and confirming or rejecting these hypotheses on the basis of observations:

- determine the hypotheses on which to focus *(focus hypotheses)*
- validate these focus hypotheses:
  - determine relevant observations on which to focus *(focus observations)*
  - perform these focus observations
  - evaluate the focus hypotheses on the basis of the observation results
  - repeat this (hypothesis validation) process
- repeat the whole process

The example diagnostic process used in this chapter is a simplified case of diagnosis of refrigerator malfunctioning. Four types of faults are considered:

- the light *bulb* is *broken*,
- there is *no power supply*,

- the *engine* is *broken*, and
- the *cooling system* is *broken*.

Only three observations can be performed:

- if the door is opened there is (no) *light*,
- (no) *noise* of the engine is heard, and
- it is (not) *cold* in the fridge.

Assumptions are:

- only situations in which one type of fault occurs are considered *(single fault assumption)*

Requirements are:

- for each situation a correct diagnosis shall be determined
- the diagnostic process shall be parsimonous: an efficient number of observations shall be performed during the diagnostic process to reach the conclusion

3 CONCEPTUAL DESIGN AND DETAILED DESIGN

A conceptual and detailed design consist of specifications of the following three types:

- process composition,
- knowledge composition,
- the relation between process composition and knowledge composition.

These three types of specifications are discussed in more detail below.

3.1 Process composition

Process composition identifies the relevant processes at different levels of (process) abstraction, and describes how a process can be defined in terms of lower level processes.

3.1.1 Identification of processes at different levels of abstraction

Processes can be described at different levels of abstraction; for example, the process for the multi-agent system as a whole, processes for individual agents and the external world, processes for task-related components of individual agents. Different views can be taken: a task perspective, and a multi-agent perspective. The *task perspective* refers to the view in which the processes needed to perform an overall task are distinguished. These processes (or sub-tasks) are then *delegated* to appropriate agents and the external world. The *multi-agent perspective* refers to the view in which agents and one or more external worlds are first distinguished and then the processes within them, including agent-related processes such as management of communication, or controlling its own processes.

3.1.2 Specification of a process

The identified processes are modelled as *components*. For each process the *types of information* used as input and resulting as output are identified as well. This is modelled as *input and output interfaces* of the components.

To model the example diagnostic reasoning process, a generic model can be used consisting of a reasoning process and a process of performing observations. The reasoning process is composed of the processes hypothesis determination and hypothesis validation, where the hypothesis validation process is composed of the processes observation determination and hypothesis evaluation. Based on this generic model the following processes are identified for the example reasoning pattern: External World, and Diagnostic Reasoning, with sub-processes Hypothesis Determination and Hypothesis Validation. The process Hypothesis Validation has the two sub-processes: Observation Determination and Hypothesis Evaluation. The types of information used and produced by each of these processes are shown in Figures 3, 4 and 5.

process	*input information type*	*output information type*
Diagnostic Reasoning	Observation Result Info	Assessed Hypotheses Selected Observations
External World	Required Observations	Observation Result Info

Figure 3. Interface information types within the Top Level

As shown in Figure 3 three information types are distinguised for the process as a whole. The diagnostic reasoning process requires information on specific (focus) observations that are selected (Selected Observations) and receives this information (Observation Result Info) as input. Acquisition of information in the external world, is focussed on the basis of information on the observations to be performed (Required Observations) and provides the results (Observation Result Info).

Figure 4 depicts the input and output information types of the processes within the diagnostic reasoning process. The process of determining on which hypotheses to focus, Hypothesis Determination, uses information on which hypotheses have already been assessed (Assessed Hypotheses), which hypotheses have already been in focus (Selected Hypotheses) and results of observations that have been performed. The result is a list of one or more hypotheses which may be used to focus the diagnostic process (Selected Hypotheses).

process	*input information type*	*output information type*
Hypothesis Determination	Assessed Hypotheses Selected Hypotheses	Selected Hypotheses
Hypothesis Validation	Observation Result Info Focussed Hypotheses	Assessed Hypotheses Selected Observations

Figure 4. Interface information types within Diagnostic Reasoning

The process of validating one or more hypotheses, Hypothesis Validation, uses information on the hypotheses on which to focus (Focussed Hypotheses) and results of observations (Observation Result Info). During validation, often a need for specific information is identified (Selected Observations). Once the validation process has been completed the results—hypotheses that have been assessed (Assessed Hypotheses)—are available as output.

process	*input information type*	*output information type*
Observation Determination	Focussed Hypotheses Observation Information	Selected Observations
Hypothesis Evaluation	Target Domain Hypotheses Assumption Domain Info Domain Info	Epistemic Domain Hypotheses Epistemic Domain Info Domain Info Domain Hypotheses

Figure 5. Interface information types within Hypothesis Validation

Figure 5 depicts the information types used and produced by the processes needed to validate hypotheses. To determine which observations to perform, Observation Determination, information is needed on the hypotheses on which to focus (Focussed Hypotheses) and the available information on observations (Observation Information). The results of this process are a list of one or more observations to be performed (Selected Observations). The process Hypothesis Evaluation, involves evaluating one or more hypotheses (Target Domain Hypotheses) on the basis of information on observations performed (Assumption Domain Info). The result is an evaluation of the hypotheses (Epistemic Domain Hypotheses) on which evaluation focussed.

3.1.3 *Specification of abstraction levels*

The identified levels of process abstraction are modelled as *abstraction/specialisation relations* between components at adjacent levels of abstraction: components may be *composed* of other components or they may be *primitive*. Primitive components may be either reasoning components (based on a knowledge base), or, alternatively, components capable of performing tasks such as calculation, information retrieval, optimisation, et cetera.

For the example diagnostic reasoning process the process abstraction levels are depicted in Figure 6. The identification of processes at different abstraction levels results in specification of components that can be used as building blocks, and of a specification of the sub-component relation, defining which components are a sub-component of a which other component. The distinction of different process abstraction levels results in process hiding.

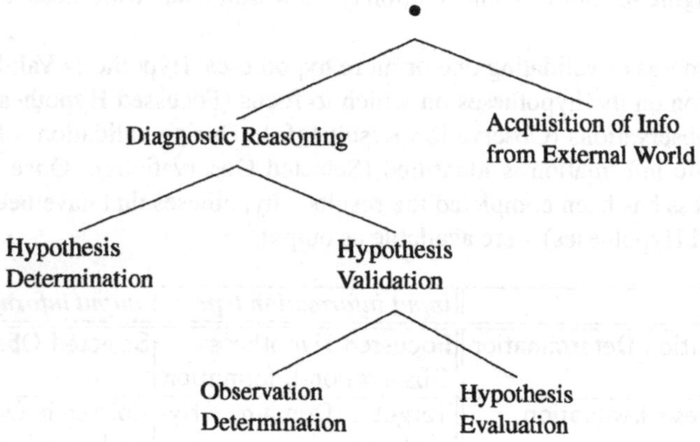

Figure 6. Process abstraction levels in a strategic information gathering process

3.1.4 *Composition*

The way in which processes at one level of abstraction are composed of processes at the adjacent lower abstraction level is called *composition*. This composition of processes is described by the possibilities for *information exchange* between processes (*static view* on the composition), and *task control knowledge* used to control processes and information exchange (*dynamic* view on the composition).

3.1.5 *Information exchange*

A specification of information exchange defines which types of information can be transferred between components and the *information links* by which this can be achieved.

Within each of the components information *private links* are defined to transfer information from one component to another. In addition, *mediating links* are defined to transfer information from the input interfaces of encompassing components to the input interfaces of the internal components, and to transfer information from the output interfaces of the internal components to the output interface of the encompassing components.

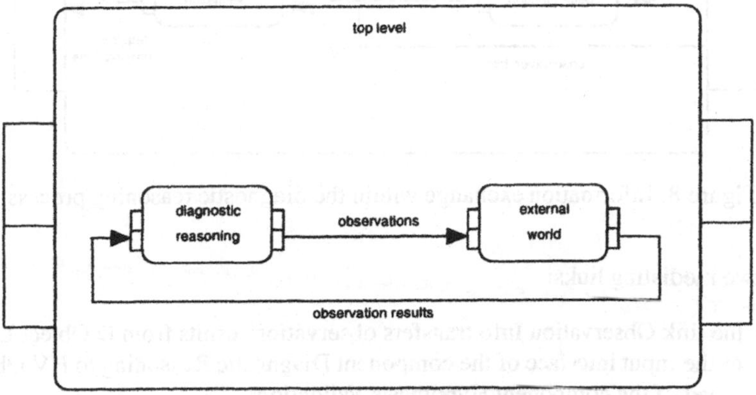

Figure 7. Information exchange within the top level process

As shown in Figure 7 two private links are defined at the top level: Observations and Observation Results. The link Observations transfers the list of observations to be performed (Selected Observations) from the lower level (called D Object Level) of the output interface of the component Diagnostic Reasoning to the lower level (called EW Object Level) of the input interface of the component External World. The link Observation Results transfers results of observations (Observation Result Info) from EW Object Level of the output interface of the component External World to D Object Level of the input interface of the component Diagnostic Reasoning.

In Figure 8 four information links defined for the component Diagnostic Reasoning are depicted: two private and two mediating. The private links:

- the link Hypotheses transfers Selected Hypotheses from the lower level (called HD Object Level) of the component Hypothesis Determination to the lower level (called HV Object Level) of the component Hypothesis Validation to become the hypotheses on which the diagnostic process is to focus,

- the link Assessments transfers Assessed Hypotheses from level HV Object Level of the output interface of the component Hypothesis validation to HD object level of the input interface of the component Hypothesis Determination.

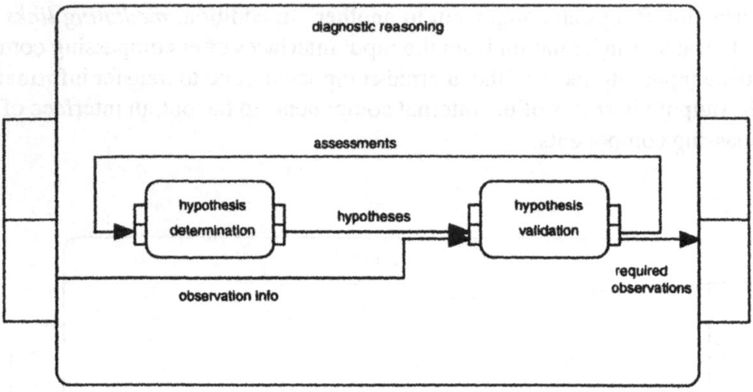

Figure 8. Information exchange within the diagnostic reasoning process

The two mediating links:

- the link Observation Info transfers observation results from D Object Level of the input interface of the component Diagnostic Reasoning to HV Object Level of the component Hypothesis Validation,

- the link Required Observations transfers selected observations from HV Object Level of the component hypothesis validation to D Object Level of the output interface of the component Diagnostic Reasoning.

Within the component Hypothesis Validation, see Figure 9, one private link and five mediating links are defined. The private link:

- the link Performed Obs transfers epistemic domain information (on the results of observations) from HE Meta-Level of the output interface of the component Hypothesis Evaluation to HD Object Level of the input interface of Observation Determination

The five mediating links:

- the link Focus Hyp To OD transfers the hypotheses to be validated, Focussed Hypotheses, from HV Object Level of the input interface of the component Hypothesis Validation to OD Object Level of the component Observation Determination,

- the link Focus Hyp To HE transfers the hypotheses to be validated, Focussed Hypotheses, from HV Object Level of the input interface of the component Hypothesis Validation to HE Meta-Level of the component Hypothesis Evaluation,

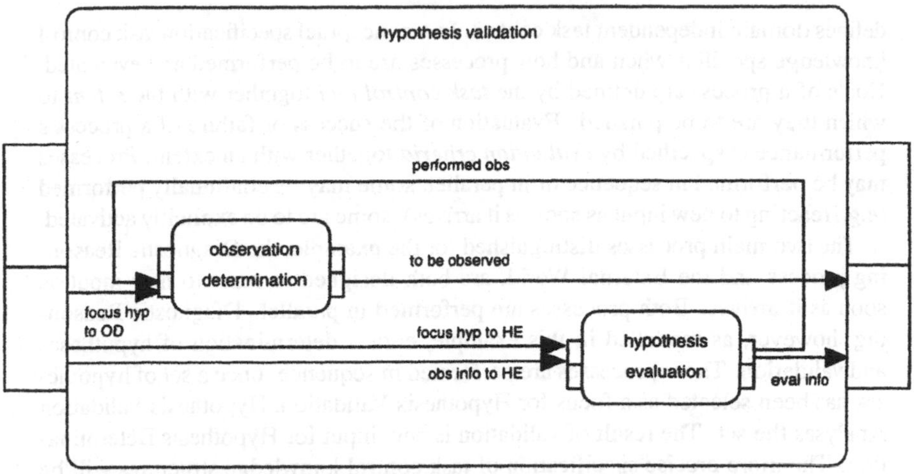

Figure 9. Information exchange within Hypothesis Validation

- the link Obs Info To HE transfers the results of observations, Observation Result Info, from HV level of the input interface of the component Hypothesis Validation to HE Meta-Level of the input interface of the component Hypothesis Evaluation,
- the link To Be Observed transfers Selected Observations from OD Object Level of the output interface of the component Observation Determination to HV Object Level of the output interface of the component Hypothesis Validation,
- the link Eval Info transfers the result of Hypothesis Evaluation, namely Assessed Hypotheses from HE Meta-Level of the component Hypothesis Evaluation to HV Object Level of the output interface of the component Hypothesis Validation.

3.1.6 *Task control knowledge*

Components may be activated sequentially or they may be continually capable of processing new input as soon as it arrives (*awake*). The same holds for information links: information links may be explicitly activated or they may be awake. *Task control knowledge* specifies under which conditions which components and information links are active (or made awake). Evaluation criteria, expressed in terms of the evaluation of the results (success or failure), provide a means to guide further processing.

Conceptual specification of task control is often initially partial: some knowledge is available about necessary sequencing of processes, more detailed knowledge needs to be acquired. The detailed specification of task control knowledge, defines domain independent task control. In a conceptual specification task control knowledge specifies when and how processes are to be performed and evaluated. Goals of a process are defined by the *task control foci* together with the *extent* to which they are to be pursued. Evaluation of the success or failure of a process's performance is specified by *evaluation criteria* together with an extent. Processes may be performed in sequence or in parallel, some may be continually performed (e.g., reacting to new input as soon as it arrives), some are to be explicitly activated.

The two main processes distinguished for the example, the Diagnostic Reasoning process and the External World, are both designed to react to new input as soon as it arrives. Both processes are performed in parallel. Diagnostic Reasoning, however, as modelled in this example, entails determination of hypotheses and validation. These processes are performed in sequence: once a set of hypotheses has been selected as a focus for Hypothesis Validation, Hypothesis Validation Analyses the set. The result of validation is new input for Hypothesis Determination. The more precise specification of task control knowledge structures will be addressed in Section 3.2.

3.2 Knowledge composition

Knowledge composition identifies the knowledge structures at different levels of (knowledge) abstraction, and describes how a knowledge structure can be defined in terms of lower level knowledge structures. The knowledge abstraction levels may correspond to the process abstraction levels, but this is not often the case; often the matrix depicted in Figure 2.2 shows more than a one to one correspondence between process abstraction levels and knowledge abstraction levels.

3.2.1 Identification of knowledge structures at different abstraction levels

The two main structures used as building blocks to model knowledge are: *information types* and *knowledge bases*. Knowledge structures can be identified and described at different levels of abstraction. At the higher levels the details can be hidden. The resulting levels of knowledge abstraction can be distinguished for both information types and knowledge bases.

3.2.2 Information types

An information type defines an *ontology* (lexicon, vocabulary) to describe objects or terms, their sorts, and the relations or functions that can be defined on these objects. Information types are defined as signatures (sets of names for sorts, objects, functions, and relations) for order-sorted predicate logic. Information types can be specified in graphical form or in formal textual form. For the example diagnostic reasoning task the following information types have been specified in textual form. The attribute information types is used to import other information types. Via the attribute meta-descriptions all atoms that can be built by the referenced information type are imported in the indicated sort. For example,

```
meta-descriptions
                    domain_hypotheses      :      HYPOTHESIS;
```

specifies that all atoms that can be formed using the information type domain_hypotheses are included as objects or terms in the sort HYPOTHESIS. This language construct is used to define object-meta relations in the reasoning process.

Generic information types on hypotheses:

```
information type hypotheses_sorts
    sorts       HYPOTHESIS;
end information type

information type meta_domain_hypotheses
    information types   hypotheses_sorts;
    meta-descriptions
        domain_hypotheses  :    HYPOTHESIS;
end information type

information type assessed_hypotheses
    information types  meta_domain_hypotheses;
    relations  tried, rejected, confirmed   :  HYPOTHESIS;
end information type

information type selected_hypotheses
    information types  meta_domain_hypotheses;
    relations
        to_be_validated,
        has_been_focus    :   HYPOTHESIS;
        subhypothesis  :  HYPOTHESIS * HYPOTHESIS;
end information type
```

Generic information types on symptoms:

```
information type symptoms_sorts
    sorts      SYMPTOM;
end information type

information type meta_domain_symptoms
    information types symptoms_sorts;
    meta-descriptions
        domain_symptoms  :    SYMPTOM;
end information type
```

information type selected_observations
  information types meta_domain_symptoms, meta_domain_hypotheses
  relations to_be_observed : SYMPTOM;
    relevant_observation_for : SYMPTOM * HYPOTHESIS;
end information type

information type observation_results
  information types domain_symptoms;
end information type

information type value_signs
  sorts SIGN
  objects pos, neg : SIGN;
end information type

information type observation_information
  information types meta_domain_symptoms , value_signs;
  relations observed : SYMPTOM * SIGN;
end information type

Domain-specific information types on hypotheses:

information type domain_hypotheses
  information types hypotheses_sort;
  objects fridge_problem, electricity_problem, cooling_problem,
    no_power_supply, broken_bulb, broken_engine, broken_cooling_system:
      HYPOTHESES;
end information type

Domain-specific information types on symptoms:

information type domain_symptoms
  information types symptoms_sort;
  objects light, cold, noise: SYMPTOM;
end information type

3.2.3 Knowledge bases

A knowledge base defines a part of the knowledge that is used in one or more of the processes. Knowledge bases use ontologies defined in information types. Which information types are used in a knowledge base defines a relation between information types and knowledge bases. In a detailed design, knowledge bases are specified in order-sorted predicate logic form, normalised to a classical rule format (implications between conjunctions of literals). Knowledge bases can also be specified in conceptual pre-formal manners, for example in graphical forms.

Graphical representations of knowledge bases for the example diagnostic reasoning task are as follows. In Figure 10 the causal relations between faults and observations are depicted.

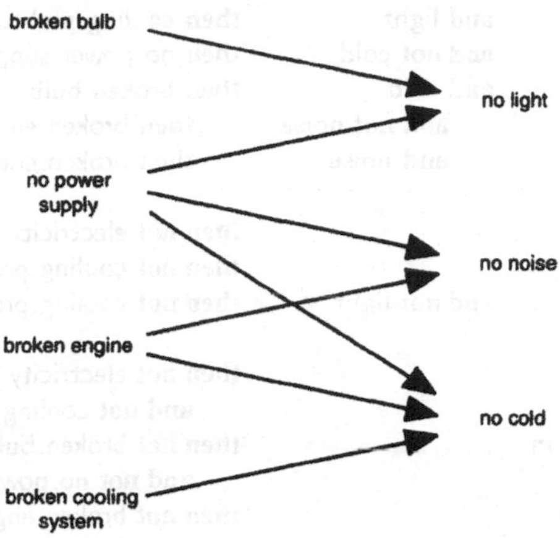

Figure 10. The causal relations between faults and observations

As mentioned in the problem description, for simplicitys sake, only single faults are considered. If any one of the faults occurs, there is a *fridge problem*. If the fridge either has a broken bulb or there is no power supply, the fridge is said to have an *electricity problem*. If either the engine or the cooling system is broken, the fridge is said to have a *cooling problem*. In Figure 11 this taxonomy of types of problems is depicted.

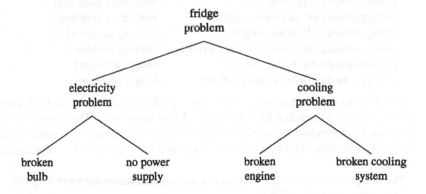

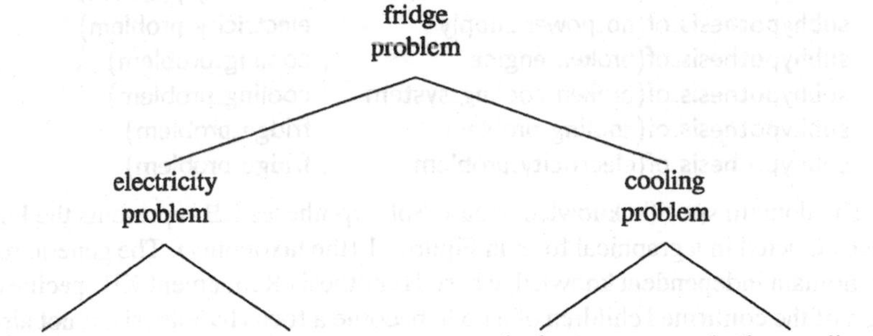

Figure 11. Taxonomy of types of fridge problems

The object level domain knowledge can be modelled in a *causal* (i.e., faults imply observations, as depicted in Figure 10) or *anti-causal* (i.e., observations imply faults) manner. For reasons of presentation, the anti-causal manner is chosen. The knowledge base Anti Causal Fridge Knowledge consists of:

if fridge_problem **and not** light **then** electricity_problem
if fridge_problem **and not** cold
and light **then** cooling_problem
if electricity_problem **and not** cold **then** no_power_supply
if electricity_problem **and** cold **then** broken_bulb
if cooling_problem **and not** noise **then** broken_engine
if cooling_problem **and** noise **then** broken_cooling_system

if light **then not** electricity_problem
if cold **then not** cooling_problem
if not cold **and not** light **then not** cooling_problem

if not fridge_problem **then not** electricity_problem
and not cooling_problem
if not electricity_problem **then not** broken_bulb
and not no_power_supply
if not cooling_problem **then not** broken_engine
and not broken_cooling_system

The component Hypothesis Determination uses the following knowledge bases:

Hypothesis Refinement KB (*domain independent knowledge*)
if confirmed(H:HYPOTHESIS)
and subhypothesis_of(H1:HYPOTHESIS, H:HYPOTHESIS)
and not tried(H1:HYPOTHESIS)
then focus_hypothesis(H1:HYPOTHESIS)

Subhypotheses KB (*domain specific knowledge*)
subhypothesis_of(broken_bulb , electricity_problem)
subhypothesis_of(no_power_supply , electricity_problem)
subhypothesis_of(broken_engine , cooling_problem)
subhypothesis_of(broken_cooling_system , cooling_problem)
subhypothesis_of(cooling_problem , fridge_problem)
subhypothesis_of(electricity_problem , fridge_problem)

The domain specific knowledge base Subhypotheses KB represents the knowledge depicted in a graphical form in Figure 11 (the taxonomy). The generic rule in the domain independent knowledge base Hypothesis Refinement KB specifies that each of the confirmed children of a node become a focus hypothesis, if not already validated.

The component Observation Determination uses the following knowledge bases:

Generic Obs Determination KB *(domain independent knowledge)*
if focus_hypothesis(H:HYPOTHESIS)

```
    and relevant_observation_for(S:SYMPTOM, H:HYPOTHESIS)
then relevant_observation(S:SYMPTOM)

if relevant_observation(S:SYMPTOM)
    and not observed(S:SYMPTOM, pos)
    and not observed(S:SYMPTOM, neg)
then to_be_observed(S:SYMPTOM)
```

Obs Relevance KB (*domain specific knowledge*)

```
relevant_observation_for(light   , electricity_problem)
relevant_observation_for(cold    , cooling_problem)
relevant_observation_for(cold    , broken_bulb)
relevant_observation_for(cold    , no_power_supply)
relevant_observation_for(noise   , broken_engine)
relevant_observation_for(light   , broken_engine)
relevant_observation_for(noise   , broken_cooling_system)
```

These knowledge bases specify that the observations that are relevant for at least one of the focus hypotheses are selected, unless already performed.

3.2.4 *Composition of knowledge structures*

Information types can be composed of more specific information types, following the principle of compositionality discussed above. Similarly, knowledge bases can be composed of more specific knowledge bases. The compositional structure is based on the different levels of knowledge abstraction that are distinguished, and results in information and knowledge hiding.

3.2.5 *Composition of information types*

The relations between the information types distinguished above and generic information types distinguished for diagnosis, are depicted below in Figures 12 and 13. The information types Selected Hypotheses, Assessed Hypotheses and Focussed Hypotheses, see Figure 12, refer to the information type Meta Domain Hypotheses. The information type Meta-Domain Hypotheses refers to the information types Hypotheses Sorts and Domain Hypotheses. Note in this respect the meta-object distinction between Meta-Domain Hypotheses and Domain Hypotheses.

In a similar way Predicted Symptoms, Observation Information and Selected Obervations refer to Meta-Domain Symptoms, see Figure 13. The information types Predicted Symptoms and Observation Information both refer to Value Signs.

The information types Predicted Symptoms, Selected Observations and Observation Information all refer to the information type Meta-Domain Hypotheses. Information type Observation Results refers to Domain Symptoms. Only Domain Symptoms and Observation Results are on the same object level, all others are on the same meta-level.

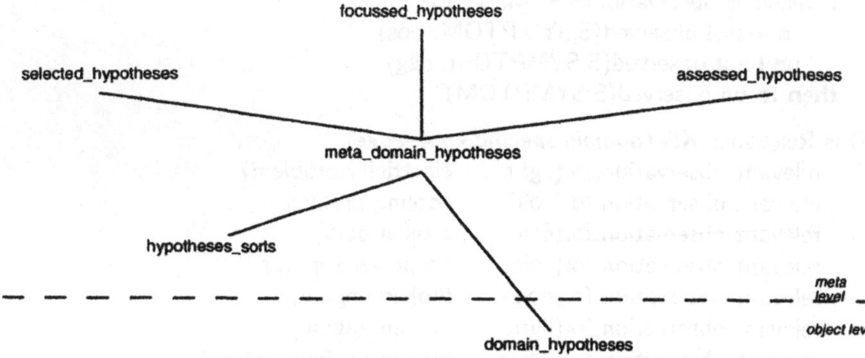

Figure 12. Composition of information types: hypotheses

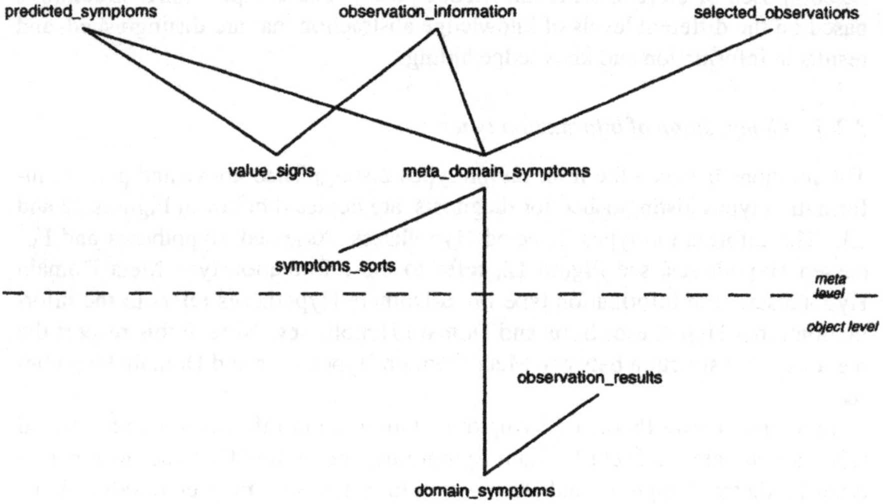

Figure 13. Composition of information types: observations

3.2.6 Composition of knowledge bases

The knowledge base Hypothesis Determination KB is composed of the generic knowledge base Hypothesis Refinement KB and the domain-specific knowledge base Subhypotheses KB. The knowledge base Observation Determination KB is composed of the generic knowledge base Generic Obs Determination KB and the domain-specific knowledge base Obs Relevance KB.

3.2.7 Task control knowledge

Processes at different abstraction levels can have different degrees of autonomy. For example, to constrain behaviour the following forms of control can occur:

- *Fully decentralised control*
  Processes at the lowest process abstraction levels are all autonomous (i.e., not constrained by control from the higher levels) and the behaviour of the processes at the higher abstraction levels emerges on the basis of the behaviour of the processes at the lower levels (and the composition relation defined by the information links).
- *Fully centralised control*
  The top level process has control knowledge that constrains the behaviour of the processes at the lower level of process abstraction (e.g., by prescribing a sequence in which they have to be active), which in their turn have control knowledge that constrains the processes at the levels below, and so on.
- *Centralised top level, decentralised lower level control*
  The top level has centralised control of the processes at the level of process abstraction level (just) below it and processes at the lowest levels of process abstraction are autonomous.
- *Decentralised top level, centralised lower level control*
  Control is minimal at the top level (e.g., agents are completely autonomous), but at the lower levels (within the agents) more centralised forms of control are used to obtain coherent behaviour (the agents themselves have control over their activities). This form of control is often used within multi-agent systems.

Control can be modelled within DESIRE in a variety of ways, including all types of control mentioned above. Whether or not control knowledge is modelled (and in which form) to constrain the behaviour of the sub-processes can be decided independently for each of the process abstraction levels. Together with the information links, the control knowledge defines the process composition relation.

Task control knowledge specifies control of a component. A component can be in three *states*: it can be active, awake, or idle. A component in state idle is not in the process of doing anything. An active or awake component is actively

extent	*to be derived*
all p	all possible targets
every	every target
any	any target
any new	any target not previously derived

Figure 14. Extents

pursuing a strategy defined by its *task control focus* and *extent*. A task control focus defines the focus of the process: processes can have different foci, specified as task control foci. Dynamic selection of a task control focus, is one of the ways in which a process can be controlled. To each task control focus a set of *targets* can be associated. These are the outputs the process tries to achieve. An *extent* specifies to which extent targets associated to a task control must be derived: see Figure 14.

If an *active* component *succeeds* or *fails* to derive the targets specified in its task control focus to a given extent, it becomes idle. An *awake component* is a component that is continually capable of deriving new information. Its task control focus and extent are used to focus the reasoning process in the same way as the task control focus and extent are used to focus the reasoning in active components. Links can be either awake (in which case information is transferred as soon as it has become available), uptodate (in which case information has just been transferred) or idle. A link or component in state awake remains in this state, also if nothing new can be done (in this way they are stand-by). A component in state active and link in state uptodate becomes idle as soon as no new information can be derived or transferred (termination). A component or link in state awake reacts to arriving information immediately (event-driven). A component or link in state idle does not react to arriving information. It only can react if its state is explicitly changed to awake or active, respectively uptodate.

The names of task control foci are part of the public task information in a component specification. The *initial task control focus* and the *initial extent* can be specified (the initial task information, as part of private task information) as well as which targets initially are associated to which task control focus (*initial targets*): initial kernel information, as part of private kernel information. A target specifies whether it is aimed to confirm, determine or reject a specific output atom. Evaluation criteria can be specified to assess a components results. These may be the same as task control foci, or they may be different. Also to evaluation criteria targets can be associated. Different evaluation criteria can be used to determine the status of a components process, each with different implications: depending on which evaluation criteria have been successfully achieved, or failed, one or more different components may be activated (made active or awake).

In the diagnostic reasoning example the component Diagnostic Reasoning has initial information about its task control focus diagnose fault and its extent, namely all p. This implies that the component will succeed if all possible targets associated to the task control focus diagnose fault have been derived. The definition of this information is included in the definition of Diagnostic Reasoning's private kernel information as initial kernel information: this specifies that the output atoms the component aims to determine are all atoms of the form diagnosis(H: HYPOTHESIS) for some instantiation of H: HYPOTHESIS for the task control focus diagnose fault. This is specified by the expression: diagnosis(H : HYPOTHESIS) : confirm, where confirm is the *target type*. Note that targets are specified one level higher than the output atoms to which they refer: target information is not information about the world but about the component's process: a target describes meta-information expressed by a meta-atom of the form target(diagnose_fault, diagnosis(H:HYPOTESIS), confirm).

Instances of task control knowledge of the diagnostic reasoning example are used to illustrate task control specifications. Continual activity of both the component Diagnostic Reasoning and the component External World, and the two links Observations and Observation results, is specified as task control knowledge of the component Top Level.

if start
then next_component_state(diagnostic_reasoning, **awake**)
  and **next_component_state**(external_world, **awake**)
  and **next_link_state**(observations, **awake**)
  and **next_link_state**(observation_results, **awake**)

The component Diagnostic Reasoning is made awake: the system is to continually react to new information in its effort to confirm all possible hypotheses. The External World is assumed to be continually awake, capable of executing an observation as soon as it arrives. Information provided by one of the two components is immediately transferred to the other by the respective link which is awake.

An example of *task control knowledge* within the component Diagnostic Reasoning is as follows:

if start
then next_component_state(hypothesis_determination, **active**)
  and next_link_state(assessments, **uptodate**)
  and next_task_control_focus(hypothesis_determination, determine_hypos)
  and next_extent(hypothesis_determination, **any_new**)

This rule states that when activated (in the sense of active or awake), the component Diagnostic Reasoning activates its sub-component Hypothesis Determination with the state active, with the task control focus determine hypos and extent any new. The last two specifications about task control focus and extent can be left out, as they never change during the process; therefore they also can be specified as initial task information. Another example of task control knowledge of the component Diagnostic Reasoning is:

if component_state(hypothesis_determination, **idle**)
  and previous_component_state(hypothesis_determination, **active**)
  and evaluation(hypothesis_determination, hypos_determined, **any, succeeded**)
then next_component_state(hypothesis_validation, **awake**)
  and next_link_state(hypotheses, **uptodate**)

This statement specifies that if the component Hypothesis Determination terminates and succeeds in deriving any target associated to hypos determined (an evaluation criterion), then, at the next point in time, the component Hypothesis Validation, has state **awake**. In addition the link between Hypothesis Determination and hypothesis validation, has become **uptodate**.

Task control knowledge constrains the process states of components. Typically, in the current version of DESIRE task control knowledge is specified according to the pattern

previous state & current state ⇒ next state

This form (actually a form of executable temporal logic; cf. ([Barringer *et al.*, 1996])), enables direct computation of the next control state from the current and previous control state. In the near future, extensions of this format (e.g., by adding intermediate information types) will be investigated.

3.3 Relation between process composition and knowledge composition

Each process in a process composition uses knowledge structures. Which knowledge structures (information types and knowledge bases) are used for which processes is defined by the relation between process composition and knowledge composition. The cells within the matrix depicted in Figure 2 define these relations.

For the example diagnostic reasoning task the knowledge base related to the primitive component Hypothesis Evaluation is Anticausal Fridge KB. The knowledge base related to Hypothesis Determination is Hypothesis Determination KB; the knowledge base related to Observation Determination is Observation Determination KB.

4 DESIGN RATIONALE

One of the parts of the *design rationale* describes verification of the relevant properties of the designed system in relation to the design requirements identified in the problem description. Also the assumptions under which these properties hold are made explicit in the design rationale. Important design decisions are specified, together with some of the alternative choices that could have been made and the arguments in favour of and against the different options. At the operational level the design rationale includes decisions based on operational considerations, such

atoms	*possible world states*						
light	false	false	false	true	true	true	true
noise	false	true	false	false	true	true	false
cold	true	true	false	false	false	true	true
fridge problem	true	true	true	true	true	false	false
electricity problem	true	true	true	false	false	false	false
cooling problem	false	false	false	true	true	false	false
broken bulb	true	true	false	false	false	false	false
no power supply	false	false	true	false	false	false	false
broken engine	false	false	false	true	false	false	false
broken cooling system	false	false	false	false	true	false	false

Figure 15. Possible world states for the example domain

as the choice to implement a parallel process on one or more machines, depending on the available capacity.

For verification, the possible states of the world (world situations) taken into account are depicted in Figure 15.

It is easy to verify that the anti-causal knowledge is *correct* with respect to the possible world states depicted in Figure 15, in the sense that for given observation results, conclusions drawn by means of this knowledge are true in the world state in which the observations were performed. Moreover, the knowledge is *decisive* in the sense that if it is known whether there is a fridge problem, and sufficient observation information is available, for each of the causes it can be derived whether or not it is true (one of the variants of completeness; see [Treur and Willems, 1994]).

A trivial approach to diagnosis would be to perform all possible observations and then draw a conclusion on the hypotheses (e.g., using the knowledge base above). For practical applications, such a trivial approach is not satisfactory: many observations would have to be performed that do not contribute to the solution. The challenge in diagnostic reasoning mainly lies in the question which strategy can be followed to obtain observation information, sufficient to determine which hypotheses are true and which are false, but as economically as possible.

One strategy for the strategic reasoning processes, described in the literature is called *hierarchical classification*. This strategy uses a taxonomy of types of problems (also called *abstract hypotheses*) of different levels of abstraction like the one used in the example model for diagnostic reasoning. The strategy of hypothesis determination is to first determine hypotheses at the highest level in the taxonomy, and depending on which of these abstract hypotheses are confirmed (by validation), proceed by selection of more specific hypotheses. Thus, if an abstract hypothesis is confirmed, then the next hypotheses to focus on are its children in the taxonomy. The *specific hypotheses* for the diagnostic process as a whole are the hypotheses at the bottom of the taxonomy.

The abstract hypotheses in the taxonomy play two intermediary roles during the diagnostic process:

- they serve as intermediate results in the object level reasoning (e.g., using the knowledge base above), and
- they play an important strategic role within the Hypothesis Determination task (meta-level reasoning about focussing the diagnostic process).

A confirmed abstract hypothesis defines a sub-taxonomy (below), that can be further pursued by selecting its children (sub-hypotheses in the taxonomy) as the next hypotheses to be validated. The children of non-confirmed hypotheses are not further pursued. In this way each of the intermediate outcomes of the reasoning process (a confirmed abstract hypothesis) implies control of the direction of search. This hierarchical strategy of hypothesis determination is more economical than treating all specific hypotheses in sequence. Analysing all specific hypotheses, for example, for a binary tree of depth n entails validation of $\mathbf{2n}$ specific hypotheses, whereas for hierarchical classification following the taxonomy from top to bottom, two hypotheses per level need to be analysed, entailing validation of only $\mathbf{2n}$ hypotheses for the whole process.

5 THE DESIRE SOFTWARE ENVIRONMENT

The DESIRE design method is supported by the DESIRE software environment. This environment includes tools to support all phases of design. Graphical editors, for example, support specification of conceptual and detailed design of processes and knowledge. A detailed design is a solid basis to develop an operational implementation in any desired environment. An implementation generator supports prototype generation of both partially and fully specified models. The code generated by the implementation generator can be executed in an execution environment, which runs on different platforms: UNIX-based, LINUX based and Windows-based systems.

6 CONTROL BY DYNAMIC TARGETS AND DYNAMIC ASSUMPTIONS

Control of reasoning processes can be imposed in different manners. One option is related to the process composition and uses task control knowledge as presented above to control the global phases (specified by different components) in a reasoning process. Besides this global form of control of reasoning, also control at a more detailed level is possible. An option for more fine-grained control is to dynamically generate goals (targets) for the reasoning process within a (primitive) component, thus supporting dynamic control of limited reasoning within a component. Another option for more fine-grained control is to dynamically generate additional presuppositions (assumptions) for the reasoning process in a component. In this section the control of reasoning by dynamic targets and by dynamic assumptions is discussed.

6.1 Control by dynamic generation of targets

The example of fridge diagnosis is used to illustrate the use of dynamic targets to control reasoning patterns in a fine-grained manner.

6.1.1 First example trace

The first example reasoning trace that is considered is depicted in Figure 16. The trace starts when the system is informed that there are problems with the fridge (1). This information is transferred to the component Hypothesis Determination (2), where, based on the taxonomy of hypotheses, two (abstract) hypotheses to be validated (electricity problem, cooling problem) are determined (3). These focus hypotheses are transferred to the component Hypothesis Validation (4), and within this component to both the components Observation Determination and Hypothesis Evaluation (5). In the latter component they serve as targets: the reasoning process within the component is limited to deriving these outputs only; however, in the beginning there is not enough observation information available to derive any of these targets. In Observation Determination, based on the hypotheses in focus, two observations to be performed (to observe light, cold) are determined (6), which are transferred to the output interface of Hypothesis Validation (7), and from there to the output interface of Diagnostic Reasoning (8).

The observations that are to be performed, are transferred to the component External World (9), where they are actually performed (10). The observation results (light, not cold) are transferred to the component Diagnostic Reasoning (11), and within this component to Hypothesis Validation (12) and further down to the lower process abstraction level in Hypothesis Evaluation (13). Given the observation information (light, not cold), this component (which has electricity problem and cooling problem as targets) is able to derive that one of the abstract hypotheses (cooling problem) is true (14). This information is transferred to the component Hypothesis Determination, as is the information that both hypotheses in focus have seen validated; based on this updated input, revision takes place, which leads to the retraction of the earlier derived conclusions on hypotheses to be validated (15).

After this revision, new (this time specific) hypotheses to be validated (broken engine and broken cooling system) are derived, based on the taxonomy of hypotheses (16), which again are transferred to Hypothesis Validation (17), and from there to the lower process abstraction levels of Hypothesis Evaluation and Observation Determination, where revision takes place (18). In observation determination a new observation to be performed (to observe noise) is found (19). Again, it is

Figure 16. First example trace description

time points	1 2 3 4 5 6 7 8 9 10 11 12 13 14 15 16
Diagnostic Reasoning	[ ] ; [confirmed(fridge_problem)] [light, not cold] ; [confirmed(fridge_problem), selected_observation(light), selected_observation(cold)] [ ] ; [confirmed(fridge_problem), selected_observation(light), selected_observation(cold)]
Hypothesis Determination	[confirmed(fridge_problem)] [confirmed(fridge_problem), validated(electricity_problem), validated(cooling_problem), confirmed(cooling_problem)] [confirmed(fridge_problem), to_be_validated(electricity_problem), to_be_validated(cooling_problem)] [confirmed(fridge_problem), validated(electricity_problem), validated(cooling_problem), confirmed(cooling_problem), to_be_validated(broken_engine), to_be_validated(broken_ cooling_system)]
Hypothesis Validation	[ ] ; [focus_hypothesis(electricity_problem), focus_hypothesis(cooling_problem), confirmed(fridge_problem)] [light, not cold] ; [focus_hypothesis(electricity_problem), focus_hypothesis(cooling_problem), confirmed(fridge_problem), selected_observation(light), selected_observation(cold)] [ ] ; [focus_hypothesis(electricity_problem), focus_hypothesis(cooling_problem), confirmed(fridge_problem), selected_observation(light), selected_observation(cold)]
Observation Determination	[focus_hypothesis(electricity_problem), focus_hypothesis(cooling_problem)] [focus_hypothesis(electricity_problem), focus_hypothesis(cooling_problem), selected_observation(light), selected_observation(cold)]
Hypothesis Evaluation	[fridge_ problem]; [ ] [fridge_problem]; [target(he_tcf, electricity_problem,det), target(he_tcf, cooling_problem, det)] [light, not cold, fridge_problem]; [target(he_tcf,electricity_problem, det)] [target(he_tcf,cooling_problem, det)] [light, not cold, fridge_problem, cooling_problem]; [target(he_tcf, electricity_problem det), target(he_tcf, cooling_problem, det), true(cooling_problem)]
External World	[ ] ; [target(ew_tcf, light, det), target(ew_tcf, cold, det)] [light, not cold] ; [target(ew_tcf, light, det), target(ew_tcf, cold, det)]

time points	17 18 19 20 21 22 23 24 25 26 27 28 29
Diagnostic Reasoning	[light, not cold, not noise]; [confirmed(fridge_problem) confirmed(cooling_problem) selected_observation(noise)
	[light, not cold]; confirmed(fridge_problem), confirmed(cooling_problem), selected_observation(noise)
	[light, not cold, not noise]; confirmed(fridge_problem), confirmed(cooling_problem), confirmed(broken_engine) selected_observation(noise)
Hypothesis Determination	
Hypothesis Validation	[light, not cold] ; [focus_hypothesis(broken_engine), focus_hypothesis(broken_cooling_system), selected_observation(light), selected_observation(cold), confirmed(fridge_problem), confirmed(cooling_problem)]
	[light, not cold, not noise]; [focus_hypothesis(broken_engine), focus_hypothesis(broken_cooling_system), selected_observation(noise), confirmed(fridge_problem), confirmed(cooling_problem)]
	[light, not cold]; [focus_hypothesis(broken_engine), focus_hypothesis(broken_cooling_system), selected_observation(noise), confirmed(fridge_problem), confirmed(cooling_problem)]
	[light, not cold, not noise]; [focus_hypothesis(broken_engine), focus_hypothesis(broken_cooling_system), selected_observation(noise), confirmed(fridge_problem), confirmed(cooling_problem), confirmed(broken_engine)]
Observation Determination	[focus_hypothesis(broken_engine), focus_hypothesis(broken_cooling_system), observed(light), observed(cold)]
	[focus_hypothesis(broken_engine), focus_hypothesis(broken_cooling_system), observed(light), observed(cold), selected_observation(noise)]
Hypothesis Evaluation	[light, not cold, fridge_problem, cooling_problem]; [target(he_tcf, broken_engine, det), target(he_tcf, broken_cooling_system, det), true(cooling_problem)]
	[light, not cold, not noise, fridge_problem, cooling_problem]; [target(he_tcf, broken_engine, det), target(he_tcf, broken_cooling_system, det), true(cooling_problem)]
	[light, not cold, not noise, fridge_problem, cooling_problem, broken_engine]; [target(he_tcf, broken_engine, det), target(he_tcf, broken_cooling_system, det), true(cooling_problem),true(broken_engine)]
External World	[light, not cold]; [target(ew_tcf, noise, det)]
	[light, not cold, not noise]; [target(ew_tcf, noise, det)]

transferred to the output interface of Hypothesis Validation (20), and from there to the output interface of Diagnostic Reasoning (21).

The new observation to be performed, is transferred to the component External World (22), where it is actually performed (23). The observation result (not noise) is transferred to the component Diagnostic Reasoning (24), and further down to Hypothesis Validation (25) and Hypothesis Evaluation (26). This time this component is able to derive that one of the specific hypotheses (broken engine) is true (27). This result is transferred up to the output interfaces of Hypothesis Validation (28) and Diagnostic Reasoning (29), respectively.

6.1.2 Second example trace

The second example reasoning trace considered in this chapter is depicted in Figure 17. Steps (1) to (9) are exactly the same steps as for the first trace. A difference occurs at step (10). This time the observations results are (not light and cold). From (11) to (18) a similar pattern as in the first trace is followed: after transfer of the observation results (12, 13), within the component Hypothesis Evaluation, which has electricity problem and cooling problem as its targets, the truth of the abstract hypothesis electricity problem is derived from the observation information (not light, cold) (14). This information is transferred to Hypothesis Determination entailing revision (15). After this the new (specific) hypotheses to be validated broken bulb and no electricity supply are derived, based on the taxonomy of hypotheses (16). These focus hypotheses are transferred down to the lower process abstraction level of Hypothesis Evaluation and Observation Determination, where revision takes place (17, 18). A difference occurs at (19). In Observation Determination no new observation to be performed is found (19). At the same time, the component Hypothesis Evaluation is able to derive one of the specific hypotheses in focus (which, actually, are targets of Hypothesis Evaluation), namely broken bulb, without further observations. Therefore, this evaluation result is transferred up in the component hierarchy to Hypothesis Validation (20) and Diagnostic Reasoning (21), after which the diagnostic process stops.

Note that within the component Hypothesis Evaluation, the necessary information to derive broken bulb was already available at step (13). However, at that time broken bulb was not a target of this component, and therefore it was not derived. Only after the targets of Hypothesis evaluation had been changed and included broken bulb (18), was it actually derived (19). This shows how targets dynamically influence the reasoning behaviour of the system.

Figure 17. Second example trace description

time points	1 2 3 4 5 6 7 8 9 10 11 12 13 14 15 16
Diagnostic Reasoning	[ ]; [confirmed(fridge_problem)] [not light, cold]; [confirmed(fridge_problem), selected_observation(light), selected_observation(cold)] [ ]; [confirmed(fridge_problem), selected_observation(light), selected_observation(cold)]
Hypothesis Determination	[confirmed(fridge_problem)] [confirmed(fridge_problem), validated(electricity_problem), validated(cooling_problem), confirmed(electricity_problem)] [confirmed(fridge_problem), to_be_validated(electricity_problem), to_be_validated(cooling_problem)] [confirmed(fridge_problem), validated(electricity_problem), validated(cooling_problem), confirmed(electricity_problem), to_be_validated(broken_bulb), to_be_validated(no_power_supply)]
Hypothesis Validation	[ ] ; [focus_hypothesis(electricity_problem), focus_hypothesis(cooling_problem), confirmed(fridge_problem)] [not light, cold]; [focus_hypothesis(electricity_problem), focus_hypothesis(cooling_problem), confirmed(fridge_problem), selected_observation(light), selected_observation(cold)] [ ] ; [focus_hypothesis(electricity_problem), focus_hypothesis(cooling_problem), confirmed(fridge_problem), selected_observation(light), selected_observation(cold)]
Observation Determination	[focus_hypothesis(electricity_problem), focus_hypothesis(cooling_problem)] [focus_hypothesis(electricity_problem), focus_hypothesis(cooling_problem), selected_observation(light), selected_observation(cold)]
Hypothesis Evaluation	[fridge_problem]; [ ] [fridge_problem]; [target(he_tcf, electricity_problem,det), target(he_tcf, cooling_problem, det)] [fridge_problem, not light, cold]; target(he_tcf,electricity_problem, det), target(he_tcf,cooling_problem, det)] [not light, cold, fridge_problem, electricity_problem]; [target(he_tcf, electricity_problem det), target(he_tcf, cooling_problem, det), true(electricity_problem)]
External World	[ ] ; [target(ew_tcf, light, det), target(ew_tcf, cold, det)] [not light, cold]; [target(ew_tcf, light, det), target(ew_tcf, cold, det)]

time points	17	18	19	20	21
Diagnostic Reasoning					[not light, cold]; [confirmed(fridge_problem), confirmed(electricity_problem), confirmed(broken_bulb)]
Hypothesis Determination					
Hypothesis Validation	[not light, cold]; [focus_hypothesis(broken_bulb), focus_hypothesis(no_power_supply), selected_observation(light), selected_observation(cold), confirmed(fridge_problem), confirmed(electricity_problem)]				
				[not light, cold]; [focus_hypothesis(broken_bulb), focus_hypothesis(nopowersupply), selected_observation(light), selected_observation(cold), confirmed(fridge_problem), confirmed(electricity_problem) confirmed(broken_bulb)]	
Observation Determination		[focus_hypothesis(broken_bulb), focus_hypothesis(no_power_supply), observed(light), observed(cold)]			
		[focus_hypothesis(broken_bulb), focus_hypothesis(no_power_supply), observed(light), observed(cold)]			
Hypothesis Evaluation		[not light, cold, fridge_problem, electriciti_problem]; [target(he_tcf, broken_bulb, determine), target(he_tcf, no_power_supply, determine), true(electricity_problem)]			
			[not light, cold, fridge_problem, electricity_problem, broken_bulb]; [target(he_tcf, broken_bulb, determine), target(he_tcf, no_power_supply, determine), true(electricity_problem),true(broken_bulb)]		
external world					

6.2 *Control by dynamic generation of assumptions*

How to acquire and handle beliefs is an important but not simple task of an agent. In practice, information acquisition is often defeasible. Information acquired earlier may be found to be incorrect, and has to be retracted or revised. To support such processes the status of information can be modelled as well, for example whether or not a fact was assumed, or observed, or which agent communicated the fact. The decision of an agent to actually believe acquired information may depend on an estimation of the degree to which the source of the information is trustworthy. Therefore different beliefs of an agent may have a different status, and the agent has to be prepared to revise its beliefs in the light of newly acquired information anyway. A possible pattern of strategic reasoning is the following:

- identify the *required* information
- determine the *method to acquire* the required information; for example, one of
  - derive information in a *deductive* manner from available information
  - determine an appropriate *assumption*, and make this assumption
  - acquire additional information by *observation* (in interaction with the world)
  - acquire additional information by *communication* (ask another agent)
- *apply* the chosen method for information acquisition
- *verify* the obtained information in the light of other available information
- *integrate* the new information in the available information

In this section a reasoning method in which assumptions are dynamically added and retracted (sometimes called hypothetical reasoning), is discussed. The reasoning method is illustrated by a simple example of diagnostic reasoning on malfunctioning cars. Reasoning with and about assumptions entails deciding about a set of assumptions to be assumed for a while (reasoning *about* assumptions), and deriving which facts are logically implied by this set of assumptions (reasoning *with* assumptions). The derived facts may be evaluated; based on this evaluation some of the assumptions may be rejected and/or a new set of assumptions may be chosen (reasoning *about* assumptions). As an example, if an assumption was chosen, and the facts derived from this assumption contradict information obtained from a different source (e.g., by observation), the assumption may be rejected and the converse may be assumed.

Reasoning with and about assumptions is a reflective reasoning method. It proceeds by the following alternation of object level and meta-level reasoning, and upward and downward reflection:

- inspecting the information currently available (epistemic upward reflection),
- determining a set of assumptions (meta-level reasoning),
- assuming this set of assumptions (downward reflection of assumptions),
- deriving which facts follow from this assumed information (in the object level reasoning)
- inspecting the information currently available (epistemic upward reflection),

- evaluating the derived facts (meta-level reasoning)
- deciding to reject some of the assumptions and/or to choose a new set of assumptions based on this evaluation (meta-level reasoning).

and so on

As an example, if an assumption 'a is true' has been chosen, and the facts derived from this assumption contradict information that is obtained from a different source, the assumption 'a is true' may be rejected and the converse 'a is false' may be assumed. This reasoning pattern also occurs in diagnostic reasoning based on causal knowledge (discussed below).

6.2.1 Car diagnosis based on causal knowledge

In this section a simple diagnostic reasoning pattern on car malfunctioning is analysed. The causal knowledge on the domain of cars depicted in Figure 18 is used:

- if the battery is empty

  then the lights do not work

  and the car will not start
- if the sparking-plugs are tuned up badly

  then the car will not start

The causal knowledge could be easily extended, but for reasons of presentation the knowledge is kept limited.

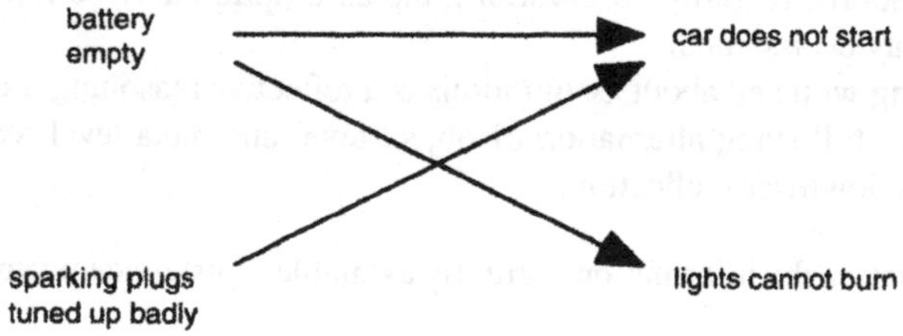

Figure 18. Causal knowledge for car diagnosis

At the meta-level the diagnostic reasoning process focusses on finding out whether an empty battery can be excluded as the cause of the problems. The following (simplified) meta-knowledge is used to reason about hypotheses: to propose hypotheses on which to focus, and to reject them if possible:

- if it has been observed that 'the car does not start'

  and it is not known whether the hypothesis 'the battery is empty' holds

  then 'the battery is empty' is an adequate hypothesis on which to focus

- if it has been observed that 'the car does not start'

  and it is true that 'the battery is non-empty'

  and it is not known whether the hypothesis 'the sparking-plugs are tuned up badly' holds

  then 'the sparking-plugs are tuned up badly' is an adequate hypothesis on which to focus

- if the focus is on a hypothesis X

  and assuming X it has been derived that the observable Y is the case

  and it has been observed that Y is not the case,

  then the hypothesis X should be rejected

In the following section the generic model behind this reasoning pattern is introduced.

6.2.2 A generic model for reasoning with and about assumptions

The generic model for reasoning with and about assumptions consists of four primitive components: External World, Predict Observation Results, Assumption Determination, Assumption Evaluation (see Figure 19). The first two of these components represent the object level, the last two the meta-level. The component Observation Result Prediction reasons with assumptions, the two components Assumption Determination and Assumption Evaluation reason about assumptions.

The generic model will be explained for the domain of car diagnosis. In this example, the reasoning pattern starts with the information that the car does not start. Using this information and the knowledge described above, the following reflective reasoning pattern is performed:

1. **component *Assumption Determination***
   From thc observation that the car does not start and that it is as yet unknown whether the battery is empty, draw the conclusion at the meta-level that 'battery is empty' is an adequate hypothesis on which to focus

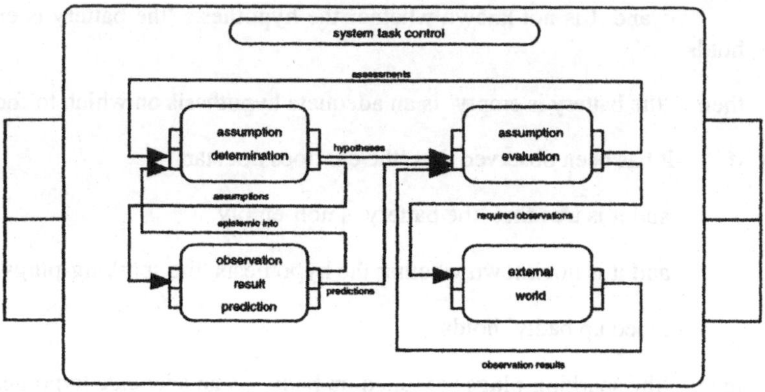

Figure 19. A generic model for reasoning with and about assumptions

2. **information link *Assumptions* (object-assumption)**
   Reflect this hypothesis downwards: introduce it at the object level as an assumption

3. **information link *Hypotheses* (object-object)**
   Transfer the hypothesis to Assumption Evaluation

4. **component *Observation Result Prediction***
   Draw the conclusion at the object level that the lights do not work

5. **information link *Predictions* (epistemic-object)**
   Reflect upwards the information that object level reasoning has predicted that the lights do not work

6. **component *Assumption Evaluation***
   Draw the conclusion at the meta-level that the observation to find out whether the lights work is a useful observation to perform

7. **information link *Observations* (object-target)**
   Reflect downwards the observation to be performed (as a target for the External World)

8. **component *External World*** Perform an observation in the external world to find out whether the lights work

9. **information link *Observation Results* (epistemic-object)**
   Reflect the observation result upwards

10. **component *Assumption Evaluation***
    At the meta-level use the observation result that the lights work, and notice that the actual observation result contradicts the prediction on the observation. Draw the conclusion that the focus hypothesis 'battery is empty' should be rejected

11. **information link *Assessments* (object-object)**

12. **component *Assumption Determination***

    and so on

Note the interaction between the two levels at points 5. and 9. (epistemic upward interaction), at point 2. (downward interaction to make an assumption), and at point 7. (downward interaction to set a target). These are the points in the traces, where interaction between levels takes place.

6.2.3 Overview of the components and information links

In this section a short overview is given of the components and information links of the generic model introduced above and their instantiations for the domain of car diagnosis.

Overview of the components

The generic model is not specified in detail. Instead, the components and information links are depicted in Figure 19, and relevant parts of the detailed design are presented.

External World

This component is used for executing observations. It has no knowledge base.

Relevant *input atoms*	(meta-level):	target(observations, A:OA, determine)
Relevant *output atoms*	(object level):	car_starts, lights_work
	(meta-level):	true(A:OA), false(A:OA), known(A:OA)

Observation Result Prediction

Based on the assumption, observations are predicted.

Relevant *input atoms*	(object level):	battery_empty, sparking_plugs_problem
	(meta-level):	assumption(A:IA, S:SIGN)
Relevant *output atoms*	(meta-level):	true(A:OA), false(A:OA), known(A:OA)

Relevant part of the *knowledge base*:

if battery_empty
then not lights_work
  and not car_starts

if sparking_plugs_problem
then not car_starts

Assumption Determination

Based on the current state of one diagnostic process, assumptions are generated.

Relevant *input atoms*:	rejected(H:HYPOTHESIS, S:SIGN), has_been_considered(H:HYPOTHESIS), observation_result(O:OBSERVATION, S:SIGN)
Relevant *output atoms*:	poss_assumption(H:HYPOTHESIS, S:SIGN)

Relevant part of the *knowledge base:*

if observation_result(car_starts, neg)
  and not has_been_considered(battery_empty)
then poss_assumption(battery_empty, pos)

if rejected(battery_empty, pos)
  and not has_been_considered(sparking_plugs_problem)
then poss_assumption(sparking_plugs_problem, pos)

if rejected(H:HYPOTHESIS, pos)
then poss_assumption(H:HYPOTHESIS, neg)

Assumption Evaluation

Based on the selected assumption, the predicted and actual observation result, an evaluation is made.

Relevant *input atoms*:	assumed(H:HYPOTHESIS, S:SIGN), predicted(O:OBSERVATION, S:SIGN), observation_result(O:OBSERVATION, S:SIGN)
Relevant *output atoms*:	rejected(H:HYPOTHESIS, S:SIGN), has_been_considered(H:HYPOTHESIS) to_be_observed(O:OBSERVATION)

Relevant part of the *knowledge base*:

if predicted(O:OBSERVATION, S:SIGN)
then to_be_observed(O:OBSERVATION)

if assumed(H:HYPOTHESIS, S:SIGN)
  and predicted(O:OBSERVATION, pos)
  and observation_result(O:OBSERVATION, neg)
then rejected(H:HYPOTHESIS, S:SIGN)

if assumed(H:HYPOTHESIS, S:SIGN)
  and predicted(O:OBSERVATION, neg)
  and observation_result(O:OBSERVATION, pos)
then rejected(H:HYPOTHESIS, S:SIGN)

if assumed(H:HYPOTHESIS, S:SIGN)
then has_been_considered(H:HYPOTHESIS)

Note that the knowledge base specified above contains domain independent knowledge and the knowledge is rather minimal: the example reasoning pattern can be generated but not much more. This could be easily extended.

Overview of the information links
The information links of the generic model for reasoning with and about assumptions are shortly described as follows.

***Assumptions* (type object-assumption)**
The truth values of instances of output atoms poss_assumption(H:HYPOTHESIS, S:SIGN) of Assumption Determination are transferred to the same truth values of input meta-facts assumption(H:IA, S:SIGN) of the component Observation Result Prediction. As a result the hypothesis to which the assumption refers can be used in this component.

***Epistemic Information* (type epistemic-object)**
The truth values of output meta-facts of the form true(O:OA) of the component Observation Result Prediction are transferred to the same truth values of object level input atoms of the component assumption evaluation of the form known_to_hold (O:OBSERVATION, pos). Similarly, the truth values of output meta-facts of the form false(O:OBSERVATION) of the component Observation Result Prediction are transferred to the same truth values of object level input atoms of the component Assumption Evaluation of the form known_to_hold(O:OBSERVATION, neg).

***Predictions* (type epistemic-object)**
The truth values of output meta-facts of the form true(O:OA) of the component Observation Result Prediction are transferred to the same truth values of object level input atoms of the component Assumption Evaluation of the form

predicted(O:OBSERVATION, pos). Similarly, the truth values of output meta-facts of the form false(O:OBSERVATION) of the component Observation Result Prediction are transferred to the same truth values of object level input atoms of the component Assumption Evaluation of the form predicted(O:OBSERVATION, neg).

***Required Observations* (type object-target)**
The truth values of the object level output atoms to_be_observed(O:OBSERVATION) of the component Assumption Evaluation are transferred to the same truth values of meta-level input atoms target(observations, O:OA, determine) of the component External World.

***Observation Results* (type epistemic-object)**
This interaction transfers truth values of output meta-facts of the form true(O:OA) (resp. false(O:OA)) of the component External World to the same truth values of object level input atoms of the component assumption evaluation of the form observation_result(O:OBSERVATION, pos) (resp. observation_result(O:OBSERVATION, neg)).

***Hypotheses* (type object-object)**
The truth values of the object level output atoms of the form poss_assumption(H:HYPOTHESIS, S:SIGN) of the component Assumption Determination are transferred to the same truth values of object level input atoms assumed(H:HYPOTHESIS, S:SIGN) of the component Assumption Valuation.

***Assessments* (type object-object)**
This interaction transfers truth values of object level output facts of the form rejected(H:HYPOTHESIS) and has_been_considered(H:HYPOTHESIS) of the component Assumption Evaluation to the same truth values of identical object level input atoms of the component Assumption Determination.

6.2.4 The dynamics of the reasoning method

Task control knowledge controls the activation of the components. For instance, first the component Assumption Determination is activated, next the component Observation Result prediction; if this succeeds (which is the case in the example trace below), then the component Assumption Evaluation is made active. After activation of the component external world to perform the observation, the component assumption evaluation is again activated. An example trace of a part of such a reasoning pattern is depicted in the example in Figure 20 starting with the initial information in the component Assumption Determination that it was observed that car starts is false and that the truth or falsity of the hypotheses battery empty and sparking plugs problem have not, as yet, been determined.

This example trace combines traces for the four components. For the object level components Observation Result Prediction and External World the information states for two levels are depicted as:

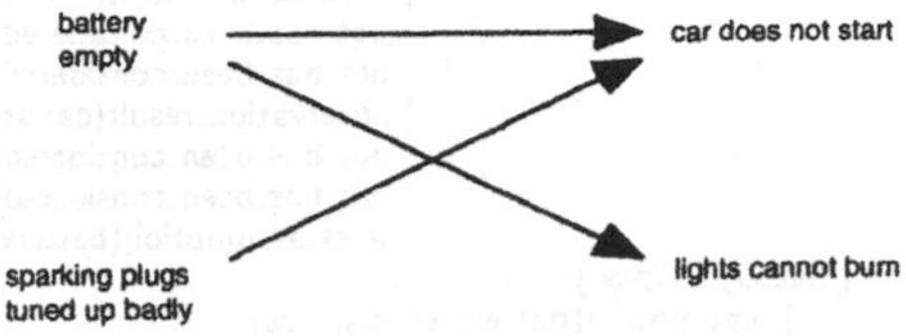

Figure 20. Trace of a hypothetical reasoning process

[⟨object level information state⟩];
[⟨meta-level information state⟩]

These meta-level information states show how the interaction between levels takes place. Their content is directly related to the content of the object level information state. For example, in the example trace, after the first activation (1)-(2) of the component Assumption Determination the information link Assumptions is made uptodate. This means that at the next moment in time the meta-statement assumption(battery_empty, pos) is true in the meta-information state of Observation Result Prediction (3). However, at the same moment the assumption is actually made: at the object level the atom battery_empty becomes true. This is an example of (downward) interaction between the levels. As a consequence the meta-atom true(battery_empty) gets truth value true in the meta-level information state (upward interaction between the levels). As soon as the information link becomes uptodate (i.e., at the next moment in time), this whole revision process is assumed to have finished. At this point (3) only the result of the process of level interaction is visible, the different steps as sketched are not recorded separately. Therefore, the interaction between the levels is considered to take place instantaneously, as soon as new information arrives as input.

The same holds in the opposite direction. Based on the assumption battery empty the component Observation Result Prediction derives that lights work is false (4). At the same time the related epistemic meta-atoms have been assigned their appropriate truth values; e.g., false(lights_work) gets truth value true, which, again, is the actual (upward) interaction between the levels. This meta-information is transferred to the component Assumption Evaluation by the information link predictions (5).

A next interaction between levels can be found after the component Assumption Evaluation has derived the atom to_be_observed(lights_work) (6). The information link Required Observations transfers this information to the meta-atom target(observations, lights_work, determine) of the component External World (7). This time the actual (downward) interaction between the levels immediately determines control of the the component. The effect of the target is that the pro-

Time point	External World	Observation Result Prediction	Assumption Determination	Assumption Evaluation
1			[observation_result(car_starts, neg), not has_been_considered(battery_empty), not has_been_considered(sparking_plugs_problem)]	
2			[observation_result(car_starts, neg), not has_been_considered(battery_empty), not has_been_considered(sparking_plugs_problem), poss_assumption(battery_empty, pos)]	
3		[battery_empty]; [assumption(battery_empty, pos), true(battery_empty)]		
4		[battery_empty, not lights_work]; [assumption(battery_empty, pos), true(battery_empty), false(lights_work)]		
5				[assumed(battery_empty, pos), predicted(lights_work, neg)]
6				[assumed(battery_empty, pos), predicted(lights_work, neg), to_be_observed(lights_work)]
7	[ ]; [target(observations, lights_work, determine)]			
8	[lights_work]; [target(observations, lights_work, determine), true(lights_burn)]			
9				[assumed(battery_empty, p predicted(lights_work, n to_be_observed(lights_wo observation_result(lights_work, p
10				[assumed(battery_empty, p predicted(lights_work, n to_be_observed(lights_wo observation_result(lights_work, p rejected(battery_empty, pos), has_been_considered(battery_empty)
11			[observation_result(car_starts, neg), has_been_considered(battery_empty), not has_been_considered(sparking_plugs_problem) rejected(battery_empty, pos)]	
12			[observation_result(car_starts, neg), has_been_considered(battery_empty), rejected(battery_empty, pos), poss_assumption(battery_empty, neg), poss_assumption(sparking_plugs_problem, pos)]	
13		[not battery_empty, sparking_plugs_problem]; [assumption(battery_empty, neg), assumption(sparking_plugs_problem, pos) false(battery_empty), true(sparking_plugs_problem)]		
14		[not battery_empty, sparking_plugs_problem, not car_starts]; [assumption(battery_empty, neg), assumption(sparking_plugs_problem, pos) false(battery_empty), true(sparking_plugs_problem), false(car_starts)]		

cess focusses on lights_work only and does not try to find truth values for other atoms. For example, if the component External World actually stands for a human observer, then the target tells him or her what to observe. If, instead, the Exterenal World is a reasoning component, then the target focuses its reasoning by a goal-directed inference strategy with the target as its goal.

The result of the observation is that lights_work becomes true in the object level information state of the component External World (8). Again, an upward level interaction from object level to meta-level takes place instantaneously: at the same moment true(lights_work) becomes true. This result is transferred to the component Assumption Evaluation (9).

A fifth example of level interaction in the example trace is the second time assumptions are made in the component Observation Result Prediction. This time the earlier made assumption that battery empty is true is retracted. This takes place because, due to revision within the component Assumption Determination, the earlier drawn conclusion poss_assumption(battery_empty, pos) is retracted, which means that it is assigned truth value unknown. This truth value is propagated through the link Assumptions: the meta-level atom assumption(battery_empty, pos) of the object level component Observation Result Prediction is assigned the truth value unknown, the effect of which is that within the same component the object level atom battery_empty is assigned the truth value unknown (downward level interaction). At the same time the meta-level atom assumption(battery_empty, neg) is provided as input, which by downward level interaction assigns the truth value false to the object level atom empty_battery (note that in this downward level interaction the truth value unknown for the object level atom empty_battery is overruled by the truth value false). Furthermore, the meta-level atom assumption(sparking_plugs_problem, pos) is provided as input, which, by downward level interaction assigns the truth value true to the object level atom sparking_plugs_problem. After all these downward level interactions, in an upward level interaction the epistemic information at the meta-level is updated, which finishes the whole level interaction and results in the state at time point (13).

The generic model for reasoning with and about assumptions or parts of it has been used to model, for example, default reasoning and reasoning based on the closed world assumption.

7 DISCUSSION

As shown in this paper compositional DESIRE models specify processes and knowledge at different levels of abstraction. Information exchange between processes and process sequencing are explicitly defined at each of the levels distinguished. Different levels of abstraction within the knowledge composition structure information types and knowledge bases. Reuse of generic models within DESIRE is supported by their transparent compositional structure. The basic principles behind compositional multi-agent system design described in this paper (process and knowledge abstraction, compositionality, reusability) are principles generally acknowledged to be of importance in both software engineering and knowledge engineering. The operationalisation of these principles within a compositional design method for multi-agent systems is, however, a distinguishing element. The method is supported by a (graphical) software environment in which all three levels of design are supported: from conceptual design to implementation. Libraries of both generic models and instantiated components, of which a few have been highlighted in this paper, support system designers at all levels of design. Generic agent models, generic task models and generic models of reasoning patterns help structure the process of system design. Formal semantics provide a basis for methods for verification — an essential part of such a method.

A number of approaches to conceptual-level specification of multi-agent systems have been recently proposed. On the one hand, general-purpose formal specification languages stemming from Software Engineering are applied to the specification of multi-agent systems (e.g., [Luck and d'Inverno, 1995] for an approach using Z). A compositional design method such as DESIRE is committed to well-structured compositional designs that can be specified at a higher level of conceptualisation than in Z or VDM and can be implemented automatically using automated prototype generators. On the other hand, new development methods for the specification of multi-agent systems have been proposed. These methods often commit to a specific agent architecture. For instance, [Kinny *et al.*, 1996] describe a language on the one hand based on the BDI agent architecture and on the other hand based on object-oriented design methods. A more in depth comparative analysis of these methods from the perspective of compositionality and the related principles presented in this paper would be interesting further research.

The compositional approach to agent design followed in this paper has some aspects in common with object oriented design methods; e.g., [Booch, 1994; Coleman *et al.*, 1994; Rumbaugh *et al.*, 1991]. However, there are differences as well. Examples of approaches to object-oriented agent specifications can be found in [Aridor and Lange, 1998; Kendall *et al.*, 1998]. A first interesting point of discussion is to what the difference is between agents and objects. Some tend to classify agents as different from objects. For example, [Jennings and Wooldridge, 1998a]) compare objects with agents on the dimension of autonomy in the following way:

> "An object encapsulates some state, and has some control over this state in that it can only be accessed or modified via the methods that the object provides. Agents encapsulate state in just the same way. However, we also think of agents as encapsulating behavior, in addition to state. An object does not encapsulate *behavior*: it has no control over the execution of methods – if an object x invokes a method m on an object y, then y has no control over whether m is executed or not – it just *is*. In this sense, object y is not autonomous, as it has no control over its own actions. In contrast, we think of an agent as having *exactly* this kind of control over what actions it performs. Because

> of this distinction, we do not think of agents as invoking methods (actions) on agents – rather, we tend to think of them *requesting* actions to be performed. The decision about whether to act upon the request lies with the recipient."

Some others consider agents as a specific type of objects that are able to decide by themselves whether or not they execute a method (objects that can say 'no') upon a received message, and that can initiate action (objects that can say 'go') without any message received.

A difference between the compositional design method DESIRE and object-oriented design methods in representation of basic functionality is that within DESIRE declarative, knowledge-based specification forms are used, whereas method specifications (which usually have a more procedural style of specification) are used in object-oriented design. Another difference is that within DESIRE the composition relation is defined in a more specific manner: the static aspects by information links, and the dynamic aspects by (temporal) task control knowledge, according to a prespecified format. A similarity is the (re)use of generic structures: generic models in DESIRE, and patterns [Alexander, 1977; Gamma *et al.*, 1995; Fowler, 1997; Grand, 1998] in object-oriented design methods, although their functionality and compositionality are specified in different manners, as discussed above.

The use of reflection principles to specify the control of complex dynamic reasoning patterns in diagnosis is described in [Treur, 1991]. The approach to diagnosis using fine-grained dynamic control of (limited) reasoning by means of dynamic generation of targets in particular was described in [Treur, 1993]. Temporal semantics for the dynamics of this reasoning pattern were introduced in [Treur, 1994]. The generic model for reasoning with and about dynamic assumptions was introduced in [Treur, 1992]. Semantics for this approach to reasoning with and about dynamic assumptions (temporal epistemic reflection) were introduced in [Hoek *et al.*, 1994]. In [Brazier *et al.*, 2000] more details about these diagnostic reasoning models can be found. In [Brazier *et al.*, 1995; Brazier *et al.*, 1996; Brazier *et al.*, 1998a; Brazier *et al.*, 1998b] a number of applicatios of DESIRE can be found. In [Brazier *et al.*, 1998c] an overview of a large number of reusable modell in DESIRE is given. Compositional verification is addressed in [Jonker and Treur, 1998].

ACKNOWLEDGEMENT

Copyright 2000. From Compositional Design and Reuse of a Generic Agent Model by F. M. T. Brazier, C. M. Jonker and J. Treur. Reporduced by permission of Taylor & Francis, Inc., http://www.routledge-ny.com

Frances M.T. Brazier, Catholijn M. Jonker and Jan Treur
Vrije Universiteit Amsterdam, The Netherlands.

BIBLIOGRAPHY

[Alexander, 1977] C. Alexander. *A Pattern Language*. Oxford University Press, 1997.

[Aridor and Lange, 1998] Y. Aridor and D.B. Lange. Agent Design Patterns: Elements of Agent Application Design. *Proc. of the Second Annual Conference on Autonomous Agents, Agents98*, ACM Press, 108–115, 1998.

[Barringer *et al.*, 1996] H. Barringer, M. Fisher, D. Gabbay, R. Owens, and M. Reynolds, (eds.). *The Imperative Future: Principles of Executable Temporal Logics*. Research Studies Press, Chichester, United Kingdom, 1996.

[Booch, 1994] G. Booch. *Object-Oriented Analysis and Design* (2nd ed.). Addison-Wesley, 1994.

[Brazier *et al.*, 1995] F.M.T. Brazier, B.M. Dunin-Keplicz, N.R. Jennings and J. Treur. Formal specification of Multi-Agent Systems: a Real-World Case. In: V. Lesser (ed.), *Proc. of the First International Conference on Multi-Agent Systems, ICMAS95*, MIT Press, Cambridge, MA, 25–32. Extended version in: *International Journal of Cooperative Information Systems*, M. Huhns, M. Singh, (eds.), special issue on Formal Methods in Cooperative Information Systems: Multi-Agent Systems, **6**, 67–94, 1997.

[Brazier *et al.*, 1996] F.M.T. Brazier, C.M. Jonker and J. Treur. Modelling Project Coordination in a Multi-Agent Framework. In: *Proceedings of the Fifth Workshops on Enabling Technology for Collaborative Enterprises, WET ICE'96*, IEEE Computer Society Press, 148–155, 1996. Extended version in *Int. Journal of Cooperative Information Systems*, **9**, 171–207, 2000.

[Brazier *et al.*, 1998a] F.M.T. Brazier, F. Cornelissen, R. Gustavsson, C.M. Jonker, O. Lindeberg, B. Polak and J. Treur. Agents Negotiating for Load Balancing of Electricity Use. In: M.P. Papazoglou, M. Takizawa, B. Krämer, S. Chanson (eds.), *Proceedings of the 18th International Conference on Distributed Computing Systems, ICDCS'98*, IEEE Computer Society Press, 622–629, 1998.

[Brazier *et al.*, 1998b] F.M.T. Brazier, C.M. Jonker, F.J. Jungen and J. Treur. Distributed Scheduling to Support a Call Centre: a Co-operative Multi-Agent Approach. In: *Proceedings of the Third International Conference on the Application of Intelligent Agents and Multi-Agent Technology*, H.S. Nwana and D.T. Ndumu (eds.), The Practical Application Company, Blackpool, 555–576, 1998. Also in: *Applied Artificial Intelligence Journal*, **13**, 65–90, 1999. Special Issue with selected papers from PAAM98.

[Brazier *et al.*, 1998c] F.M.T. Brazier, C.M. Jonker and J. Treur. Principles of Compositional Multi-agent System Development. In: J. Cuena (ed.), *Proceedings of the 15th IFIP World Computer Congress, WCC'98*, Conference on Information Technology and Knowledge Systems, IT&KNOWS'98, 347–360, 1998.

[Brazier *et al.*, 2000] F.M.T. Brazier, C.M. Jonker, J. Treur and N.J.E. Wijngaards. On the Use of Shared Task Models in Knowledge Acquisition, Strategic User Interaction and Clarification Agents. *International Journal of Human-Computer Studies*, **52**, 77–110, 2000.

[Coleman *et al.*, 1994] D. Coleman, P. Arnold, S. Bodoff, C. Dollin, H. Gilchrist, F. Hayes and P. Jeremaes. *Object-Oriented Development: the FUSION method*. Prentice Hall International: Hempel Hempstead, England, 1994.

[Fowler, 1997] M. Fowler. *Analysis Patterns: Reusable Object Models*. Addison Wesley, 1997.

[Gamma *et al.*, 1995] E.R. Gamma, R. Helm, R. Johnson and J. Vlissides. *Design Patterns: Elements of Reusable Object-Oriented Software*. Addison-Wesley, 1995.

[Grand, 1998] M. Grand. *Patterns in Java: Volume 1*. John Wiley and Sons, 1998.

[Hoek *et al.*, 1994] W. van der Hoek, J.-J. Meyer and J. Treur. Formal semantics of temporal epistemic reflection, In: L. Fribourg and F. Turini (eds.), *Logic Program Synthesis and Transformation-Meta-Programming in Logic, Proc. Fourth Int. Workshop on Meta-programming in Logic, META'94*. Lecture Notes in *Computer Science*, **883**, Springer Verlag, 332–352, 1994.

[Jennings and Wooldridge, 1998a] N.R. Jennings and M. Wooldridge. *Applications of Intelligent Agents*. In: Jennings and Wooldridge, 3–28, 1998.

[Jennings and Wooldridge, 1998b] N.R. Jennings and M. Wooldridge (eds.), *Agent Technology: Foundations, Applications, and Markets*. Springer Verlag, 1998.

[Jonker and Treur, 1998] C.M. Jonker and J. Treur. Compositional Verification of Multi-Agent Systems: a Formal Analysis of Pro-activeness and Reactiveness. In: W.P. de Roever, A. Pnueli *et al.* (eds.), *Proceedings of the International Workshop on Compositionality, COMPOS97*, Springer Verlag, 1998. Extended version in *Int. Journal of Cooperative Information Systems*, in press, 2001.

[Kendall *et al.*, 1998] E.A. Kendall, P.V. Murali Krisna, C.V. Pathak and C.B. Suresh. *Proc. of the Second Annual Conference on Autonomous Agents, Agents98*. ACM press, 1998.

[Kinny *et al.*, 1996] D. Kinny, M.P. Georgeff and A.S. Rao. A Methodology and Technique for Systems of BDI Agents. In: W. van der Velde, J.W. Perram (eds.), *Agents Breaking Away, Proc. 7th European Workshop on Modelling Autonomous Agents in a Multi-Agent World, MAAMAW'96*, Lecture Notes in AI, **1038**, Springer Verlag, 56–71, 1996.

[Luck and d'Inverno, 1995] M. Luck and M. dInverno. A formal framework for agency and autonomy. In: V. Lesser (ed.) Proc. of *The first International Conference on Multi-Agent Systems, ICMAS95*, AAAI Press, 254–260, 1995.

[Rumbaugh *et al.*, 1991] J. Rumbaugh, M. Blaha, W. Pelerlani, F. Eddy and W. Lorensen. *Object-Oriented Modelling and Design*, Prentice Hall, Eaglewoods Clifs, NJ, 1991.

[Treur, 1991] J. Treur. On the use of reflection principles in modelling complex reasoning. *International Journal of Intelligent Systems*, **6**, 277–294, 1991.

[Treur, 1992] J. Treur Interaction types and chemistry of generic task models. In: M. Linster and B. Gaines (eds.). *Proc. of the Fifth European Knowledge Acquisition Workshop, EKAW91*. GMD Studien, **211**, 390–414, 1992.

[Treur, 1993] J. Treur. Heuristic reasoning and relative incompleteness. *International Journal of Approximate Reasoning*, **8**, 51–87, 1993.

[Treur, 1994] J. Treur Temporal Semantics of Meta-Level Architectures for Dynamic Control of Reasoning. In: L. Fribourg and F. Turini (eds.), Logic Program Synthesis and Transformation-Meta-Programming in Logic, *Proceedings of the Fourth International Workshop on Meta-Programming in Logic, META'94*, Springer Verlag. Lecture Notes in *Computer Science*, **883**, 353–376, 1994.

[Treur and Willems, 1994] J. Treur and M. Willems. A logical foundation for verification. In: A.G. Cohn (ed.), Proc. of The 11th European Conference on Artificial Intelligence, ECAI'94. John Wiley & Sons, Chichester, 745–749, 1994.

LLUÍS GODO, JOSEP PUYOL-GRUART AND CARLES SIERRA

CONTROL TECHNIQUES FOR COMPLEX REASONING: THE CASE OF *MILORD II*

1 INTRODUCTION

Reasoning patterns occurring in complex problem solving tasks usually cannot be modelled by means of just a pure classical logic approach. This is due to several reasons, for instance: incompleteness of the available information, need of using and representing uncertain or imprecise knowledge, or combinatorial explosion of classical theorem proving when knowledge bases become large. To deal with these problems, *Milord II*, an architecture for Knowledge Base Systems (KBS), combines modularization techniques with both implicit and explicit control mechanisms and with an approximate reasoning component based on many-valued logics.

Roughly speaking, a Knowledge Base (KB) in *Milord II* consists of a hierarchy of modules interconnected by their export interfaces. Each module contains an Object Level Theory (OLT) and a Meta-Level Theory (MLT) interacting through a reflective mechanism (see Figure 1).

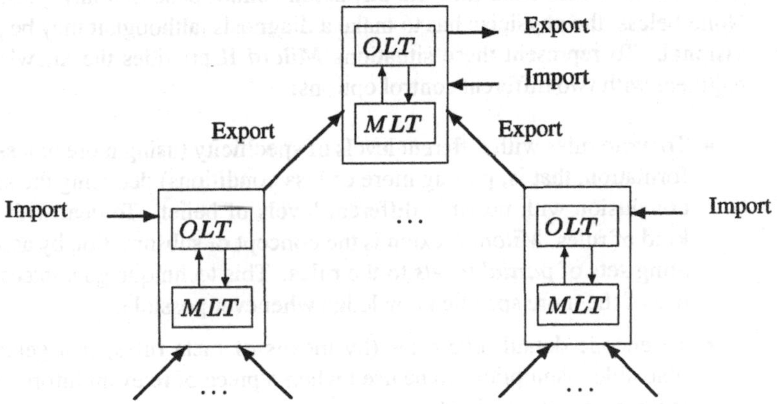

Figure 1. Structure of a *Milord II* module hierarchy.

A module can be understood as a functional abstraction between the set of components it needs as input and the type of results it can produce. From the logical point of view, *Milord II* makes use of both many-valued logic and epistemic

J.J.Ch. Meyer and J. Treur (eds.),
Handbook of Defeasible Reasoning and Uncertainty Management Systems, Vol. 7, 65–97.

meta-predicates to express the truth status of propositions. For further details in these logical topics the reader is referred to [Godo *et al.*, 1995; Puyol *et al.*, 1992; Puyol-Gruart *et al.*, 1998; Puyol-Gruart and Sierra, 1997; Sierra and Godo, 1992; Sierra and Godo, 1993].

In this chapter we focus on the control techniques used in *Milord II* that determine a KBS execution. The explicit part of the control, declarative in nature, is mainly based on a reflective approach and a declarative backtracking mechanism. In this context, reflection makes sense as a control mechanism because there is a clear separation between domain (object-level) and control (meta-level) knowledge. The basic implicit control components are a subsumption mechanism and a process of elimination of unnecessary rules, both concerning the object level.

Next we list the most usual control requirements for a KBS language together with the solutions adopted in *Milord II*.

Locality of Control: All explicit control mechanisms are specified locally to each module. This allows us to identify a module as the complete description of a problem (or subproblem). The separation between domain and control knowledge is a typical characteristic of most KBS languages to offer a clear and declarative programming style.

Specificity versus generality: To solve problems, human experts are able to reason at different levels of precision depending on the amount of data at hand. For instance, a physician cannot always gather all the relevant data to make a complete and accurate diagnosis. This is the case, for example, when a patient is in a coma and thus the physician cannot pose him any question. Nonetheless, the physician has to make a diagnosis, although it may be provisional. To represent these situations *Milord II* provides the knowledge engineer with two different control options:

- To write rules with different levels of specificity (using more or less information, that is, putting more or less conditions) deducing the same conclusion with possibly different levels of belief. To deal with this kind of rules, *Milord II* extends the concept of subsumption by associating sets of *partial labels* to the rules. This technique guarantees the use of the more specific knowledge whenever possible.
- To encode default-like rules (by means of meta-rules) that generate plausible assumptions to be used when a piece of relevant information is missing (see Section 9).

Avoidance of unnecessary work: *Milord II* takes advantage of the specialization deductive mechanism [Puyol *et al.*, 1992; Puyol-Gruart *et al.*, 1998] to eagerly detect when a rule cannot increase the certainty on a conclusion. When a rule is applied, *Milord II*'s engine decides whether other rules with the same conclusion can increase its certainty or not. If not, they are removed.

Locality of threshold: In some cases knowledge engineers are interested in programming modules whose deduced predicates are only useful if their certainty is above a minimum truth level. This is done by declaring a *threshold* local to each module. Whenever a rule gets, by specialization, a truth interval with its minimum value below the module threshold, it is removed.

Flexibility in data gathering: Given a query to a module, different strategies for the module to get an answer can be used. The different evaluation strategies of *Milord II* determine how and in which order the necessary external information is gathered.

Declarativity of Control: *Milord II* Horn-like meta-rules are used as a declarative language to implement several control actions, e.g. elimination of rules, generation of plausible assumptions, dynamic changing of the modules hierarchy or dynamic creation of modules.

The detailed description of the different implicit and explicit control mechanism of *Milord II* is structured in this Chapter as follows. In Section 2 we present a general picture of the whole control structure. Sections 3 through 8 are devoted to describe the different *Milord II* control mechanisms, that is, the object level process, the upwards reflection operation, the meta-level process, the downwards reflection operation and the communication among modules respectively. In Section 9 two reasoning tasks are implemented using some of the previously presented control mechanisms.

2 *MILORD II* OVERVIEW

A *Milord II* KB consists of a hierarchy of modules, each module containing different kinds of knowledge, structured as sketched in Figure 2.

From a logical point of view, a module is composed of an Object Level Theory (OLT) and a Meta Level Theory (MLT). The OLT is generated by a set of rules which are specified in the `Deductive Knowledge` definition. These rules are formulas belonging to the Object Level Language $\mathcal{OL}_n$, a propositional language based on many-valued semantics. Formulas of this language are of the form (r, V), where r is a Horn-like rule and V is an interval of truth values belonging to a finite and totally ordered set of values, also specified in the module declaration. Deduction in the object level language, denoted $\vdash_{\mathcal{O}}$, is mainly based on a *specialization inference rule*, a straightforward generalization of the many-valued version of Modus Ponens, which allows to simplify rules as soon as we know truth-intervals for any of their conditions. On the other hand, the MLT is generated by a set of meta-rules which are specified in the `Deductive Control` definition. These meta-rules are formulas of the Meta Level Language $\mathcal{ML}_n$, a restricted first order classical language of Horn rules. Variables in meta-rules, if any, are considered universally quantified. Deduction at the meta level, denoted by

$\vdash_{\mathcal{M}}$, is based on Modus Ponens and particularization. The overall reasoning process of a module consists on reasoning at each level and interacting between both levels. This process produces a sequence of modifications over the initial OLT and MLT. For a deeper insight of *Milord II* modules, the reader is again referred to [Godo *et al.*, 1995; Sierra and Godo, 1993].

From an operational point of view, a module can be identified with a process attached to it, used to compute values (truth intervals) for all the propositions and variables contained in its export interface. Namely, a module execution consists of the reasoning process necessary to compute the values for the propositions and variables in the export interface the user queries about. The execution of a module can possibly activate the execution of submodules in the hierarchy. These executions only interact with the parent module through the export interface of the submodules, giving formulas back as result. It is worth noticing that the interaction is made only at the object level.

```
Begin
            Hierarchy of submodules
            Import:   ...
            Export:   ...
            Deductive knowledge
                 Dictionary:  ...
                 Rules:  ...
                 Inference System:   ...
                         Truth-values:   ...
                         Connectives:  ...
                         Renaming:  ...
            end deductive
            Control knowledge
                 Evaluation Type:  ...
                 Truth Threshold:  ...
                 Deductive Control:   ...
                 Structural Control:  ...
            end control
end
```

Figure 2. Knowledge components of a module.

Conceptually, the execution of a module involves two deductive subprocesses, object and meta-level, that act as co-routines, and three operations. Two of them, *upwards reflection* and *downwards reflection*, are resume-type operations between the co-routines that, besides acting as resuming operations, modify the knowledge used by the deductive subprocesses. The third operation is the *communication* with the user and/or other modules that has the effect of adding new formulas to the object-level co-routine. Figure 3 shows the structure of a module and the relations

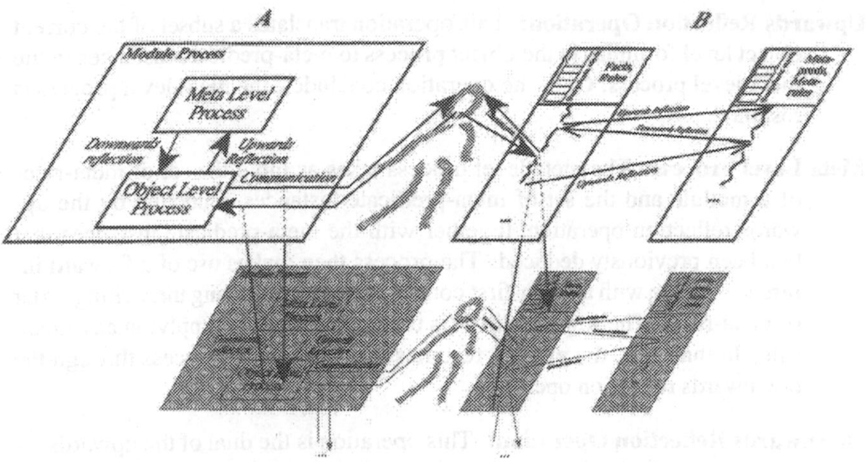

Figure 3. A: structure of the components of *Milord* module process.
B: Co-routine view of a module process.

between its components. Besides that, the module evaluation type determines in which way subprocesses and operations are combined to get the global control behaviour of a module execution.

Next we succinctly describe each one of the above mentioned processes and operations.

Object Level Process: This process uses as data the set of propositional variables and rules of the module possibly updated by the previous downwards reflection and communication operations. With this data and a goal to be solved, the task of the process is to obtain a value for the goal and potentially for other propositional variables of the module. Obtaining a value for a propositional variable can be done in one of the following ways:[1] by using the ***communication*** operation, either by querying the user (when the propositional variable is declared as ***Import***) or querying a submodule (when the propositional variable belongs to the export interface of a submodule); or by ***deduction*** when the propositional variable is the conclusion of a rule. To do so, the process follows the rule specialization algorithm with two implicit control mechanisms, namely the ***subsumption*** and the ***elimination of unnecessary rules***, and a parametric control mechanism, the ***truth-threshold rule elimination***.

The type of evaluation determines when the control is passed to the upwards reflection or to the communication operations.

[1] Actually there are other ways such as functional evaluation—in the case of propositional variables with an attached function—or constraint propagation, but they are out of the scope of this paper.

Upwards Reflection Operation: This operation translates a subset of the current object level formulas in the object process to meta-predicate instances in the meta-level process. Once the operation concludes, the meta-level process is resumed.

Meta Level Process: The meta level process takes as input the set of meta-rules of a module and the set of meta-predicate instances generated by the upwards reflection operation, together with the meta-predicate instances that had been previously deduced. The process then makes use of a forward inference engine with a depth-first control strategy, following the writing order of meta-rules. The stop condition is the impossibility of applying any meta-rule. In that case, the process resumes the object level process through the downwards reflection operation.

Downwards Reflection Operation: This operation is the dual of the upwards reflection one. It translates formulas from the meta-level process into the object level one and executes the actions determined by the meta-level process. When the translation is finished, the object level process is resumed. Special mention has to be made when an instance of the action $Assume$ is applied. In this case, as many extensions of the meta-theory MLT as elements in the argument of the $Assume$ action are generated (see Figure 4). These extensions conform a tree of MLTs. Every time an $Assume$ action is executed a new branching is added to this tree. Whenever a *Resume* action is executed, a backtracking in that tree is performed and the computation is resumed. This is how the declarative backtracking mechanism (see Section 7) is implemented.

Communication: This operation is used to add new formulas to the object-level process either from the external user or from other modules of the KB. The evaluation type determines when this operation is to be applied.

The control mechanisms determine the algorithmic behaviour of the processes themselves or just the way processes and operations are combined. The combination of the previous processes and operations is done by the explicit declaration of the evaluation strategy inside each module. In *Milord II* there are three evaluation strategies: *lazy*, *eager* and *reified*. Each one of them produces a different behaviour. On the one hand, the lazy and eager evaluation types are opposite strategies about how to obtain external data (from the user and/or from its submodules). The lazy strategy always tries to use the minimum information while the eager strategy makes use of as much information as possible.

In the following algorithmic descriptions of the evaluation strategies, OLP stands for object level process and MLP for meta level process.

Lazy: A module with lazy evaluation finds the cheapest path to compute a solution for a goal, that is, no irrelevant data will ever be gathered. The control

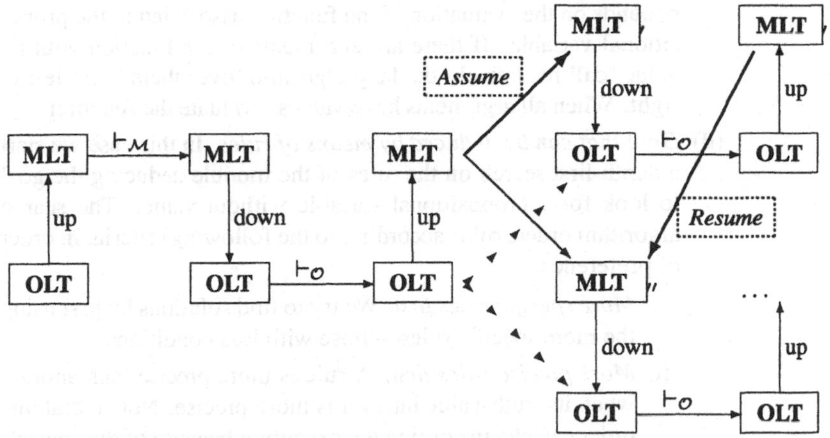

Figure 4. Module processes and operations. *Assume* and *Resume* actions.

used in the module to answer a query, in a simplified view, is a loop over first finding the next relevant propositional variable to look for a value and then specializing the deductive knowledge. This cycle is repeated until the goal is solved or no more relevant questions exist. This is the evaluation strategy used by default.

Given a query to a lazy module, the control flow of the module process is the following one:[2]

1. [*OLP*] If the goal has already a value, STOP.
2. [*OLP*] Otherwise, depending on the kind of goal, it performs one of the next steps in order to get a value for it.
    (a) *Submodule goal.* If the goal is a path to a submodule of the current module, and if that submodule is visible,[3] then call the communication operation with this goal (the communication operation will call the submodule object process to solve the goal).
    (b) *Goal belonging to the import interface.* Call the communication operation with this goal (the communication operation will query the user to give a value for the goal).

[2] A symbol between square brackets stands for the name of the active process in which the algorithm step is performed

[3] Submodules can be hidden by a refinement operation between modules [Godo and Sierra, 1994]. This kind of operation is out of the scope of this paper.

(c) *Goal with a function attribute.* Now the evaluation of the goal depends on the evaluation of the function associated to the propositional variable. If there are arguments of the function with no value, call recursively the lazy algorithm over them from left to right. When all arguments have values, evaluate the function.

(d) *Goal that can be deduced by means of rules.* In this case we start a depth-first search on the rules of the module deducing the goal to look for a propositional variable without value. The search algorithm orders rules according to the following criteria, in order of preference.

i. *More specific rules first.* We try to find solutions by first using the more specific rules—those with less conditions.

ii. *More precise rules first.* A rule is more precise than another when its truth-value interval is more precise. Notice that this order can change during the execution because of the specialization of rules.

iii. *The writing order of rules.*

To evaluate the conditions of a selected rule, the search strategy follows the writing order of the conditions (left to right), in a depth-first manner. Finally, call recursively the lazy algorithm with the above mentioned propositional variable as a subgoal.

Notice that the algorithm finally returns a path to a submodule of the current module, a propositional variable belonging to its import interface, or a propositional variable with a evaluable function associated to it, and its associated value.

3. $[OLP]$ Specialization of rules
4. $[OLP]$ Call the reification operation
5. $[MLP]$ The meta level fires all possible meta-rules.
6. $[MLP]$ Call the reflection operation
7. $[OLP]$ If the reflection operation does not modify the object level set of formulas GOTO 1, otherwise GOTO 3.

Notice that this algorithm always provides the goal with a value since, in the worst case, it will get the value *unknown*, which corresponds to the maximum imprecision interval.

Eager: An eager strategy asks the user for all the variables and propositions declared in the *Import* interface of the module and queries all the exportable propositional variables of its submodules as well.

Given a query to a module with an eager evaluation, the control flow of the module process is the following one:

1. [OLP] If the goal has already a value, STOP.
2. [OLP] Otherwise, call the communication operation as many times as necessary to get values for all imported propositional variables in their writing order.
3. Steps 3, 4, 5 and 6 of the Lazy evaluation algorithm.
4. FOR each submodule DO (Submodules are ordered by their writing order).
   (a) [OLP] call the communication operation to get values for all the submodule exportable propositional variables used in the rules or meta-rules of the module.
   (b) Steps 3, 4, 5 and 6 of the Lazy evaluation algorithm.

   END FOR
5. [OLP] If the goal has already a value, STOP.
6. Steps 3, 4, 5 and 6 of the Lazy evaluation algorithm.
7. [OLP] GOTO 4

Reified: This kind of evaluation strategy does not differ from the eager one in the way of gathering data. The main difference of a reified strategy with respect to both lazy and eager strategies is that the specialization mechanism of the object level is not used at all. Therefore, deduction is only performed at the meta-level process. The motivation behind this evaluation strategy is to provide module designers with the possibility to define meta-interpreters.

3 OBJECT LEVEL PROCESS

3.1 Object-level deduction

Milord II provides the user with approximate reasoning capabilities at the object level. The approximate reasoning mechanisms are based on the use of a finitely-valued fuzzy (or many-valued) logic. Before describing the logical deduction system, and for the sake of a better understanding, we first outline the semantics behind it.

A particular many-valued logic can be specified inside each module by defining which is the algebra of truth-values, i.e. which is the (finite) ordered set of truth-values and which is the set of logical operators associated to them. Formally speaking, a *Milord II* algebra of truth-values $A_{n,T} = \langle A_n, \leq, N_n, T, I_T \rangle$ is a finite linearly ordered residuated lattice with a negation operation. In plain words, the set of truth-values $A_n = \{\mathbf{0} = a_1 < a_2 < ... < a_n = \mathbf{1}\}$ is a chain of n elements where $\mathbf{0}$ and $\mathbf{1}$ are the booleans *false* and *true* respectively; the negation operation N_n is the involution in A_n, i.e. $N_n(a_i) = a_{n-i+1}$; the conjunction operator T is a t-norm, i.e. a binary, commutative, associative and non-decreasing operation on A_n with $\mathbf{1}$ as neutral element and $\mathbf{0}$ as null element; finally I_T is the residuum of T, i.e. defined as $I_T(a, b) = Max\{c \in A_n \mid T(a, c) \leq b\}$, and it is used to model a many-valued implication. As it is easy to notice from the above definition, any of such truth-values algebras is completely determined as soon as the set of truth-values A_n and the conjunction operator T are chosen. So, varying these two characteristics we generate a family of different multiple-valued logics. For instance, taking $T(a_i, a_j) = a_{min(i,j)}$ or $T(a_i, a_j) = a_{min(n,n-i+j)}$ we get the well-known Gödel's and Łukasiewicz's semantics (truth-tables) for finitely-valued logics [Gottwald, 1988; Gottwald, 1993; Hájek, 1998; Hájek, 1995].

In a given module, and thus for a given truth-value algebra A_n and a set of propositional variables Σ_O, the set $\mathcal{OL}_n$ of object-level formulas consists of:

- $\mathcal{OL}_n$-Atoms: $\{(p, V) \mid p \in \Sigma_O\}$
- $\mathcal{OL}_n$-Literals: $\{(p, V), (\neg p, V) \mid (p, V) \in \mathcal{OL}\text{-Atoms}\}$
- $\mathcal{OL}_n$-Rules: $\{(p_1 \wedge p_2 \wedge \cdots \wedge p_n \rightarrow q, V^*) \mid p_i$ and q are literals (atoms or negations of atoms) and $\forall i, j (p_i \neq p_j, p_i \neq \neg p_j, q \neq p_j, q \neq \neg p_j)\}$

where V and V^* are intervals of truth-values. Intervals V^* for rules are constrained to be upper intervals, i.e. of the form $[a, \mathbf{1}]$, where $a > 0$. That is, object level formulas are indeed signed formulas under the form of pairs of usual propositional formulas (restricted to be literals or rules) and intervals of truth-values.

The *semantics* is obviously determined by the connective operators of the truth-value algebra $A_{n,T}$. *Interpretations* are defined by valuations ρ mapping the (propositional) sentences to truth-values of A_n fulfilling the following conditions:[4]

$\rho(true) = \mathbf{1}$,
$\rho(\neg p) = N_n(\rho(p))$,
$\rho(p_1 \wedge \ldots \wedge p_n \rightarrow q) = I_T(T(\rho(p_1), \ldots, \rho(p_n)), \rho(q))$.

Then the *satisfaction relation* between interpretations and $\mathcal{OL}_n$-formulas is defined as

$$\rho \models_O (\varphi, V) \text{ iff } \rho(\varphi) \in V$$

and it is extended to a *semantical entailment* between sets of $\mathcal{OL}_n$-formulas and $\mathcal{OL}_n$-formulas as usual:

$$\Gamma \models_O (\varphi, V) \text{ iff } \rho \models_O (\varphi, V) \text{ for all } \rho \text{ such that } \rho \models_O A, \text{ for all } A \in \Gamma.$$

Once the semantics is clear, we come to the (syntactical) deduction system which is implemented in each module. The *Many-valued Specialisation Calculus* (**Mv-SC** for short) is defined by the following axioms:

- **A1:** $(\varphi, [0, 1])$

[4]The expression $T(r_1, r_2, r_3, \ldots)$ is the recurrent application of T as $T(r_1, T(r_2, T(r_3 \ldots)))$.

- **A2:** $(true, 1)$

and by the following inference rules:

- **Weakening**: from (φ, V_1) infer (φ, V_2), where $V_1 \subseteq V_2$
- **Not-introduction**: from (p, V) infer $(\neg p, N_n^*(V))$
- **Not-elimination**: from $(\neg p, V)$ infer $(p, N_n^*(V))$
- **Composition**: from (φ, V_1) and (φ, V_2) infer $(\varphi, V_1 \cap V_2)$
- **Specialization**: from (p_i, V) and $(p_1 \wedge \cdots \wedge p_n \rightarrow q, W^*)$ infer $(p_1 \wedge \cdots \wedge p_{i-1} \wedge p_{i+1} \wedge \cdots \wedge p_n \rightarrow q, MP_T^*(V, W^*))$

where $N_n^*([a,b]) = [N_n(b), N_n(a)]$ is the point-wise extension of N_n to intervals and $MP_T^*(V, W^*)$ is defined as follows: $MP_T^*([a,b],[c,1]) = [T(a,c), 1]$. In [Puyol-Gruart *et al.*, 1998] it is shown that this deductive system is sound with respect to the above semantics and complete for deriving $\mathcal{OL}_n$-atoms. Object-level deduction will be denoted by $\vdash_O$.

3.2 Object Level Control Mechanisms

Subsumption mechanism

When expressing the deductive knowledge of a module, experts might write different rules concluding the same propositional variable to represent the possibility of either:

- having different unrelated sets of conditions entailing that propositional variable, or
- having different sets of conditions related by an inclusion relation that may allow concluding that propositional variable (with different certainty values). Subsumption is the mechanism that ensures that only the most specific sets of conditions will be used.

The widely accepted subsumption criterion is to use always the more specific knowledge in the deductive process. This idea is made precise in the following general definition.

DEFINITION 1 (Subsumption). Given a knowledge base KB and two rules $R_1 : (A_1 \rightarrow B, \alpha_1)$ and $R_2 : (A_2 \rightarrow C, \alpha_2)$ where B and C are literals over the same propositional variable, we say that rule R_1 is more specific than rule R_2 if, taking for granted the set of formulas in KB, whenever A_1 is true A_2 is also true, that is, when

$$KB \models A_1 \rightarrow A_2.$$

In the particular multi-valued logical framework of *Milord II* this definition can be expressed as $\rho(A_1 \rightarrow A_2) = 1$, for all many-valued interpretation ρ such that $\rho \models KB$. By definition of the implication connective as a residuum, the condition $\rho(A_1 \rightarrow A_2) = 1$ is equivalently expressed as $\rho(A_1) \leq \rho(A_2)$.

This criterion can be described in terms of the set of *labels*—non deducible propositional variables needed to apply the rule—associated to each premise. Namely, it can be checked that, in the conditions of the above definition, rule R_1 is more specific than rule R_2 if for each label L_i of A_1 there is a label L_j of A_2 such that $L_i \rightarrow L_j$ is a valid formula. The condition $\models L_i \rightarrow L_j$ reduces to the inclusionship of labels $L_j \subset L_i$.

For instance consider the following set of rules:

$$\begin{cases} R_1 : a \wedge b \wedge c \wedge d \rightarrow g \\ R_2 : e \wedge f \rightarrow g \\ R_3 : c \rightarrow e \\ R_4 : a \wedge b \rightarrow f \end{cases}$$

It is easy to see that there is a subsumption relation between the rules R_1 and R_2. The set of non deducible propositional variables necessary to apply the rule R_1 is $\{a, b, c, d\}$, whilst for the rule R_2 is $\{a, b, c\}$. Therefore R_1 is more specific than R_2.

Due to the special deductive mechanism of *Milord II*, based on specialization of rules, the subsumption relation changes as deduction progresses. This is so because the specialization mechanism reduces the conditions in the premises of rules, and thus modifies the reference KB used to compute labels. Because of that, *Milord II* incorporates an algorithm that dynamically computes and completes *partial labels*, in the sense that, the set of labels can be incomplete and even labels may be incomplete.

Elimination of Unnecessary Rules

The maximum precision given to the conclusion of a rule is limited by the truth interval of the rule. Consider a rule with certainty value $[a_r, 1]$ and whose premise has been evaluated to the interval $[a_i, a_j]$. Then, the interval associated to the concluded propositional variable by the application of this rule is given by

$$MP_T^*([a_i, a_j], [a_r, 1]) = [T(a_i, a_r), 1)] = [a'_r, 1] \text{ ,where } a'_r \leq a_r.$$

This consideration leads us to the following definition.

DEFINITION 2. A rule $(A \rightarrow q, [a_r, 1])$ is *unnecessary* for a propositional variable $(q, [a_i, a_j])$ if $a_r \leq a_i$. Similarly, a rule $(A \rightarrow \neg q, [a_r, 1])$ is *unnecessary* for a propositional variable $(q, [a_i, a_j])$ if $a_r \leq N_n(a_j)$, where N_n is the negation operator.

Therefore we can easily test whether the remaining rules concluding a propositional variable are still useful or not. This is what we call the *elimination of unnecessary rules process*. The test is applied every time a rule is specialized since the specialization mechanism broadens rule intervals. This control technique allows us to save unnecessary deductions as well as unnecessary information requirements and processing.

4 META-LEVEL PROCESS

4.1 Meta-level deduction

The meta-level language $\mathcal{ML}_n$, corresponding to an object-level language $\mathcal{OL}_n$, is a restricted classical first order language. It is defined from a set Σ_{rel} of predicate symbols including predicates K, P and WK which play a special role in the reflection mechanism; a set Σ_{act} of action symbols ($inhibit\_rules$, $assume$, $resume$, *filter*, *stop* and *module*); a set Σ_{fun} of classical arithmetic function symbols; a set Σ_{con} of constants including the truth-values of A_n and object propositional variables of $\Sigma_{\mathcal{O}}$; and a set Σ_{var} of variable symbols,[5] which can be empty.

Meta-level formulas are either ground literals, in a classical sense, or rules of the type

$$\{P_1 \wedge P_2 \wedge \ldots \wedge P_n \rightarrow Q \mid P_i, Q \text{ literals }\},$$

where each variable occurring in Q must occur also in some P_i. Variables in meta-rules, if any, are considered universally quantified. Quantifiers are all outermost. Only the conclusion Q may contain action symbols.

The *semantics* of the language is the classical of first order logic. The meaning of the special predicates K, P and WK will be explained in the next subsection along with the definition of the reification rules which use them to represent object-level sentences.

Finally, the *deduction* system is based on only one (modus ponens-like) inference rule:

$$\text{from } \{P_1 \wedge P_2 \wedge \ldots \wedge P_n \rightarrow Q, P'_1, P'_2, \ldots, P'_n\} \text{ infer } Q'$$

where $P'_1, \ldots, P'_n$ are ground instances of $P_1, \ldots, P_n$ respectively, such that there exists a unifier σ for $\{P_1 \wedge P_2 \wedge \ldots \wedge P_n, P'_1 \wedge P'_2 \wedge \ldots \wedge P'_n\}$, and $Q' = \sigma Q$ is the ground instance of Q resulting from σ. The deductive system of *Milord II* meta-level is thus not complete with respect to the classical semantics we use for it. Nevertheless, the deduction mechanism based on this single inference rule is powerful enough for our modelling purposes. Meta-level Deduction will be denoted by the symbol $\vdash_{\mathcal{M}}$.

4.2 Control Actions

Control actions may affect the deductive knowledge of a module by inhibiting rules and by branching and backtracking the reasoning process. Control actions

[5] When using *Milord II* syntax variable are prefixed by \$, for instance $\$x$.

may also modify the hierarchy of a module by inhibiting modules, or creating new ones. They can also abort the execution.

Inhibit Rules: This action takes out of the OLT a particular set of rules. When we execute `inhibit_rules(pathpredid)`, all the rules containing the propositional variable *pathpredid* in their premises are removed. We can also inhibit all rules containing in their premises propositional variables related to a given one.

Assume: The argument of this action stands for an ordered list of possible assumptions to be made at the object level that can be retracted later on.

Resume: It retracts the latest assumption performed.

Filter: This action consists on inhibiting (filtering) a set of submodules of the module. This means that all the propositions p exported by the filtered submodules will be considered as being $(p, unknown)$

Stop: This is an abort action. In some cases it is necessary to abort the execution when an unrecoverable situation holds.

Module: When a meta-rule concludes an instance of *module*, for example *module(= (A, B))*, an action will be performed, at downwards reflection time, to add a submodule named A and equal to B as the last, in writing order, of the already existing submodules. B can be any allowed modular expression, in particular, the application of a generic module. Generic modules containing as control knowledge meta-rule calls to themselves are allowed. This is the way recursion can be defined inside *Milord II* [Puyol-Gruart and Sierra, 1997].

Assume and *Resume* predicates deserve special attention because they allow to define a backtracking mechanism in *Milord II* (see Section 7), useful to model hypothetical reasoning.

5 UPWARDS REFLECTION OPERATION

The upwards reflection operation translates formulas from the current OLT to the MLT in the form of meta-predicate instances. It relates a sub-theory of the OLT with the set of ground literals of the meta-language ML. The meta-predicate WK is used to relate the set of object mv-literals with the set of ground meta-literals. Given that the constant names used in the MLT are exactly the same as those used in the OLT as proposition names, the quoting functions for literals are omitted for the sake of simplicity. The same applies for the intervals of truth-values. So we will write $WK(p, V)$ instead of $WK(\lceil p \rceil, \lceil V \rceil)$. The reification rules are:

$$\frac{(p, V) \in OLT}{\vdash_{\mathcal{M}} WK(p, V)}$$

$$\frac{(p,V) \notin OLT}{\vdash_{\mathcal{M}} \neg WK(p,V)}$$

$$\frac{(p_1 \wedge p_2 \wedge \cdots \wedge p_n \rightarrow q, V) \in OLT}{\vdash_{\mathcal{M}} WK(implies(and(p_1, p_2, \cdots, p_n), q), V)}$$

The other two meta-predicates, K and P, used in the meta-level language to represent the OLT state are definable from the meta-predicate WK:

$$K(p, [a_i, a_j]) \equiv WK(p, [a_i, a_j]) \wedge \neg WK(p, [a_{i+1}, a_j]) \wedge \neg WK(p, [a_i, a_{j-1}])$$

$$P(p) \equiv WK(p, [a_2, 1])$$

5.1 *Other meta-predicates*

Upwards reflection also contains programmer defined relations between propositional variables, the threshold and the rules. Although the submodules of a module are not persistent, the initial submodules are also reified, as well as those that have been filtered.

Relations: When declaring a propositional variable in *Milord II*, it is possible to establish a relation with another propositional variable, in the same module or in a submodule. There is a set of system-defined relations used for control. Other relations are domain dependent and defined by programmers. The name of the relation used in the definition of propositional variables, corresponds to a binary meta-predicate identifier. The two arguments correspond to the name of the propositional variables being related. For instance the definition

```
p1 = name: ...
     ...
     relation: relationid p2
```

becomes the next meta-predicate instance: $relationid(p_1, p_2)$

Threshold: A certainty threshold is treated as a meta-predicate instance. There is an instance per module and one per each submodule: $threshold(a_i)$ and $threshold(submodule_j, a_k)$.

Submodules: There is a meta-predicate called *submodule* which has an instance per submodule (meta-predicate with only one argument), and an instance per sub-submodule (the same meta-predicate name but with two arguments), that is, $submodule(submodule_1)$ or $submodule(submodule_1, subsubmodule_2)$.

Filtered: Instances of this meta-predicate represent the submodules that have been filtered (removed) by meta-rules, *filtered*$(submodule)$.

6 DOWNWARDS REFLECTION OPERATION

The downwards reflection operation is responsible of making effective at the object level the consequences of the deduced meta-predicate instances.

$$\frac{K(p,V) \in MLT}{\vdash_O (p,V)}$$

The reflection operation modifies the data structure of the OLP to make it causally connected, using the terminology of Patty Maes [Maes, 1988], with the meta-predicate instances.

7 DECLARATIVE BACKTRACKING

When a meta-rule with an *Assume* action in its conclusion is applied, as many extensions of the meta-theory MLT as elements in the argument set of the *Assume* action are generated. For instance, consider the case where, in a certain moment, we have in the current object theory only the literal

$$(p,1)$$

and MLT consists of the following meta-rule:

$$\text{If } K(p,1) \text{ and } \neg P(q) \text{ then } Assume(\{(q,1),(q,0)\})$$

Suppose also that q could not be proved in OLT. Then, after the upwards reflection process, the current MLT will be the extension of the previous one with the ground literals $K(p,1)$ and $\neg P(q)$. So, now the above meta-rule can be applied, and this causes the system to obtain the conclusion $Assume(\{(q,1),(q,0)\})$. The meaning of the action *Assume* is that the elements of its argument should be assumed in different extensions of the current OLT. This is done by building a tree of MLTs, each containing a K meta-predicate instance for all the elements of the argument of *Assume*, implemented by a snapshots stack. So in this case we obtain the following two different extensions of the current MLT in Figure 5:

$$MLT_1 = MLT \cup K(q,1)$$

$$MLT_2 = MLT \cup K(q,0)$$

From now on, and until another instance of an *Assume* or *Resume* action is obtained, the existing communication (upward reflection and downward reflection) between OLT and MLT is moved to a communication between OLT and MLT_1.[6] Thus, in this case, after the downward reflection process OLT is extended with $(q,1)$.

[6]Computationally speaking, MLT_1 is managed by the same co-routine of MLT but with a new snapshot in the stack containing the state of MLT plus $Assume(\{(q,1),(q,0)\})$.

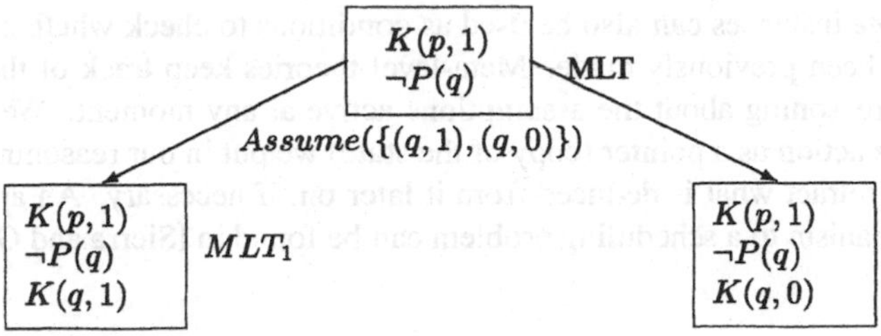

Figure 5. Meta-theories branching using the *Assume* action.

In order to backtrack in the tree of MLTs generated by successive applications of *Assume* actions, the language provides a special 0-ary predicate *Resume*. When a meta-rule concluding *Resume* is applied, we perform a backtracking in the meta-theories tree. This backtracking restores the parent MLT, and the current OLT becomes the OLT which was active at the moment the assumptions were made by the parent MLT. In the above example, backtracking from MLT_1 to MLT makes that q will not be true in the current OLT, and that immediately the communication (upward reflection and downward reflection) between OLT and MLT is moved to a communication between OLT and MLT_2 (see Figure 6).

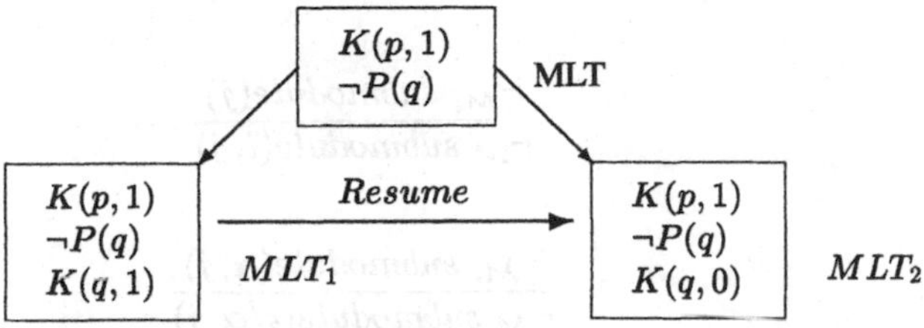

Figure 6. Backtracking using the action *Resume*.

It is worth noticing that actions *Assume* and *Resume* provide the system with a declarative backtracking mechanism, similar to the approach taken in MetaProlog [Bacha, 1988]. This declarative mechanism allows us to implement several complex reasoning patterns. For instance, consider that an assumption was made at the meta-level. Whenever a contradiction occurs in OLT afterwards, we can declaratively detect it, and then, by means of the *Resume* meta-predicate, we can move back to a previous non contradictory OLT. This can be achieved by a meta-rule, such as

If $Assume(\$y)$ and $K(\$x,())$[7] then *Resume*

[7]Notice that a contradiction in OLT occurs when OLT contains literals of the form (p,V) and

Assume instances can also be used as conditions to check whether an assumption has been previously made. Meta-level theories keep track of them to allow explicit reasoning about the assumptions active at any moment. We can see the *Assume* action as a pointer (copy of the state) we put in our reasoning process in order to retract what is deduced from it later on, if necessary. An application of the mechanism to a scheduling problem can be found in [Sierra and Godo, 1993].

8 COMMUNICATION CONTROL MECHANISMS

The last issue considered in this Chapter referring to control mechanisms in *Milord II* concerns the communication operation. The object level module process activates the communication operation to either query the user or query some of the submodules. When the operation queries the user, the result is the extension of the current OLT by a propositional variable. However, the communication from a submodule to its present parent module is governed by a set of inference rules concerning the translation between the possibly different corresponding local logics of the modules,[8] and the structural relations concerning the hierarchy. Some of these rules are shown below.

$$\frac{\vdash_{\mathcal{O}_i} (p, V)}{\vdash_{\mathcal{O}} (i/p, \mathcal{T}(V))}$$

$$\frac{\vdash_{\mathcal{M}_i} submodule(j)}{\vdash_{\mathcal{M}} submodule(i, j)}$$

$$\frac{\vdash_{\mathcal{M}_i} submodule(\alpha, j)}{\vdash_{\mathcal{M}} submodule(i/\alpha, j)}$$

$$\frac{\vdash_{\mathcal{M}_i} eval\_type(a)}{\vdash_{\mathcal{M}} eval\_type(i, a)}$$

The first rule translates object level formulas from the submodule $\mathcal{O}_i$ to the module $\mathcal{O}$. The second one informs the module $\mathcal{O}$ that its submodule $\mathcal{O}_i$ has the module $\mathcal{O}_j$ as a submodule. The third one allows us to propagate the KB structure through the module hierarchy. Finally, the fourth rule informs the module $\mathcal{O}$ of the evaluation type of its submodule $\mathcal{O}_i$.

$(\neg p, V')$ such that $V \cap N_n^*(V') = \emptyset$. In this case the literal $(p, ())$ is generated.

[8]The system allows the specification of mappings between local logics in the sense of renaming mappings $\mathcal{T} : A_n \to I(A_m)$ sending each truth-value of an algebra A_n to an interval of truth-values of another algebra A_m, extending in the obvious way to intervals of A_n.

9 EXAMPLES OF COMPLEX REASONING TASKS

9.1 Scheduling Reasoning System

This example is an excerpt of the detailed scheduling problem presented in [Sierra and Godo, 1993].

In general, to specify a complex reasoning system in *Milord II* it is necessary to define a hierarchy of modules. This hierarchy captures the usual task/subtask decomposition. However, in some cases it is necessary to iterate over a set of subtasks (for instance, *hypothesis assumption*, *evaluation*, and *revision*). The only way to perform iteration in our language is through the reification/reflection mechanism. This leads to understand the task/subtask decomposition in such cases as a particular relation between the OLTs and the MLTs.

In this scheduling example we build a module in which we associate to each variable a proposition identifier. The space of values for variables is understood in the proposed implementation as the set of truth values of a particular multiple-valued logic. Requirements are expressed as restrictions over the truth value assignments for these propositions. Solutions to the scheduling problem are then considered to be truth value assignments that fulfil the requirements.

In order to implement a scheduling task with a set of requirements to be fulfilled, two modules must be defined:

Requirements module This module will contain the requirements as meta-predicates over the propositions of the object level, i.e. restrictions over the possible values that object level propositions can take. These meta-predicates are defined in the dictionary of the module. This module also defines the particular multiple-valued logic for the object level. In the example there is no truth-values combination, so selecting connectives is irrelevant.

Design task Module This module contains the initial conditions of the problem as object level rules, and the meta-rules that perform the different subtasks of the scheduling process.

Each problem setting is determined by a number of tasks and a set of constrains among them, and requires a particular *requirements module* and a particular *design task module*. In order to be generic all *design task modules* have been programmed with the MLT in common. Thus, to build the actual module that will perform the design, it is necessary to connect this particular generic *design task module* with a concrete *requirements module*, so the former can inherit the requirements of the problem from the later. It is done in the following way, using the refinement operation:[9]

Module Example = Design : Requirements

[9] $A : B$ is a modular expression that generates a new module that results from modifying A by adding elements inherited from B, such as *dictionary* or *logic*. See for details [Puyol-Gruart and Sierra, 1997].

Requirements Module

The requirements of the scheduler under study are:

- The number of tasks to be scheduled.
- The temporal constraint relations among them.
- The number of available time points to perform the tasks.

The tasks are represented as a set of object level propositions $A_1, \ldots, A_n$, the temporal relations as meta-predicates over pairs of elements in the set $\{A_1, \ldots, A_n\}$, and the number of time points $time_1, \ldots, time_q$ as the truth-values of the logic, which will be $\{false, time_1, \ldots, time_q, true\}$.

In our particular case there are four types of constraints that we will represent by four meta-predicates: *before*, *equ*, *diff* and *notbefore*.

1. $before(x, y)$ means that activity x must occur before activity y.
2. $equ(x, y)$ means that activities x and y must occur during the same time period.
3. $diff(x, y)$ means that activities x and y must not occur in the same period.
4. $notbefore(x, y)$ means that activity x must not occur before activity y.

Design Task Module

The implementation of the heuristic search to find a solution to the scheduling problem, is done by defining a set of rules and meta-rules. Rules are responsible for the initial attachment of the whole space of values to the propositions and meta-rules are responsible for the pruning of the search space.

For each proposition A_i representing a scheduling activity a rule like

R00i if true then conclude A_i is $time_1$

has to be written in order to define the initial possible truth-values for the propositions, i. e. the interval $[time_1, true]$. These intervals represent the root node of the search space. In general the truth-value of the rules determine the initial time point to start the scheduling of the corresponding activity. So initial conditions of the problem can be stated just modifying the certainty values of these rules.

At the meta-level, for each possible constraint violation a meta-rule is written, having as premise a set of conditions that are true when a particular constraint is violated, and as conclusion "how" to restrict the set of possible values for one activity in such a way that the violation is solved. That is, the meta-rule cuts off a set of children states. Meta-rules can be of two types depending on the violation:

1. Meta-rules that restrict the possible values of propositions in such a way that there is no need for backtracking, i.e. only one child remains. This is the case of requirements of type *before*, *equ* and *notbefore*. These meta-rules use the K meta-predicate in their conclusion. An example of such a meta-rule for the *before* requirement is:

   M002 **if** before($x,$y) **and K**($x,**int**($z,true))
   and K($y, **int**($w,true)) **and ge**($z,$w)
   then conclude K($y, **int**(**suc**($z),true))

2. Meta-rules that perform branching, i.e. two children remain. This is the case of constraints of type "diff". These meta-rules have the action *Assume* in their conclusion. Example:

   M005 **if** diff($x,$y) **and K**($x,**int**($z,true))
   and K($y,**int**($z,true)) **then**
   Assume(**list**(($x,**int**(suc($z),true)), ($y,**int**(suc($z),true))))

   A special meta-rule is also needed to detect when no solution is found, and then in that case to backtrack. This situation can be detected when a proposition gets the interval $[true, true]$ as follows:

   M001 **if K**($x,**int**(true,true)) **then Resume**

   When the search space is exhausted and no solution is found, this situation reflects that the set of constraints is inconsistent.

Code of the example

Here we present the complete code of the scheduling module for a set on 4 tasks to be scheduled (see Figure 7). The set of constraints is specific for each example test.[10]

To use the scheduler in a test example we define a module *requirements* which contains the meta-predicates defining the relations between the propositions at the object level, and the set of truth-values representing the admissible time points (see Figure 8).

Now, as said before, using the inheritance property of the operator ":", we define the module that performs a scheduling of four tasks, module *scheduler_test*, with a particular set of requirements, defined in *requirements*.

[10] In the Meta-rules, it is possible to use some system-defined meta-predicates such as: ge (greater or equal), lt (lower than), gt (greater than). The meta-predicates ge, lt, gt can be applied over the order of the truth-values of the local logic, or over the real numbers. It is also possible to perform operations on top of the truth-values, such as suc (successor function), that have to be understood in the context of an ordered set of truth-values.

```
Module Scheduler =
Begin
    Export A1, A2, A3, A4
    Deductive knowledge
        Rules:
            R001 if true then conclude A1 is t1
            R002 if true then conclude A2 is t1
            R003 if true then conclude A3 is t1
            R004 if true then conclude A4 is t1
        Inference system:
            Truth values = (false, t1, t2, t3, true)
    end deductive
    Control knowledge
        Evaluation type: eager
        Deductive control:
        ;; If a propositional variable gets the maximum value
        ;; no solution can be found.
        M001 if K($x,int(true,true))
            then Resume
        ;; X before Y.
        M002 if before($x,$y) and K($x,int($z,true))
            and K($y, int($w,true)) and ge($z,$w)
            then conclude K($y, int(suc($z),true))
        ;; X equal Y
        M003 if equ($x,$y) and K($x,int($z,true))
            and K($y,int($w,true)) and gt($z,$w)
            then conclude K($y, int($z,true))
        ;; X not before Y
        M004 if notbefore($x, $y) and K($x,int($z,true))
            and K($y,int($w,true)) and lt($z, $w)
            then conclude K($x, int($w,true))
        ; X different Y
        M005 if diff($x,$y) and K($x,int($z,true))
            and K($y,int($z,true)) then
            Assume(list(($x,int(suc($z),true)), ($y,int(suc($z),true))))
    end control
end
```

Figure 7. Scheduler module declaration.

```
Module Requirements =
Begin
    Export A1, A2, A3, A4
    Deductive knowledge
        Dictionary:
        Predicates:
            A1 = Name: "A1" Type: many-valued
                      Relation: notbefore A4
            A2 = Name: "A2" Type: many-valued
            A3 = Name: "A3" Type: many-valued
            A4 = Name: "A4" Type: many-valued
                      Relation: before A2
                      Relation: before A3
                      Relation: diff A1
        Inference system:
            Truth values = (false, t1, t2, t3, true)
    end deductive
    Control knowledge
        Evaluation type: eager
    end control
end
```

Figure 8. Requirements module declaration.

Module Scheduler_test = scheduler : requirements

Now, let's see the execution trace of the module *Scheduler_test*:

1. Initially OLT deduces the interval [t1,true] for all the propositions A1-A4 by means of rules R001-R004. Upwards reflection operation introduces the following set of predicates into MLT:

   $$\mathbf{K}(\mathrm{A}_i,\mathbf{int}(\mathrm{t1},\mathrm{true})),\ \text{for } i = 1, 2, 3, 4.$$

   For the sake of simplicity, in the following we will consider only the minimum value of the interval of truth-values of propositions. This initial situation and all the meta-level processes are represented in Tables 1 to 3.

Table 1. First assumption.

←	**1**	→		←	**2**	→		←	**3**	→
t1	t2	t3		t1	t2	t3	**Assume**	t1	t2	t3
A1				A1				A1		
A2			*M002*	⇒	A2				A2	
A3			*M002*	⇒	A3				A3	
A4				A4			*M005*	⇒	A4	

2. Table 1 represents a part of the meta-level process until the first assumption is generated. Meta-rule M002 is used two times considering the relations "A4 before A2" and "A4 before A3", increasing the value of the propositional variables A2 and A3.

3. Meta-rules M003 and M004 cannot be fired. Given the relation "A4 diff A1" and that A4 and A1 have the same value, an assumption is produced by meta-rule M005:

   Assume(**list**((A4,**int**(t2,true)), (A1,**int**(t2,true))))

   first the value t2 is assumed for A4.

4. Similarly to the previous meta-level process, Table 2 represents meta-rule actions until a new assumption is performed. M002 increase again the values of A2 and A3.

5. Now a matching occurs for meta-rule M004 because of the relation "A1 notbefore A4", producing a new value for A1.

6. Given that A4 and A1 have the same value, a new assumption is performed by meta-rule M005:

   Assume(**list**((A4,**int**(t3,true)), (A1,**int**(t3,true))))

   first the value t3 is assumed for A4.

Table 2. Second assumption.

	←	4	→		←	5	→		←	6	→
	t1	t2	t3		t1	t2	t3	**Assume**	t1	t2	t3
	A1			*M004*	⇒	A1				A1	
M002		⇒	A2				A2				A2
M002		⇒	A3				A3				A3
		A4				A4		*M005*		⇒	A4

7. Table 3 represents a new cycle of the meta-level process until a resume operation is performed. Now meta-rule M002 increase again the value of A2, making it equal to *true* (represented as * in the table).

Table 3. Resume.

	←	7	→		←	8	→
	t1	t2	t3	**Resume**	t1	t2	t3
		A1			*M001*	⇒	A1
M002		⇒	*				A2
			A3				A3
			A4		*M001*	A4	⇐

8. Given that the value for A2 is true, a matching is possible for meta-rule M001 producing a resume operation. Remember the last assumption:

   Assume(**list**((A4,**int**(t3,true)), (A1,**int**(t3,true))))

   now the value t3 is assumed for A1, and A4 return to the previous value, t2.

   Now no meta-rules can be applied, and the solution is found.

9. Finally downwards reflection operation assigns to the propositions of OLT the solution result:

   {(A1,[t3,true]),(A2,[t3,true]),(A3,[t3,true]),(A4,[t2,true])}

Notice that if the we invert the relation "A4 diff A1" in the code of *requirements module* an equivalent solution is obtained without any assumption.

9.2 *A General Method for solving a class of Default Reasoning problems*

In this section we describe, through an example, a simple approach implemented in *Milord II* to tackle some of the usual problems in defeasible reasoning, such as inheritance, irrelevance and specificity, in a restricted propositional framework.

Consider the following well-known set of defeasible rules:

$$B \to F, P \to B, P \to \neg F$$

The intended behaviour of this set of rules is to infer $\neg F$ given P, to infer F given B, and to infer B given P.

The implementation in *Milord II* makes use of three modules. The module *Penguin* (see Figure 9) defines an object level component with the following characteristics:

```
Module Penguin =
Begin
    Import P, B
    Export F
    Deductive knowledge
        Dictionary:
            Predicates:
                B = Name: "Bird" Type: many-valued
                P = Name: "Penguin" Type: many-valued
                F = Name: "Flies" Type: many-valued
                FPos = Name: "Flies+" Type: many-valued
                    Relation: supports F
                FNeg = Name: "Flies-" Type: many-valued
                    Relation: distracts F
        Rules:
            R001 If B then conclude FPos is d
            R002 If P then conclude B is d
            R003 If P then conclude FNeg is d
        Inference system:
            Truth values = (0, dd, d, 1)
            Conjunction = min
            Modus ponens = Truth table:
                                    ((0  0   0   0)
                                     (0  0   0   dd)
                                     (0  0   dd  d)
                                     (0  dd  d   1))
    End Deductive
End
```

Figure 9. Module Penguin.

- For those propositional variables with contradictory default conclusions a couple of extra propositional variables (in this case FPos and FNeg) that will accumulate the evidence for the particular sign coming from eventually different deductive paths. These propositional variables are related through two relations named *supports* and *distracts*.
- Default rules are written as object level rules with truth-value *d*, one degree below the maximum one (see next point). Notice that all three rules in the

example are considered as default rules, although rules $P \rightarrow B$ and $P \rightarrow \neg F$ could be considered as well as *strict* rules and have attached maximum truth-value 1.

- A *local logic* with as many truth-values as the maximum path in the deductive trees associated to exportable propositional variables. In the current example we take as $\{0 < dd < d < 1\}$ as truth-value set. The combination of conditions (conjunction declaration) is done with the *min* operator. So, shorter paths win. The truth table for the *modus ponens* operation[11] corresponds to the so-called *Łukasiewicz* t-norm and has the characteristic of "counting" the number of applied defaults because it makes the minimum interval value decrease by one term (look at the third column or row in the table).

- At the end of a deductive process, by specialization, the value of the coupled propositional variables is an interval with the minimum value as low as the maximum number of default rules applied to get it.

```
Module Default_interpreter =
Begin
    Control knowledge
        Evaluation type: eager
        Deductive control:
        M001 If K($x, int($min1, $max1)) and supports($x, $y)
                and distracts($z, $y) and K($z, int($min2, $max2))
                and gt($min1, $min2))
                then conclude K($y, int(1,1))
        M002 If K($x, int($min1, $max1)) and supports($x, $y)
                and distracts($z, $y) and K($z, int($min2, $max2))
                and lt($min1, $min2))
                then conclude K(not($y), int(1,1))
        Structural control:
        M001 If K($x, $cert) and supports($x, $y)
                and distracts($z, $y) and K($z, $cert)
                then abort
    End control
End
```

Figure 10. Default Interpreter module.

The *Default interpreter* module (see Figure 10) contains a generic control able to manage any module containing default rules written in the way outlined in the

[11] In this example the algebra of truth-values is different for that defined in Section 3, here defining explicitly the *modus ponens* operator.

module *Penguin*. The connection between the default interpreter module and any containing default rules is done as in the declaration of the module *Solution* (see Figure 11). The *Union module operation*, constructs a new module from two other modules by performing the union component by component. In this example from the modules *Penguin* and *Default_interpreter*. Whenever the union is not feasible an error is raised, for example, when trying to make the union of two modules with different *local logics*. Once connected, the *Default interpreter* module and the *Penguin* module, the new module acts as a module having the deductive knowledge of module *Penguin* and the control knowledge of module *Default_interpreter*. So the execution of *Solution* acts in the following way, because of the eager interpretation defined in *Solution*:

Module Solution = **Union** (Penguin, Default_interpreter)

Figure 11. Solution module.

- P and B are queried to the user. Let suppose P is true.
- An upwards reflection step is performed. Nothing can be deduced.
- $FPos$ and $FNeg$ are deduced. $Fpos$ will have an interval with a minimum value lower than $FNeg$ because two rules were necessary to deduce it. Given $(P, [1, 1])$ and using modus ponens we get $(FPos, [dd, 1])$ and $(FNeg, [d, 1])$.
- An upwards reflexion is performed with $K(FPos, int(dd, 1))$ and $K(FNeg, int(d, 1))$. M002 is applied and $K(not(F), int(1, 1))$ is concluded.
- Downwards reflection produces: $(F, [0, 0])$. As expected penguins don't fly.

9.3 A Legal Problem: Default Reasoning

This example is borrowed from Brewka [Brewka, 1994] and it is based on the next statements :

- According to Uniform Commercial Code (UCC) a security interest in goods is perfected by taking possession of the collateral.
- According to Ship Mortgage Act (SMA) security interest in a ship may only be perfected by filing a financing statement.
- UCC is state law, SMA federal law. UUC is more recent than SMA.
- The principle Lex Posterior gives precedence to newer laws.

- The principle Lex Superior gives federal law precedence over state law.
- Miller has possession of a certain ship but did not file a financing statement.

We are interested in formalizing this example in such a way that we can answer negatively the question *Is Millers' security interest perfected?*

In [Brewka, 1994] default logic is enriched by allowing to represent priorities among defaults and reasoning about them. In this formalism, defaults are referenced by a unique name identifier, and preferences among defaults are encoded by a strict partial order, noted <, in the set of default names. This preferences are used then to eliminate all those Reiter extensions which are incompatible with the priority information they contain.

Using Brewka's approach, the previous statements are represented as follows.

Defaults:

$UCC : possession \rightarrow perfected$
$SMA : ship \wedge \neg financial\text{–}statement \rightarrow \neg perfected$
$LP(d_i, d_j) : more\text{–}recent(d_i, d_j) \rightarrow d_i < d_j$
$LS(d_i, d_j) : federal\text{–}law(d_i) \wedge state\text{–}law(d_j) \rightarrow d_i < d_j$

Propositional variables and relations:

$possession$
$ship$
$\neg financial\text{–}statement$
$more\text{–}recent(UCC, SMA)$
$federal\text{–}law(SMA)$
$state\text{–}law(UCC)$

The set of Reiter extensions would be:

$$E_1 = Th(W \cup perfected, UCC < SMA)$$
$$E_2 = Th(W \cup \neg perfected, UCC < SMA)$$
$$E_3 = Th(W \cup perfected, SMA < UCC)$$
$$E_4 = Th(W \cup \neg perfected, SMA < UCC)$$

The only extensions compatible with the priority information defined so far are E_1 and E_2. If we add the next priority information

$$LS(x, y) < LP(y, x)$$

the conflict is solved in favour of E_4.

In Figure 12 an implementation of this example in *Milord II* is presented.[12] Let us comment the more relevant aspects of the code.

[12]The meaning of `set_of_instances(var1, expression, var2)` is the following: given an `expression` containing the variable `var1`, the variable `var2` will be bound to a list containing all the instances of `var1` that make the `expression` true.

```
Module Legal =
Begin
    Import possession, ship, financial_statements
    Export perfected
    Deductive knowledge
        Dictionary:
        Predicates:
        Perfected = Name: "Perfected" Type: many-valued
        Possession = Name: "Possession" Question: "Possession?"
                    Type: boolean
        Ship = Name: "Ship" Question: "Ship?" Type: boolean
        fin_stat = Name: "Financial Statements" Type: boolean
                Question: "Financial Statements?"
        SMA = Name: "SMA" Type: boolean Relation: law federal
        UCC =   Name: "UCC" Type: boolean
                Relation: law state Relation: more_recent SMA
        Federal = Name: "Federal" Type: class
        State = Name: "State" Type: class
        Rules:
        R001 if possession then conclude UCC is s
        R002 if ship and no(fin_stat) then conclude SMA is s
    End deductive
    Control Knowledge
        Evaluation type: eager
        Deductive control:
        M001 if more_recent($y, $z) then conclude LP($y,$z)
        M002 if law($y, federal) and law($z, state)
            then conclude LS($y, $z)
        M003 if LS($x, $y) and LP($y, $x)
            then conclude prefered(LS($x, $y), LP($y, $x))
        M004 if more_recent($y, $z) and
                set_of_instances($x, prefered($x,LP($y,$z)), $list)
                and equal($list,nil) then conclude prefered($y, $z)
        M005 if law($y,federal) and law($z, state) and
                set_of_instances($x,prefered($x,LS($y,$z)), $list)
                and equal($list,nil) then conclude Prefered($y, $z)
        M006 if K(UCC, int(s,s)) and
                set_of_instances($x,conj(prefered($x, UCC),
                    K($x, int(s, s))), $list) and equal($list, nil)
                and no(K(not(perfected), int(s,s)))
            then conclude K(perfected, int(s,s))
        M007 if K(SMA, int(s,s)) and
                set_of_instances($x,conj(prefered($x, SMA),
                    K($x, int(s, s))), $list) and equal($list, nil)
                and no(K(perfected, int(s,s)))
            then conclude K(not(perfected), int(s,s))
    End control
End
```

Figure 12. Legal module declaration.

- Object level rules are used to model the verification of the conditions of UCC and SMA laws. When they are verified, rules deduce object level predicates named UCC and SMA with the value *s* (for *sure*), meaning *true* in the many-valued logic used by default in *Milord II*.
- Most important elements are in the meta-level. The first five meta-rules (M001-M005) model, in a straightforward manner, the LP and LS preference criteria.
- The last two meta-rules model the defaults. For example M006 says: If it is known that the conditions of the UCC law are fulfilled, UCC with truth-value *true*, and there are no laws preferred to UCC with their conditions fulfilled, and it is not known the negation of the propositional variable *perfected*, then the propositional variable *perfected* can be assumed.
- The eager evaluation mechanism starts by querying the user about *possession*, *ship* and *financial-statements* in this order, then applies, if possible, the object level rules, upwards reflect, and deduces at the meta-level. Let us follow a trace:

1. Possession? *true*
2. Ship? *true*
3. Financial statements? *false*
4. OL deduction gets: $(ucc, [s, s])$ and $(sma, [s, s])$
5. Upwards reflection produces: $K(ucc, int(s, s))$, $K(sma, int(s, s))$, $more\_recent(ucc, sma)$, $law(sma, federal)$, $law(ucc, state)$, and other irrelevant meta-predicates.
6. M001 applies getting: $LP(ucc, sma)$
7. M002 applies getting: $LS(sma, ucc)$
8. M003 applies getting: $preferred(LS(sma, ucc), LP(ucc, sma))$
9. M004 fails, $\$list$ is not *nil*
10. M005 applies getting: $preferred(sma, ucc)$
11. M006 fails, $\$list$ is not *nil*
12. M007 applies getting: $K(not(perfected), int(s, s))$
13. No more meta-rule applies
14. Downwards reflection produces: $(not(perfected), [s, s])$
15. No object level deductions are possible. STOP

10 CONCLUSIONS

It is often the case that reasoning patterns occurring in complex problem solving tasks cannot be modelled (or at least it may turn very cumbersome) by means of a pure logical approach. Extra-logical mechanisms may be of great help in such situations if correctly used in suitable contexts. In this paper we have described the control techniques successfully used in the *Milord II* system. The most remarkable feature is its declarative control which is modelled by a meta-level approach, based on reflection techniques and equipped with a declarative backtracking mechanism. The use of reflection techniques, together with an (implicit) subsumption mechanism at the object level, has been proved specially well suited to tackle the problem of incompleteness of knowledge. As a final remark, it is interesting to notice that, although particular to this system, most of the considered techniques can be of general interest for a variety of multi-language logical architectures (e.g. multi-agent systems).

11 ACKNOWLEDGEMENTS

We acknowledge the Spanish Comisión Interministerial de Ciencia y Tecnología (CICYT) for its continued support through projects: ACRE (CAYCIT 836/86), SPES (880J382), ARREL (TIC92–0579–C02–01) and SMASH (TIC96–1038–C04–01). This research has also been partially supported by the Esprit Basic Research Action number 3085 (DRUMS) and by the European Community project MUM (Copernicus 10053). The definition of this language has profited from ideas coming from many people, among which we would like to thank Francesc Esteva, Ramon López de Màntaras, Jaume Agustí and Don Sannella. Parts of the software have been developed by Josep Lluís Arcos. We also thank the experts that have developed applications based on *Milord II*, Albert Verdaguer, Miquel Belmonte, Marta Domingo, Pilar Barrufet, Lluís Murgui and Ferran Sanz.

This material was previously published in *Future Generation Computer Systems Journal*, pp. 157–172, 1996, and is reproduced with the kind permission of Elsevier.

Lluís Godo, Josep Puyol-Gruart and Carles Sierra
IIIA, Artificial Intelligence Research Institute, Catalonia, Spain.

BIBLIOGRAPHY

[Bacha, 1988] H. Bacha. Metaprolog design and implementation. In *Proceedings of fifth International Conference on Logic Programming*, 1988.

[Brewka, 1994] G. Brewka. Reasoning about preferences in consistency-based nonmonotonic logics. In *Notes of the Workshop on reasoning about incosistency and partiality in dynamic contexts*, 1994.

[Godo and Sierra, 1994] L. Godo and C. Sierra. Knowledge base refinement in *Milord II*. In *Proceedings of 14 th. IMACS World Congress*, 1994.

[Godo *et al.*, 1995] L. Godo, W. van der Hoek, J.J. Ch. Meyer, and C. Sierra. Many-valued Epistemic States. Application to a Reflective Architecture: Milord II. In B. Bouchon-Meunier, R.R. Yager, and L.A. Zadeh, editors, *Advances in Intelligent Computing*, volume 945 of *Lecture Notes in Computer Science*, pages 440–452. Springer-Verlag, 1995.

[Gottwald, 1988] S. Gottwald. *Mehrwertige Logik*. Akademie-Verlag, Berlin, 1988.

[Gottwald, 1993] S. Gottwald. *Fuzzy Sets and Fuzzy Logic*. Vieweg, 1993.

[Hájek, 1995] P. Hájek. Fuzzy logic from the logical point of view. In M. Bartošek, J. Staudek, and J. Wiedermann, editors, *SOFSEM'95: Theory and practice of informatics*, volume 1012 of *Lecture Notes in Computer Science*, pages 31–49. Springer-Verlag, Milovy, Czech Republic, 1995.

[Hájek, 1998] P. Hájek. *Metamathematics of fuzzy logic*. Kluwer, 1998.

[Maes, 1988] P. Maes. *Meta-Level Architectures and Reflection*, chapter Issues in computational reflection, pages 21–35. North Holland, 1988.

[Puyol *et al.*, 1992] J. Puyol, L. Godo, and C. Sierra. A specialisation calculus to improve expert system communication. In *Proceedings of the 10th European Conference on Artificial Intelligence, ECAI'92*, pages 144–148, 1992.

[Puyol-Gruart *et al.*, 1998] J. Puyol-Gruart, L. Godo, and C. Sierra. Specialisation calculus and communication. *International Journal of Approximate Reasoning (IJAR)*, 18(1/2):107–130, 1998.

[Puyol-Gruart and Sierra, 1997] J. Puyol-Gruart and C. Sierra. Milord II: a language description. *Mathware and Soft Computing*, 4(3):299–338, 1997.

[Sierra and Godo, 1992] C. Sierra and L. Godo. Modularity, uncertainty and reflection in Milord II. In *Proceedings of 1992 IEEE International Conference on Systems, Man and Cybernetics*, pages 255–260, 1992.

[Sierra and Godo, 1993] C. Sierra and L. Godo. *Formal Specification of Complex Reasoning Systems*, chapter Specifyinf simple schedulling tasks in a reflective and modular architecture, pages 199–232. Ellis Horwood, 1993.

ADAM KELLETT AND MICHAEL FISHER

COORDINATING HETEROGENEOUS COMPONENTS USING EXECUTABLE TEMPORAL LOGIC

1 INTRODUCTION

Concurrent METATEM is a programming language for reactive systems [Fisher, 1993] that has been shown to be particularly useful in representing and developing multi-agent systems [Fisher, 1995]. It is based on the combination of two complementary elements: the direct execution of temporal logic specifications providing the behaviour of an individual object [Fisher, 1996]; and a concurrent operational model in which such objects execute asynchronously, communicate via broadcast message-passing, and are organized using a grouping mechanism [Fisher, 1994].

In this paper we consider the extension of Concurrent METATEM to act as a coordination language. We propose an extension to the existing operational model which allows Concurrent METATEM objects to act as controlling processes for agents written in a subject language. Our aim is to extend the functionality of Concurrent METATEM and provide a consistent link between formal specification using temporal logic and implemented systems.

Temporal logic is now widely used in the specification and verification of concurrent and reactive systems. This development has come about largely through the failure of conventional system development approaches to provide adequate techniques for the formal specification of such systems. As a specification language, temporal logic provides the expressiveness necessary for complex problems, and is declarative in nature. What differentiates it from other approaches is its independence from execution sequences, expressing instead the properties required of a system. It this way many problems such as mutual exclusion, deadlock and livelock can be expressed in a very natural manner.

Concurrent METATEM programs are, in effect, executable specifications. Programs express the properties required to be true throughout an execution. Using the *Imperative Future* paradigm [Barringer *et al.*, 1996], a program execution is a forward chaining process which dynamically reflects changes over time in a temporal model. This imperative approach means that an execution not only assesses the validity of events against the properties required, but actively take steps to ensure they as satisfied. If, for example, the event X occurring means that event Y should happen sometime in the future, then the execution of a program expressing this property will actively endeavour to make Y true as soon as possible.

Though implementations of the language have been successfully used to demonstrate the applicability of this approach to executable temporal logic, the lack of functionality has so far restricted the domain of suitable applications. With the

J.J.Ch. Meyer and J. Treur (eds.),
Handbook of Defeasible Reasoning and Uncertainty Management Systems, Vol. 7, 99–112.

model presented here for the integration of a subordinate language, we provide a framework in which larger application development can be undertaken. In this paper, we first present an overview of temporal logic and Concurrent METATEM. Our approach to Concurrent METATEM as a coordination language is then described. Finally, we illustrate how the languages will interact and the manner in which their behaviour can be utilized with a number of examples.

2 TEMPORAL LOGIC

Temporal logic can be seen as classical logic extended with various modalities representing temporal aspects of logical formulae [Emerson, 1990]. The propositional and first-order temporal logics we use (called PTL and FTL) are based on a linear, discrete model of time. Thus, time is modeled as an infinite sequence of discrete states, with an identified starting point, called 'the beginning of time'. Classical formulae are used to represent constraints *within* states, while temporal formulae represent constraints *between* states. As formulae are interpreted at particular states in a sequence, operators which refer to both the past and future are required. Examples of such operators are given below[1].

$\Diamond\varphi$ is satisfied now if φ is satisfied *sometime* in the future (often termed *eventualities*).

$\Box\varphi$ is satisfied now if φ is satisfied *always* in the future.

$\varphi\,\mathcal{U}\,\psi$ is satisfied now if ψ is satisfied from now *until* a future moment when ψ is satisfied.

$\bigcirc\varphi$ is satisfied now if φ is satisfied at the next moment in time.

$\varphi\,\mathcal{S}\,\psi$ is satisfied now if ψ was satisfied in the past and φ was satisfied from that moment until (but not including) the present moment.

'$\blacklozenge$' is the past-time analogue of '$\Diamond$', while '$\blacksquare$', is the past-time analogue of '$\Box$'.

$\bullet\varphi$ is satisfied if there was a last moment in time and, at the moment, φ was satisfied.

start is only satisfied at the beginning of time.

---

[1]The full syntax and semantics of this temporal logic can be found in, for example, [Fisher, 1993].

2.1 Separated Normal Form

As an agent's behaviour is represented by a temporal formula, we can transform this formula into Separated Normal Form (SNF) [Fisher, 1997a]. This not only removes many of the temporal operators, but also translates the formula into a set of *rules* suitable for execution. Each of these rules is of one of the following forms.

$$\mathbf{start} \quad \Rightarrow \quad \bigvee_{j=1}^{r} m_j \qquad \text{(an \textit{initial} rule)}$$

$$\bigcirc \bigwedge_{i=1}^{q} k_i \quad \Rightarrow \quad \bigvee_{j=1}^{r} m_j \qquad \text{(a \textit{step} rule)}$$

$$\bigcirc \bigwedge_{i=1}^{q} k_i \quad \Rightarrow \quad \Diamond l \qquad \text{(a \textit{sometime} rule)}$$

where each k_i, m_j or l is a literal. Note that the left-hand side of each *initial* rule is a constraint only on the *first* state, while the left-hand side of the other rules represent constraints upon the previous state. The right-hand side of each step rule is simply a disjunction of literals referring to the current state, while the right-hand side of each sometime rule is a single eventuality (i.e., '$\Diamond$' applied to a literal).

3 CONCURRENT METATEM

The motivation for the development of Concurrent METATEM [Fisher, 1993] has been provided from many areas. Being based upon executable logic, it can be utilized as part of the formal specification and prototyping of reactive systems. In addition, as it uses *temporal*, rather than classical, logic the language provides a high-level programming notation in which the dynamic attributes of individual components can be concisely represented [Barringer *et al.*, 1995]. This, together with its use of a novel model of concurrent computation, ensures that it has a range of applications in distributed and concurrent systems [Fisher, 1994].

Concurrent METATEM is a programming language comprising two distinct aspects:

1. the fundamental behaviour of a single agent is represented as a temporal formula and animation of this behaviour is achieved through the direct execution of the formula [Fisher, 1996];

2. agents are placed within an operational framework providing both asynchronous concurrency and broadcast message-passing.

While these aspects are, to a large extent, independent, the use of *broadcast* communication provides a natural link between them as it represents both a flexible communication model for concurrent objects [Birman, 1991] and a natural interpretation of distributed deduction [Fisher, 1997b]. Thus, these features together provide an coherent and consistent programming model within which a variety of reactive systems can be represented and implemented.

3.1 Agents

The basic elements of Concurrent METATEM are agents. These are considered to be encapsulated entities, executing independently, and having complete control over their own internal behaviour. There are two elements to each agent: its *interface definition* and its *internal definition*. The definition of which messages an agent recognizes, together with a definition of the messages that an agent may itself produce, is provided by the interface definition for that particular agent. The internal definition of each agent is provided by a temporal formula.

In order to animate the behaviour of an agent, we choose to execute its temporal specification directly [Fisher, 1996]. Execution of a temporal formula corresponds to the construction of a model for that formula and, in order to execute a set of SNF rules representing the behaviour of a Concurrent METATEM agent, we utilize the *imperative future* [Gabbay, 1989; Barringer *et al.*, 1996] approach. This evaluates the SNF rules at every moment in time, using information about the history of the agent in order to constrain its future execution. Thus, as the aim of execution is to produce a model for a formula, a *forward-chaining* process is employed. The underlying (sequential) METATEM language [Barringer *et al.*, 1995] exactly follows this approach.

The operator used to represent the basic temporal indeterminacy within the SNF rules is the *sometime* operator, '$\Diamond$'. When $\Diamond\varphi$ is executed, the system must try to ensure that φ *eventually* becomes true. As such eventualities might not be able to be satisfied immediately, we must keep a record of the unsatisfied eventualities, retrying them as execution proceeds. It should be noted that the use of temporal logic as the basis for the computation rules gives an extra level of expressive power over the corresponding classical logics. In particular, operators such as '$\Diamond$' give us the opportunity to specify future-time (temporal) indeterminacy. Transformation to SNF allows us to capture these expressive capabilities concisely.

As an example of a simple set of rules which form a fragment of an agent's description, consider the following.

$$
\begin{array}{rcl}
\textbf{start} & \Rightarrow & \neg\texttt{moving} \\
\bullet\texttt{go} & \Rightarrow & \Diamond\texttt{moving} \\
\bullet(\texttt{moving} \wedge \texttt{go}) & \Rightarrow & \texttt{overheat} \vee \texttt{fuel}
\end{array}
$$

Here, we see that `moving` is false at the start of execution and, whenever go is true in the last moment in time, a commitment to eventually make `moving` true is made. Similarly, whenever both go and `moving` are true in the last moment in time, then either `overheat` or `fuel` must be made true.

3.2 Concurrency and Communication

It is fundamental to our approach that all agents are (potentially) concurrently active. In particular, they may be asynchronously executing. Each agent, in executing its temporal formula, independently constructs its own temporal sequence. Within Concurrent METATEM, a mechanism is provided for communication between separate agents which simply consists of a partition of each agent's predicates into those controlled by the agent and those controlled by its environment. To fit in with this logical view of communication, whilst also providing a flexible and powerful message-passing mechanism, *broadcast* message-passing is used to pass information between agents. Here, when an agent sends a message it does not send it to a specified *destination*, it merely sends it to its environment where it can be received by *all* other agents. Although broadcast is the basic mechanism, both multicast and point-to-point message-passing can be defined on top of this [Fisher, 1994]. Finally, the default behaviour for a message is that if it is broadcast, then it will *eventually* be received at all possible receivers. Also note that, by default, the order of messages is not preserved, though such a constraint can be added, if required.

3.3 Applications and Implementation

The combination of executable temporal logic, asynchronous message-passing and broadcast communication provides a powerful and flexible basis for the development of reactive systems. Concurrent METATEM is being utilized in the development of a range of applications in areas from distributed artificial intelligence [Fisher, 1994], concurrent theorem-proving [Fisher, 1997b], agent societies [Fisher and Wooldridge, 1995], and transport systems [Finger *et al.*, 1992]. A survey of some of the potential applications of the language is given in [Fisher, 1994],

The current implementation is based upon the compilation of programs into an automata representation [Kellett, to appear]. This is interpreted by a standard execution mechanism. Replacing earlier versions of the language which directly interpreted program formulae, the new implementation serves to reduce the cost of evaluation of PTL and FTL formulae by pre-determining the potential temporal models generated during execution. Both compiler and execution mechanism are written in C++.

4 COORDINATION

As a declarative language, Concurrent METATEM provides an efficient mechanism for specifying the behavioural properties of an application. The language has been applied so far to the areas of *reactive* [Fisher, 1993] and *multi-agent systems* [Fisher, 1995]. In these domains the language has demonstrated the ability to formally specify complex properties in a natural manner. Used as a language for implementing multi-agent systems, Concurrent METATEM has provided an effective approach to specifying both *reactive* and *pro-active* behaviours. Pro-active applications, with the ability to influence their environment, have been demonstrated with a variety of cooperative and competitive behaviours [Fisher and Wooldridge, 1995].

The premise for our representation of Concurrent METATEM as a coordination language is that it provides a formal mechanism for defining the interaction between independent software modules. As a coordination language, the use of temporal logic, utilizing the declarative mechanisms of rules and constraints, provides a highly expressive formal mechanism for the abstract definition of compositional properties. While purely reactive systems are suitable for many applications, it has been proposed that pro-active behaviour is necessary to provide a framework for full coordination [Andreoli *et al.*, 1994]. Using the ability of Concurrent METATEM to specify such properties allows this facility to be directly reflected in the execution of a coordinated system.

4.1 Coordination Schema

A framework for a coordination language must support both *coordinators* and *participants* [Andreoli *et al.*, 1994]. In our approach we regard each agent in a Concurrent METATEM program as a coordinator for some independent software process. To support our model, we define the participants of an application to be object-based software processes. Each object is continuously active and executes independently. The encapsulation of participants into active agent frameworks provides the interface between the two languages (see Figure §1). Each coordinator can only directly access operations provided by its own participant, and the services of other agents can only be accessed through the Concurrent METATEM coordinators controlling their behaviour.

An agent in a Concurrent METATEM program consists of a formula comprising atoms; either propositions and predicates, related through classical and temporal logic operators. The formula is structured as a set of rules, specifying properties which must be satisfied at every moment in time. For our approach, some subset of the atoms of an agent are associated with the methods of a controlled participant. This abstract view of software components allows us to use the reasoning mechanism of Concurrent METATEM in order to define behaviours and relationships between these entities using the facilities of classical and temporal logic. A method in the underlying object is executed when its abstract representation is made true in the current state of the temporal model generated by the Concurrent METATEM agent. In the following example, a method associated with the proposition `process` is executed after the receipt of a request from the environment:

$$\bullet\texttt{request} \;\Rightarrow\; \texttt{process}$$

As participants must also be able to return data to the coordinator, some subset of

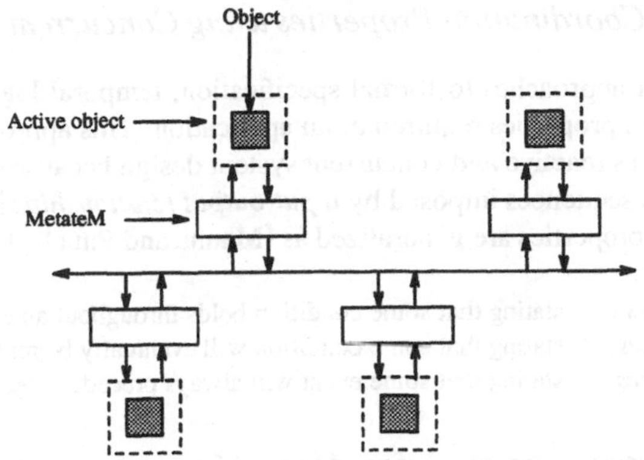

Figure 1. Coordination Schema

the terms in a Concurrent METATEM program are used to indicate the completion of a method. These are treated in the same way as environment predicates received from other Concurrent METATEM agents and are assigned a value of true in the current state of a temporal model when the associated message from the participant has been received. Extending the above example, the proposition `return` indicates the completion of `process` and specifies the subsequent behaviour:

$$\bigcirc\texttt{return} \;\Rightarrow\; \texttt{respond}$$

To define the formal relationship between terms representing methods and return values we include an additional rule specifying their behaviour. This uses the eventuality modality, $\Diamond$, to indicate that the initiation of the method `process` will cause the proposition `return` to be true at sometime in the future[2] :

$$\bigcirc\texttt{process} \;\Rightarrow\; \Diamond\texttt{return}$$

While propositions can be used to indicate the initiation or completion of a method, first-order predicates allow a transfer of data with the participant. The terms of a predicate are associated with the arguments of a method. Similarly for return values, predicate variables hold the results of a methods execution. For example:

$$\bigcirc\texttt{request(X,Y)} \;\Rightarrow\; \texttt{process(X,Y)}$$
$$\bigcirc\texttt{return(W,Z)} \;\Rightarrow\; \texttt{respond(W,Z)}$$

The association between call and return value is performed where necessary by association of an identifier with a term referenced in both predicates.

[2]Though formally defining the relationship between method and return value, such rules are not required in determining the behaviour of coordinators.

4.2 Specifying Coordination Properties using Concurrent METATEM

As opposed to other approaches to formal specification, temporal logic is based upon the definition of properties required of an application. This approach is beneficial in areas such as reactive and concurrent system design because of the freedom from execution sequences imposed by *input-output relationships* [Lynch and Tuttle, 1988]. Such properties are generalized as [Manna and Pnueli, 1988]:

Invariance properties :	stating that some condition holds throughout an execution.
Eventuality properties :	stating that some condition will eventually be realized.
Precedence properties :	stating that some event will always precede another.

Invariance properties, sometimes referred to as *safety* properties, define the parameters which an application must always satisfy to ensure correct execution. *Eventuality* properties, also referred to as *liveness* properties, specify that a 'goal' of a system will at some time be satisfied. *Precedence* properties define sequences of execution required by an application.

In Concurrent METATEM programs any or all of these categories may be used to define the behaviour of an agent. In general safety properties are used to constrain the execution to a set of acceptable circumstances. For example, the following rule defines that both `read` and `write` may not be true at the same time:

$$\bullet\mathbf{true} \Rightarrow \neg\mathtt{read} \vee \neg\mathtt{write}$$

Because of the execution mechanism used by Concurrent METATEM, all precedence properties are also members of this class[3]. We distinguish them here as rules which specify sequences of execution rather than general restrictions. For example:

$$\bullet\mathtt{request} \Rightarrow \mathtt{read}$$

Given these two categories we can define an application in terms of a set of requirements which must be satisfied including immutable sequences of actions. The third category of properties, that of eventualities, is used to apply to this framework a goal directed behaviour. Though properties associated with eventualities are specified as *satisfied at sometime in the future*, because of the implementation of Concurrent METATEM the more strict interpretation of *satisfied as soon as possible* is applied. This allows us to use eventualities to select the most profitable sequence of actions from the range of possibilities falling within the constraints applied by other rules. In the following example, a safety property specifies that two methods `read` and `write` may never be executed concurrently. Two eventuality properties define that when requests for these services are received, they are satisfied as soon

[3]The full definition of a Concurrent METATEM agent is given as $\Box \bigwedge_i R_i$ for a set of rules R_i of the form *past* ⇒ *future*.

as possible:

```
●true  ⇒  ¬read ∨¬write
●get   ⇒  ◇read
●put   ⇒  ◇write
```

5 APPLICATIONS

In this section we demonstrate some of the features of Concurrent METATEM as a coordination language by defining a number of examples of its use. These applications are simplified for demonstration purposes and do not incorporate formal specification of the relationship between call and return parameters.

5.1 Goal Directed Scheduling

This application demonstrates how Concurrent METATEM utilizes goal directed behaviours to provide dynamic processing capabilities. A Concurrent METATEM agent schedules requests from the environment to utilize its participant in processing data. Request are accepted in the form `request(X)` and `urgent(X)` where `X` represents an argument for the method `process`. Completion of the method is signaled by the receipt of the predicate `return(X)`. The first rule directs that requests must be processed serially by using the *since* operator to specify that, if a return value has not been received since processing began, then another request may not be initiated. Both normal and urgent requests are scheduled to be satisfied at sometime in the future. The implementation of ◇ directs that this will occur as soon as possible. Urgent requests are given higher priority by specifying that no normal request can be processed until all outstanding urgent request have been completed (see Figure §2).

```
scheduler()
      in :  request, urgent
     out :  result
 ●(¬return(X) S process(X)) ⇒ ¬process(Y)
 ●request(X) ⇒ ◇process(X)
 ●urgent(X) ⇒ ◇process(X)
 ●((¬process(X) S request(X))∧(¬process(Y) S urgent(Y))⇒¬process(X)
 ●return(X) ⇒ result(X)
```

Figure 2. Goal Directed Scheduler

5.2 Distributed Problem Solving

In this example we show how the coordination of a number of processes collaborating in a distributed application can be achieved using Concurrent METATEM (see Figure §3). The program consists of four processes, performing operations on common data. The Concurrent METATEM agents, `stage1, stage2` and `stage3` control processes which may execute in parallel. The agent `controller` performs the final stage of an operation which utilizes the results of the other processes.

An operation begins with a message, `initiate`, broadcast to all agents. Because each *stage* agent is independent, each may immediately satisfy the predicates `process` and utilize the participant methods. The result of these operations is returned to Concurrent METATEM in the predicate `return(X)` which is distributed by the agents as `stage1, stage2` and `stage3`.

When the final stage agent, `final`, receives an `initiate(X)` message, it indicates that an operation is in progress by specifying `answer(X)` will be true at some time in the future. The satisfaction of this predicate is prevented until the final stage operation, `evaluate`, is performed. As the initiation of this participant method requires the results of each the other processes (`stage1, stage2` and `stage3`), the predicate `answer(X)` cannot be satisfied until all stages have been completed. To coordinate the results of stages 1,2 and 3, the *until* operator is used to preserve the data recieved until all agents have produced an answer. When a value is returned from the `evaluate` method, the result of the combined operation is distributed with the predicate `result`.

5.3 Distributed Transaction Processing

This example demonstrates a simple mechanism for the implementation of meta-level transactions (see Figure §4). Three agents are used to access independent databases and perform updates upon them. A fourth agent implements a commit procedure which ensures that the operation was successful in all three databases before confirming changes. A failure in any of the three agents results in the cancellation of changes to the other two.

A message `modify(X,Y)` is distributed to all agents to initiate an operation. Each database agent activates a participant method, `update`, to perform changes on receipt of this signal. Results from these operations are delivered in the predicate `return` on completion of the first stage of an update. The result `return(X,0)` indicates a failure of the database to perform the operation and causes a predicate `fail` to be distributed to all agents. Database agents receiving this message activate a participant method undo to rollback the update operation.

Successful updates cause the messages `commit1, commit2` and `commit3` to be distributed by the agents. The controller will signal that and update is successful when each database has reported that it is ready to commit. The predicate go is distributed to indicate that changes can be made permanent. A `fail` message causes the requests to commit to be discarded.

```
stage1()
    in :   initiate
   out :   stage1

○initiate(X) ⇒ process1(X)
○return(X,Y) ⇒ stage1(X,Y)

stage2()
    in :   initiate
   out :   stage2

○initiate(X) ⇒ process2(X)
○return(X,Y) ⇒ stage2(X,Y)

stage3()
    in :   initiate
   out :   stage3

○initiate(X) ⇒ process3(X)
○return(X,Y) ⇒ stage3(X,Y)

final()
   in :   initiate, stage1, stage2, stage3
  out :   result

○initiate(X) ⇒ ◇answer(X)
○¬evaluate(W,X,Y,Z) ⇒ ¬answer(W)
○stage1(X,Y) ⇒ stage1(X,Y) U answer(W)
○stage2(X,Y) ⇒ stage2(X,Y) U answer(W)
○stage3(X,Y) ⇒ stage3(X,Y) U answer(W)
○(stage1(X,J) ∧ stage2(X,K) ∧ stage3(X,L)) ⇒ evaluate(X,J,K,L)
○return(X,Y) ⇒ result(X,Y)
```

Figure 3. Distributed Problem Solving

```
database1()
      in :   update, go, fail
     out :   commit1, fail
 ●modify(X,Y) ⇒ update1(X,Y)
 ●return(X,Y) ⇒ commit1(X)
 ●return(X,0) ⇒ fail(1,X)
 ●(fail(X,Y) ∧ X<>1) ⇒ undo(Y)
 ●go(X) ⇒ confirm(X)

database2()
      in :   update, go, fail
     out :   commit2, fail
 ●modify(X,Y) ⇒ update2(X,Y)
 ●return(X,Y) ⇒ commit2(X)
 ●return(X,0) ⇒ fail(1,X)
 ●(fail(X,Y) ∧ X<>2) ⇒ undo(Y)
 ●go(X) ⇒ confirm(X)

database3()
      in :   update, go, fail
     out :   commit3, fail
 ●modify(X,Y) ⇒ update3(X,Y)
 ●return(X,Y) ⇒ commit3(X)
 ●return(X,0) ⇒ fail(1,X)
 ●(fail(X,Y) ∧ X<>3) ⇒ undo(Y)
 ●go(X) ⇒ confirm(X)

controller()
       in :   commit1, commit2, commit3, fail
      out :   go
 ●commit1(X) ⇒ commit1(X) U (go(X) ∨ fail(X,Y))
 ●commit2(X) ⇒ commit2(X) U (go(X) ∨ fail(X,Y))
 ●commit3(X) ⇒ commit3(X) U (go(X) ∨ fail(X,Y))
 ●(commit1(X) ∧ commit2(X) ∧ commit3(X)) ⇒ go(X)
```

Figure 4. Transaction Processing

6 CONCLUSIONS AND FUTURE WORK

Concurrent METATEM, when used as a coordination language, can provide a formal mechanism for defining the interaction between independent software modules. The use of temporal logic gives a highly expressive notation for the abstract definition of compositional properties. The association between coordinated system and temporal formulae (potentially) allows the verification of coordination properties via temporal theorem-proving.

In this paper we have proposed an extension to Concurrent METATEM to support the coordination of underlying software processes. The benefits of using the expressive power of temporal logic to define the integration of a coordinated system have been described. We suggest that the ability of the language to define pro-active behaviours, demonstrated with multi-agent systems, can provide significant advantages for coordination. In our ongoing work, Concurrent METATEM is being developed to support larger applications, full implementation of the extensions described above is being developed and the integration of dynamic agent creation is being undertaken.

7 ACKNOWLEDGEMENT

An earlier version of this paper appeared in *Coordination Languages and Models*, published by Springer-Verlag in Lecture Notes in Computer Science 1282, and is reproduced here with their kind permission.

Adam Kellett
BJSS Ltd, UK.

Michael Fisher
Department of Computer Science, University of Liverpool, UK.

BIBLIOGRAPHY

[Barringer *et al.*, 1995] H. Barringer, M. Fisher, D. Gabbay, G. Gough, and R. Owens. METATEM: An Introduction. *Formal Aspects of Computing*, 7(5):533–549, 1995.

[Barringer *et al.*, 1996] H. Barringer, M. Fisher, D. Gabbay, R. Owens, and M. Reynolds, editors. *The Imperative Future: Principles of Executable Temporal Logics*. Research Studies Press, Chichester, United Kingdom, 1996.

[Birman, 1991] K. P. Birman. The Process Group Approach to Reliable Distributed Computing. Techanical Report TR91-1216, Department of Computer Science, Cornell University, July 1991.

[Emerson, 1990] E. A. Emerson. Temporal and Modal Logic. In J. van Leeuwen, editor, *Handbook of Theoretical Computer Science*, pages 996–1072. Elsevier, 1990.

[Finger *et al.*, 1992] M. Finger, M. Fisher, and R. Owens. METATEM at Work: Modelling Reactive Systems Using Executable Temporal Logic. In *Sixth International Conference on Industrial and Engineering Applications of Artificial Intelligence and Expert Systems (IEA/AIE)*, Edinburgh, U.K., June 1993. Gordon and Breach Publishers.

[Fisher, 1994] M. Fisher. A Survey of Concurrent METATEM — The Language and its Applications. In *First International Conference on Temporal Logic (ICTL)*, Bonn, Germany, July 1994. (Published in *Lecture Notes in Computer Science*, volume 827, Springer-Verlag).

[Fisher, 1995] M. Fisher. Representing and Executing Agent-Based Systems. In M. Wooldridge and N. R. Jennings, editors, *Intelligent Agents*. Springer-Verlag, 1995.

[Fisher, 1996] M. Fisher. An Introduction to Executable Temporal Logics. *Knowledge Engineering Review*, 11(1):43–56, March 1996.

[Fisher and Wooldridge, 1995] M. Fisher and M. Wooldridge. A Logical Approach to the Representation of Societies of Agents. In N. Gilbert and R. Conte, editors, *Artificial Societies*. UCL Press, 1995.

[Fisher, 1993] M. Fisher. Concurrent METATEM — A Language for Modeling Reactive Systems. In *Parallel Architectures and Languages, Europe (PARLE)*, Munich, Germany, June 1993. (Published in *Lecture Notes in Computer Science*, volume 694, Springer-Verlag).

[Fisher, 1997a] M. Fisher. A Normal Form for Temporal Logic and its Application in Theorem-Proving and Execution. *Journal of Logic and Computation*, 7(4), August 1997.

[Fisher, 1997b] M. Fisher. An Open Approach to Concurrent Theorem-Proving. In *Parallel Processing for Artificial Intelligence III*. North-Holland, 1997.

[Gabbay, 1989] D. Gabbay. Declarative Past and Imperative Future: Executable Temporal Logic for Interactive Systems. In B. Banieqbal, H. Barringer, and A. Pnueli, editors, *Proceedings of Colloquium on Temporal Logic in Specification*, pages 402–450, Altrincham, U.K., 1987. (Published in *Lecture Notes in Computer Science*, volume 398, Springer-Verlag), 1989.

[Kellett, to appear] A. Kellett. *Implementation Techniques for Concurrent* METATEM. PhD Thesis. In preparation.

[Lynch and Tuttle, 1988] N. A. Lynch and M. R. Tuttle. An Introduction to Input/Output Automata. Technical Report MIT/LCS/TM-373, Laboratory for Computer Science, Massachusetts Institute of Technology, November 1988.

[Andreoli *et al.*, 1994] Jean-Marc Andreoli, Herve Gallaire, and Remo Pareschi. Rule Based Object Coordination. *Object-Based Models and Languages for Concurrent Systems* ed. P.Ciancarini, O.Nierstrsz, A.Yonezawa. LNCS 924. Springer-Verlag. 1994.

[Manna and Pnueli, 1988] Zohar Manna and Amir Pnueli. The Anchored Version of the Temporal Framework In *Linear Time, Branching Time and Partial Order Logics and Models for Concurrency*. LNCS 354. Springer-Verlag. 1988.

FRANCES M.T. BRAZIER, CATHOLIJN M. JONKER, JAN TREUR

COMPOSITIONAL DESIGN AND REUSE OF A GENERIC AGENT MODEL

1 INTRODUCTION

The term agent has become popular, and has been used for a wide variety of applications, ranging from simple batch jobs and simple email filters, to mobile applications, to intelligent assistants, and to large, open, complex, mission critical systems (such as systems for air traffic control). Some of the key concepts in agent technology lack universally accepted definitions. In particular, there is only partial agreement on what an agent is. For example, simple batch jobs are termed agent because they can be scheduled in advance to perform tasks on a remote machine, mobile applications are termed agent because they can move themselves from computer to computer, and intelligent assistants are termed agents because they present themselves to human users as believable characters that manifest intentionality and other aspects of a mental state normally attributed only to humans. Besides this variety in different appearances of agents, applications of agents often are concentrated on specific implementations of agents (often in Java). Often the only precise description of an agent is its implementation code, which is dependent on the chosen implementation platform. Therefore, existing agent architectures are often only comparable in an informal manner. A principled design description of an agent at a conceptual and logical level lacks, which makes it difficult to compare agents from different applications.

As agents show a variety of appearances, perform a multitude of tasks, and their abilities vary significantly [Nwana, 1996; Nwana and Ndumu, 1998], attempts have been made to define what they have in common. In [Wooldridge and Jennings, 1995b; Wooldridge and Jennings, 1995c] the weak notion of agent is introduced; this is often used as a reference. This notion is explained in more detail in Section 2; a number of primitive concepts relevant for this type of agent are identified in Section 3. During the design of such agents, these concepts have to be incorporated, and a number of generic agent processes can be identified; for example relating to interaction with the world or to social behaviour with respect to other agents.

To obtain a unified, formally defined conceptual but implementation-independent description, Section 5 describes the compositional design of a generic agent model (GAM) at a conceptual and logical level, in which generic agent concepts and processes related to the weak agent notion are predefined. This generic agent model abstracts from specific application domains; by refinement (specialisation and instantiation) it can be (re)used as a core design for a large variety of

J.J.Ch. Meyer and J. Treur (eds.),
Handbook of Defeasible Reasoning and Uncertainty Management Systems, Vol. 7, 113–163.

agent types and application domains. The model was designed on the basis of experiences in applications to, among others, monitoring, diagnosis and restoration of an electricity network (Brazier, Dunin-Keplicz, Jennings and Treur, 1995) and negotiation for load balancing of electricity use (Brazier, Cornelissen, Gustavsson, Jonker, Lindeberg, Polak and Treur, 1998). The compositional development method used to design this agent model, DESIRE, is briefly introduced in Section 4. To illustrate reuse of this agent model, an application with co-operative information gathering agents is described in more depth in Section 6. Section 7 discusses how the model GAM can be used to obtain a unified, and thus comparable, description at the level of design of a large variety of agent architectures occurring in the literature. In Section 8 the paper concludes with a discussion on design and reuse of this generic agent model.

2 AGENT NOTIONS

The weak notion of agent was introduced in [Wooldridge and Jennings, 1995b; Wooldridge and Jennings, 1995c] and is often used as a reference in the literature (see also [Jennings and Wooldridge, 1998]).

2.1 Weak Notion of Agent

The weak notion of agent is a notion that requires the behaviour of agents to exhibit at least the following four types of behaviour:

- Autonomous behaviour
- Responsive behaviour (also called reactive behaviour)
- Pro-active behaviour
- Social behaviour

Autonomy relates to control: although an agent may interact with its environment, the processes performed by an agent are in full control of the agent itself. [Jennings and Wooldridge, 1998] define *autonomous* behaviour as:

> ... the system should be able to act without the direct intervention of humans (or other agents) and should have control over its own actions and internal state.

This means that an agent can only be requested to perform some action, and, as [Jennings and Wooldridge, 1998] state:

> The decision about whether to act upon the request lies with the recipient.

Examples of autonomous processes are: process control systems (e.g., thermostats, missile guiding systems, and nuclear reactor control systems), software deamons (e.g., one that monitors a user's incoming email and obtains their attention by displaying an icon when new, incoming email is detected), operating systems.

Many processes that exhibit autonomous behaviour are being termed agents. However, if such agents do not exhibit flexible behaviour, they are not, in general, considered to be intelligent agents. An intelligent agent is defined in [Jennings and Wooldridge, 1998] to be a computer system that is capable of flexible autonomous actions in order to meet its design objectives. Intelligence requires flexibility with respect to autonomous actions, meaning that intelligent agents also exhibit responsive, social, and pro-active behaviour.

An agent exhibits *responsive* (or *reactive*) behaviour if it reacts or responds to new information from its environment. Jennings and Wooldridge define responsive behaviour as follows:

> Agents should perceive their environment (which may be the physical world, a user, a collection of agents, the Internet, etc.) and respond in a timely fashion to changes that occur in it.

A barometer is a simple example of a system that exhibits responsive behaviour: It continually receives new information about the current air pressure and responds to this new information by adjusting its dial.

Pro-active behaviour is defined by [Jennings and Wooldridge, 1998] as follows:

> Agents should not simply act in response to their environment, they should be able to exhibit opportunistic, goal-directed behaviour and take the initiative where appropriate.

Pro-active behaviour is the most difficult of the required types of behaviour for an agent defined according to the weak agent notion. For example, pro-active behaviour can occur simultaneously with responsive behaviour. It is possible to respond to incoming new information in an opportunistic manner according to some goals. Also initiatives can be taken in response to incoming new information from the environment, and thus this behaviour resembles responsive behaviour. However, it is also possible to behave pro-actively when no new information is received from the environment. This last behaviour can by no means be called responsive behaviour. A more elaborate comparison between responsive behaviour and pro-active behaviour can found in [Jonker and Treur, 1998a; Jonker and Treur, 1998b].

An agent exhibits *social behaviour* if it communicates and co-operates with other agents. Jennings and Wooldridge define social behaviour as follows:

> Agents should be able to interact, when they deem appropriate, with other artificial agents and humans in order to complete their own problem solving and to help others with their activities.

An example of an agent that exhibits social behaviour is a car: it communicates with its human user by way of its dials (outgoing communication dials: speed, amount of fuel, temperature) and its control mechanisms (incoming communication control mechanisms: pedals, the steering wheel, and the gears). It co-operates with its human user, e.g., by going in the direction indicated by the user, with the speed set by that user.

2.2 *Other Notions of Agent*

Agents can also be required to have additional characteristics. In this section three of these characteristics are discussed: adaptivity, pro-creativity, and intentionality.

Adaptivity is a characteristic that is vital in some systems. An adaptive agent learns and improves with experience. This behaviour is vital in environments that change over time in ways that would make a non-adaptive agent obsolete or give it no chance of survival. This characteristic is modelled often in simulations of societies of small agents, but also, for example, in adaptive user interface agents.

Pro-creativity is of similar importance to find agents that satisfy certain conditions. The chance of survival is often measured in terms of a fitness function. This characteristic is modelled often in simulations of societies of small agents (see the literature in the area of Artificial Life). A computer virus is a very infamous form of a pro-creative agent.

According to [Denett, 1987] an *intentional system* is an entity

> ... whose behaviour can be predicted by the method of attributing beliefs, designs and rational acumen.

Mentalistic and intentional notions such as ***beliefs, desires, intentions, commitments, goals, plans, preference, choice, awareness***, may be assigned to agents. The ***stronger notion of agenthood*** in which agents are described in terms of this type of notions provides additional metaphorical support for the design of agents.

3 PRIMITIVE AGENT CONCEPTS

The notions of agenthood discussed in Section 2 are highly abstract notions. In order to design agents, it is necessary to be familiar with a number of primitive agent concepts. These primitive concepts serve as an ontology or vocabulary to express analyses and designs of applications of agents and multi-agent systems. Two classes of primitive notions are distinguished: those used to describe the behaviour of agents in terms of their external (or public) states and interactions (Section 3.1), and those used to describe the behaviour of agents in terms of their internal (or private) states, and processes (Section 3.2). In Section 3.3, to illustrate the concepts, an example agent is discussed in terms of these concepts: an elevator.

3.1 External primitive concepts

Two types of interaction of an agent with its environment are distinguished, depending on whether the interaction takes place with an agent or with something else (called an *external world*), for example a database, or the material world. For each of these two types of interaction specific terminology is used.

3.1.1 Interaction with the external world

Two primitive types of interaction with the external world are distinguished. The first type of interaction, *observation*, changes the information the agent has about the world, but does not change the world state itself, whereas the second type, *performing an action*, does change the world state, but does not change the information the agent has about the world. Combinations of these primitive types of interaction are possible; for example, performing an action, and observing its results.

Observation

In which ways is the agent capable of observing or sensing its environment? Two types of observation can be distinguished: the agent passively receives the results of observations without taking any initiative or control to observe (*passive observation*), or the agent actively initiates and controls which observations it wants to perform; this enables the agent to focus its observations and limit the amount of information acquired (*active observation*).

Execution of actions in the external world

An agent may be capable of making changes to the state of its environment by initiating and executing specific types of actions.

3.1.2 Communication with other agents

Two directions of communication are distinguished, which can occur together: *outgoing communication* (is the agent capable of communicating to another agent; to which ones?), and *incoming communication* (is the agent capable of receiving communication from another agent; from which ones?).

3.2 Internal primitive concepts

A description in terms of the external primitive concepts abstracts from what is inside the agent. In addition to descriptions of agents in terms of the external concepts, often descriptions in terms of internal concepts are useful. The following internal primitive agent concepts are distinguished.

World and Agent Models

An agent may create and maintain information on (a model of) external *world* based on its observations of that world, on information about that world communicated by other agents, and its own knowledge about the world. The agent may also create and maintain information on (models of) *other agents* in its environment based on its observations of these agents as they behave in the external world, on information about these agents communicated by other agents, and knowledge about the world.

Self Model and History

Some agents create and maintain information on (a model of) their own characteristics, internal state, and behaviour. Or the agent creates and maintains a history of the world model, or agent models, or self model, or own and group processes.

3.2.1 Goals and Plans

To obtain pro-active, goal-directed behaviour, an agent often represents, generates, and uses explicit goals and its own plans of action in its processing.

3.2.2 Group Concepts

Besides individual concepts, often agents use group concepts that allow it to co-operate with other agents. For example, *joint goals*: is the agent capable of formulating or accepting and using goals for a group of agents, i.e., goals that can only be achieved by working together? Or *joint plans*: is the agent capable of representing, generating, and using plans of action for joint goals, i.e., involving which actions are to be performed by which agents in order to achieve a certain joint goal? Also *commitments* to joint goals and plan, *negotiation protocols* and *strategies* can be useful group concepts for agents, depending on their role and function.

3.3 An Example Analysis

These agent concepts introduced in Section 3.1 and 3.2 are illustrated by an example: an elevator is analysed from the agent perspective using these basic concepts; see Tables 1 through 3 below and their motivation.

3.3.1 The elevator in terms of external primitive concepts

Observation

No reasoning is performed as to when observations are to be performed. However, an elevator is capable of receiving *passive* observation results on the presence of objects between the doors (an optical sensor), the total weight of its contents, and

Table 1. Elevator: External primitive concepts

I. External primitive concepts	elevator
Interaction with the world	
observations	
passive observations	presence of objects between doors (optically) total weight its position
active observations	presence of objects between the doors (mechanically)
performing actions	moving opening and closing doors
Communication with other agents	
incoming communication	from users in the elevator: where they want to go (pushing button in elevator) from users outside: where they want to be picked up (pushing button outside elevator)
outgoing communication	to users in the elevator: where we are (display) there is overweight (beep) to users outside: where is the elevator (display) in which direction it moves (display)

possibly, its position in the building (at which floor). Besides it is able to perform *active* observation: the presence of objects between the doors (a mechanical sensor which is moved in the door opening just ahead of the doors themselves).

Performing actions

It performs actions in the world like moving itself (and people) vertically from one position to another and opening and closing doors.

Incoming communication

The elevator receives communication from users by buttons that have been pressed (providing information about the floor to which they wish to be transported).

Table 2. Elevator: Internal primitive concepts

II. Internal primitive concepts	elevator
A. *World Model*	the current floor, max load, current load
B. *Agent Models*	a user wants to be picked up from floor X a user wants to go to floor Y
C. *Self Model*	when maintenance is next due
D. *History*	when maintenance was last performed
E. *Goals*	to go to the floor X to pick up somebody to go to the floor X to deliver somebody
F. *Plans*	the order in which the required floors are visited sometimes: the speed that is taken
G. *Group Concepts*	
Joint goals	With other elevators to transport people and goods as efficiently as possible
Joint plans	Some elevators are capable of distributing the work
Commitments	The elevators then commit to their part of the work
Negotiation protocol	To reach a good distribution, they may have to negotiate
Negotiation strategies	To reach a good distribution, they may have to negotiate

Outgoing communication

The elevator communicates to a user by lighting buttons (information on the floor) and sounding beeps (information about overload).

3.3.2 The elevator in terms of internal primitive concepts

World and Agent Models

Elevators need to know *world information* on which floor they are on. They may maintain this knowledge themselves based on the actions (going two floors up, going one floor down) they perform. Another possibility is that the elevator immediately observes where it is, then it would not need to maintain a world state. Furthermore, the elevator needs to know how much weight its physical self is capable of transporting. The *agent information* of the user goals (where they want to go) may be maintained as well.

Self Model and History

The agent might have an explicit representation of when its physical form needs maintenance. It does not need to know what actions it previously performed to perform its current task. It might have an explicit representation of when it has last received maintenance.

Goals and Plans

Most modern elevators make use of the explicit goals (adopted from the goals communicated by the users). The goals are used to determine which actions to perform. They may even make plans for reaching these goals: determine the order of actions, for example when one of the users has the goal to be at a highter floor and another on a lower floor.

Group Concepts

The elevator co-operates with its users. Sometimes the elevator can also co-operate with other elevators so that they could strategically distribute themselves over the floors. *Joint goals*: The goals adopted from the goals communicated by the users are joint goals (joint with the users), and sometimes even joint with the other elevators. *Joint plans*: Modern elevators are capable of distributing the work load, and thus of making joint plans. *Commitments*: To achieve the joint goals an elevator must commit to its part of the work as specified in the joint plans. *Negotiation protocols*: To make a joint plan, the elevators may negotiate as to which elevator goes where. Negotiation is only possible if a negotiation protocol is followed. *Negotiation strategies*: To make a joint plan, the elevators may negotiate as to which elevator goes where. Negotiation is only possible if each elevator has at least one strategy for negotiation.

Table 3. Elevator: Types of behaviour

III. Types of behaviour	elevator
Autonomy	yes
Responsiveness	in reaction to user requests in immediate reaction to observed objects between the doors
Pro-activeness	taking the initiative to go to a normally busy floor, if empty and not being called by a user
Social behaviour	co-operation with users, and, sometimes, with other elevators
Own adaptation and learning	often not possible

3.3.3 Types of behaviour of the elevator

Autonomy

As soon as it is activated, no system or human is controlling an elevator's machinery, and (normally) it is not switched off and on by the user. The fact that it responds to the immediate stimuli of buttons being pressed is not the same as being controlled. The elevator has full control of its motor, doors, and lights.

Pro-activeness

The most simple elevators stay where they are (some take the initiative to close their doors) when no longer in use, but more intelligent elevators go to a strategic floor (e.g., the ground floor).

Reactiveness

The elevator reacts to the immediate stimuli of buttons pressed, therefore, it shows reactive behaviour. Furthermore, elevators often show delayed-response behaviour in picking up people. People often have to wait for the elevator as the elevator picks up people on other floors, however, the elevator does not forget a signal and will, eventually, come to the requested floor.

Social behaviour

The elevator co-operates with users and, sometimes, with other elevators.

Own adaptation and learning

Simple elevators are not capable of adjusting their own behaviour to new situations, nor are they capable of learning. However, it is possible to conceive of more intelligent elevators that can learn the rush hours for the different floors.

4 COMPOSITIONAL DEVELOPMENT OF MULTI-AGENT SYSTEMS

The example multi-agent system described in this paper has been developed using the compositional development method DESIRE for multi-agent systems (DEsign and Specification of Interacting REasoning components); for the underlying principles, see [Brazier *et al.*, 1998c], for a real-world case study, see [Brazier *et al.*, 1995]. The development of a multi-agent system is supported by graphical design tools within the DESIRE software environment. Translation to an operational system is straightforward; the software environment includes implementation generators with which formal specifications can be translated into executable code of a prototype system. In DESIRE, a design consists of knowledge of the following three types: process composition, knowledge composition, the relation between process composition and knowledge composition. These three types of knowledge are discussed in more detail below.

4.1 Process Composition

Process composition identifies the relevant processes at different levels of (process) abstraction, and describes how a process can be defined in terms of (is composed of) lower level processes. Processes can be described at different levels of abstraction; for example, the process of the multi-agent system as a whole, processes defined by individual agents and the external world, and processes defined by task-related components of individual agents. The identified processes are modelled as *components*. For each process the *input and output information types* are modelled. The identified levels of process abstraction are modelled as *abstraction/specialisation relations* between components: components may be *composed* of other components or they may be *primitive*. Primitive components may be either reasoning components (i.e., based on a knowledge base), or, components capable of performing tasks such as calculation, information retrieval, optimisation. These levels of process abstraction provide process hiding at each level. The way in which processes at one level of abstraction are composed of processes at the adjacent lower abstraction level is called *process composition*. This composition of processes is described by a specification of the possibilities for *information exchange* between processes (*static view* on the composition), and a specification of *task control knowledge* used to control processes and information exchange (*dynamic view* on the composition).

4.2 Knowledge Composition

Knowledge composition identifies the knowledge structures at different levels of (knowledge) abstraction, and describes how a knowledge structure can be defined in terms of lower level knowledge structures. The knowledge abstraction levels may correspond to the process abstraction levels, but this is often not the case.

The two main structures used as building blocks to model knowledge are: *information types and knowledge bases*. Knowledge structures can be identified and described at different levels of abstraction. At higher levels details can be hidden. An *information type* defines an ontology (lexicon, vocabulary) to describe objects or terms, their sorts, and the relations or functions that can be defined on these objects. Information types can logically be represented in order-sorted predicate logic. A *knowledge base* defines a part of the knowledge that is used in one or more of the processes. Knowledge is represented by formulae in order-sorted predicate logic, which can be normalised by a standard transformation into rules. Information types can be composed of more specific information types, following the principle of compositionality discussed above. Similarly, knowledge bases can be composed of more specific knowledge bases. The compositional structure is based on the different levels of knowledge abstraction distinguished, and results in information and knowledge hiding.

4.3 Relation between Process and Knowledge Composition

Each process in a process composition uses knowledge structures. Which knowledge structures are used for which processes is defined by the relation between process composition and knowledge composition.

4.4 Generic Models and Reuse

Instead of designing each and every new agent application from scratch, an existing generic model can be used. Generic models can be distinguished for specific types of agents, of specific agent tasks and of specific types of multi-agent organisation. The use of a generic model in an application structures the design process: the acquisition of a conceptual model for the application is based on the generic structures in the model. A model can be generic in two senses:

- generic with respect to the *processes* or *tasks*
- generic with respect to the *knowledge structures*.

Genericity with respect to processes or tasks refers to the level of process abstraction: a generic model abstracts from processes at lower levels. A more specific model with respect to processes is a model within which a number of more specific processes are distinguished, at a lower level of process abstraction. This type of refinement is called *specialisation*. Genericity with respect to knowledge refers to levels of knowledge abstraction: a generic model abstracts from more specific knowledge structures. Refinement of a model with respect to the knowledge in specific domains of application, is refinement in which knowledge at a lower level of knowledge abstraction is explicitly included. This type of refinement is called *instantiation*.

In Section 5 a generic agent model for weak agency is presented. The application for co-operative information gathering agents presented in Section 6 of this paper is an instantiation of this generic agent model. *Reuse* as such, reduces the time, expertise and effort needed to design and maintain system designs. Which components, links and knowledge structures from the generic model are applicable in a given situation depends on the application. Whether a component can be used immediately, or whether instantiation, modification and/or specialisation is required, depends on the desired functionality. Other existing (generic) models can be used for specialisation of a model; existing knowledge structures (e.g., ontologies, thesauri) can be used for instantiation. Which models and structures are used depends on the problem description: existing models and structures are examined, rejected, modified, specialised and/or instantiated in the context of the problem at hand.

5 THE GENERIC AGENT MODEL: GAM

The characteristics of weak agency and the primitive agent concepts, introduced in Sections 2 and 3, provide a means to reflect on the tasks an agent needs to be able to perform. Pro-activeness and autonomy are related to the primitive concepts self model, goals, and plans. Reactivity and social ability are related to the primitive concepts world model, agent models, history, communication with other agents, and interaction with the external world. The ability to communicate with other agents and to interact with the external world often relies on the knowledge an agent has of the world and other agents.

The design of the generic agent model (GAM) in a compositional approach entails consideration of the processes and knowledge an agent needs to perform and the composition of related components and knowledge structures.

5.1 Process composition

Process composition within the generic agent model identifies the processes within an agent at the highest level of abstraction, and the manner in which they are composed to obtain the agent process (composition relation). Section 5.1.1 identifies the processes and their levels of abstraction. In Section 5.1.2 their interface information types are identified. The way in which these processes are composed is defined by information links and task control knowledge. Sections 5.1.3 (information links) and 5.1.4 (task control) address this composition relation.

5.1.1 Processes at different levels of abstraction

Identification of a process includes its abstraction level and its interface information types. The processes modelled within the generic agent model are depicted in

Figure 1. The processes involved in controlling an agent (e.g., determining, monitoring and evaluating its own goals and plans) but also the processes of maintaining a self model are the task of the component own process control. The processes involved in managing communication with other agents are the task of the component agent interaction management. Maintaining knowledge of other agents' abilities and knowledge is the task of the component maintenance of agent information. Comparably, the processes involved in managing interaction with the external (material) world are the task of the component world interaction management. Maintaining knowledge of the external (material) world is the task of the component maintenance of world information. The specific task for which an agent is designed (for example: design, diagnosis, information retrieval), is modelled in the component agent specific task. Existing (generic) task models may be used to further structure this component. In addition, a component co-operation management may be distinguished for all tasks related to social processes such as co-operation in a project, or negotiation. This component is not discussed in this paper, but is addressed elsewhere in more detail [Brazier *et al.*, 1997b].

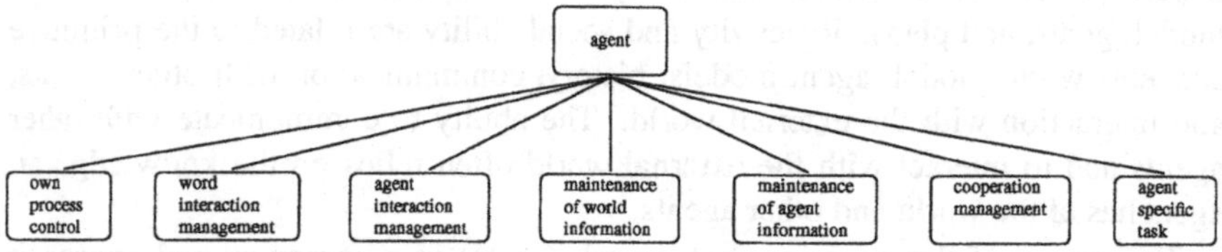

Figure 1. Processes at the two highest process abstraction levels within the agent

The four characteristics of weak agency discussed in Section 2 are related to these components in the following sense. Perception of the environment is performed by world interaction management (managing the perception process), maintenance of world information and maintenance of agent information (representation of perception information obtained from the environment). Actions in the world are managed by world interaction management. Social actions are managed by the tasks agent interaction management and cooperation management. The task cooperation management is not explained further in this paper. Performing the agent's processes is initiated and co-ordinated by the task own process control; thus the agent's autonomous and pro-active behaviour is modelled.

5.1.2 *Interface information types*

A number of generic information types can be distinguished for the input and output of the generic agent model (based on external concepts) and for the generic processes within the agent (based on internal concepts).

Interface information types of the agent

An agent capable of communication with other agents may receive incoming communication info and may send outgoing communication info. Moreover, the agent may observe and perform actions in the external (material) world. The information type observation info models the observations that are to be performed in the component external world. The information type observation result info models the incoming results of observations. The information type action info models the actions the agent performs. In Table 4 an overview of the agent's interface information types is specified, based on the external primitive agent concepts.

The information types that express communication information are composed of information types on the subject of communication, and an information type to specify the agent from, or to whom, the communication is directed.

Table 4. Specification of interface information types of the agent

process	*input information types*	*output information types*
agent	incoming communication info observation result info	outgoing communication info observation info action info

Interface information types of components within the agent

The interface information types of the components within the agent are based on the internal primitive agent concepts; these interface information types are listed in Table 5. Within the agent component, the component own process control uses belief information on other agents and the external (material) world, as input. This information is modelled in the information type belief info which is composed of belief info on world and belief info on agents. The output of the component own process control includes the agent's characteristics (modelled in the information type own characteristics), used by the components agent interaction management and world interaction management. In addition to this information type, the component agent interaction management also receives the incoming communication received by the agent (and forwarded directly to the component agent interaction management), modelled in the input interface in the information type incoming communication info, and world and agent information, modelled in the input infor mation type belief info. The output generated by the component agent interaction management includes the output for the agent as a whole (outgoing communication info), extended with maintenance info which is composed of maintenance info on agents and maintenance info on world (communicated information on the world and other agents that needs to be maintained).

Table 5. Specification of interface information types within the generic agent model

process	*input information types*	*output information types*
own process control	belief info	own characteristics
agent interaction management	incoming communication info own characterist belief info	outgoing communication info maintenance info
world interaction management	observation result info own characteristics belief info	observation info action info maintenance info
maintenance of	agent info agent information	agent info
maintenance of world information	world info	world info

The component maintenance of agent information receives new information on other agents (the agent's beliefs on other agents) in its input interface. These beliefs on other agents are made available to other components in the output interface of the component maintenance of agent information. Likewise the component world interaction management receives the agent's characteristics in the input information type own characteristics, observation results received by the agent (and forwarded directly to the component world interaction management) in the input interface type observation result info, and information the agent has about the world and agents in the information type belief info. The output generated by the component world interaction management includes the output for the agent as a whole (action info, observation info), extended with maintenance info (information obtained from observation of the world and other agents that needs to be maintained).

The component maintenance of world information receives new information on the world (the agent's beliefs on the world) in its input interface. Beliefs on the world are available in the output interface of the component maintenance of world information.

5.1.3 *Composition relation: information exchange*

Information exchange within the agent is specified by the information links listed in Table 6, and depicted in Figure 2.

Table 6. Specification of information exchange in table format

	from		to	
information link	*process*	*information type*	*process*	*information type*
communicated info	agent	incoming communication info	agent interaction management	incoming communication info
info to be communicated	agent interaction management	outgoing communication info	agent	outgoing communication info
observation results to wim	agent	observation result info	world interaction management	observation result info
observations and actions	world interaction management	observation info action info	agent	observation info action info
communicated world info	agent interaction management	maintenance info on world	maintenance of world info	assumption world info
communicated agent info	agent interaction management	maintenance info on agents	maintenance of agent info	assumption agent info
observed world info	world interaction management	maintenance info on world	maintenance of world info	assumption world info
observed agent info	world interaction management	maintenance info on agents	maintenance of agent info	assumption agent info
world info to aim	maintenance of world information	epistemic world info	agent interaction management	belief info on world
agent info to aim	maintenance of agent information	epistemic agent info	agent interaction management	belief info on agents
world info to wim	maintenance of world informatio	epistemic world info	world interaction management	belief info on world
agent info to wim	maintenance of agent information	epistemic agent info	world interaction management	belief info on agents
own process info to aim	own process control	own characteristics	agent interaction management	own characteristics
own process info to wim	own process control	own characteristics	world interaction management	own characteristics
own process info to mwi	own process control	own characteristics	maintenance of world information	target world info
own process info to mai	own process control	own characteristics	maintenance of agent information	target agent info
world info to opc	maintenance of world info	epistemic world info	own process control	belief info on world
agent info to opc	maintenance of agent info	epistemic agent info	own process control	belief info on agents

Observation results are transferred through the information link observation result info to wim from the agent's input interface to the component world interaction management. In addition, this component receives belief information from the component maintenance of world information through the information link world info to wim, and the agent's characteristics from the component own process control through the link own process info to wim. The selected actions and observations (if any) are transferred to the output interface of the agent through the information link observations and actions.

The component maintenance of world information receives meta-information on observed world information from the component world interaction management, through the information link observed world info and meta-information on communicated world information (through the link communicated world info) from the component agent interaction management. Epistemic information from maintenance of world information, epistemic world info, is transferred to input belief info on world of the components world interaction management, agent interaction management and own process control, through the information links world info to wim, world info to aim and world info to opc.

Comparably the component maintenance of agent information receives meta-information on communicated information from the component agent interaction management, through the information link communicated agent info and meta-information on observed agent information (through the link observed agent info) from the component world interaction management. Epistemic information, epistemic agent info, is output of the component maintenance of agent information, becomes input belief info on agents of the components world interaction management, agent interaction management and own process control, through the information links agent info to wim, agent info to aim and agent info to opc.

5.1.4 Composition relation: task control

Task control at the highest process abstraction level within the agent is simple: all components and links are made awake when the agent is awakened, which means that they all process (in an asynchronous manner) information as soon as it arrives.

5.2 Knowledge composition

A number of generic knowledge structures, in particular information types, can be distinguished: application domain independent knowledge structures which can be instantiated for a particular domain of application.

Information types provide the *ontology* with which knowledge used in the processes can be expressed. Information types provide the ontology (or lexicon, or vocabulary) for the languages used in one (or more) components, knowledge bases and information links. In information type specifications the following concepts are used: sorts, objects, relations, functions, and meta-descriptions. Furthermore, information types can be composed from other information types. Each concept

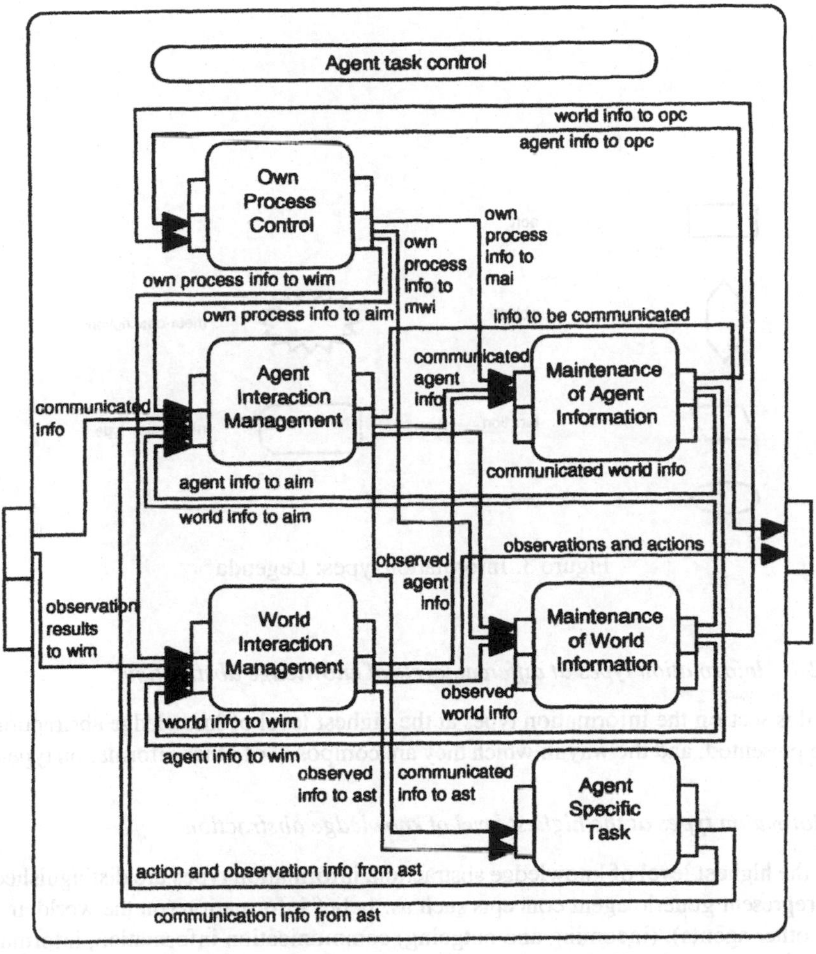

Figure 2. Information exchange at the highest process abstraction level within the agent

is represented graphically, see Figure 3. The icon for information types is used as depicted in this Figure 3 (containing only the name of an information type), but also as depicted in Figure 4 containing the sorts, object, functions, relations, and meta-descriptions used in the design of that information type.

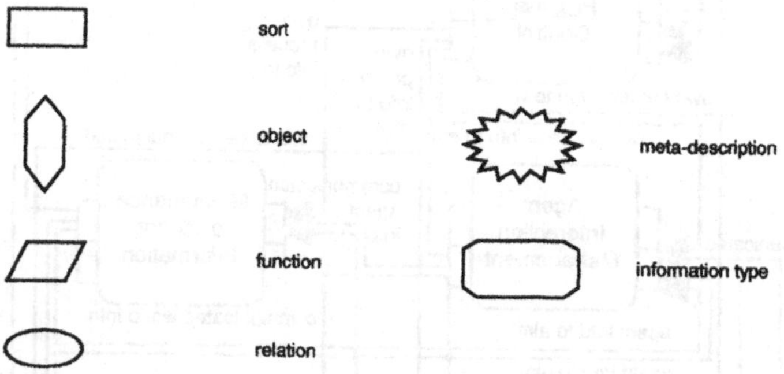

Figure 3. Information types: Legenda

5.2.1 *Information types at different levels of knowledge abstraction*

In this section the information types at the highest level of knowledge abstraction are presented, and the way in which they are composed of other information types.

Information types at the highest level of knowledge abstraction

At the highest level of knowledge abstraction, information types are distinguished to represent generic agent concepts such as: belief information (on the world and on other agents), (incoming and outgoing) communication information, information on observation (information on observations to be performed and obtained observation results), action information, information on the agent's characteristics, information to be remembered. These notions (abstracting from lower levels of knowledge abstraction), are modelled by the information types listed in Table 7.

Composition relations between information types

Each of the information types in Table 7 is composed of information types at a lower level of knowledge abstraction. Two of the information types (belief info

Table 7. Information types at the highest level of knowledge abstraction

information type	*short explanation*
belief info	information on the beliefs of the agent (information the agent has on the world and on other agents)
incoming communication info	information on communication the agent has received from another agent
outgoing communication info	information on communication the agent has decided to perform
observation info	information on observations the agent has decided to perform
observation result info	information on the observation results the agent has obtained
action info	information on the actions the agent has decided to perform
own characteristics	information on the agent's characteristics
maintenance info	information to be remembered by the agent

and maintenance info) are composed of two more specific information types: one for information on the world and one for information on other agents. All information types are (either directly or indirectly) composed of (1) generic information types and (2) domain specific information. Generic information types are fully specified within the generic model. Domain specific information types are defined by references; they are instantiated for a specific domain of application. For example, the information type action info is composed of the generic information type actions to be performed and the domain specific information type domain actions (see Figure 4). The specific actions for a given domain of application are not specified within the generic model.
In a similar manner:

- the information type observation info is composed of the generic information type obs to be performed and the domain specific information type domain meta-info.

- the information type observation result info is composed of the generic information types observation results and truth indication, and the domain specific information type domain meta-info

- the information type incoming communication info is composed of the generic information types incoming communication and truth indication, and the domain specific information type domain meta-info.

- the information type outgoing communication info is composed of the generic

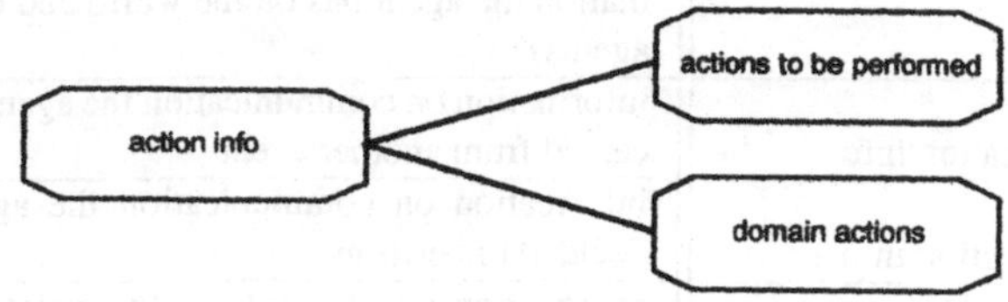

Figure 4. Action info as a composition of a generic and domain specific information type

information types outgoing communication and truth indication, and the domain specific information type domain meta-info

The information type domain meta-info is composed of world meta-info, agent meta-info and meta-info hierarchy. The information type world meta-info is a meta-description of the information type world info, using the sort world info element, as shown in Section 5.2. Similarly, the information type agent meta-info is a meta-description of the information type agent info using the sort agent info element. The information type meta-info hierarchy defines the sorts world info element and agent info element to be sub-sorts of the sort info element (see also Figure 7 in Section 5.2).

The information types maintenance info on world and maintenance info on agents are composed of two generic information type (maintenance on world, resp. maintenance on agents and truth indication) and a domain specific information type (world meta-info, resp. agent meta-info). Comparable information type compositions have been defined for belief information. The information type own characteristics is composed of the generic information type agent characteristics and the domain specific information type domain agent characteristics. Finally, the standard meta-information types assumption info, epistemic info, required info and target info are used to define (by composition) specific variants information types for the given world information and agent information separately. The information type meta-input agent info is a meta-description of the information type agent info using the sort for input atoms IA; the other variants of meta-information types are defined similarly.

5.2.2 *Generic information types*

The information types world meta-info and agent meta-info include meta-descriptions of the information types world info and agent info using the sort world info element and agent info element, respectively. Note that within the generic model the information types world info and agent info are only references. They can be instantiated for a specific domain of application.

Generic information types for observations and actions

The generic information type observation results enables the agent to express statements on observation results. In applications the observations can be *passive*: without taking any initiative, the agent automatically receives the observation results from the external world, or, *active*: observations initiated by the agent; the agent decides to do a specific observation and transfers this decision to perform an observation to the external world. After receipt of this selected observation the world executes this observation and transfers observation results back to the agent. The decision of an agent to perform an active observation, for example depends on its own goals (*pro-active observation behaviour*) or on requests of other agents (*reactive observation behaviour*). Using the generic information type obs to be performed, the observations selected by the agent are expressed by the relation to be observed (see Figure 5). The generic information type truth indication defining the sort sign and the objects pos and neg in this sort, is also used in the information type observation results.

Using these information types it is possible to make statements about the process of observation of the state of the world in contrast to statements about the world. It is possible for the statement 'my observation result is that the pressure is high' to be true, while in the world state 'the pressure is high' is false. For example, a sensor could give the wrong information. Similarly, it could also be the other way around: the statement 'the pressure is high' could be true in the world state, while the statement 'my observation result is that the pressure is high' is false, simply because it was not observed. Note also that 'I did not observe that the pressure is high' means something different from 'I observed that the pressure is not high'. A statement of the form 'my observation result is that the pressure is high' cannot be expressed using the information type that describes the world. For example, the statement 'the pressure is high' is not adequate. Therefore, another structure is necessary to express statements about statements. Statements about statements are called meta-level statements. The statements that form the subjects of such *meta-level* statements are called *object level* statements. The generic information type actions to be performed enables the agent to reason about actions; see Figure 6.

Generic information types for communication

A social agent is able to receive incoming communication and to generate outgoing communication. The generic information types for communication are depicted in Figure 7.

With these information types it is possible to make statements about the process of communication (in contrast to, for example, statements about the world). It is possible for the statement 'I was told that the pressure is high' to be true, while in

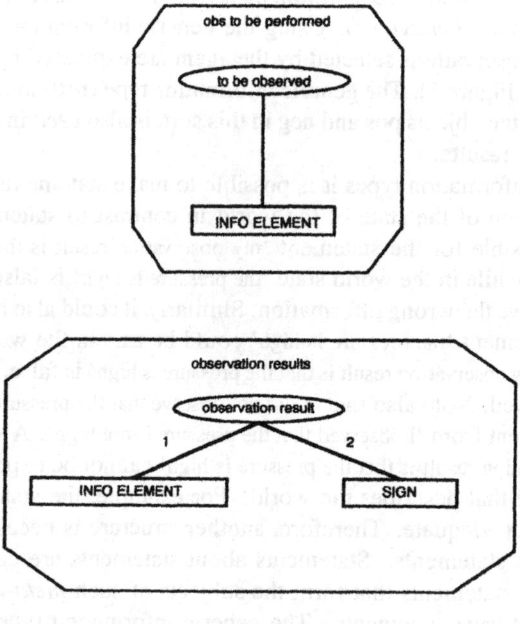

Figure 5. Generic information types on observation

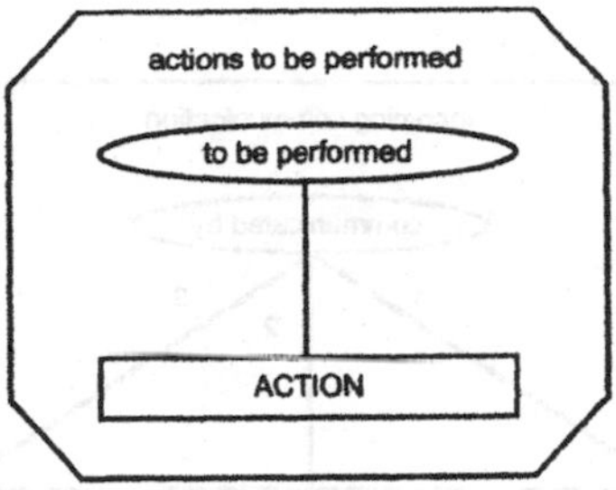

Figure 6. Generic information type: actions to be performed

the world state 'the pressure is high' is false: the other agent may simply not tell the truth. It could also be the other way around: the statement 'the pressure is high' could be true in the world state, while the statement 'sombody told me that the pressure is high' is false, simply because nobody told me. Note also that 'he did not tell me that the pressure is high' does not mean the same as 'he told me that the pressure is not high'. Similar to statements about observation, statements about communication are meta-level statements.

Generic information types for internal information

The information communicated to the agent may be used to extend or update an agent's beliefs on the world or on other agents. The information received is analysed, selected and prepared to be stored as information either on the world or on other agents; the related information types are depicted in Figure 8.

The generic information type beliefs can be used to maintain information on the world and other agents; see Figure 9.

The generic information type agent characteristics can be used to express meta-information about the agent's characteristics in an explicit, declarative manner.

Standard meta-information types

Generic standard meta-information types on assumption information and epistemic information are included. The sort IA models the input atoms of the component in which this information type is used. Similarly the sort IOA models the input and output atoms. Target information expresses on which output atoms (modelled by the sort OA) a component can focus. A target type expresses whether the focus is on confirmation (truth value true) or rejection (truth value false) of an information element, or just on determination of its truth value. The information type

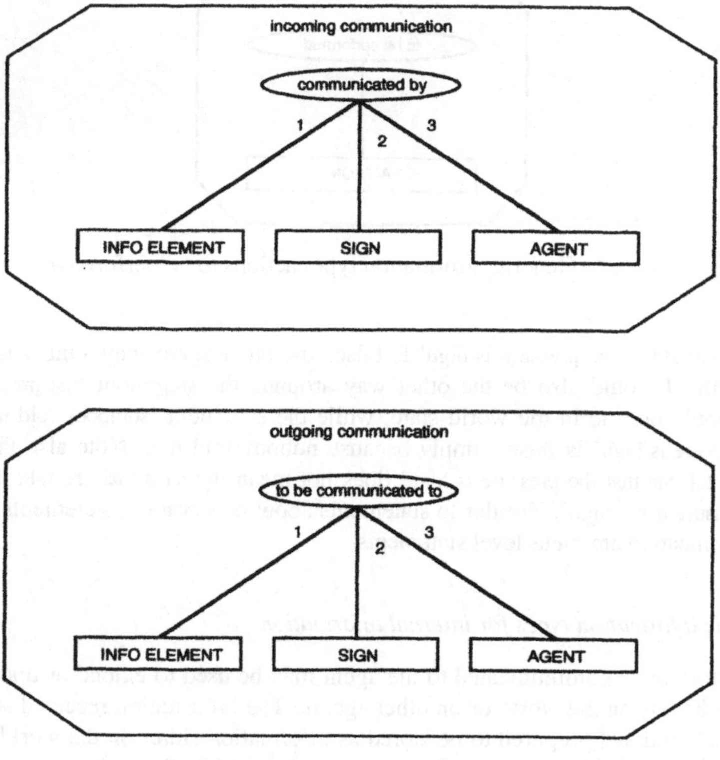

Figure 7. Generic information types on communication

required info specifies the input atoms needed to derive target atoms. This meta-information makes it possible to focus the reasoning process: to provide input needed to derive the targets.

A generic standard meta-information type is of a form named by meta-input ⟨information-type-name⟩, meta-output ⟨information-type-name⟩ and meta-interface ⟨information-type-name⟩. These information types are meta-descriptions of the information type named, using sort IA, OA, or IOA, respectively. Note that all standard information types as described are pre-defined and as such known in any component. They do not need to be explictly specified, but can be used in information links.

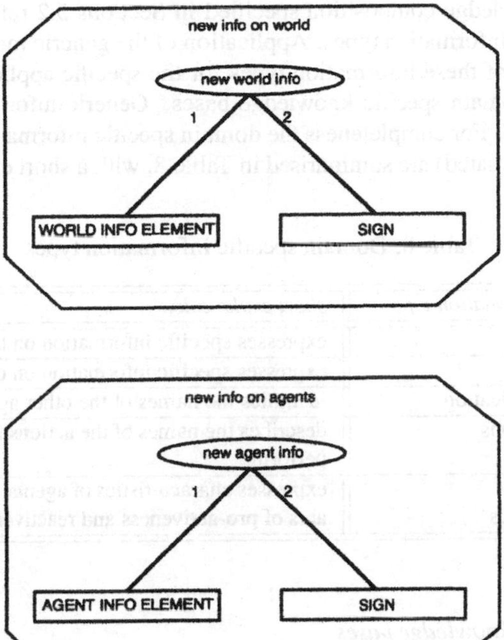

Figure 8. Generic information types: maintenance on agents, maintenance on world

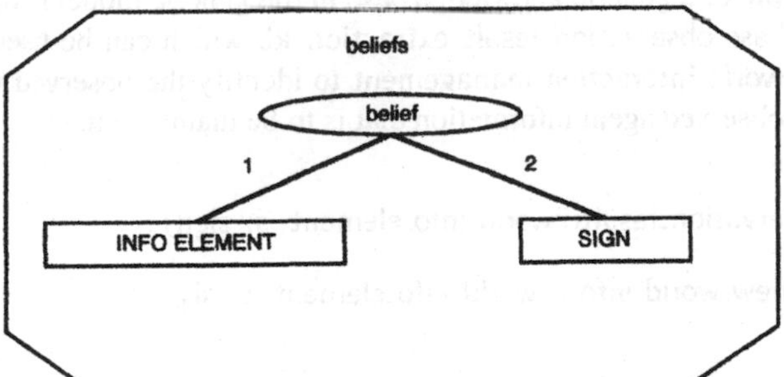

Figure 9. Generic information type: beliefs

5.2.3 Domain specific information types

Within the knowledge composition specified in Sections 5.2 references occur to domain specific information types. Application of the generic model concentrates on instantiation of these information types for the specific application domain at hand, and on domain specific knowledge bases. Generic information types can simply be reused. For completeness the domain specific information types (which need to be instantiated) are summarised in Table 8, with a short explanation.

Table 8. Domain specific information types

specific information type	*short explanation*
world info	expresses specific information on the world
agent info	expresses specific information on other agents
agent identification	identifies the names of the other agents
domain actions	describes the names of the actions the agent can perform
domain agent characteristics	expresses characteristics of agents, such as variants of pro-activeness and reactiveness

5.2.4 Generic knowledge bases

Often the knowledge to be used for a specific application strongly depends on the application domain. However, sometimes parts of the knowledge can be formulated in a more generic, domain independent manner, which makes reuse possible in domains with similar characteristics. These generic knowledge bases are available to be used in the agent model. They may be (re)used in a specific application depending on their relevance. If during an application of the generic model to a specific domain, the knowledge is applicable and relevant, it can be reused. If it is not relevant, it simply can be left out.

An example of a generic knowledge base in the generic model is the following knowledge base observation result extraction kb which can be used within the component world interaction management to identify the observed world information and observed agent information that is to be maintained:

if observation_result(i:world_info_element, s: sign)

then new_world_info(i: world_info_element, s: sign);

if observation_result(i:agent_info_element, s: sign)

then new_agent_info(i:agent_info_element, s: sign);

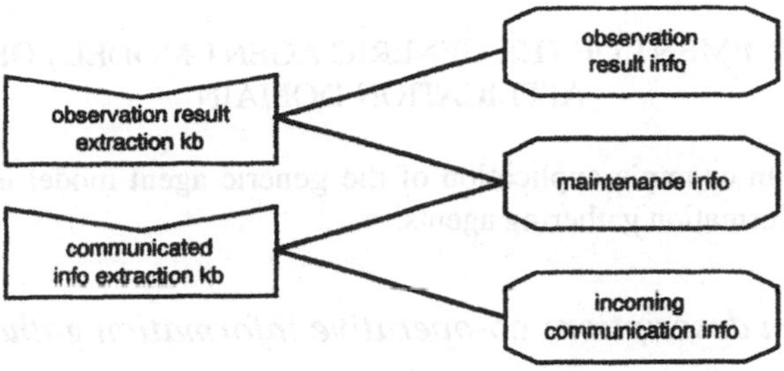

Figure 10. Relation between generic knowledge bases and information types

This generic knowledge expresses that the agent blindly trusts its own observations. In applications the knowledge can be refined, for example by adding conditions. Similarly the generic knowledge base communicated info extraction kb is part of the generic model. This knowledge base may be used within the component agent interaction management to identify the communicated world information and communicated agent information that needs to be maintained:

if communicated_by(i:world_info_element, s: sign, A:AGENT)

then new_world_info(i: world_info_element, s: sign);

if communicated_by(i:agent_info_element, s: sign, a:agent)

then new_agent_info(i:agent_info_element, s: sign);

This generic knowledge expresses that the agent blindly trusts what other agents communicate. In applications this knowledge can be refined, for example by adding conditions.

5.2.5 Relations between knowledge bases and information types

The knowledge bases defined in Section 5.2 are related to information types depicted in Figure 9.

5.3 Relations between process and knowledge composition

The generic information types described in this section are all used in interfaces of components. The relations between the two generic knowledge bases introduced in Section 5.2 and processes in which they occur is straightforward: observation result extraction kb is used within component world interaction management, and communicated info extraction kb within agent interaction management.

6 REFINEMENT OF THE GENERIC AGENT MODEL FOR AN APPLICATION DOMAIN

In this section an example application of the generic agent model is presented: co-operative information gathering agents.

6.1 Problem description: co-operative information gathering

This example multi-agent system consists of two agents that can each gather partial information on the world, but can only draw further conclusions by combining their individual information.

6.1.1 The domain

The application is as follows. Assume two agents A and B start a small project: they have to do some investigation and make up a report on some topic. Each of the agents has access to useful sources of information, but which information differs for the two agents. By co-operation they can benefit from the exchange of information that is only accessible to the other agent. If both types of information are combined, conclusions can be drawn that would not have been achievable for each of the agents separately. Co-operation may fail for a number of reasons. For example one of the agents, say A, may not be pro-active in its individual search for information. This may be compensated if the agent B is pro-active in asking the other agent for information, but then at least A has to be reactive (and not entirely inactive in information search). Another reason for failure is that one of the agents may not be willing to share its acquired information with the other agent. Yet another reason for failure may be that although both agents are active in searching and exchanging information, neither of them is able to combine different types of information and deduce new conclusions.

To make the example more precise: the example multi-agent model is composed of three components: two information gathering agents A and B and a component W representing the external world. Each of the agents is able to acquire partial information about the external world (by observation). Each agent's own observations are insufficient to draw conclusions of a desired type, but the combined information of both agents is sufficient: they have to co-operate to be able to draw conclusions. Therefore communication is required; the agents can communicate their own observation results and requests for observation information of the other agent. For reasons of presentation, this, by itself quite common situation for co-operative information agents, is materialised in the following more concrete form. The world situation consists of an object that has to be classified. One agent can

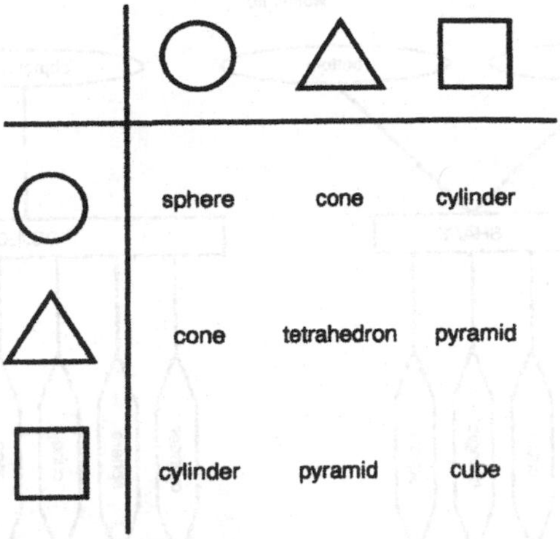

Figure 11. Object classification knowledge

only observe the bottom view of the object, the other agent the side view. By exchanging and combining observation information they are able to classify the object. In the example interview protocol presented below, two experts in the field of classification of three-dimensional objects are studied; agent *A* has done this job for almost twenty-five years, and agent *B* only started a year ago. Their daily work consists of observing three-dimensional objects, and trying to identify the nature of these objects. They need to co-operate to be successful, because they each can only see one side of the object. Agent *B* can only see the bottom, and *A* can only see one of the sides. They need to combine their two two-dimensional views to come to a correct conclusion about the object, using the knowledge depicted in Figure 10.

6.1.2 The requirements

Based on the generic agent model described in Section 5, some variants of agents that can play the role of Arnie and Bernie are designed. The variants of agents can differ in some of their characteristics; an agent may or may not be *pro-active*, in the sense that it takes the initiative to:

- perform observations
- communicate its own observation results to the other agent
- ask the other agent for its observation results
- determine the classification of the object (by reasoning)

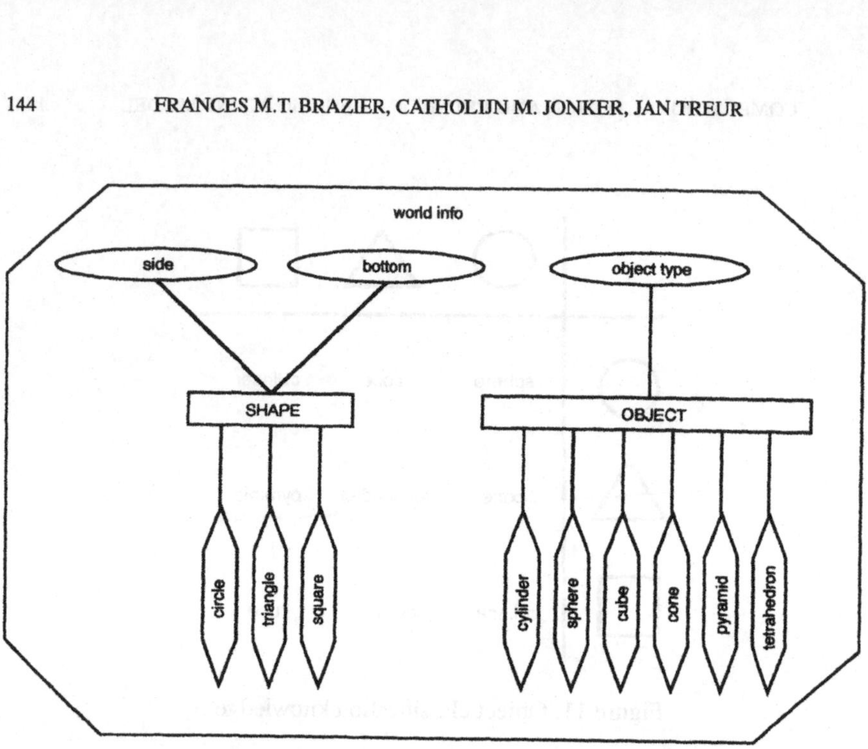

Figure 12. Instantiation of the information type world info

Moreover, it may be *reactive* to the other agent in the sense that it responds to a request for observation information:

- by communicating its observation result as soon as they become available
- by starting to observe for the other agent upon request

These agent characteristics can be represented explicitly as facts in the agent's component own process control. By varying these facts, different variants of this agent can be defined. Of course, the impact of these explicitly specified characteristics needs to be specified in the model. For example, if an agent has the characteristic that it always takes the initiative to communicate its observation results as soon as they are acquired, then the agent needs to behave accordingly, but if the agent does not have this characteristic, then the agent need not behave this way. This requires an adequate interplay between the component own process control and the component agent interaction management within the agent, and adequate knowledge within the component agent interaction management.

6.2 *An agent model for co-operative information gathering*

In this section the generic agent model introduced in Section 5 is applied to the application domain described in Section 6.1. Reusing a generic model entails that instantiations are made for a number of domain specific information types of the model. However, also some (preferably minor) extensions or modifications of the model are often made. For example, in this domain of application a component for the agent-specific task (named object classification) and some information links are added. In this section first the information types are discussed (Section 6.2), and next the knowledge bases (Section 6.2). Finally, the model is slightly extended by adding an information link from agent interaction management to world interaction management and information links from own process control and maintenance of world information to object classification (Section 6.2).

6.2.1 *Domain specific information types*

The information types needed to model the example of co-operative information gathering agents are the instantiations of the domain specific information types of the generic model and a few additional domain specific information types.

Instantiations of domain specific information type of the generic model

In Section 5.2 the domain specific information types are listed: world info, agent info, agent identification, domain actions, domain agent characteristics. For some of these information types domain specific instantiations are needed. The information types agent info and domain actions can be left empty in this domain, as the agents do not perform actions. In Figure 11 the instantiation of the information type world info is modelled. Six different types of objects form the sort object. The two-dimensional shapes that can be observed form the sort shape. The two perspectives are modelled by the relations side and bottom. Finally, the classification of the type of object is expressed by the relation object type.

The agent characteristics are taken from Section 3.2. An agent can be pro-active with respect to taking the initiative to observe, to inform the other agent if information is available, to request information from the other agent, and to reason in order to draw a conclusion on the object classification. It can be reactive with respect to providing the other agent with available information upon request and observation for the other agent, if the requested information is not yet available. The instantiation of the information type domain agent characteristics is depicted in Figure 12. The information type world meta-info is used in domain agent characteristics.

To distinguish communicated information in requests and information provision, functions and relations requested and info are defined in additional information types.

6.2.2 *Domain knowledge*

In this section the domain specific knowledge bases are discussed in the context of the component in which they are used.

Object classification knowledge

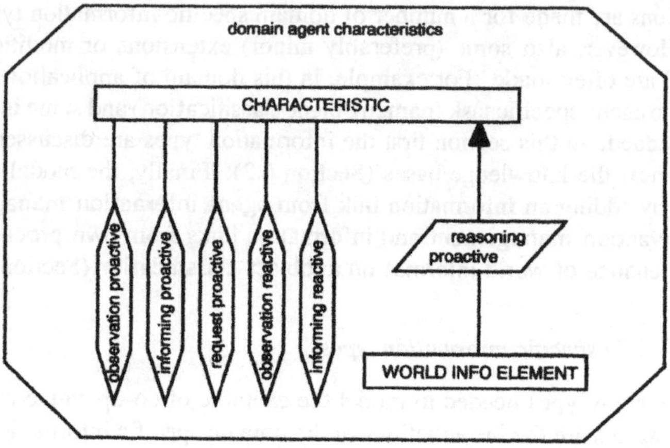

Figure 13. Instantiation of the information type domain agent characteristics

The knowledge used to classify the object based on available observation information can easily be taken from the table depicted in Figure 15:

```
if bottom(circle)   and side(circle)   then object_type(sphere);
if bottom(circle)   and side(square)   then object_type(cylinder);
if bottom(square)   and side(circle)   then object_type(cylinder);
if bottom(circle)   and side(triangle) then object_type(cone);
if bottom(triangle) and side(circle)   then object_type(cone);
if bottom(square)   and side(square)   then object_type(cube);
if bottom(square)   and side(triangle) then object_type(pyramid);
if bottom(triangle) and side(square)   then object_type(pyramid);
if bottom(triangle) and side(triangle) then object_type(tetrahedron);
```

It is assumed that objects are placed in the correct orientation. For example, viewed from the bottom a cylinder is always a circle. Note that there is no situation in which the conclusion can be drawn on the basis of one observation only.

World interaction knowledge

As the agents in the domain do not perform actions, world interaction focusses entirely on observation. No passive observations exist in the domain. There are two reasons to actively perform an observation: the agent may be pro-active (expressed by the first knowledge element below) or reactive (expressed by the second knowledge element) with respect to observation. Note that an observation is only selected if no information is available.

```
if own_characteristic(observation_proactive)
   and not belief(side(X:SHAPE), pos)
   and not belief(side(X:SHAPE), neg)
then to_be_observed(side(X:SHAPE));

if own_characteristic(observation_reactive)
   and requested(side(X:SHAPE))
   and not belief(side(X:SHAPE), pos)
   and not belief(side(X:SHAPE), neg)
then to_be_observed(side(X:SHAPE));
```

Actually, this knowledge base is meant for one of the agents. For the other agent side must be replaced by bottom.

Communication knowledge

The component agent interaction management makes use of knowledge to analyse incoming communication, and to generate outgoing communication.

Knowledge to analyse incoming communication

Generic knowledge needed to analyse incoming information is defined in the generic model; see Section 6.2 above. This knowledge identifies the information on the world that is to be maintained. However, in line with the communication differentiation added in this example model, a more sensitive treatment is preferred. The first knowledge element below expresses that the information provided by the other agent is identified as world information that is to be maintained. The second knowledge element identifies the information requested. The choice is made to only use this information in the component world interaction management, and to not maintain this information separately.

```
if communicated_by(info(I:WORLD_INFO_ELEMENT), S: SIGN, A:AGENT)
then new_world_info(I: WORLD_INFO_ELEMENT, S: SIGN);

if communicated_by(request(I:WORLD_INFO_ELEMENT), S: SIGN, A:AGENT)
then requested(I:WORLD_INFO_ELEMENT);
```

Knowledge to generate outgoing communication

Whether or not the agent actively communicates information to other agents depends on its own characteristics. If an agent is pro-active with respect to information provision, the first knowledge element below is applicable:

```
if own_characteristic(informing_proactive)
   and belief(I: WORLD_INFO_ELEMENT, S: SIGN)
then to_be communicated_to(info(I: WORLD_INFO_ELEMENT), S: SIGN,bernie);
```

If an agent is reactive in informing the other agent upon request then the second and third knowledge element are relevant:

if own_characteristic(informing_reactive)
  and communicated_by(requested(I: WORLD_INFO_ELEMENT),
    pos, A:AGENT)
  and belief(I: WORLD_INFO_ELEMENT, S: SIGN)
then to_be_communicated_to(info(I: WORLD_INFO_ELEMENT),
    S: SIGN,A:AGENT);

if own_characteristic(observation_reactive)
  and communicated_by(requested(I: WORLD_INFO_ELEMENT),
    pos, A:AGENT)
  and belief(I: WORLD_INFO_ELEMENT, S: SIGN)
then to_be_communicated_to(info(I: WORLD_INFO_ELEMENT),
    S: SIGN,A:AGENT);

The fourth knowledge element is applicable for pro-active behaviour with respect to requesting:

if own_characteristic(request_proactive)
  and not belief(bottom(S:SHAPE), pos)
  and not belief(bottom(S:SHAPE), neg)
then to_be_communicated_to(requested(bottom(S:SHAPE)), pos, bernie);

Own process control knowledge

The knowledge base for the component own process control contains meta-information that defines the agent character. For each agent the own process control knowledge is defined by a sub-set of the following set of meta-facts

own_characteristic(observation_proactive)
own_characteristic(observation_reactive)
own_characteristic(informing_proactive)
own_characteristic(informing_reactive)
own_characteristic(request_proactive)
own_characteristic(reasoning_proactive(object_type(O:OBJECT)))

Each sub-set defines a specific type of agent (the possibility of having represented the negation of an own characteristic is not considered). For example, the empty sub-set defines a totally apathic agent: it does nothing except maintain the information it receives. The complete set defines a fully pro-active and reactive agent.

6.2.3 *Addition of information links*

Three information links are added to the model. One of these links takes care of requests. The management of requests from the other agent and information provision to the other agent could be modelled as an additional agent specific task. However, because the management is rather simple, the choice has been made to have the two components agent interaction management and world interaction management take care of all request management. To this end, the information link requests is added to transfer requests from agent interaction management to world interaction management. The information types requests and world meta-info are used in this link, both at the source and destination. Furthermore, two information links are added to connect the agent specific task object classification. One information link is used to transfer the information from maintenance of world information to object classification. The other information link is used to transfer information of the form

own_characteristic(reasoning_proactive(object_type(o:object)))

from own process control to the information that the output atom object_type(o: object) is a target of the component object classification.

Table 9. Some of the outcomes of two co-operative information gathering agents

agent *B* / agent *A*	obs proactive reas proactive	obs proactive reas proactive inf proactive	obs proactive reas proactive inf proactive req proactive	obs reactive	obs reactive reas proactive	inf reactive obs proactive
obs proactive reas proactive	—	*A*	*A*	—	—	—
obs proactive reas proactive inf proactive	*B*	*A, B*	*A, B*	—	—	—
obs proactive reas proactive inf proactive req proactive	*B*	*A, B*	*A, B*	*A*	*A, B*	*A*
obs reactive	—	—	*B*	—	—	—
obs reactive reas proactive	—	—	*A, B*	—	—	—
inf reactive obs proactive	—	—	*B*	—	—	—

6.3 The behaviour of co-operative information gathering agents

The behaviour of the co-operative information gathering agents strongly depends on their characteristics. The number of sub-sets of the set of six agent characteristic facts in Section 6.2 is 64. Whether or not an agent succeeds in classification of the object also depends on the behaviour of the other agent. In principle it is possible to create a 64 by 64 matrix to identify the behaviour of all 4096 combinations of two agents. For practical reasons, only a small subset of such combinations is discussed in this section. Table 9 indicates which of the agents will be able to classify the object for 36 combinations of two agents.

The table shows that two pro-active but purely individualistic agents (both observation pro-active and reasoning pro-active) will never find a solution. Nevertheless, if one of these agents is also social in communicating its observation results (informing pro-active), the other agent (but not the agent itself) will find a solution. A fully pro-active agent will find a solution as soon as its partner is observation reactive, or informing reactive and observation pro-active, or informing pro-active and observation pro-active. An observation reactive and reasoning pro-active agent will find a solution if the other agent is request pro-active, observation pro-active and informing pro-active. Agents that both are only reactive in communication will not succeed. These are only some of the possibilities. A more complete analysis of the conditions under which one of the two or both agents will find a solution can be found in [Jonker and Treur, 1998a].

7 COMPARISON WITH EXISTING AGENT ARCHITECTURES AND APPLICATIONS

In the agent literature, various agent architectures can be found, often specialised to a particular type of application. The design of most of these agent architectures is not formally specified in detail; usually they are only available in the form of an implementation, and at the conceptual level some informal pictures and natural language explanations. In general, the aim for the development of these agent architectures in the first place is to have a working piece of software for a specific type of application. The design of the generic agent model GAM introduced in this paper has a different aim. The generic agent model GAM was meant as a unified design model for weak agency, formally specified in an implementation- and domain-independent manner at a high level of abstraction. A success criterion for this aim is the possibility to specialise and instantiate the agent model GAM to obtain conceptual, formal specifications of design models for a variety of (implemented, but not formally specified) agent types and agent behaviours. Thus a unified design description is obtained which enables comparison of these agent architectures at a conceptual but yet formally defined level. Evaluation of this aim has taken place for two different groups of agent architectures:

- *agent architectures for new applications* designed, after an informal analysis, as a formally specified refinement of GAM
- *existing agent architectures*, developed for specific applications without formal specification of a design model; in the context of the research reported here they have been reverse-engineered at a conceptual design level using the structure of GAM

Evaluation for the first of these two groups of agent architectures has shown that GAM is an adequate means to design specific types of agents, given a variety of requirements imposed by specific application domains. Evaluation for the second group of requirements shows that GAM is an adequate means for reverse engineering, to obtain unified, comparable formal descriptions of different types of existing agent architectures. For a summarizing overview, see Table 10.

7.1 *Applications designed on the basis of GAM*

The following types of agents tuned to specific application domains have been developed using (refinements of) the structure of GAM to obtain a formally specified design model.

Simulated animal behaviour

Instantiations of the generic agent model GAM have been designed to fulfill the requirements imposed by purely reactive, delayed response, pro-active goal-directed, and social animal behaviour, as identified in the literature on animal behaviour; e.g., see [Vauclair, 1996]. Within the model for purely reactive behaviour, only one component is instantiated to model the associations between observations and actions used in the direct interaction with the world. For the model with delayed response behaviour, a separate component for memory (maintenance of world information) was instantiated, in addition to world interaction management. For pro-active behaviour, also the component own process control was instantiated, to represent specific agent characteristics and to generate goals. To obtain a model for a specific type of social behaviour, in addition, the components maintenance of agent information (where the pick order between the animals is represented) and agent interaction management (to generate and interpret growling) have been instantiated. For more details, see [Jonker and Treur, 1998b].

Negotiating agents to achieve load balancing of electricity use

The application to load balancing of electricity use by means of a flexible form of one-to-many negotiation was made in co-operation with Swedish electricity industry. A precursor of the generic agent model GAM was used to develop this application. Within this application, the component cooperation management has a more complex refinement to address the evaluation and generation of bids. Also the components own process control (representing agent characteristics that have impact on the negotiation, and decisions to start or stop a negotiation process) and agent interaction management (to transfer the bids to the other agents) are present in an instantiated form. The component AST was instantiated to the task 'determine balance of predicted use'. For more details, see [Brazier *et al.*, 1998a].

Table 10. Overview of refinements of GAM to designs for various agent architectures

	WIM	AIM	MWI	MAI	OPC	CM	AST
animal behaviour	observation action generation	growling	memory	pick order	goal generation	—	—
negotiating agents	—	bids transfer	—	—	decision to negotiate or stop	composed negotiation model	determine balance
information agents	inspecting the WWW	interests and provided information	information on (info) objects	profile information	pro-activeness and reactiveness characteristics	—	models for strict and soft matching
deliberate normative agents	—	communication about norms	information on world	MAI info about norms of other agents; MSI info about norms in society	composed of: norm management, strategy management, goal management, plan management	—	—
electricity transportation management	—	refined to Generate Outgoing, Receive Incoming	World Model	Acquaintance Models	refined for Monitor Incoming Data, Evaluate Process State	—	Diagnostic Process Model, Analysis Model of Incoming Data, Planning and Monitoring Model
cooperative agent architecture	—	—	—	—	refined for Monitoring, Planning and Control of own activities	refined for Project Generation, Project Monitoring	—
BDI	Beliefs on other agents	Commitment transfer	Beliefs on world	Commitments of others	refined for Handling Beliefs, Desires and Intentions	—	—

	WIM	AIM	MWI	MAI	OPC	CM	AST
society simulation	refined for Observation Information Interpretation, Action Execution Preparation	—	—	—	refined for Own Resource Management, Own Characteristics, Goal Determination, Plan Determination	—	—
Touring Machines	Reactive Layer	Reactive Layer	Environment Model	Agent Models	refinement for Planning Layer, Control Rules	—	—
INTERRAP	Sensors, Actors, Behaviour-Based Layer	Communication, Behaviour-Based Layer	World Model	Social Model	refinement for Mental Model, Local Planning Layer	Social Mental Model, Cooperative Planning Layer	—
ZEUS	—	Mailbox, Message Handler	Resource Database	Acquaintance Models	refined for Planner and Scheduler, Task/Plan Database, Execution Monitor	Coordination Engine	—
ADEPT	—	CM	—	AM	SAM SM	IMM	SEM

Personal information agents and information brokering agents at the World Wide Web

For different applications of information agents in a World Wide Web context, agent models have been developed on the basis of GAM. First, an instantiation of GAM has been designed to serve as an information broker agent. This broker agent model has instantiations of all components of GAM. For example, within maintenance of world information information on the objects of the brokering is maintained (i.e., meta-information of the brokered information objects), and within maintenance of agent information, (interest) profiles of users and other agents are created and maintained. Within the agent specific task different matching forms have been specified. Within the instantiated component world interaction management it is specified how the agent can observe tags with meta-information in a HTML page at a given Website. In [Jonker and Treur, 1998c], the broker agent model, and an application to a Personal Assistant to support researchers in the exchange of scientific papers is described. Moreover, it is described how the information broker agent model can support its own maintenance by installing at run-time new ontologies and knowledge bases communicated to the agent by maintenance agents (instantiation of own process control). In [Jonker and Treur, 1999], a multi-agent architecture of an intelligent Website is introduced, based on (a number of instantiations of) the information broker agent model, and illustrated for the domain of a department store. Here the information agents play the role of servants at the Website, who are able to have an informed dialogue with visitors of the Website, tailored to the background and needs of the visitor. In [Jonker *et al.*, 1999] an application of this architecture to a Website for employees of an insurance company is described.

Agents in social simulation applications based on deliberate normative behaviour

To simulate societies in which agents can behave in a deliberate normative manner, a model has been developed for a deliberate normative agent [Castelfranchi *et al.*, 1999]. This type of agent has explicit mental representations of norms, which are interpreted operationally as (meta-)goals for its own behaviour. The deliberation also incorporates deciding about when to follow a norm and when to violate it. The model has been designed as a refinement of GAM in the following manner. Besides components for maintenance of world information and maintenance of agent information, also a component maintenance of society information is added. In this component the norms distinguished in the society are maintained. Society information could have been represented within maintenance of agent information as a specific, global form of agent information; however it was decided that it is more natural to include a separate component for this 'Society Model' to make society norms more explicitly visible as distinct from personal norms of specific agents. Other components reused are agent interaction management, world interaction management and own process control. The latter component is refined into four sub-components: norm management, goal management, plan management, and strategy management. In the first of these components decisions on (personal) norm adoption are made. The adopted norms are operationalised within strategy management in terms of control of the goal management and plan management processes.

7.2 *Reverse engineering of existing agent architectures and applications*

A number of existing applications have been reverse-engineered at a conceptual design level using the structure of GAM as a starting point for refinement. The generic model GAM has been refined to obtain a formally specified design description of the following types of agents.

- *monitoring, diagnostic and restoration agents in electricity transportation management*

The multi-agent system for electricity transportation management developed in the ARCHON project was one of the first operational real-world applications of agent technology [Cockburn and Jennis, 1995]; [Jennings *et al.*, 1996]. It is currently running on-line in a control room in the North of Spain. An electricity transportation network carries electricity from generation sites to the local networks where it is distributed to customers. Managing this network is a complex activity which involves a number of different subprocesses: monitoring the network, diagnosing faults, and planning and carrying out maintenance when such faults occur. The application involves two co-operating diagnostic agents, a monitoring agent, and a restoration agent. The reverse engineering application of GAM to ARCHON can be found in [Brazier *et al.*, 1995]. All of the agents maintain a World Model, which clearly can been obtained as an instantiation of the component maintenance of world information in GAM. Moreover, they maintain information about the other agents in the system in socalled Acquaintance Models, obtained as an instantiation of GAM's component maintenance of agent information. Furthermore, Monitor Incoming Data and Monitor Process State were obtained as an instantiation of own process control in GAM. The agent-specific task component AST was instantiated to obtain the different specialisations of the agents: it is refined to a complex diagnostic model for the diagnosis agents, to a model for monitoring disturbances and the progress of restoration processes for the monitor agent, and to a model for restoration planning for the restoration agent. More details can be found in [Brazier *et al.*, 1995].

- *co-operative agents based on joint intentions*

In [Jennings, 1995] an informally described multi-agent model for cooperative problem solving is proposed. Essential elements of this model are the dynamic organisation and management of joint activities, susceptive to change due to unexpected events. As described, the model only provides a restricted amount of detail to support analysis, modelling and implementation of co-operative agents in specific domains. In [Brazier *et al.*, 1997b] it is described how a formal design model of this cooperative agent architecture has been made as a refinement of GAM. Within this model monitoring, planning, control of own activities, and monitoring, planning, allocation, and communication of activities with other agents are explicitly distinguished. To obtain this cooperative agent model both the components own process control (for the monitoring, planning and control of own activities) and the component cooperation management (for the monitoring, planning, allocation and communication about activities involving others) have been refined to more complex, composed components; see [Brazier *et al.*, 1997b] for more details. Application of this model to Call Center support is described in [Brazier *et al.*, 1998b].

- *BDI-agents*

The well known BDI architecture [Rao and Georgeff, 1991], and its predecessor PRS [Georgeff and Lansky, 1987], is organised around the notions beliefs, desires, and intentions. How the generic agent model GAM can be refined to obtain a formally specified design model of the BDI-architecture, can be found in [Brazier *et al.*, 1999]. The beliefs on the environment (the world and the other agents) are maintained within the components maintenance of world information and maintenance of agent information. The desires and intentions are represented within a refinement of component own process control, which in this case has a more complex, compositional structure, based on components belief determination, desire determination and intention and commitment determination. The latter component is composed of components goal determination, and plan determination, which, in turn are composed of intended goal determination and committed goal determination, resp. intended plan determination and committed plan determination. For more details, see [Brazier *et al.*, 1999].

- *agents in social simulation experiments*

In [Cesta *et al.*, 1996] experiments are reported with which social theories are tested by simulating interaction between different types of simple agents (i.e., agents with limited knowledge and capabilities). Four types of agents are distinguished on the basis of their social characteristics: social agents, parasite agents, solitary agents and selfish agents. The effect of an agent's social characteristic on interaction with other agents is measured by simulating agent behaviour in a situation in which 30 agents try to survive on a 15 * 15 grid in which 60 pieces of food are continually available in random positions. An agent's welfare is measured on the basis of its energy level. The end result of a simulation is the number of agents that survive in a given society of agents, given the energetic value of the food available. Agents do not communicate explicitly but implicitly: a hungry agent changes colour, and this can be seen by other agents. Agents' social characteristics are assumed to be static. An agent does not change from being, for example, selfish to social. The implications of agents' social characteristics for its behaviour is as follows. A solitary agent will always search for food, regardless of its internal energy level. Likewise, a parasite agent will always look for help. A selfish agent will look for help only if it is in danger, otherwise it searches for food. A social agent will also look for help if it is in danger. If it is in a hungry state, it will search for food. If it is in a normal state, then it will search for food if no help-seeking agents are seen. Otherwise, the social agent will give food to one of the help-seeking agents nearby.

The experiments reported in [Cesta *et al.*, 1996] have been replicated and extended by reverse engineering based on GAM. The refinement of the generic model GAM to obtain the four types of agents was performed on the basis of the informal, textual descriptions provided by [Cesta *et al.*, 1996]. The only components within the generic agent model, applicable to these small agents, are the components own process control and world interaction management. The component own process control is composed of four components: own resource management, own characteristics, goal determination, and plan determination. The component own resource management receives information about its current energy level and the resources it has consumed, with which it determines its new energy level. On the basis of information the component goal determination receives about its own social characteristics and its own energy level, it determines the goals the agent is to pursue: for example to find food, or to look for help. The component own characteristics receives information on the agent's energy level from the component own resource management. This information is used to determine the agent's next state (e.g., hungry, normal or in danger). The component plan determination receives information (1) from the component own characteristics, namely the agent's current state, (2) from the component goal determination, namely which goals are to be pursued and (3) from outside the component, namely the current state of the world. With this information the component plan determination determines which actions to take in the external world.

The component world interaction management interprets information it receives from the external world, and transforms information about actions to be taken in the external world into specifications for actions which the external world can execute. Two components are defined to perform these tasks: the component observation information interpretation and the component action execution preparation. For more details, see [Brazier *et al.*, 1997a].

- *Touring Machines, INTERRAP, ZEUS, and ADEPT*

The remainder of this section discusses how the generic agent model GAM can be refined to obtain a formally specified design model for four other existing agent architectures: Touring Machines [Ferguson, 1992], INTERRAP [Müller *et al.*, 1995]; [Müller, 1996], ZEUS [Nwana *et al.*, 1998], and ADEPT [Jennings *et al.*, 1996].

The Touring Machines architecture described in [Ferguson, 1992] distinguishes three layers: a reactive layer, a planning layer, and a modelling layer; all layers process concurrently. The reactive layer can be formally specified as an instantiation of the the components world interaction management and agent interaction management in the generic agent model GAM. If reactions on combined input from observation and communication have to be modelled, two information links between world interaction management and agent interaction management are added for direct information exchange, avoiding modelling this information as beliefs. The planning layer can be specified as a refinement of component own process control; also the Control Rules are part of this refinement of own process control. The modelling layer can be obtained by instantiation of the components maintenance of world information and maintenance of agent information, where models of the agent's environment are maintained. The specific approach to control by Control Rules (in the form of Censors and Suppressors) entails that all incoming and outgoing information has to be filtered by the Control Rules within own process control. This means that, although in principle all layers are meant to be connected independently to the outside world, in order to do the filtering, in practice these connections come together in the Control Rules component within own process control. This confirms analyses of this agent architecture available in the literature; e.g., see [Müller, 1996].

Within the INTERRAP architecture [Müller *et al.*, 1995; Müller, 1996], the following components play a role: *World Interface* (Sensors, Communication, and Actors), *Agent KB* (Social Model (SM), Mental Model (MM), World Model (WM)), *Agent Control Unit* (Cooperative Planning Layer (CPL), Local Planning Layer (LPL), Behaviour-Based Layer (BBL)). A formal design specification of the World Interface can be obtained as an instantiation of the components agent interaction management (communication) and world interaction management (sensors, actors) within GAM. A design specification of Agent KB's Social Model can be obtained as an instantiation of the component maintenance of agent information and the World model of maintenance of world information. The Mental Model can be obtained as a refinement within own process control, as far as mental concepts referring to the agent itself are concerned. If also mental concepts such as joint intentions are involved, these can be included within cooperation management. The Local Planning Layer can be obtained as a refinement of own process control, the Cooperative Planning Layer of cooperation management, and the Behaviour-Based Layer of the components agent interaction management and world interaction management. The INTERRAP model has a much richer structure than the generic agent model GAM, especially in control aspects. Control differs from the Touring Architecture in that only the Behaviour-Based Layer is connected to the outside world, and the Local Planning Layer (within own process control) becomes involved as soon as the Behaviour-Based Layer indicates that the situation is assessed as beyond its competence. Similarly, own process control can indicate that the situation is beyond its (individual) competence and involve the Cooperative Planning Layer (in cooperation management).

For the refinement of GAM this means that it is specified that the appropriate control information is exchanged between world interaction management and agent interaction management, own process control and cooperation management.

The ZEUS architecture distinguishes: Mailbox, Message Handler, Co-ordination Engine, Execution Monitor, Acquaintance Model, Planner and Scheduler, Task/Plan Database, Resource Database. The Mailbox and the Message Handler together can be formally specified as a specialisation and instantiation of the component agent interaction management within GAM. The Co-ordination Engine can be obtained as a refinement of the component cooperation management. The Execution Monitor with the Planner and Scheduler, and the Task/Plan Database together can be specified as a specialisation and instantiation of the component own process control. The Acquaintance Model can be obtained as an instantiation of component maintenance of agent information. Although interaction with the External World is not explicitly modelled within a ZEUS agent, the Resource Database may include some of this information.

The architecture ADEPT (Advanced Decision Environment for Process Tasks; see [Jennings *et al.*, 1996]) represents business processes by a hierarchy of cooperative agents. The hierarchy ensures that communication overhead between agents and the autonomy of the agents are balanced. Within this model, agents have the following modules: a communication module, an interaction management module (IMM), a situation assessment module (SAM), a service execution module (SEM), a self model (SM), acquaintance models (AM). These modules have been specified as a refinement of GAM as follows: the module IMM as a refinement of the component cooperation management, the modules SAM and SM as components within a specialisation of the component own process control, the module SEM can clearly be described as a specialisation of the component maintenance of agent information.

8 DISCUSSION

This section, first summarizes the process of designing and reusing a generic model, on the basis of the generic agent model GAM. Next, current and future research issues are discussed.

8.1 Designing a generic model

The generic agent model GAM was not designed from scratch. Conceptual analysis of agent capabilities and characteristics is the main motivation for the components distinguished in the generic agent model. These components have been distinguished in agent models in different domains of application. Example agent models for the applications described in [Brazier *et al.*, 1995]; [Brazier *et al.*, 1998a], based on a precursor of GAM were an important input for the process of designing the generic agent model in more detail. Further generic structures were extracted from these example models and combined, leaving out domain specific elements.

In a number of cases a choice had to be made. Some other information types could have been included as well. The more structures are included, the more support is given when reusing the generic model. However, this only holds for applications for which the generic structures are relevant: the richer a generic model is, the more restrictive is its scope of application. Since the generic model GAM has been designed to be a very widely applicable model, the choice has been made to limit the number of structures included. As discussed in Section 7, more specialised agent models have been developed as well: for example, a generic model for BDI-agents, in which the component own process control is refined [Brazier *et al.*, 1999], and a generic model for co-operation, in which both the components own process control and co-operation management are refined [Brazier *et al.*, 1997b].

8.2 *Reusing a generic model*

The scope of applicability of the generic agent model GAM covers a variety of application domains, as discussed in Section 7. As the generic model was constructed to subsume a large number of applications, it should not be difficult to reuse the generic model in similar application domains. This paper shows in more depth how the generic agent model can be applied to another application domain: co-operative information gathering agents. As a first step in the reuse of GAM the domain specific knowledge structures were instantiated: domain specific information types and knowledge bases. The information types domain actions and agent information were not considered to be relevant for this application, so these information types remained empty. In fact, the component maintenance of agent information was not used at all and could have been removed. One of the generic knowledge bases in the generic agent model could be reused (observation result extraction kb). Another generic knowledge base (communicated info extraction kb) was replaced by a more specific knowledge base. Moreover, knowledge bases were added to generate communication and observation.

A second step was the addition of two new information types to handle requests for information or observation. Finally, a third step was to add an information link to transfer requests from agent interaction management to world interaction management. The process of reusing a generic model as summarised above has realistic characteristics. In general, if a suitable generic model is available, during the design process:

- most but not all parts of the generic model can be reused as is
- parts that are not used are modified, remain empty or can be removed
- some additional knowledge structures may be needed and added

- some additional information links may be needed and added maybe
- some additional components are needed and added or modified.

The example process of reusing the generic agent model GAM discussed in this paper shows almost all of these characteristics.

8.3 Current and Future Research

Current research focuses on requirements engineering and verification for agent systems, and on applications to information brokering agents and Electronic Commerce. Within requirements engineering the aim is to obtain appropriate informal, semi-formal and formal representations of functional or behavioural properties of a multi-agent system, of the agents within a multi-agent system and of components within an agent. A first proposal can be found in [Herlea *et al.*, 1999]. Requirements specifications can be expressed in generic forms and reused in conjunction with generic models such as GAM. Compositional verification is an approach to establish that behavioural properties of a multi-agent system hold, given properties of agents and of their components; e.g., see [Jonker and Treur, 1998a].

Frances M.T. Brazier, Catholijn M. Jonker and Jan Treur
Vrije Universiteit Amsterdam, The Netherlands.

BIBLIOGRAPHY

[Brazier *et al.*, 1995] F.M.T. Brazier B. Dunin-Keplicz, N.R. Jennings and J. Treur. Formal specification of Multi-Agent Systems: a Real-World Case. In: V. Lesser (ed.), *Proceedings of the First International Conference on Multi-Agent Systems, ICMAS-95*, MIT Press, Cambridge, MA, 25–32, 1995. Extended version in: *International Journal of Cooperative Information Systems*, M. Huhns and M. Singh, (eds.), special issue on Formal Methods in Cooperative Information Systems: Multi-Agent Systems, **6**, 67–94, 1997.

[Brazier *et al.*, 1997a] F.M.T. Brazier, P.A.T. van Eck and J. Treur. Modelling a Society of Simple Agents: From Conceptual Specification to Experimentation. In: R. Conte, R. Hegselmann and P. Terna (eds.), *Simulating Social Phenomena, Proc. of the International Conference on Computer Simulations and Social Sciences, ICCS&SS'97*, Lecture Notes in Economics and Mathematical Systems, **456**, Springer-Verlag, Berlin, 103–107, 1997. Extended version in *Journal of Applied Intelligence*, **14**, 161–178, 2001.

[Brazier *et al.*, 1997b] F.M.T. Brazier, C.M. Jonker and J. Treur. Formalisation of a cooperation model based on joint intentions. In: J.P. Müller, M.J. Wooldridgeand and N.R. Jennings (eds.), Intelligent Agents III, *Proc. of the Third International Workshop on Agent Theories, Architectures and Languages, ATAL'96*, Lecture Notes in AI, **1193**, Springer Verlag, 141–155, 1997.

[Brazier *et al.*, 1998a] F.M.T. Brazier, F. Cornelissen, R. Gustavsson, C.M. Jonker, O. Lindeberg, O., B. Polak and J. Treur. Agents Negotiating for Load Balancing of Electricity Use. In: M.P. Papazoglou, M. Takizawa, B. Krmer, S. Chanson (eds.), *Proceedings of the 18th International Conference on Distributed Computing Systems, ICDCS'98*, IEEE Computer Society Press, 622–629, 1998.

[Brazier *et al.*, 1998b] F.M.T. Brazier, C.M. Jonker, F.J. Jüngen and J. Treur. Distributed Scheduling to Support a Call Centre: a Co-operative Multi-Agent Approach. Applied *Artificial Intelligence Journal*, **13**, 65–90, 1999. H.S. Nwana and D.T. Ndumu (eds.), Special Issue on Multi-Agent Systems.

Earlier shorter version in: H.S. Nwana and D.T. Ndumu (eds.), *Proceedings of the Third International Conference on the Practical Application of Intelligent Agents and Multi-Agent Technology, PAAM'98*. The Practical Application Company Ltd, 555–576, 1998.

[Brazier *et al.*, 1998c] F.M.T. Brazier, C.M. Jonker and J. Treur. Principles of Compositional Multi-agent System Development. In: J. Cuena (ed.), *Proceedings of the 15th IFIP World Computer Congress, WCC'98, Conference on Information Technology and Knowledge Systems, IT&KNOWS'98*, To be published by IOS Press.

[Brazier *et al.*, 1999] F.M.T. Brazier, B. Dunin-Keplicz, J. Treur and L.C. Verbrugge. Modelling Internal Dynamic Behaviour of BDI agents. In: J.-J. Ch. Meyer and P.Y. Schobbes (eds.), *Formal Models of Agents* (Selected papers from final ModelAge Workshop). Lecture Notes in AI, **1760**, Springer Verlag, 36–56, 1999.

[Castelfranchi *et al.*, 1999] C. Castelfranchi, F. Dignum, C.M. Jonker and J. Treur. Deliberate Normative Agents: Principles and Architecture. In: N.R. Jennings and Y. Lesperance (eds.), *Intelligent Agents VI. Proc. of the Sixth International Workshop on Agent Theories, Architectures and Languages, ATAL'99*. pp. 364–378. Lecture Notes in *AI* vol. 1757, Springer Verlag, 2000.

[Cesta *et al.*, 1996] A. Cesta, M. Micelli and P. Rizzo. Effects of different interaction attitudes on a multi-agent system performance. In: W. van de Velde and J.W. Perram (eds.) Agents Breaking Away. *Proc. 7th Eur. Workshop on Modelling Autonomous Agents in a Multi-Agent World, MAAMAW'96*. Lecture Notes in *Artificial Intelligence*, **1038**, Springer-Verlag, 128–138, 1996.

[Cockburn and Jennis, 1995] D. Cockburn and N. R. Jennings. ARCHON: A Distributed Artificial Intelligence System for Industrial Applications. In: G. M. P. O'Hare and N. R. Jennings (eds.), *Foundations of Distributed Artificial Intelligence*, Wiley & Sons, 319–344, 1995.

[Denett, 1987] D. Dennett. *The Intentional Stance*, MIT Press, Cambridge, MA, 1987.

[Ferguson, 1992] A.I. Ferguson. *Touring Machines: An Architecture for Dynamic, Rational, Mobile Agents*. Ph.D. Thesis. Computer Laboratory, University of Cambridge, UK, 1992.

[Georgeff and Lansky, 1987] M.P. Georgeff and A.L. Lansky. Reactive Reasoning and Planning. *Proc. of the National Conference of the American Association for AI, AAAI'87*. Morgan Kaufman, 1987.

[Herlea *et al.*, 1999] D.E. Herlea, C.M. Jonker, J. Treur and N.J.E. Wijngaards. Specification of Behavioural Requirements within Compositional Multi-Agent System Design. In: F.J. Garijo and M. Boman (eds.), Multi-Agent System Engineering, *Proceedings of the 9th European Workshop on Modelling Autonomous Agents in a Multi-Agent World, MAAMAW'99*. Lecture Notes in *AI*, **1647**, Springer Verlag, Berlin, 8–27, 1999.

[Jennings, 1995] N.R. Jennings. Controlling Cooperative Problem Solving in Industrial Multi-Agent Systems using Joint Intentions. *Artificial Intelligence Journal*, **74**(2), 1995.

[Jennings *et al.*, 1996] N.R. Jennings, J. Corera, I. Laresgoiti, E. H. Mamdani, F. Perriolat, P. Skarek and L. Z. Varga. Using ARCHON to develop real-word DAI applications for electricity transportation management and particle accelerator control, *IEEE Expert - Special issue on Real World Applications of DAI*, 1996.

[Jennings *et al.*, 1996] N.R. Jennings, P. Faratin, T. Norman, T.J. O'Brien, P. Wiegand, M.E. Voudouris, J.L. Alty, T. Miah and E.H. Mamdani. ADEPT: Managing Business Processes using Intelligent Agents. In: *Proc. BCS Expert Systems 96, Conference (ISIP Track)*, Cambridge, UK, 5–23, 1996.

[Jennings and Wooldridge, 1998] N.R. Jennings and M. Wooldridge. Applications of Intelligent Agents. In: [Jennings and Wooldridge, 1998], 3–28, 1998.

[Jennings and Wooldridge, 1998] N.R. Jennings and M. Wooldridge (eds.). *Agent Technology: Foundations, Applications, and Markets*. Springer Verlag, 1998.

[Jonker *et al.*, 1999] C.M. Jonker, R.A. Lam and J. Treur. A Multi-Agent Architecture for an Intelligent Website in Insurance. In: *Proceedings of the Third International Workshop on Cooperative Information Agents, CIA'99*. Lecture Notes in *AI*, Springer Verlag, 1999. Extended version in *Journal of Applied Intelligence*, **15**, 7–24, 2001.

[Jonker and Treur, 1998a] C.M. Jonker and J. Treur. Compositional Verification of Multi-Agent Systems: a Formal Analysis of Pro-activeness and Reactiveness. In: W.P. de Roever, H. Langmaack and A. Pnueli (eds.), *Proceedings of the International Workshop on Compositionality, COMPOS'97*. Lecture Notes in *Computer Science*, **1536**, Springer Verlag, 350–380, 1998. Extended version in *Int. Journal of Cooperative Information Systems*, in press, 2002.

[Jonker and Treur, 1998b] C.M. Jonker and J. Treur. Agent-based Simulation of Reactive, Pro-active and Social Animal Behaviour. In: J. Mira, A.P. del Pobil, and M. Ali (eds.), Methodology and Tools in Knowledge-Based Systems, *proceedings of the 11th International Conference on Industrial and*

Engineering Applications of AI and Expert Systems, IEA/AIE'98, vol. I, Lecture Notes in *AI*, **1415**, Springer Verlag, 584–595, 1998. Extended version in *Journal of Applied Intelligence*, **15**, 83–115, 2001.

[Jonker and Treur, 1998c] C.M. Jonker and J. Treur. Compositional Design and Maintenance of Broker Agents. In: J. Cuena (ed.), *proceedings of the 15th IFIP World Computer Congress, WCC'98, Conference on Information Technology and Knowledge Systems, IT&KNOWS'98*, 319–332, 1998.

[Jonker and Treur, 1999] C.M. Jonker and J. Treur. Information Broker Agents in Intelligent Websites. In: I. Imam, Y. Kodratoff, A. El-Dessouki and M. Ali (eds.), Multiple Approaches to Intelligent Systems, *proc. of the 12th International Conference on Industrial and Engineering Applications of AI and Expert Systems, IEA/AIE'99*. Lecture Notes in *AI*, **1611**, Springer Verlag, 430–439, 1999. Extended version in *Journal of Applied Intelligence*, **15**, 7–24, 2001.

[Müller, 1996] J.P. Müller. The Design of Intelligent Agents: a Layered Approach. Lecture Notes in *AI*, **1177**, Springer Verlag, 1996.

[Müller *et al.*, 1995] J.P. Müller, M. Pischel and M. Thiel. Modelling reactive behaviour in vetically layered agent architectures. In: [Wooldridge and Jennings, 1995a], 261–276, 1995.

[Nwana, 1996] H.S. Nwana. Software Agents: an Overview. *Knowledge Engineering Review*, **11**(3), 205–244, 1996.

[Nwana and Ndumu, 1998] H.S. Nwana and D.T. Ndumu. A Brief Introduction to Software Agent Technology. In: [Jennings and Wooldridge, 1998], 29–47, 1998.

[Nwana *et al.*, 1998] H.S. Nwana, D.T. Ndumu and L.C. Lee. ZEUS: An Advanced Tool-Kit for Engineering Distributed Multi-Agent Systems. In: *Proceedings of the Third International Conference on the Application of Intelligent Agents and Multi-Agent Technology*, H.S. Nwana and D.T. Ndumu (eds.), The Practical Application Company, Blackpool, 377–391. Also in *Applied AI*, **13**, 129, 1998.

[Rao and Georgeff, 1991] A.S. Rao and M.P. Georgeff. Modeling rational agents within a BDI architecture. In: R. Fikes and E. Sandewall (eds.), *Proceedings of the Second Conference on Knowledge Representation and Reasoning*, Morgan Kaufman, 473–484, 1991.

[Vauclair, 1996] J. Vauclair. *Animal Cognition*, Harvard Univerity Press, Cambridge, Massachusetts, 1996.

[Wooldridge and Jennings, 1995a] M. Wooldridge and N. R. Jennings, eds. *Intelligent Agents, Proc. of the First International Workshop on Agent Theories, Architectures and Languages, ATAL'94*, Lecture Notes in *Artificial Intelligence*, **890**, Springer Verlag, Berlin, 1995

[Wooldridge and Jennings, 1995b] M.J. Wooldridge and N.R. Jennings. Agent theories, architectures, and languages: a survey. In: [Wooldridge and Jennings, 1995a], 1–39, 1995.

[Wooldridge and Jennings, 1995c] M.J. Wooldridge and N.R. Jennings. Intelligent Agents: Theory and practice. In: *Knowledge Engineering Review*, **10**(2), 115–152, 1995.

PART IIIA

FORMAL ANALYSIS:

GENERAL APPROACHES

FRANCES BRAZIER, PASCAL VAN ECK AND JAN TREUR

SEMANTIC FORMALISATION OF EMERGING DYNAMICS OF COMPOSITIONAL AGENT SYSTEMS

1 INTRODUCTION

Multi-agent systems often are heterogeneous systems composed of different types of autonomous agents. Each of these agents may be based on a specific design specification, and may have its own semantics. Global dynamics at the level of the entire system emerges from the behaviours of the agents separately, and the manner in which the agents interact. Since the behaviour of each of the agents is specified in a local manner, independent of the global multi-agent system structure, an important semantical question is how global dynamics can be defined, given this variety of heterogeneous individual agents semantics. The question addressed in this chapter is how different behaviours independently defined at the level of individual agents can be composed to obtain emergent global multi-agent system behaviour, without assuming a uniform global semantic model for the system as a whole, and, in particular, without assuming a uniform global time frame.

The basic element for the approach introduced in this chapter is the semantic description of each of the agents by a (local) set of traces over a local time frame. The dynamics of a multi-agent system is described by *multitraces*: structured multisets of agent traces. Interaction between agents constrains the set of combinations of traces that model possible dynamics of the overall system. These constraints are specified by (temporal) compatibility relations which model the structure of the overall multi-agent system. Temporal compatibility relations are relations between traces of two different agents, and model their interaction: only agent traces that respect compatibility can be part of the multi-agent systems overall behaviour. The semantic structure presented in this chapter allows characterisations of what it means to respect compatibility. Only agent traces that are compatible with traces of agents with which they interact can be part of a description of the dynamics of the multi-agent system.

The semantic structure provides a basis for the formalisation of emergence of global dynamics from interaction between different forms of local dynamics. The model defines local dynamics without any reference to global elements, not even to a global clock. Given the independently defined local dynamics, and interaction between different local dynamics, the emerging global dynamics are defined in terms of the local dynamics, without assuming a uniform global semantic model. In particular, no total time ordering is enforced between events occurring internally in different agents, as would be the case for a global clock.

J.J.Ch. Meyer and J. Treur (eds.),
Handbook of Defeasible Reasoning and Uncertainty Management Systems, Vol. 7, 167–196.

In general, an agent's behaviour results from internal dynamics within the agent, such as, for example, reasoning on the basis of beliefs, desires and intentions [Bratman, 1987]. Many semantic approaches leave open how the internal dynamics of agents are described. Those that do address to a certain extent the internal dynamics of agents restrict their scope to one specific agent architecture, for example the BDI-architecture [Rao and Georgeff, 1991]. In the semantic approach introduced in this chapter, a unifying perspective is taken: the internal architecture of an agent and its dynamics are defined in terms of components and interaction between components. A formal semantic structure of component dynamics is presented in which a number of concepts are identified to describe agents and the manner in which their behaviour can be defined on the basis of their component dynamics. A component maintains certain information, which can change over time and which determines its current component state. As for agents, the dynamics of a component is described by traces of consecutive component states. As the state of a component is determined only by the information contents of this component (and not by anything outside the component), the traces that describe a component's dynamics are completely local. For a specific agent, compatibility for multitraces taken from sets of traces of its components and the set of traces of the agent itself, constrains the agent behaviour, given its internal structure.

The approach introduced here covers both (1) at the level of the multi-agent system as a whole, how the overall system dynamics emerges from the agent behaviours, and (2) at the level of individual agents, how agent behaviour results from the agent's internal component structure and dynamics. Locality is one important concept in the semantic structure developed in this chapter: the global state of a compositional agent system is not defined. Moreover, it is assumed that each agent and each component is described in its own local language. Dynamic compatibility is the second important concept, which defines constraints on local behaviours in order to obtain global behaviour.

The two basic concepts, components and interaction between components, are used both to describe the internal structure and dynamics of an agent, as well as the structure and dynamics of a multi-agent system external to agents. Thus, a uniform formal semantic model is obtained that covers both types of dynamics and their relations, and is generic with respect to the specific architecture chosen for the agents.

After describing the basic concepts in Section 2, the semantic structure of component dynamics is formalised in Section 3. Section 4 describes compatibility relations in more detail. In Section 5, a comparison with other frameworks is presented. Results are discussed in Section 6.

2 COMPOSITIONAL AGENT SYSTEMS: BASIC CONCEPTS

The basic assumption adopted in this chapter is that multi-agent systems are modelled as *compositional systems*, i.e. systems consisting of components. The starting point is to focus on a multi-agent system as a *system*. A system is often defined as a connected collection of parts. As the semantic structure focuses on multi-agent systems *dynamics*, in this chapter, a system is seen as a coherent collection of *processes and entities that execute processes (agents and the environment)*. It is understood that any pattern of change within a system constitutes a process.

A multi-agent system is then modelled as a compositional system in the following way:

- In the domain that has to be modelled, processes are identified. Moreover, these processes can either be classified as deliberation processes or environmental processes. Usually, a number of processes are distinguished as subprocesses of other processes, or as subprocesses of subprocesses, and so on;

- In the multi-agent system, (active) entities are identified (the agents and the environment) that execute the processes. An agent or the environment can execute more than one process simultaneously

- The multi-agent system is modelled as a compositional system, which is achieved by representing each process as a component. This component encapsulates both the information used by the process and the process (computation) itself. Subprocesses of a process are represented as subcomponents of the component that represents that process. At the highest level of abstraction, a compositional system that models a multi-agent system consists (solely) of one component for each agent. At lower levels of abstraction, all subprocesses distinguished in the multi-agent system are represented by subcomponents of either one of the agent's components. As a consequence, the structure of a multi-agent system (in terms of the agents that constitute the multi-agent system) is thus represented in the compositional system.

- Dynamic relationships between processes in the multi-agent system are represented by interaction between the components that represent these processes.

Each process in the multi-agent system is represented by a component. Thus, the approach is uniform: components (and interaction between components) are used to represent processes at the multi-agent system level as well as on the level of individual agents. (See [Brazier *et al.*, 2001] for a further discussion of the representation of multi-agent systems.)

Thus, components and interaction are the basic, uniform concepts of the semantic structure. Before formally defining the semantic structure in Section 3, first these concepts are discussed informally.

2.1 *Basic Concepts: Components and Interaction*

The semantic structure developed in this chapter distinguishes two basic concepts. The first concept is the concept of a *component*, of which two aspects are represented in the semantic structure. A component is a locus of information and computation that can (only) interact with other components via a well-defined pair of interfaces. In general, a component stores information, which can change over time. The current information contents of a component determines its information state, or *state* for short, which is the first aspect of a component. This information state is only accessible via the interfaces of the component. In the semantic structure developed in this chapter, each component has an input and output interface. The input interface is that part of a component's state that is used as input for the (computational) processes of the component. The output interface is the part of a component's state that is used for the output of the component. This interface can be used to make results of services provided by the component accessible to other components. Changes in the output part of a component's state are visible to other components. The second aspect of the notion of a component is its *structure*: a component may itself be composed of other components. These components are called the *subcomponents* of the component. (In this chapter, the term 'subcomponent' always refers to direct descendants of a component, not including their own subcomponents, and so on.) A third aspect of the notion of a component is the *specification language used for the description of the behaviour of a component*. However, this aspect is not represented in the semantic structure.

The second concept is the concept of *interaction*, of which two aspects are represented in the semantic structure. Interaction is modelled as information exchange between two (possibly the same) components. The first aspect of interaction is the communication channel used for interaction, which is called the *information link*. This aspect also comprises the *state* of the link, and the components to which the information link is connected. The state of a link is determined by, e.g., the state of the process of interaction. (E.g., whether interaction is enabled or currently being performed.) The second aspect of interaction is the declaration of the information exchange that models the interaction. This declaration is called the *information mapping*. From a conceptual point of view, an information mapping declares that certain state transitions in a component are related to certain state transitions in another component. As an example, a state transition of a certain component which leads to a state in which a new result is available as output may be related to a state transition of another component which leads to a state in which this result is available as input. From an operational point of view, an information mapping declares the communication performed to establish the semantic relationship. (For example, the new result is communicated.)

2.2 *Dynamics of Compositional Agent Systems: Locality and Temporal Compatibility*

The behaviour of a single component is modelled by a set of information state traces, which are (e.g., linear or branching) structures of consecutive information states of the component. The behaviour of a compositional agent system is modelled by a set of multitraces, which are structures consisting of such traces. Consider a system with two components, A and B, and an information link between A and B. The specification of the behaviour of B states that B is able to provide a service 's' that places new output in the output interface of B. To request this service, component A places the identifier 's' of the service in its output interface, which is then (in this case automatically) relayed to component B by the information link.

For both components, three sets of traces can be distinguished, as depicted in Figure 1 (in which traces are assumed to be linear). The first set is the set of traces consisting of all possible combinations of states of a component. This set includes traces that, in practice, can never be acquired, because they do not fulfil the (internal) specification of the behaviour of the component. For instance, the set of all possible traces of component B includes traces in which after receipt of the request for a service 's' in the input interface, the result is never placed in the output interface of B. A subset of the set of all possible traces is the set of local component traces: those traces that could, in principle, be acquired because they fulfil the internal specification of component behaviour. The second set of traces is called *local component traces* to emphasise the fact that they are only part of a component's behaviour from a purely local point of view, in which constraints imposed by interaction with other components are not taken into account. For instance, a trace in which, after receipt of service request 's', the result of this service is placed in the output interface, is a local component trace. However, this trace may not be an actual behaviour of component B in a compositional agent system: it is only an actual behaviour if component A generates a request for service 's' and if this request is actually transferred to B. Therefore, in a structure that models the behaviour of the system consisting of components A and B and interaction between A to B, such constraints should be represented explicitly.

The third set of traces is the subset of the set of local component traces that take constraints imposed by information exchange into account. A local component trace of a component that interacts with another component is part of the *actual overall behaviour* of the system only if it respects constraints imposed by interaction. This is modelled by temporal compatibility relations between traces of interacting components. For instance, a trace of component A in which event 's' is in A's output interface is compatible with traces in which event 's' is in B's input interface (and the result of service 's' is eventually generated at its output interface). Only compatible local component traces are part of multitraces that model the behaviour of compositional agent systems. Only local component traces that respect interaction are compatible.

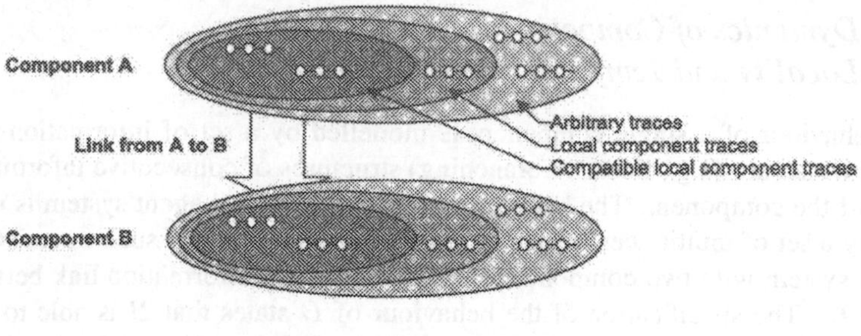

Figure 1. Sets of component traces

2.3 Running Example Multi-Agent System

To illustrate the semantic structure developed in this chapter, an example system in the area of intelligent Internet applications is used. The system is a compositional multi-agent system in which broker agents act as intermediaries between agents that provide information (on arbitrary resources) and agents that use this information. There are three agent roles in the example: broker, provider (of resources) and user. (Agents in the system may play more than one role, for instance a broker agent may play the role of a user with respect to another broker agent, but this is not further explored in this chapter.) Using the concepts developed in Section 2.1, the example system is modelled as a compositional agent system. The highest level of the composition is depicted in Figure 2. In this figure, the components labelled user_1 and user_2 represent agents that play the role of users with respect to the brokering agent. (The components labelled user_1 and user_2 may, for example, be human agents, their personal digital agents, or web browser applications serving as graphical user interfaces to interact with human agents. The precise nature of these components is not important.) The component labelled broker represents an agent that plays the role of a broker agent. The components labelled provider_1 and provider_2 represent agents playing the role of provider agents. (Like user agents, the precise nature of provider agents is not important.)

The intended function of a broker agent is as follows. A user agent communicates to the broker agent that he/she is interested in information on a resource (i.e., a research chapter, a WWW page, a product sold by means of e-commerce). A provider agent communicates descriptions of resources available to him/her to the broker agent, preferably using an internet standard such as the Resource Description Framework (RDF), see [Lassila, 1998], KQML [Finin *et al.*, 1997], or FIPA-ACL [O'Brien and Nichol, 1998]. The broker agent matches interests of users with information provided by the provider agents, in conformance with a number of requirements. Due to space restrictions, only one requirement that is placed on the behaviour of a broker agent is mentioned in this chapter: once a broker agent receives a query from a user, information matching the query has to be communicated to the user at the next moment in time if this information is known

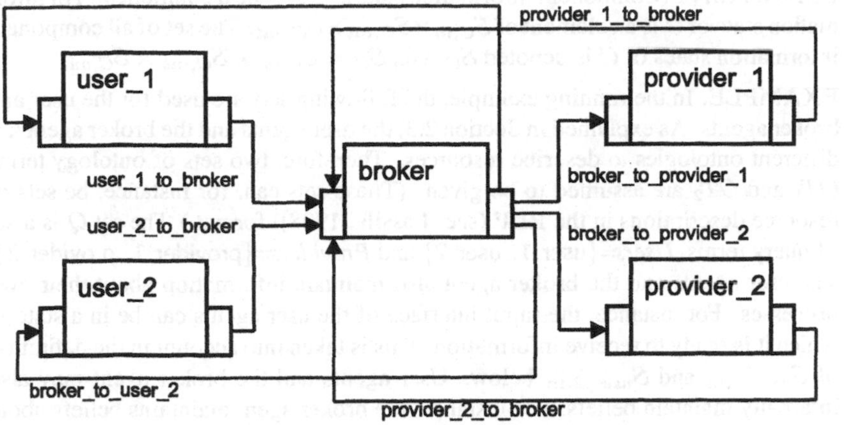

Figure 2. Information broker agent system

to the broker, or some time in the future in all other cases. However, many more requirements can be distinguished [Jonker and Treur, 1998].

3 SEMANTIC FORMALISATION

The formalisation of the semantic structure is divided in two parts: the formalisation of components and interaction, and the formalisation of behaviour.

3.1 Components and interaction

Two aspects of the notion of a component are of importance: state and structure. These aspects are formalised in Section 3.1.1 and Section 3.1.3, respectively. In Section 3.1.2, the concept of interaction is formalised.

Component State

For the first aspect of the notion of a component, its state, it is assumed that a set of components (or, more precisely, component identifiers or names) is given. Elements of this set are typically denoted by capitals C, D, etc. For each component a dynamically changing information state is defined, which consists of input, internal and output substates. It is assumed that for each component C, three sets of substates $\mathcal{S}_{C,in}$, $\mathcal{S}_{C,int}$ and $\mathcal{S}_{C,out}$ are given (for the input, internal and output substates, respectively), without further commitment to the contents of these sets. The (overall) state of a component C is composed of elements of the sets $\mathcal{S}_{C,in}$, $\mathcal{S}_{C,int}$ and $\mathcal{S}_{C,out}$ as follows:

DEFINITION 1 (Component information state). Let C be a component. An information state of C is an element of $\mathcal{S}_{C,in} \times \mathcal{S}_{C,int} \times \mathcal{S}_{C,out}$. The set of all component information states of C is denoted $\mathcal{S}_C$, i.e., $\mathcal{S}_C = \mathcal{S}_{C,in} \times \mathcal{S}_{C,int} \times \mathcal{S}_{C,out}$.

EXAMPLE. In the running example, the following sets are used for the user and broker agents. As explained in Section 2.3, the user agents and the broker agent use different ontologies to describe resources. Therefore, two sets of ontology terms OT_1 and OT_2 are assumed to be given. (These sets can, for instance, be sets of resource descriptions in the RDF (see [Lassila, 1998]) format.) The set Q is a set of query terms, *Users*={user_1, user_2} and *Providers*={provider_1, provider_2}. The user agents and the broker agent also maintain information about their own processes. For instance, the input interface of the user agents can be in a state in which it is ready to receive information. This is taken into account in the definition of $\mathcal{S}_{user\_1,in}$ and $\mathcal{S}_{user\_2,in}$ below. User agents and the broker agent may also internally maintain beliefs. For instance, the broker agent maintains beliefs about matches between user queries and information available via the providers.

- $\mathcal{S}_{user\_1,in} = \mathcal{S}_{user\_2,in} = \{\text{communicated\_by}(t,\text{broker}) \mid t \in OT_1\} \cup \{\text{ready\_for\_information}\}$,
  $\mathcal{S}_{user\_1,int} = \mathcal{S}_{user\_2,int} = \{\emptyset\}$
  $\mathcal{S}_{user\_1,out} = \mathcal{S}_{user\_2,out} = \{\text{to\_be\_communicated\_to}(q,\text{broker}) \mid q \in Q\}$.
- $\mathcal{S}_{broker,in} = \{\text{communicated\_by}(q,u) \mid q \in Q \text{ and } u \in Users\} \cup \{\text{communicated\_by}(t,p) \mid t \in OT_2 \text{ and } p \in Providers\}$,
  $\mathcal{S}_{broker,int} = \{\text{belief}(\text{match}(t,q)) \mid t \in OT_2 \text{ and } q \in Q\}$
  $\mathcal{S}_{broker,out} = \{\text{to\_be\_communicated\_to}(t,u) \mid t \in OT_2 \text{ and } u \in Users\} \cup \{\text{just\_communicated\_to}(t,u) \mid t \in OT_2 \text{ and } u \in Users\}$.

In this example, states are identified by elements of e.g. $\mathcal{S}_{user\_1,in}$ such as communicated_by(t,broker), where t is an element from the set of ontology terms OT_1. The elements of sets such as $\mathcal{S}_{user\_1,in}$ resemble propositions about states. However, in this example, elements of sets such as $\mathcal{S}_{user\_1,in}$ are (unique) names of states. These names have no internal structure.

The information state of a component changes over time. Sequences of information states, called traces, are used to model the dynamics.

DEFINITION 2 (Time frame). A time frame is a pair $TF = \langle T; < \rangle$,[1] where T is a set of time points and $<$ is a strict partial order on T. Moreover, $<$ is connected, i.e. $\forall t \in T : \exists t' \in T : t < t' \vee t' < t$. There is one element, $\perp \in T$, for which there is no $t \in T$ such that $t < \perp$.

This definition enables various types of time frames with different properties to be used for different component such as, for instance, branching time frames and dense time frames. In the remainder of this chapter, linear time frames are assumed.

[1] In this chapter, tuples are delimited by angular brackets and their elements are separated by semicolons.

DEFINITION 3 (Local traces). Local traces are defined as follows:

- A local component trace of a component C for a time frame $TF = \langle T; < \rangle$ is a pair $LT_C = \langle TF; V \rangle$, where V is a function $V : T \to \mathcal{S}_C$.

- The set of all local component traces of a component C is denoted $\mathcal{LT}_C$.

As described in Section 2.2, local component traces are used to model the behaviour of components. For each component C, a subset of $\mathcal{LT}_C$ is distinguished which contains all local component traces that are alternative behaviours of C, given the specification of the behaviour of C. Formally:

DEFINITION 4 (Local component behaviour). Let C be a component. A local component behaviour of $Beh_{loc}(C)$ of C is a set of local component traces of C.

To clarify the use of the qualification 'local' in the name of the notion defined above, a distinction is made between two views on the concept of location. On the one hand, a component has a specific location in the hierarchical structure of components, their subcomponent, the subcomponents of subcomponents and so on. This structure is determined by how processes relate to one another in terms of the function of the subprocesses. In this conception of the location of a component, a subcomponent is 'close', or 'local' to its parent component. On the other hand, there is also a 'physical' distribution of components (and the processes they represent) over processes, which are spatially divided. It is *not* assumed that the subcomponents of a component all exist at the same physical location. Thus, subcomponents of a component are not necessarily local to their parent component with respect to the 'physical' location of a compositional system. In this chapter, the qualifications 'local' and 'global' always refer to the physical distribution of a compositional system.

The notion of local component behaviour is called 'local' because the set $Beh_{loc}(C)$ is not constrained by non-local phenomena such as interaction (not even with its subcomponents). A set $Beh_{loc}(C)$ is thus independent of any (hierarchical) structure of components in which C occurs. In other words, the set $Beh_{loc}(C)$ can contain local component traces that are not possible behaviour when interaction is taken into account, for instance because these local component traces depend on information residing in other components.

EXAMPLE. The running example uses the set of natural numbers together with the usual ordering as a time frame. Local traces of the broker agent should satisfy the requirement mentioned at the end of Section 2.3. This requirement can be formalised using a temporal logic formula as follows (where P means 'sometime in the past', X means 'next' and F means 'sometime in the future', and $q \in Q$, $t \in OT_2, u \in Users$ and $p \in Providers$):

$$
\begin{aligned}
&(\,\mathsf{communicated\_by}\ (q,u) \wedge \mathsf{belief}\ (\mathsf{match}(t,q))) \rightarrow \\
&((\mathsf{Pcommunicated\_by}(t,p) \rightarrow \mathsf{Xto\_be\_communicated\_to}(t,u)) \vee \\
&(\mathsf{F}(\mathsf{communicated\_by}(t,p) \rightarrow \mathsf{Fto\_be\_communicated\_to}(t,u))))
\end{aligned}
$$

A local component trace is represented as a sequence of component states with the sets of input, internal and output propositions separated by bars. As an example of a local component trace in $Beh_{loc}(broker)$, consider the following trace (where res_1 represents a resource description and query_1 represents a query). The trace uses the natural numbers with the usual order as the time frame.

$lt_{broker} =$ ∅ | belief(match(res_1,query_1)) | ∅ →
communicated_by(query_1,user_1)|belief(match(res_1,query_1))| ∅ →
communicated_by(res_1,provider_1)|belief(match(res_1,query_1))| ∅ →
∅ |belief(match(res_1,query_1))|to_be_communicated_to(res_1,user_1)→
∅ |belief(match(res_1,query_1))|just_communicated_to(res_1,user_1)

Interaction

The beginning of this section mentions the structure of a component as its second aspect. This structural aspect describes compositions of components in terms of the subcomponents and interactions that constitute the components. To formally define this structural aspect, first the formalisation of interaction (the second concept distinguished in the semantic structure) is introduced. Two aspects of this concept are distinguished: the information link used for interaction and its state, and the information mapping.

An information link transmits information from one component (called source component or *domain* of the link) to another, or possibly the same component (called the destination component or *co-domain* of the link). Information links are first-class citizens; they are of the same standing as components.

It is assumed that a set of links *Lnk* (or, more precisely, link identifiers or names) is given. An element of this set is typically denoted with the capital I. The state of a link is determined by (1) the state of information transmission as an activity: for instance, a link can be busy exchanging information, it can be waiting for new information to exchange, it can be enabled or disabled, and (2) the contents of the link, e.g. messages in transit. The semantic structure does not enforce a commitment with respect to the contents of link states. Instead, for each information link I, a non-empty set of information link states S_I is assumed to be given.

DEFINITION 5 (Link information state). Let I be an information link. An *information state* of I is an element of a set S_I.

EXAMPLE. In the running example, a link called broker_to_user_1 exists between broker and user_1. The set of link information states of this link is defined as follows (where t is a set of ontology terms):

$$S_{broker\_to\_user\_1} = \{\mathsf{awake\_and\_empty}, \mathsf{active\_and\_contents}(t) \mid t \in OT_2\}.$$

The information state of a link also changes over time, and this is likewise modelled by traces of link states as follows:

DEFINITION 6 (Information link trace). An *information link trace* of an information link I for a time frame $TF = \langle T; < \rangle$ is a pair $LT_I = \langle TF; V \rangle$, where V

is a total function $V : T \to S_I$. The set of all link traces of a link from D to C is denoted $\mathcal{LT}_I$.

The way in which the behaviour of a link is described by a set of information link traces, as witnessed by the following definition, is similar to the way in which local component behaviour is defined by a set of local component traces:

DEFINITION 7 (Local link behaviour). Let I be an information link. The *local link behaviour* of I is a set $Beh_{loc}(I)$ of information link traces.

EXAMPLE. A possible information link trace for the link broker_to_user_1 is:

$$lt_{broker\_to\_user\_1} = \mathsf{awake\_and\_empty} \to \mathsf{active\_and\_contents(res\_1)} \to \mathsf{awake\_and\_empty} \to \ldots$$

In this trace, the link is first awake (ready to transmit information) and there are no messages in transit. At the second time point, the link is busy (actively transmitting information, and a message t is in transit. At the third point in time, the message is delivered. The link is empty again and ready to transmit information.

The domain and co-domain of a link are formally represented as follows:

DEFINITION 8 (Domain and co-domain). Let I be an information link. Two components or links, called the *domain* and *co-domain*, are related to I. This is denoted by two functions, *dom,cdom*: *Lnk*$\to$ *Comp*$\cup$ *Lnk*. A link I with $dom(I) = S_1$ and $cdom(I) = S_2$ is called a link from S_1 to S_2.

Depending on the relation between the two end points of a link, six kinds of interaction can be distinguished:

- Interaction between components that are subcomponents of the same component (called private interaction),
- Interaction from the input interface of a component to the input interface of one of its subcomponents (called import mediating interaction),
- Interaction from the output interface of a subcomponent to the output interface of its parent component (called export mediating interaction),
- Interaction from the input interface of a component to the output interface of the same component (called cross-mediating interaction),
- Interaction from a component to an information link between two other components that are all subcomponents of the same component (called link modifier interaction), and
- Interaction from an information link between two other components that are all subcomponents of the same component to a third component (called link monitoring interaction).

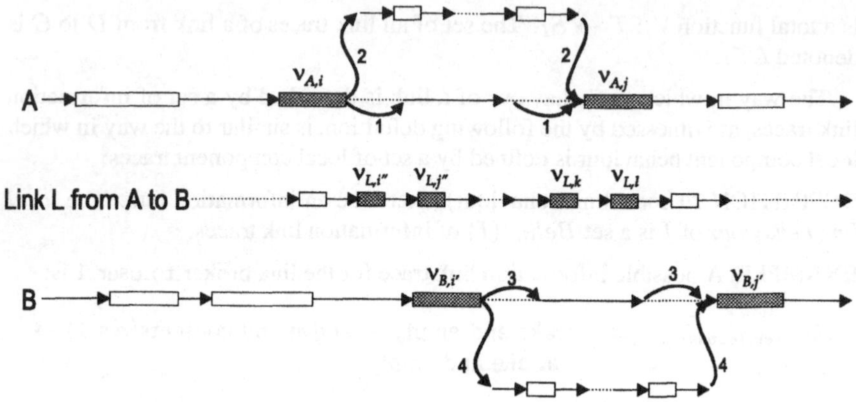

Figure 3. Interaction

The second aspect of the notion of interaction is the declaration of the information relation within interaction. This aspect is called the *information mapping*. The definition of information mappings refers to the six different kinds of interaction distinguished above.

DEFINITION 9 (Information mapping). Let I and I' be information links. An information mapping for I is a relation defined as follows:

- $\lambda_I \subseteq (\mathcal{S}_{dom(I),out} \times \mathcal{S}_{dom(I),out}) \times \mathcal{S}_I^4 \times (\mathcal{S}_{cdom(I),in} \times \mathcal{S}_{cdom(I),in})$, if I is a link for private interaction, or
- $\lambda_I \subseteq (\mathcal{S}_{dom(I),in} \times \mathcal{S}_{dom(I),in}) \times \mathcal{S}_I^4 \times (\mathcal{S}_{cdom(I),in} \times \mathcal{S}_{cdom(I),in})$, if I is a link for import mediating interaction, or
- $\lambda_I \subseteq (\mathcal{S}_{dom(I),out} \times \mathcal{S}_{dom(I),out}) \times \mathcal{S}_I^4 \times (\mathcal{S}_{cdom(I),out} \times \mathcal{S}_{cdom(I),out})$, if I is a link for export mediating interaction, or
- $\lambda_I \subseteq (\mathcal{S}_{dom(I),in} \times \mathcal{S}_{dom(I),in}) \times \mathcal{S}_I^4 \times (\mathcal{S}_{cdom(I),out} \times \mathcal{S}_{cdom(I),out})$, if I is a link for cross-mediating interaction, or
- $\lambda_I \subseteq (\mathcal{S}_{dom(I),out} \times \mathcal{S}_{dom(I),out}) \times \mathcal{S}_I^4 \times (\mathcal{S}_{cdom(I)} \times \mathcal{S}_{cdom(I)})$, if I is a link for link modifier interaction, or
- $\lambda_I \subseteq (\mathcal{S}_{dom(I)} \times \mathcal{S}_{dom(I)}) \times \mathcal{S}_I^4 \times (\mathcal{S}_{cdom(I),in} \times \mathcal{S}_{cdom(I),\in})$, if I is a link for link monitoring interaction.

The intended meaning of an information mapping is explained with reference to Figure 3. In Figure 3, interaction by a link L from a component A to B is depicted. (Thus, $dom(L) = A$ and $cdom(L) = B$.) Four states of the link, two of the domain and two of the co-domain are distinguished. These eight states correspond to an element $\langle\langle \nu_{A,i}; \nu_{A,j}\rangle; \langle \nu_{L,i''}; \nu_{L,k}; \nu_{L,l}\rangle; \langle \nu_{B,i'}; \nu_{B,j'}\rangle\rangle \in \lambda_L$. This element states that (the numbers between parentheses refer to the explanation below):

- *If* component A reaches state $\nu_{A,i}$ (1), *and* link L is in state $\nu_{L,i''}$ (2), *and* component B is in state $\nu_{B,i'}$ (3),
- *then* component B should reach state $\nu_{B,j'}$ (4) as one of the successors of $\nu_{B,i'}$ (5), and one of the following states of A should be $\nu_{A,j}$ (6),
- *and* state $\nu_{L,j''}$ should be the first state of L in which the information is in transit, *and* state $\nu_{L,k}$ should be the last state of L in which the information is in transit, *and* state $\nu_{L,l}$ should be the first state of L in which the information is no longer in transit (7).

Parts (1) and (4) form the most important part of this expression. Parts (1) and (4) show that in the semantic structure, information transmission is characterised in terms of states of the components that exchange information. Condition (2) enables the expression of a requirement on the state of the information link to enable transmission, e.g., the expression that the link must be in an enabled state $\nu_{L,i}$. Condition (3) enables expression of an enabling condition for receipt imposed by the destination component. Parts (5) and (6) reflect commitments to non-blocking receive and send, respectively. Part (7) enables expression of the dynamics of transmission as a process. The sequence of states can, for instance, be used to describe that a message is put in the link (change from $\nu_{L,i''}$ to $\nu_{L,j''}$) and later taken from it (change from $\nu_{L,k}$ to $\nu_{L,l}$). An information mapping does not, in general, need to be functional in all of its arguments, and is therefore formalised using a relation.

EXAMPLE. The information mapping of the link broker_to_user_1 is defined as follows (where *trans* is a function from OT_2 to OT_1 that translates ontology terms in OT_2 to OT_1, which is assumed to be given):

$$\begin{aligned}\lambda_{\text{broker\_to\_user\_1}} = \{&\langle\langle \text{to\_be\_communicated\_to}(t,\text{user\_1}); \text{just\_communicated\_to}(t,\text{user\_1})\rangle;\\ &\langle \text{awake\_and\_empty}; \text{active\_and\_contents}(t); \text{active\_and\_contents}(t);\\ &\text{awake\_and\_empty}\rangle; \langle \text{ready\_for\_information};\\ &\text{communicated\_by}(t',\text{broker})\rangle\rangle \mid t \in OT_2, t' \in OT_1 \text{ and } t' = \textit{trans}(t)\}.\end{aligned}$$

This information mapping specifies that if in broker's behaviour, there is a state to_be_communicated_to(t,user_1), and the state of link broker_to_user_1 is awake_and_empty, and in user_1's behaviour, there is a state ready_for_information, then a transition of user_1's input interface state to the state communicated_by (t',broker) exists, for resource description terms such that $t' = \textit{trans}(t)$. Moreover, in broker's behaviour, one of the successor states is the state just_communicated_to(t,user_1), and the state of link broker_to_user_1 changes to active_and_contents(t) and then back to awake_and_empty. (To keep the example simple, it is assumed that link broker_to_user_1 can only transmit a message if no other messages are in transit.)

The definition of an information mapping only refers to states in the state sets of the two components, not to traces of the components or to any temporal notion such as one information state being a successor of another state. As a consequence, an information mapping does not specify that interaction actually takes place. The actual exchange of information, which necessarily refers to the information states occurring in component traces, is modelled by compatibility relations in Section 3.2.

In the above example, different sets of resource description terms are used to describe the state sets of the components. The option to use different languages is an advantage of the locality principle. Information mappings can define relations between these languages. (In the example, an abstract translation function *trans* is used.)

Compositional Structure

Finally, the structural aspect of components and links can be defined. The structural aspect is very general: starting from a set of components (or, more precisely, component names) and a set of interactions (represented by the names of the information links that model these interactions), arbitrary hierarchical composition structures, called ***structure hierarchies***, can be described. A structure hierarchy is defined as follows:

DEFINITION 10 (Structure hierarchy). A structure hierarchy SH is a tuple $\langle Comp; Lnk; \circ\!\!\prec; dom; cdom\rangle$, where:

- *Comp* is a finite set of component names;

- *Lnk* is a finite set of information link names such that $Comp \cap Lnk = \varnothing$;

- $\circ\!\!\prec \subseteq (Comp \cup Lnk) \times Comp$ (the ***hierarchy relation***) such that $\circ\!\!\prec$ defines a forest: a finite, non-empty collection of trees. For all pairs $\langle I; C\rangle \in \circ\!\!\prec$ such that $I \in Lnk$, I must be a leaf. The reflexive closure of $\circ\!\!\prec$ is denoted $\underline{\circ\!\!\prec}$;

- $dom, cdom : Lnk \rightarrow Comp \cup Lnk$ are total functions (the domain and co-domain functions) such that for all $I \in Lnk$, if $dom(I) = S_1$ and $cdom(I) = S_2$, then either

  - $S_1, S_2 \in Comp$ and there is a $P \in Comp$ such that $S_1, S_2 \circ\!\!\prec P$ (private link), or

  - $S_1, S_2 \in Comp$ and $S_2 \circ\!\!\prec S_1$ (import mediating link), or

  - $S_1, S_2 \in Comp$ and $S_1 \circ\!\!\prec S_2$ (export mediating link), or

  - $S_1, S_2 \in Comp$ and there is a $P \in Comp$ such that $I, S_1, S_2 \circ\!\!\prec P$ and $S_1 = S_2$ (cross-mediating link), or

- $S_1 \in Comp$ and $S_2 \in Lnk$ and there are $S_3, S_4, P \in Comp$ such that $S_3 \neq S_1, S_4 \neq S_1$, $P \neq S_i$ and $S_i \circ\!\!\prec P$ for $i = 1, \ldots, 4$ and $dom(S_2) = S_3$ and $cdom(S_2) = S_4$ (link modifier link), or
- $S_1 \in Lnk$ and $S_2 \in Comp$ and there are $S_3, S_4, P \in Comp$ such that $S_3 \neq S_2, S_4 \neq S_2$, $P \neq S_i$ and $S_i \circ\!\!\prec P$ for $i = 1, \ldots, 4$ and $dom(S_1) = S_3$ and $cdom(S_1) = S_4$ (link monitoring link).

A structure hierarchy is called a structure hierarchy for a component $C \in Comp$ iff $\circ\!\!\prec$ defines a collection of exactly one tree (i.e., $\circ\!\!\prec$ is connected, formally $\forall C_1 C_2 \in Comp : C_1 \circ\!\!\prec C_2 \lor C_2 \circ\!\!\prec C_1$ and C is the root of the tree defined by $\circ\!\!\prec$, thus $\neg\exists C' \in Comp : C \circ\!\!\prec C'$.

In the subcomponent relation, $C_1 \circ\!\!\prec C_2$ denotes that C_1 is a subcomponent of C_2 in SH. A component C is called *primitive* in $SH = \langle Comp; Lnk; \circ\!\!\prec; dom; cdom \rangle$ iff there is no $C' \in Comp$ such that $C' \circ\!\!\prec C$. Otherwise, it is called *composed* in SH. The set of primitive components in a structure hierarchy SH is defined as follows: $Prim(SH) = \{C \mid \neg\exists C' \in Comp : C' \circ\!\!\prec C\}$. The leaves in a structure hierarchy (that is, the components C in the structure hierarchy for which there is no C' such that $C' \circ\!\!\prec C$) are by definition primitive in SH.

Different perspectives on a set of components and a set of links can be described with different structure hierarchies. This is possible because it is not assumed that only one structure hierarchy exists for a given set of components and a given set of links, nor is there any commitment with respect to whether components are composed or primitive. For instance, in a certain stage in the analysis or development of a multi-agent system, certain components are considered to be primitive: it is assumed that they do not have subcomponents. In another stage, these components may be made composed. The qualifiers 'composed' and 'primitive' are thus relative to a given structure hierarchy, and so is the notion of subcomponent.

In general, a structure hierarchy for a component C not only contains subcomponents of C, but also subcomponents of these subcomponents, and so on. Therefore, a structure hierarchy itself comprises more than the compositional structure (the third aspect of a component distinguished by the semantic structure) of a component C. A structure hierarchy consisting of C, its subcomponents and links is called the *composition structure* of C. Formally:

DEFINITION 11 (Composition structure).

- A *composition* structure for a component C is a structure hierarchy $CS = \langle Comp; Lnk; \circ\!\!\prec; dom; cdom \rangle$ for C such that for all $S \in Comp \cup Lnk$, $S \circ\!\!\prec C$.

- Let $SH = \langle Comp; Lnk; \circ\!\!\prec; dom; cdom \rangle$ be a structure hierarchy and let $C \in Comp$ be a component. The *composition structure* $CS(C, SH)$ of C with respect to SH is the structure hierarchy $\langle Comp'; Lnk'; \circ\!\!\prec'; dom'; cdom' \rangle$ where:

  - $Comp' = \{C' \in Comp \mid C' \underline{\circ\!\!\prec} C\}$;

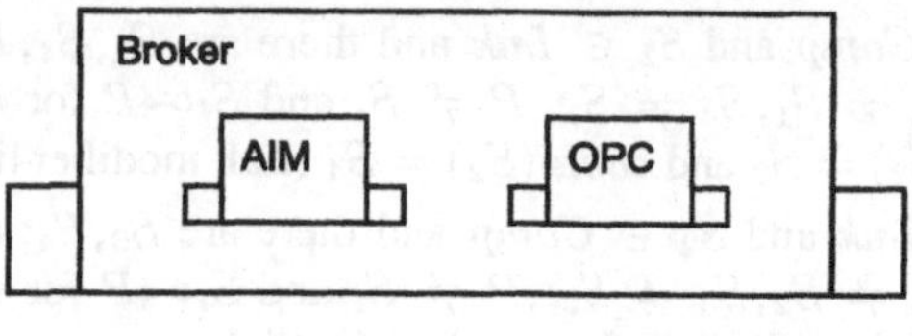

Figure 4. The broker agent

- $Lnk' = \{I \in Lnk \mid I \circ\!\!\prec C\}$;
- $S \circ\!\!\prec' C \Leftrightarrow S \circ\!\!\prec C, S \in Comp' \cup Lnk'$ and $C \in Comp'$;
- For all $I \in Lnk'$, $dom'(I) = dom(I)$;
- For all $I \in Lnk'$, $cdom'(I) = cdom(I)$.

If, for a structure hierarchy SH for C it holds that $SH = CS(C, SH)$, then SH itself is a composition structure for C. For a given component C and structure hierarchy SH, the composition structure $CS(C, SH) = \langle Comp'; Lnk'; \circ\!\!\prec'; dom'; cdom' \rangle$ is unique. The set of subcomponents of C with respect to $SH = \langle Comp; Lnk; \circ\!\!\prec; dom; cdom \rangle$ is denoted $Subc(C, SH) = \{C' \in Comp \mid C' \circ\!\!\prec C\}$. The set of links of C with respect to SH is denoted $Lnk(C, SH) = \{I \in Lnk \mid I \circ\!\!\prec C\}$. The set of subcomponents of C with respect to SH and links '*inside*' C is denoted $SLC(C, SH)$: $SLC(C, SH) = \{C\} \cup Subc(C, SH) \cup Lnk(C, SH)$. (The abbreviation 'SLC' stands for '*Subc*, *Lnk* and the Component itself', but also for 'slice'.) If a component C is a primitive component according to a structure hierarchy SH for C, then, according to the definition, $Subc(C, SH) = \varnothing$.

EXAMPLE. To illustrate structure hierarchies, assume the broker agent itself consists of the components depicted in Figure 4. Moreover, the components depicted in Figure 2 are considered to be subcomponents of a component representing the overall system, which is called toplevel. (Due to space restrictions, a simplified picture of the broker agent's architecture is used. Moreover, the function of these components is not discussed. See [Jonker and Treur, 1998] for an extensive treatment of these issues.) The following structure hierarchy can be used to analyse user_1 together with agent broker: $sh = \langle Comp; Lnk; \circ\!\!\prec; dom; cdom \rangle$, with:

- $Comp = \{$toplevel, user_1, broker, AIM, OPC$\}$;
- $Lnk = \{$user_1_to_broker, broker_to_user_1$\}$;
- $\circ\!\!\prec = \{\langle$AIM; broker$\rangle, \langle$OPC; broker$\rangle, \langle$user_1; toplevel$\rangle, \langle$broker; toplevel$\rangle\}$;
- $dom = \{\langle$user_1_to_broker; user_1$\rangle, \langle$broker_to_user_1; broker$\rangle\}$;
- $cdom = \{\langle$user_1_to_broker; broker$\rangle, \langle$broker_to_user_1; user_1$\rangle\}$.

3.2 *Behaviour of compositional agent systems*

As explained in Section 1, local component traces themselves do not describe the behaviour of a compositional agent system. Given the definition of structure hierarchies, the stage is now set to define multitraces, which are used to describe the behaviour of such compositions. In multitraces, dependencies between components imposed by interaction are modelled by temporal compatibility relations. The focus of this section is on how compatibility relations are employed to determine which combinations of local component traces of a component and its subcomponents constitute its actual behaviour. As a consequence, while compatibility relations are defined in this section, the discussion of various properties of compatibility relations is postponed until Section 4. A compatibility relation for an information link named I between a component D and a component C is defined as follows:

DEFINITION 12 (Compatibility relation). A compatibility relation *for a link I is a relation* $\mathcal{CR}_I \subseteq LT_{dom(I)} \times LT_I \times LT_{cdom(I)}$.

The behaviour of a component is defined in terms of multitraces. A multitrace is a collection of local component traces of a set of components, indexed by a structure hierarchy. (The view on behaviour developed in this section is relative to a given structure hierarchy and collection of compatibility relations.) An indexed set can also be seen as a function, which is the view taken in the formal definition below:

DEFINITION 13 (Multitrace). Let $SH = \langle Comp; Lnk; \circ\!\!\prec; dom; cdom \rangle$ be a structure hierarchy. A multitrace $(mt_S)_{S \in Comp \cup Lnk}$ for SH is a function $mt : Comp \cup Lnk \to \bigcup_{S \in Comp \cup Lnk} \mathcal{LT}_S$ such that for all $S \in Comp \cup Lnk$, $mt(S) \in \mathcal{LT}_S$. The set of all multitraces for SH is denoted MT_{SH}. Typically, an element of MT_{SH} is denoted μ. The element of a multitrace μ with index P is denoted μ_P.

The subcomponent relation $\circ\!\!\prec$ present in the index set of a multitrace induces a hierarchical structure on the collection of local component traces indexed by this index set in the multitrace. Only those traces that are compatible with each other are part of multitraces that model compositional behaviour. This is captured by the definition of a compatible multitrace as follows:

DEFINITION 14 (Compatible multitrace). Let $SH = \langle Comp; Lnk; \circ\!\!\prec; dom; cdom \rangle$ be a structure hierarchy and let $\gamma = (\gamma_i)_{i \in Lnk}$ be a collection of compatibility relations. A multitrace μ for SH is compatible for γ iff the following property holds:

$$\forall I \in Lnk : \langle \mu_{dom(I)}; \mu_I; \mu_{cdom(I)} \rangle \in \gamma_I.$$

If a component C is primitive according to a structure hierarchy SH for C, and there are no links from C to itself, (thus $SH = \langle \{C\}; \varnothing; \varnothing; \varnothing; \varnothing \rangle$, then every multitrace for SH is compatible.

Compatible multitraces are used to model the behaviour of compositional agent systems. In fact, three views on the behaviour of compositional systems are distinguished, called the black box, white box and glass box views. From a bird's-eye view, the structures that constitute these views consist of local component traces and link traces of a (possibly composed) component itself, possibly its subcomponents and its links, and possibly their subcomponents and links, and so on. All three views on behaviour developed in this section are relative for a given structure hierarchy and collection of compatibility relations. In a *black box view*, the behaviour of C is defined as a set of local component traces of C only. As a consequence, in the black box view the behaviour of subcomponents and links is not visible (although their behaviour is taken into account in the definition of the black box view to determine which local component traces of C constitute behaviour if information exchange is taken into account). In a *white box view*, the behaviour of a component not only consists of local component traces of this component, but also of local component traces of its subcomponents. In a *glass box view*, the behaviour of a component consists of local component traces of this component, of local component traces of its subcomponents and of local component traces of the subcomponents of these subcomponents, and so on.

The White Box View

The first of the three views presented is the white box view. The three views on the behaviour of a component are defined relative to a structure hierarchy and the behaviour of specific other components in the structure hierarchy. Thus, the three views can be seen as composition operators that define behaviour of a composition of components in terms of the behaviour of the constituents of the composition. Consequentially, if the behaviour of the components in the structure hierarchy to which each view is relative, is not correct, then the behaviour defined by each of the views is also incorrect. The behaviour of a component S of C to which a view on the behaviour of C is relative need not be $Beh_{loc}(S)$. However, it is required that the behaviour of such a set S is a subset of $Beh_{loc}(S)$.

The white box view on the behaviour of a component differs from the other two views with respect to the kind of structure hierarchy considered. On the one hand, the black box and glass box views are both relative to an arbitrary structure hierarchy. The definitions of these two views refer to compatible multitraces for the relevant structure hierarchy. Consequently, the black box and glass box views may take the behaviour of arbitrary components into account by choosing an appropriate structure hierarchy. On the other hand, the white box view is relative to a composition structure CS, which is not an arbitrary structure hierarchy. As a consequence, the definition of the white box view on the behaviour of a composed component C can only refer to the behaviour of the subcomponents of C. (As stated in Section 2.1, in this chapter the term 'subcomponent' always refers to *direct* subcomponent of a component C.)

In addition to a composition structure CS, the white box view is relative to a collection of compatibility relations γ and to the behaviour of the subcomponents and links (the elements of $SLC(C, SH)\backslash\{C\}$). The white box view on the behaviour of a component C consists of a set of structures (multitraces) each of which consists of local component traces of C, its subcomponents and links. Because of nondeterminism, which gives rise to different alternative behaviours, a component can have more than one multitrace. (Each multitrace contains one behaviour alternative of a specific component.) Therefore, the white box view consists of a set of multitraces.

DEFINITION 15 (Component behaviour, white box view). Let C be a component, let CS be a non-empty composition structure for C, let γ be a collection of compatibility relations and let $(\Gamma_S)_{S \in SLC(C,CS)}$ be a collection of sets of traces such that for all $S \in SLC(C, CS), \Gamma_S \in Beh_{loc}(S)$. The white box view on the behaviour of C, $Beh_{WB}(C, CS, \gamma, (\Gamma_S))$, with respect to CS, γ and (Γ_S) is the set of compatible multitraces $\mu \in MT_{CS}$ of C such that for each subcomponent or link S of C, the local trace of S in μ is an element of Γ_S. Formally:

$$\begin{aligned} Beh_{WB}(C, CS, \gamma, (\Gamma_S)) = \quad & \{\mu \mid \mu \in MT_{CS} \text{ is compatible for } \gamma \text{ and} \\ & \forall S \in SLC(C, CS) : \mu_s \in \Gamma_S\}. \end{aligned}$$

This definition of component behaviour provides a white box view on the behaviour of a component in the sense that each multitrace in $Beh_{WB}(C, CS, \gamma, (\Gamma_S))$ not only contains a local component trace of C, but also local component and link traces of the subcomponents and links of C. However, these local component and link traces themselves do not contain information of their subcomponents and so on, recursively. Nevertheless, as is the case with the black box view on the behaviour of C, the requirement on the local component traces that constitute $Beh_{WB}(C, CS, \gamma, (\Gamma_S))$ ensures that the combinations of local component traces and link traces that constitute the elements of $Beh_{WB}(C, CS, \gamma, (\Gamma_S))$ take constraints imposed by information transmission into account.

If a component C is primitive according to a composition structure CS for C, (thus $CS = \langle\{C\}; \varnothing; \varnothing; \varnothing; \varnothing\rangle$ and $Subc(C, CS) = Lnk(C, CS) = \varnothing$), then for every collection of compatibility relations γ, it holds that $Beh_{WB}(C, CS, \gamma, (\Gamma_S)) = \{\mu \in MT_{CS} \mid \mu_C \in Beh_{loc}(C)\}$ (with $(\Gamma_S) = \varnothing$).

The Black Box View

The second of the three views presented is the black box view. Similar to the other two views, the black box view is defined relative to a structure hierarchy, a collection of compatibility relations and the behaviour of specific other components in the structure hierarchy. In contrast to the white box view, but similar to the glass box view, the black box view is relative to an arbitrary structure hierarchy. (The white box view is relative to a specific type of structure hierarchy: a composition structure.)

In addition to an arbitrary structure hierarchy SH, the black box view on the behaviour of a component C is defined relative to a collection of compatibility relations γ and the behaviour of its subcomponents and links (the elements of $SLC(C, SH)\backslash\{C\}$).

DEFINITION 16 (Component behaviour, black box view). Let C be a component, let SH be a non-empty structure hierarchy for C, let γ be a collection of compatibility relations and let $(\Gamma_S)_{S \in SLC(C,SH)}$ be a collection of sets of traces such that for all $S \in SLC(C, SH)$, $\Gamma_S \in Beh_{loc}(S)$. The black box view $Beh_{BB}(C, SH, \gamma, (\Gamma_S))$ for C with respect to SH, γ and (Γ_S) on the behaviour of C is the subset of $Beh_{loc}(C)$ such that each local component trace in this subset is part of a compatible multitrace for SH that is based on the given traces of the subcomponents and links of C. Formally:

$$\begin{aligned} Beh_{BB}(C, SH, \gamma, (\Gamma_S)) = \; & \{\mu_C \mid \mu \in MT_{SH} \text{ is compatible for } \gamma, \\ & \forall S \in SLC(C, SH) : \mu_S \in \Gamma_S \text{ and} \\ & \forall I \in Lnk \text{ such that } dom(I) = cdom(I) = \\ & \qquad C : \mu_I \in Beh_{loc}(I)\}. \end{aligned}$$

The definition given above provides a black box view in the sense that the behaviour of a component C is a set consisting of local component traces of C itself only, and not of its subcomponents and links. As the definition of local component traces given above indicates, only local information is recorded in a local component trace of a component. In particular, information of subcomponents is *not* recorded in the local component trace of the encompassing component and is thus not visible in the black box view on component behaviour as defined above.

As stated in the beginning of Section 3.2, the behaviour of a composed component is, in general, not a set of *arbitrary* combinations of local component traces of its subcomponents and link traces, because only combinations that take constraints imposed by information transmission into account can be considered to represent behaviour of the component. The requirement on the local component traces that constitute $Beh_{BB}(C, SH, \gamma, (\Gamma_S))$ ensures that the combinations of local component traces from which the elements of $Beh_{BB}(C, SH, \gamma, (\Gamma_S))$ are taken, take constraints imposed by information transmission into account. This is ensured because $Beh_{BB}(C, SH, \gamma, (\Gamma_S))$ depends on the existence of a compatible multitrace, and the elements of this compatible multitrace are themselves part of the behaviour of the subcomponents, to which the definition of the black box view is relative.

To determine the black box view on the behaviour of a component C, only the behaviour of the subcomponents and links of C needs to be given. This also holds if, according to the related structure hierarchy, these subcomponents are composed. It suffices to only take the behaviour of the subcomponents into account, (and not of their subcomponents), because it is assumed that the behaviour given for these subcomponents takes constraints imposed by their subcomponents into account. However, if this is not the case, the behaviour as defined by the black box view is incorrect.

If a component C is primitive, according to a structure hierarchy SH for C, and there are no links from C to itself, (thus $SH = \langle\{C\};\varnothing;\varnothing;\varnothing;\varnothing\rangle$, and $Subc(C,SH) = Lnk(C,SH) = \varnothing$), then for every collection of compatibility relations γ, it holds that $Beh_{BB}(C,SH,\gamma,(\Gamma_S)) = Beh_{loc}(C)$ (with $(\Gamma_S) = \varnothing$). However, $Beh_{BB}(C,SH,\gamma,(\Gamma_S))$ is *not* equal to $Beh_{WB}(C,SH,\gamma,(\Gamma_S))$ because $Beh_{WB}(C,SH,\gamma,(\Gamma_S))$ is a set of multitraces, while $Beh_{BB}(C,SH,\gamma,(\Gamma_S))$ is a set of local component traces.

The Glass Box View

As is the case for the black box view, the glass box view is relative to a structure hierarchy SH, a collection of compatibility relations γ and the behaviour of specific subcomponents (in this case, the primitive subcomponents). The glass box view on the behaviour of a component C is defined by imposing three requirements on the set of multitraces μ for a structure hierarchy $SH = \langle Comp; Lnk; \multimap\!\!<; dom; cdom\rangle$ for C: (i) they must be compatible, (ii) for all components C' in *Comp*that are primitive in SH, the local component trace $\mu_{C'}$ must be an element of the given sets of traces and (iii) for all components and links I in *Lnk*, the link trace μ_I must be an element of $Beh_{loc}(I)$. Formally:

DEFINITION 17 (Component behaviour, glass box view). Let C be a component, let $SH = \langle Comp; Lnk; \multimap\!\!<; dom; cdom\rangle$ be a non-empty structure hierarchy for C, let γ be a collection of compatibility relations and let $(\Gamma_S)_{S\in Prim(SH)}$ be a collection of sets of traces for the primitive components in SH. The *glass box view* $Beh_{GB}(C,SH,\gamma,(\Gamma_S))$ for C with respect to SH,γ and (Γ_S) on the behaviour of C is the subset of the set MT_{SH} of multitraces for SH such that for all $\mu \in Beh_{GB}(C,SH,\gamma,(\Gamma_S))$ it holds that:

- μ is compatible for γ;
- $\forall C' \in Prim(SH)\backslash\{C\} : \mu_{C'} \in \Gamma_{C'}$ and
- $\forall S \in Comp \cup Lnk : \mu_S \in Beh_{loc}(S)$.

EXAMPLE. To illustrate the previous definition, an element of Beh_{GB}(toplevel, $sh,\gamma,(\Gamma_S)$), where sh is the structure hierarchy presented in the previous example, is presented (for some γ which is left implicit). Such an element is a multitrace, which is represented here by a function mt. Due to space restrictions, it is impossible to present local component traces for all components in sh. Instead, the multitrace's hierarchical structure induced by sh is illustrated in Figure 5, with a pictorial representation of the elements of mt.

In Figure 5, the local component trace associated with broker is the local component trace lt presented in Section 3.1. Formally: $mt(broker) = lt$.

The glass box view is the most complete view on the behaviour of a component: the other two views can be expressed in terms of the glass box view. In the rest

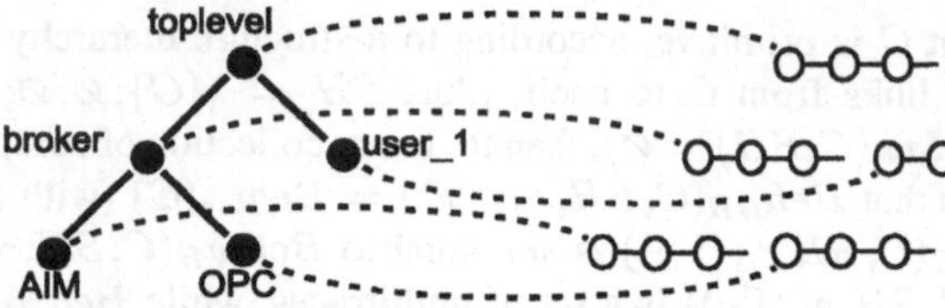

Figure 5. An element of Beh_{GB}(toplevel, $sh, \gamma, (\Gamma_S)$)

of this section, three propositions are presented that relate the three views on the behaviour of a component. These propositions show how the white box view and the black box view can be expressed in terms of the glass box view by *restricting* the multitraces (in the usual sense of restricting the domain of a function) in the glass box view to a subset of the components and links in a structure hierarchy of C. A restriction of a multitrace to a set S of components and links is denoted $\mu_{|S}$ and is called a *restricted multitrace*. The set of all multitraces restricted by a set S is denoted $MT_{|S}$.

The first proposition shows how the white box view and the glass box view are related. On the one hand, the proposition shows how the white box view can be expressed in terms of the glass box view by restricting multitraces. On the other hand, the proposition shows how the glass box view on the behaviour of a composed component can be constructed from the white box views on the behaviour of this component, its subcomponents and their subcomponents, and so on. The proposition assumes that a structure hierarchy SH for a component C is given and states that if a multitrace for this structure hierarchy satisfies a specific requirement for the restriction of this multitrace to the subcomponents of C, and their subcomponents, and so on, then this multitrace is an element of the glass box view on the behaviour of C. The definition of the proposition involves the following technical issues:

- Figure 6 shows a structure hierarchy for a composed component C consisting of six composed components (grey ovals), thirteen primitive components (white ovals) and four links (small white boxes). For each composed component C', the composition structure $CS(C', SH)$ is enclosed in a solid line. Within component S_1, as an example also $SLC(S_1, SH)\setminus\{S_1\}$ is depicted by a dashed line. The '=' between $CS(S_1, SH)$ and $SLC(S_1, CS(S_1, SH))$ is put between quotes because $CS(S_1, SH)$ is a structure hierarchy and $SLC(S_1, CS(S_1, SH))$ is a set. The proposition below expresses the white box view in terms of the glass box view on the behaviour of C' by restricting multitraces for SH to $SLC(C', CS(C', SH))$. To construct the glass box view from the white box views on the behaviour of each C', the proposition assumes that for each component C' in SH (composed and primitive), the restriction of a multitrace for SH to $SLC(C', CS(C', SH))$ is an element of the white box view on the behaviour of C'.

- The glass box view constructed from the white box views is relative to a collection of sets of local component and link traces of the primitive components in SH, as indicated by the definition of the glass box view on the behaviour of a composed component. Such a collection of sets $(\Delta_S)_{S \in Prim(SH)}$ is assumed to be given.

- Each of the white box views from which the glass box view is constructed, is itself relative to a collection of sets of local component and link traces, as indicated by the definition of the white box view on the behaviour of a component. These collections are taken from multitraces for SH as follows. Let μ be a multitrace for SH. For each composed component C' in SH, a collection of sets of local component and link traces $(\Gamma_S)_{S \in SLC(C', CS(C', SH)) \backslash \{C'\}}$ is defined as follows: $\Gamma_S = \Delta_S$ if S is a primitive component in SH, or $\Gamma_S = \{\mu_S\}$ otherwise. The collections of sets of traces for each composed component in SH are themselves grouped as a collection of collections $\Gamma\Gamma$ indexed by the composed components in SH. (Figure 6 might help to obtain an overview of the index sets involved.)

PROPOSITION 18. *Let $SH = \langle Comp; Lnk; \circ\!\!\prec; dom; cdom \rangle$ be a structure hierarchy for a composed component C. Let $\mu \in MT_{SH}$ be a multitrace for SH such that for all $I \in Lnk$, $\mu_S \in Beh_{loc}(I)$. Let γ be a collection of compatibility relations and let $(\Delta_S)_{S \in Prim(SH)}$ be a collection of sets of local component traces for the primitive components in SH. Define*

$$\Gamma\Gamma = ((\Gamma_S)_{S \in SLC(C', CS(C', SH)) \backslash \{C'\}})_{C' \in Comp \backslash Prim(SH)}$$

such that for all $C' \in Comp \backslash Prim(SH)$ and for all $S \in SLC(C', CS(C', SH)) \backslash \{C'\}$,

$$(\Gamma\Gamma_{C'})_S = \begin{cases} \Delta_s & \text{if } S \in Prim(SH), \\ \{\mu_S\} & \text{otherwise.} \end{cases}$$

Then the following equivalence holds:

$$\begin{array}{ll} & \text{for all } \; C' \in Comp : \mu_{|SLC(C', CS(C', SH))} \in Beh_{WB} \\ & \qquad (C', CS(C', SH), \gamma, \Gamma\Gamma_{C'}) \\ \Leftrightarrow & \quad \mu \in Beh_{GB}(C, SH, \gamma, (\Delta_S)_{S \in Prim(SH)}). \end{array}$$

This proposition is important for compositional verification of compositional systems [Engelfriet *et al.*, 1999], because this proposition shows that the semantic structure supports proving global properties of a system from local properties. This topic is not further discussed in this chapter.

The next proposition shows how the black box view can be expressed in terms of the glass box view. As stated before, all three views are, among others, relative to sets of traces of specific subcomponents. In the context of the propositions presented below, these sets of traces to which the black box view is relative, are

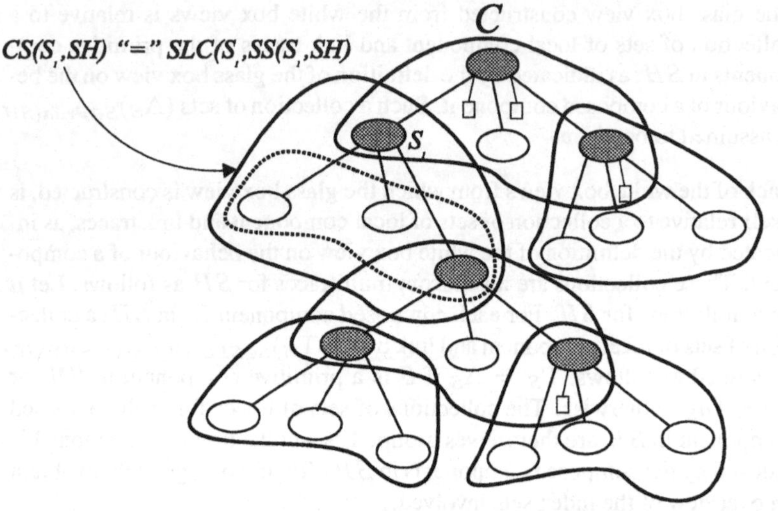

Figure 6. Composing the glass box view

defined in terms of the glass box view from which the black box view is generated. This is done by defining a collection of sets $(\Delta_S)_{S \in SLC(C,SH)\backslash\{C\}}$ such that for each $S \in SLC(C,SH)\backslash\{C\}, \Delta_S = \{\mu_S \mid \mu \in Beh_{GB}(C,SH,\gamma,(\Gamma_S))\}$. In the proposition below, the black box view is taken relative to (Δ_S), which (only) consists of sets of traces for the subcomponents of C in SH taken from the glass box view. The glass box view itself is defined relative to a collection of sets $(\Gamma_S)_{S \in Prim(SH)\backslash\{C\}}$ of traces of *the primitive components* in SH. In general, the primitive components in a structure hierarchy SH for C and the subcomponents of C are distinct.

PROPOSITION 19. *Let C be a composed component, let SH be a non-empty structure hierarchy for C, let γ be a collection of compatibility relations and let $(\Gamma_S)_{S \in Prim(SH)\backslash\{C\}}$ be a collection of sets of traces for the primitive components in SH. Define $(\Delta_S)_{S \in SLC(C,SH)\backslash\{C\}}$ such that for all $S \in SLC(C,SH)\backslash\{C\}, \Delta_S = \{\mu_S \mid \mu \in Beh_{GB}(C,SH,\gamma,(\Gamma_S))\}$. Then:*

$$Beh_{BB}(C,SH,\gamma,(\Delta_S)) = \{\mu_C \in Beh_{loc}(C) \mid \mu \in Beh_{GB}(C,SH,\gamma,(\Gamma_S))\}.$$

The last proposition presented in this section shows how the three views relate to each other in the case of primitive components:

PROPOSITION 20. *Let C be a primitive component, let $SH = \langle\{C\}; Lnk; \varnothing; dom; cdom\rangle$ be a structure hierarchy for C, let γ be a collection of compatibility relations and let μ be a multitrace for SH such that for all $I \in Lnk$, $\mu_I \in$*

$Beh_{loc}(I)$. Then:

$$\mu \in Beh_{GB}(C, SH, \gamma, \varnothing) \Leftrightarrow \mu \in Beh_{WB}(C, CS(C, SH), \gamma, \varnothing) \Leftrightarrow \mu_C \in Beh_{BB}(C, SH, \gamma, \varnothing).$$

4 COMPATIBILITY RELATIONS

Section 3.2 defines a compatibility relation for a link between a component D and a component C as a relation on the set of local component traces of D, the link itself, and C. As stated in Section 3.2, compatibility relations are used to model interaction. However, Section 3.2 does not define how compatibility relations model different properties of interaction, such as whether interaction is synchronous or asynchronous. This section defines properties of compatibility relations that reflect such interaction properties. Applications of the semantic structure may or may not choose to adopt these properties.

Compatibility relations are closely related to information mappings. An information mapping for a link I between a component D and a component C is defined on the sets of states S_D and S_C of D and C and thus does not necessarily refer to states that actually occur in D or C. A compatibility relation for I, however, is defined on the set of local component traces of D and C, and thus refers to states that are part of the actual behaviour of D and C. A compatibility relation should relate traces from the sets of local component traces that respect interaction as specified by the information mapping for I and according to the interaction properties adopted. Several properties for compatibility relations are made formal in the following definitions. The formalisation refers to the notion of *transmission octets*, which is defined as follows: (where $in(\nu)$ maps a state $\nu = \langle \nu_{in}; \nu_{int}; \nu_{out} \rangle$ to ν_{in}, and analogously for $int(\nu)$ and $out(\nu)$. Moreover, a state $V(i)$ of a trace $LT = \langle TF; V \rangle$ of a component C is denoted $\nu_{C,i}$)

DEFINITION 21 (Transmission octet). Let LT_A, LT_L, and LT_B be three local traces of components A and B and a link L with information mapping λ_L such that $A = dom(L)$ and $B = cdom(L)$. Let $\nu_{A,i}$ and $\nu_{A,j}$ be two states in LT_A, let $\nu_{L,i''}, \nu_{L,j''}, \nu_{L,k}$ and $\nu_{L,l}$ be four states in LT_L and let $\nu_{B,i'}$ and $\nu_{B,j'}$ be two states in LT_B. The octet $\langle\langle \nu_{A,i}; \nu_{A,j} \rangle; \langle \nu_{L,i''}; \nu_{L,j''}; \nu_{L,k}; \nu_{L,l} \rangle; \langle \nu_{B,i'}; \nu_{B,j'} \rangle\rangle$ is a *transmission octet* with respect to LT_A, LT_L and LT_B iff either:

- $\langle\langle out(\nu_{A,i}); out(\nu_{A,j}) \rangle; \langle \nu_{L,i''}; \nu_{L,j''}; \nu_{L,k}; \nu_{L,l} \rangle; \langle in(\nu_{B,i'}); in(\nu_{B,j'}) \rangle\rangle \in \lambda_L$, for the case that L is a private link, and
- $\langle\langle in(\nu_{A,i}); in(\nu_{A,j}) \rangle; \langle \nu_{L,i''}; \nu_{L,j''}; \nu_{L,k}; \nu_{L,l} \rangle; \langle in(\nu_{B,i'}); in(\nu_{B,j'}) \rangle\rangle \in \lambda_L$, for the case that L is an import mediating link, and
- $\langle\langle out(\nu_{A,i}); out(\nu_{A,j}) \rangle; \langle \nu_{L,i''}; \nu_{L,j''}; \nu_{L,k}; \nu_{L,l} \rangle; \langle out(\nu_{B,i'}); out(\nu_{B,j'}) \rangle\rangle \in \lambda_L$, for the case that L is an export mediating link, and

- $\langle\langle in(\nu_{A,i}); in(\nu_{A,j})\rangle; \langle \nu_{L,i''}; \nu_{L,j''}; \nu_{L,k}; \nu_{L,l}\rangle; \langle out(\nu_{B,i'}); out(\nu_{B,j'})\rangle\rangle \in \lambda_L$, for the case that L is a cross-mediating link, and
- $\langle\langle out(\nu_{A,i}); out(\nu_{A,j})\rangle; \langle \nu_{L,i''}; \nu_{L,j''}; \nu_{L,k}; \nu_{L,l}\rangle; \langle \nu_{B,i'}; \nu_{B,j'}\rangle\rangle \in \lambda_L$, for the case that L is a link modifier link, and
- $\langle\langle \nu_{A,i}; \nu_{A,j}\rangle; \langle \nu_{L,i''}; \nu_{L,j''}; \nu_{L,k}; \nu_{L,l}\rangle; \langle in(\nu_{B,i'}); in(\nu_{B,j'})\rangle\rangle \in \lambda_L$, for the case that L is a link monitoring link.

Properties of compatibility relations can be defined using the notion of a transmission octet. Two properties are presented below. The asynchronous and synchronous interaction properties are based on the notion of dependence between states, which is itself defined in terms of transmission octets. Dependence is a binary relation on the union of the sets of states of the domain and co-domain of a link. A state $\nu_{B,j'}$ depends on a state $\nu_{A,i}$ if either $A = B$ and $i < j$ according to A's local time ordering, or if there is a state $\nu_{C,m}$ such that $\nu_{B,j'}$ depends on $\nu_{C,m}$ and $\nu_{C,m}$ depends on $\nu_{A,i}$, or if $A = dom(L)$ and $B = cdom(L)$ for a link L and $\nu_{B,j'}$ contains the result of interaction with $\nu_{A,i}$. The latter requirement can be expressed formally in terms of a transmission octet, as is shown in the following definition. To keep the presentation concise, it is assumed that the state sets of all components are disjoint. Moreover, it is assumed that in a specific trace, all states are different. (Both assumptions can be fulfilled by colouring states of a component C with the name of their component and the point in time at which they occur.) The union of the state sets of all components and links in a structure hierarchy is denoted $\mathtt{S}_{SH}$.

DEFINITION 22 (Dependence). Let LT_A, LT_L and LT_B be three local traces of components A and B and a link L such that $A = dom(L)$ and $B = cdom(L)$. Let $\nu_{A,i}$ and $\nu_{A,j}$ be two states in LT_A, let $\nu_{L,i''}, \nu_{L,j''}, \nu_{L,k}$ and $\nu_{L,l}$ be four states in LT_L and let $\nu_{B,i'}$ and $\nu_{B,j'}$ be two states in LT_B. The dependence relation $\rightarrow_s \subseteq \mathtt{S}_{SH} \times \mathtt{S}_{SH}$ for LT_A, LT_L and LT_B is defined as follows: $\nu_{A,i} \rightarrow_s \nu_{B,j'}$ iff either

1. $A = B$ and $i < j$, or
2. $\langle\langle \nu_{A,i}; \nu_{A,j}\rangle; \langle \nu_{L,i''}; \nu_{L,j''}; \nu_{L,k}; \nu_{L,l}\rangle; \langle \nu_{B,i'}; \nu_{B,j'}\rangle\rangle$ is a transmission octet for
   LT_A, LT_L, and LT_B and $i < j$, and $i' < j'$, or
3. There is a state $\nu_{C,m}$ of a component or link $C \in Comp \cup Lnk$ such that $\nu_{A,i} \rightarrow_s \nu_{C,m}$ and $\nu_{C,m} \rightarrow_s \nu_{B,j}$.

DEFINITION 23 (Asynchronous interaction property). Let $\mathtt{CR}_I$ be a compatibility relation for a link I. For this compatibility relation the asynchronous interaction property holds iff for each $\langle LT_{dom(I)}; LT_I; LT_{cdom(I)}\rangle \in \mathtt{CR}_I$: the dependence relation $\rightarrow_s$ for $LT_{dom(I)}, LT_I$, and $LT_{cdom(I)}$ is a partial order.

The dependence relation $\rightarrow_s$ is similar to Lamport's [1986] "happens before" relation that defines a temporal order without assuming the existence of global time. (However, Lamport's notion is defined in terms of events and does not support locality.) If $\rightarrow_s$ is a partial order, it cannot contain cycles, so there is no chain of states that all have to occur before their predecessor. This characterisation of asynchronous interaction is also presented in [Charron-Bost *et al.*, 1996] in terms of events. A more precise evaluation of this property cannot be presented here.

The formalisation of the synchronous interaction property is based on a construction that is analogous to the notion of dependence.

DEFINITION 24 (Synchronous interaction property). Let LT_A, LT_L and LT_B be three local component or link traces of components A and B and a link L such that $A = dom(L)$ and $B = cdom(L)$. Let $\nu_{A,i}$ and $\nu_{A,j}$ be two states in LT_A, let $\nu_{L,i''}, \nu_{L,j''}, \nu_{L,k}$ and $\nu_{L,l}$ be four states in LT_L and let $\nu_{B,i'}$ and $\nu_{B,j'}$ be two states in LT_B.

- The relation $K \subseteq \mathrm{S}_{SH} \times \mathrm{S}_{SH}$ for LT_A, LT_L, and LT_B is defined as follows: $K = \{\langle \nu_{A,i}; \nu_{B,j'}\rangle \; \langle\langle \nu_{A,i}; \nu_{A,j}\rangle; \langle \nu_{L,i''}; \nu_{L,j''}; \nu_{L,k}; \nu_{L,l}\rangle; \langle \nu_{B,i'}; \nu_{B,j'}\rangle\rangle$ is a transmission octet for LT_A, LT_L, and LT_B and $i < j, i' < j'\}$;
- The relation R is defined as $R = \{\langle \nu_{S,i}; \nu_{S,j}\rangle \mid (S = A$ or $S = B)$ and $i < j\} \cup K$;
- Let K^{-1} be the set $\{\langle \nu_{B,j'}; \nu_{A,i}\rangle \mid \langle\langle \nu_{A,i}; \nu_{A,j}\rangle; \langle \nu_{L,i''}; \nu_{L,j''}; \nu_{L,k}; \nu_{L,l}\rangle; \langle \nu_{B,i'}; \nu_{B,j'}\rangle\rangle$ is a transmission octet for LT_A, LT_L, and LT_B and $i < j, i' < j'\}$.
- Let $R' = (R \cup K^{-1})^* \backslash (K \cup K^{-1})$, where R^* denotes the transitive closure of R;
- Let CR_I be a compatibility relation for a link I. For this compatibility relation the synchronous interaction property holds iff the following condition holds for each $\langle LT_{dom(I)}; LT_I; LT_{cdom(I)}\rangle \in \mathrm{CR}_I$: the relation R' for $LT_{dom(I)}, LT_I$, and $LT_{cdom(I)}$ is a partial order.

The construction $R' = (R \cup K^{-1})^* \backslash (K \cup K^{-1})$ is adapted from a proof in [Charron-Bost *et al.*, 1996] in which an equivalent characterisation of synchronous interaction in terms of events is given. From the synchronous interaction property, it can be proven that an arbitrary state ν of D or C happens before (in Lamport's sense) a state in which information is made available in D if and only if it happens before a state in which this information is received by C. Likewise, an arbitrary state ν of D or C happens after (in Lamport's sense) a state in which information is made available in D if and only if it happens after a state in which this information is received by C. Thus, at the moment of synchronous interaction, components C and D have the same past and future. The definitions of these properties demonstrate how compatibility relations relate information mappings and local component traces, how properties of interaction can be captured and how the (global, event-based) notions of [Lamport, 1986; Charron-Bost *et al.*, 1996] are applicable within a framework that supports locality.

5 COMPARISON WITH OTHER APPROACHES

The formal semantic structure for dynamics of compositional agent systems presented in this chapter can serve as a formal foundation for compositional modelling frameworks such as DESIRE: their semantics can be expressed as a mapping from such a language or framework to the concepts that constitute the semantic structure presented in this chapter. A semantics for an earlier version of DESIRE in which, in contrast to this chapter, global time and sequential processing are assumed, is presented in [Gavrila and Treur, 1994; Brazier *et al.*, 1999].

A related area of research is that of co-ordination languages; e.g., [Garlan and le Métayer, 1997]. Many approaches to developing semantics for coordination languages are based on process calculi such as CCS [Milner, 1989] or the π-calculus [Milner *et al.*, 1992]. One of the differences between such approaches and the approach proposed in this chapter is that process calculi are event-based (events, or labelled transitions, are first-class citizens, while state is implicit), while the approach presented in this chapter is state-based, to capture the fact that one of the characteristics of agents is that agents have a persistent internal state; e.g., [Bradshaw, 1997].

Another state-based approach is provided by the Concurrent MetateM [Fisher and Wooldridge, 1997]. Concurrent MetateM's semantic view of dynamic behaviour is based on global time (discrete in [Fisher and Wooldridge, 1997], but a version of the semantics for dense time is also available) and a single, global language, contrary to the approach presented in this chapter.

The concepts locality and compatibility are also used by [Giunchiglia and Ghidini, 1997] to model contextual deduction. Their model, however, does not cover dynamics of behaviour.

6 DISCUSSION

The semantic approach introduced in this chapter provides a transparent basis for the formalisation of emergence of global dynamics of a multi-agent system from interaction between different forms of local dynamics of the agents. Similarly, the approach describes how the behaviour of an agent is obtained from the dynamics of components within the agent. The semantic structure defines local behaviour without any reference to global elements, not even to global time. Given the independently defined local dynamics, the model transparently defines interaction between different local dynamics, and the emerging global dynamics.

If no global clock is assumed (which is the only possibility in relativistic systems and which is often beneficial when modelling widely distributed systems, see [Pratt, 1986]), then no global temporal order can be defined whatsoever in the usual sense: a view in terms of dependence is the best possibility. The definition of temporal compatibility introduced above amounts to a theory of dependence. In particular, the definition of composed component behaviour abstracts from (i) the actual time spent by components between two states according to some observer and (ii) synchronisation characteristics and buffering of communication between components. A non-local view in terms of dependence is, on the one hand, sufficient for many purposes, while, on the other hand, it is the best possibility if one does not want to assume a global clock. A view in terms of dependence is sufficient for the following reasons. If a global clock is assumed, then a dependence ordering implies a global temporal ordering, in the real physical reality as well as in any conceivable computer implementation [Lamport, 1986]. However, a real temporal ordering is only required if (local) state transitions can have side effects not explicitly modelled as interaction with other components in the system. Thus, a view in terms of dependence is sufficient, if such side effects are modelled explicitly (which is the case for the semantic structure presented in this chapter).

Due to space restrictions, a number of extensions of the semantic structure that have been explored, have been omitted in this chapter, including; e.g., characterisations of different properties of interaction in terms of compatibility relations, and an extended notion of information links. Future research will among others focus on investigating the more precise relationship with, e.g., Lamport's theory of synchronisation [Lamport, 1986].

ACKNOWLEDGEMENTS

The authors are indebted to Joeri Engelfriet for comments on a preliminary version of this chapter.

Frances Brazier, Pascal van Eck and Jan Treur
Vrije Universiteit Amsterdam, The Netherlands.

BIBLIOGRAPHY

[Bradshaw, 1997] J. Bradshaw. An Introduction to Software Agents. In *Software Agents*, J. Bradshaw, ed. pp. 3–46. MIT Press, 1997.

[Bratman, 1987] M. E. Bratman. *Intentions, Plans and Practical Reason*. Harvard University Press, 1987.

[Brazier *et al.*, 2001] F. M. T. Brazier, C. M. Jonker and J. Treur. Compositional Multi-agent System Design: Dynamics and Control. This *volume*, 2001.

[Brazier *et al.*, 1999] F. M. T. Brazier, J. Treur, N. J. E. Wijngaards and M. Willems. Temporal semantics of task models and problem solving methods. *Data and Knowledge Engineering*, **29**, 17–42, 1999.

[Charron-Bost *et al.*, 1996] B. Charron-Bost, F. Mattern and G. Tel. Synchronous, asynchronous, and causally ordered communication. *Distributed Computing*, **9**, 173–191, 1996.

[Engelfriet *et al.*, 1999] J. Engelfriet, C. M. Jonker and J. Treur. Compositional verification of multi-agent systems in temporal multi-epistemic logic. In *Intelligent Agents V, Proc. of the Fifth International Workshop on Agent Theories, Architectures and Languages, ATAL'98*, J. P. Mueller, M. P. Singh and A. S. Rao, eds. pp. 177–194. Lecture Notes in AI, vol. 1555, Springer Verlag, 1999. Also this *volume*.

[Finin *et al.*, 1997] T. Finin, Y. Labrou and J. Mayfield. KQML as an agent communication language. In *Software Agents*, J. Bradshaw, ed. MIT Press, 1997.

[Fisher and Wooldridge, 1997] M. Fisher and M. Wooldridge. On the formal specification and verification of multi-agent systems. *International Journal of Cooperative Information Systems*, **6**, 37–65, 1997.

[Garlan and le Métayer, 1997] D. Garlan and D. le Métayer, eds. *Coordination Languages and Models. Proceedings of the Second International Conference, COORDINATION '97*. Lecture Notes in Computer Science, volume 1282, Springer-Verlag, 1997.

[Gavrila and Treur, 1994] I. S. Gavrila and J. Treur. A formal model for the dynamics of compositional reasoning systems. In *Proc. 11th European Conference on Artificial Intelligence, ECAI'94*, A. G. Cohn, ed. pp. 307–311. Wiley and Sons, 1994.

[Giunchiglia and Ghidini, 1997] F. Giunchiglia and C. Ghidini. Local models semantics, or contextual reasoning = locality + compatibility. In *AAAI Fall 1997 Symposium on Context in Knowledge Representation and Natural Language*. MIT, Cambridge, MA.

[Jonker and Treur, 1998] C. M. Jonker and J. Treur. A generic architecture for broker agents. In *Proceedings of the Third International Conference on the Practical Application of Intelligent Agents and Multi-Agent Technology, PAAM'98*, H. S. Nwana and D. T. Ndumu, eds. pp. 623–624. The Practical Application Company Ltd, 1998.

[Lamport, 1986] L. Lamport On interprocess communication, part I–basic formalism. *Distributed Computing*, **1**, 77–85, 1986.

[Lassila, 1998] O. Lassila. Web metadata: A matter of semantics. *IEEE Internet Computing*, **2**, 1998. See also http://www.w3.org/RDF/.

[Milner, 1989] R. Milner. *Communication and Concurrency*. International Series in Computer Science. Prentice Hall, 1989.

[Milner *et al.*, 1992] R. Milner, J. Parrow and D. Walker. A calculus of mobile processes, parts I and II. *Journal of Information and Computation*, **100**, 1–40 and 41–77, 1992.

[O'Brien and Nichol, 1998] P. D. O'Brien and R. C. Nichol. FIPA—towards a standard for software agents. *BT Technology Journal*, **16**, 51–59, 1998.

[Pratt, 1986] V. R. Pratt. Modeling concurrency with partial orders. *International Journal of Parallel Programming*, **15**, 33–71, 1986.

[Rao and Georgeff, 1991] A. S. Rao and M. P. Georgeff. Modeling rational agents within a BDI architecture. In *Proceedings of the Second Conference on Knowledge Representation and Reasoning*, R. Fikes and E. Sandewall, eds. pp. 473–484, Morgan Kaufman, 1991.

CARLES SIERRA, LLUÍS GODO, RAMON LÓPEZ DE MÁNTARAS AND MARIA MANZANO

DESCRIPTIVE DYNAMIC LOGIC AND ITS APPLICATION TO REFLECTIVE ARCHITECTURES

1 INTRODUCTION

In Artificial Intelligence, Multi-Language Logical Architectures, *MLA* for short, (MC [Giunchiglia and Serefini, 1994; Giunchiglia and Traverso, 1991; Giunchiglia *et al.*, 1993], BMS [Tan, 1992], ML^2 [Balder *et al.*, 1993], $DESIRE$ [Treur, 1992], $MILORD - II$ [Agusti *et al.*, 1991; Agusti *et al.*, 1994; Sierra and Godo, 1993], $OMEGA$ [Attardi and Simi, 1995], FOL [Weyhrauch, 1980]) are particular types of architectures, used to build knowledge-based systems, that play a major role in dealing with complex reasoning patterns, such as those involved in non-monotonic reasoning, scheduling or planning. Despite the fact that many commonalities can be intuitively found [F. Van Harmelen *et al.*, 1993], there was a lack of formal frameworks to compare and describe them. These architectures are based on the use of several logical languages to define local theories (or meta-theories acting upon theories) that influence/modify each other. These influences are modeled by complex control patterns of the reasoning flow between system units (also called modules or contexts) containing different knowledge theories evolving in time. Furthermore, the control patterns are often dynamically changed at run time. Therefore, when trying to define a formal framework to describe multi-language architectures it is mandatory that such a framework be able to model this dynamic behaviour.

Dynamic logic [Goldblatt, 1992; Harel, 1984; Pratt, 1976] has been traditionally used to describe and compare dynamic systems. Particularly, it has been used to deal with computational systems, understanding computations of programs as dynamic state changes. The aim of this kind of logic is the study of the mathematical properties of programs, and their behaviour. The construction of tools to reason about programs and the discovery of the key concepts involved in this reasoning process are the long term research in the field.

In this paper we present an extension of our previous work [Sierra *et al.*, 1995] proposing a particular theory in propositional dynamic logic called *Descriptive Dynamic Logic* (*DDL* from now on) in such a way that a Knowledge Base built within a concrete multi-language logical architecture, will be mapped into a theory in *DDL*. The atomic *DDL*-formulas will be taken as the *quotations* of the formulas of the set of languages of the architecture upon which the Knowledge Base is built. The key point is what a state and a program are meant to be in this descriptive dynamic logic. In the dynamic logic terminology, a state is a complete description of which atoms are true and which atoms are false. Thus an state for *DDL* will

J.J.Ch. Meyer and J. Treur (eds.),
Handbook of Defeasible Reasoning and Uncertainty Management Systems, Vol. 7, 197–220.

be a complete description of which formulas belong to the local theory of each unit in the Knowledge Base in a given moment of its execution. A change in a particular local theory of a unit will occur by the action of its local deductive system. Since programs in dynamic logic are understood in terms of transitions between states, programs in *DDL* will represent deductions. In particular, the set of atomic programs will correspond to the set of possible elementary deductive steps in the deduction systems (intra or inter-unit) of the architecture, while compound programs will represent any possible control reasoning flow combining as many atomic programs as necessary. Finally, which kind of changes of local theories in units are allowed is a matter of the particular computational characteristics of each architecture. Our approach will allow to describe such characteristics in terms of compound *DDL*-formulas, and, on the whole, to check computational properties of a knowledge bases built upon a *MLA* by making proofs in suitable theories in *DDL*.

Summarizing, the basic idea of our approach is to map multi-language knowledge based systems into *DDL* by representing:

1. elementary computational steps of Knowledge Bases as atomic programs in *DDL*,

2. local theories of the Knowledge Bases as formulas of *DDL*, and

3. operational semantics of Knowledge Bases as (non-logical) axioms in *DDL*.

Our goal is to show that execution of a Knowledge Base can be made equivalent to deduction in *DDL*. In this context *DDL* can be understood as a formal basis to describe multi-language architectures, as well as a specification language for Knowledge Bases.

The paper is structured as follows. Section 2 contains a brief remainder of Propositional Dynamic Logic. Section 3 presents a formalization of Multi-Language Logical Architectures and related notions. In sections 4-6 a particular theory in dynamic logic called *DDL* is proposed to model *MLA*s. A complete axiomatization of it is also provided. Finally, in Section 7 the mapping of Multi-Language Knowledge-bases into *DDL* is exemplified with the description of two different *reflective* architectures which are a particular case of Multi-Language architectures.

2 BACKGROUND ON PROPOSITIONAL DYNAMIC LOGIC

Propositional Dynamic Logic is a powerful program logic used as a metalanguage for talking about computer programs. A program can be seen as a dynamic object, that is, an object able to make the computer pass from one state (the content of the memory registers used by the program) to another. Due to the state change, the truth values of the formulas describing the state also change. The objective of the logic of programs is to create a logical basis to be able to express our reasoning about computer programs. The correctness of the programs, the property of halting, etc. are among the properties of programs that we may be interested to express in a logic of programs. It is also interesting to formalize general common features of programs such as WHILE, REPEAT, or IF THEN constructs. Furthermore, it is also useful to have a deductive calculus in which to verify our reasoning about programs. *PDL* provides all that and uses modal logic as its basis because modal logics allow to express changes in truth values due to changes of states. Modal models and in particular Kripke models suit *PDL* perfectly. The universe of the Kripke structure is now a universe of states. To each program we associate an accessibility relation in such a way that a pair of states (s, t) is in that relation if and only if there is a computation of the program transforming the state s into the state t. Finally, as in modal logic, each formula is interpreted as a set of states. Since we conceive a program as a binary relation between initial and final states, we associate an accessibility relation to every program, thus having a multimodal language. PDL is not only multimodal, but there are also operations that can be performed on programs. In fact, this is the most important feature of *PDL* and its axioms are not just describing the accessibility relation in itself, as it is the case with modal logics, but the operations that our program operators perform in them.

Next we briefly review the basic concepts of propositional dynamic logic *PDL*, for a detailed description the reader is referred to [Goldblatt, 1992; Harel, 1984; Manzano, 1996].

General Syntax for PDL Given a set of propositional atomic variables Φ_0 and atomic programs Π_0, the set Φ of compound formulas and the set Π of compound programs of PDL are defined as:[1]

1. $true \in \Phi$, $false \in \Phi$, $\Phi_0 \subseteq \Phi$,
2. if $A, B \in \Phi$ then $\neg A \in \Phi$ and $(A \vee B) \in \Phi$,
3. if $A \in \Phi$ and $\alpha \in \Pi$ then $\langle\alpha\rangle A \in \Phi$,
4. $\Pi_0 \subseteq \Pi$,
5. if $\alpha \in \Pi$ and $\beta \in \Pi$ then $(\alpha; \beta) \in \Pi$, $(\alpha \cup \beta) \in \Pi$ and $\alpha^* \in \Pi$,
6. if $A \in \Phi$ then $A? \in \Pi$

$[\alpha]A$ is the usual modal abbreviation for $\neg\langle\alpha\rangle\neg A$. Also $\wedge, \rightarrow$ and $\leftrightarrow$ are abbreviations with the standard meaning.

General Semantics for PDL The semantics of PDL is defined relative to a structure M of the form $M = (W, \tau, \rho)$, where W is a set of states, τ a mapping $\tau : \Phi \longrightarrow 2^W$ assigning to each formula A the set of states in which A is true, and a mapping $\rho : \Pi \longrightarrow 2^{W \times W}$ which assigns to each program a set

[1]Notation convention: we will use $p, q, \ldots$, to denote atomic propositional variables; $A, B, \ldots$ to denote arbitrary PDL formulas; $\alpha, \beta, \ldots$ to denote arbitrary programs. For sets of formulas of *PDL* we use $\Gamma, \Delta, \ldots$

of pairs (s,t) representing transitions between states. More concretely, the mappings τ and ρ are defined as follows:

$$\begin{array}{ll}
\tau(true) & = W \\
\tau(false) & = \emptyset \\
\tau(\neg A) & = W - \tau(A) \\
\tau(A \vee B) & = \tau(A) \cup \tau(B) \\
\tau(\langle\alpha\rangle A) & = \{s \in W \mid \exists t, (s,t) \in \rho(\alpha) \text{ and } t \in \tau(A)\} \\
\rho(\alpha;\beta) & = \{(s,t) \mid \exists u, (s,u) \in \rho(\alpha) \text{ and } (u,t) \in \rho(\beta)\} \\
\rho(\alpha \cup \beta) & = \rho(\alpha) \cup \rho(\beta) \\
\rho(\alpha^*) & = \{(s,t) \mid \exists s_0, \ldots s_k \text{ such that } s = s_0, s_k = t \text{ and} \\
 & \quad (s_{i-1}, s_i) \in \rho(\alpha) \text{ for all } 1 \leq i \leq k\} \\
\rho(A?) & = \{(s,s) \mid s \in \tau(A)\}
\end{array}$$

Henceforth, we shall denote by $\mathcal{C}^{PDL}$ the class of the above standard structures for *PDL*. As for notions of satisfiability and validity we write:

- $(M,s) \models A$, saying that A is *true* in s, iff $s \in \tau(A)$,
- $(M,s) \models \Gamma$, iff $(M,s) \models A$ for all $A \in \Gamma$.
- $M \models A$, saying that A is M–valid, iff $(M,s) \models A$ for every s in M,
- $M \models \Gamma$ iff $M \models A$ for all $A \in \Gamma$.
- $\models_{\mathcal{PDL}} A$, saying that A is *valid*, iff A is M–valid for every $M \in \mathcal{C}^{\mathcal{PDL}}$.

Furthermore, a formula A is said to be a *global logical consequence* of a set of formulas Γ if for any structure $M \in \mathcal{C}^{\mathcal{PDL}}$, we have that $M \models A$ whenever $M \models \Gamma$, in which case we write $\Gamma \models^G_{\mathcal{PDL}} A$. A formula A is said to be a *local logical consequence* of a set of formulas Γ if for any structure M and state s, we have that $(M,s) \models A$ whenever $(M,s) \models \Gamma$; in which case we simply write $\Gamma \models_{\mathcal{PDL}} A$.

Axiomatic of PDL The next set of axioms define the propositional dynamic logic [Harel, 1984]:

(A1) All instances of tautologies of the propositional calculus
(A2) $\langle\alpha;\beta\rangle A \leftrightarrow \langle\alpha\rangle\langle\beta\rangle A$
(A3) $\langle\alpha \cup \beta\rangle A \leftrightarrow (\langle\alpha\rangle A \vee \langle\beta\rangle A)$
(A4) $\langle\alpha^*\rangle A \leftrightarrow (A \vee \langle\alpha\rangle\langle\alpha^*\rangle A)$
(A5) $\langle A?\rangle B \leftrightarrow A \wedge B$
(A6) $[\alpha^*](A \to [\alpha]B) \to (A \to [\alpha^*]B)$
(A7) $[\alpha](A \to B) \to ([\alpha]A \to [\alpha]B)$

The set of theorems of *PDL*, denoted by $\vdash_{PDL}$, is defined as the set of axioms above plus the theorems that can be obtained from the following inference rules applied to other theorems:

(MP) from $\vdash_{PDL} A$ and $\vdash_{PDL} A \to B$ infer $\vdash_{PDL} B$ (Modus Ponens)
(G) from $\vdash_{PDL} A$ infer $\vdash_{PDL} [\alpha]A$ (Generalization)

It has been shown (see for instance [Goldblatt, 1992]) that $\vdash_{PDL}$ completely axiomatizes the set of valid formulas of the class of models $\mathcal{C}^{\mathcal{PDL}}$, i.e. it holds that $\vdash_{PDL} A$ iff $\models_{\mathcal{PDL}} A$. The notion of proof from a set of formulas Γ is defined as follows: $\Gamma \vdash_{PDL} A$ iff $\exists\gamma_1, \dots \gamma_n \in \Gamma$ such that $\vdash_{PDL} \gamma_1 \wedge \cdots \wedge \gamma_n \to A$.

3 MULTI-LANGUAGE LOGICAL ARCHITECTURES

The basic notion we want to express by a "multi-language logical architecture" is that of a logical system that provides as basic tools:

1. a set of possibly different logical languages,
2. a set of inference rules within each language,
3. a set of inference rules between some of these languages,
4. the possibility of defining a set of units containing theories written in a particular language together with a particular set of inference rules, and
5. a set of possible interconnections of units through bridge rules, respecting some topological criteria, e.g. a binary tree structure, an acyclic graph, etc.

In order to make easier understanding the formalization, in the next definitions we introduce, at different levels of abstraction, three concepts related to the above notion of Multi-Language knowledge-based system. Namely, what we call *Multi-Language Logical Architecture* (in the highest level of abstraction) represents the most general characteristics of our target computational systems. Then, a *Multi-Language Knowledge-Based Structure* specifies a subclass of those systems by fixing its components, that is the set of units, their languages, the inference rules among them and the topology. Finally, a *Multi-Language Knowledge-Based System* is the specification of a concrete system obtained by filling the units of a Multi-Language Knowledge-Based Structure with particular domain theories.

DEFINITION 1. A **Multi-Language Logical Architecture** is a 4-tuple MLA $= (L, \Delta, S, T)$, where:

1. $L = \{L_j\}_{j \in J}$ is a set of *finite* logical languages.
2. $\Delta = \bigcup_{j_1, j_2 \in J} \Delta_{j_1,j_2}$ is a set of (instances of) inference rules between pairs of languages, where $\Delta_{j_1,j_2} \subseteq 2^{L_{j_1}} \times L_{j_2}$. In particular, when $j_1 = j_2$, Δ_{j_1,j_2} denotes a set of inference rules of the corresponding language; otherwise it denotes a set of bridge rules between two different languages.

3. S is a finite set of symbols to identify units.
4. T is the set of possible topologies. Each topology is determined by a set of directed links between symbols from S, i.e T is a subset of $2^{S \times S}$.

Notice that we focus only on finite languages as it is the usual case in Knowledge bases where some limitative rules are imposed on the generation of formulas. This fact will be essential in our approach because it will allow to express (finite) big conjunctions in *DDL* involving all the formulas of a language.

Notice also that It would be possible to further generalize the above definition of architecture by allowing inference rules with premises in different languages, that is, having $\Delta = \{\Delta_{j_1,j_2,\ldots,j_n}\}_{j_1,j_2,\ldots,j_n \in J}$, where

$$\Delta_{j_1,j_2,\ldots,j_n} \subseteq 2^{\bigcup_{i=1}^{n-1} L_{j_i}} \times L_{j_n}$$

DEFINITION 2. A **Multi-Language Knowledge-Based Structure** *MKB-ST* for a given *MLA* is a 5- tuple *MKB-ST* = (*MLA*, U, M_L, M_Δ, B) where:

1. *MLA* = (L, Δ, S, T) is a Multi-Language Logical Architecture
2. $U = \{u_k\}_{k \in K}$ is a set of unit identifiers, i.e. U is a subset of S
3. M_L assigns a language to each unit identifier, i.e. $M_L : U \longrightarrow L$.
4. M_Δ assigns a set of inference rules to each unit identifier, i.e. $M_\Delta : U \longrightarrow \bigcup_{i \in J} 2^{\Delta_{ii}}$ such that if $M_L(u) = L_j$, for some $j \in J$, then $M_\Delta(u) \subseteq \Delta_{jj}$
5. B is a mapping that assigns a set of directed bridge rules to pairs of different units, i.e. $B : U \times U \longrightarrow \bigcup_{i,j \in J} 2^{\Delta_{ij}}$, such that:
   (i) if $u_1 \neq u_2$, $M_L(u_1) = L_i$ and $M_L(u_2) = L_j$ then $B(u_1, u_2) \subseteq \Delta_{ij}$
   (ii) $B(u, u) = \emptyset$, for any $u \in U$
   (iii) $\{(u_1, u_2) \mid B(u_1, u_2) \neq \emptyset\} \in T$, that is, the topology of *MKB-ST* is in accordance with the allowed topologies in *MLA*

Notice that in this definition, even in the case where $u_1 \neq u_2$, $B(u_1, u_2)$ can be empty, denoting that unit u_1 has no (directed) link with the unit u_2. In this way, a unit u_1 is connected to a unit u_2 whenever $B(u_1, u_2) \neq \emptyset$.

DEFINITION 3. A **Multi-Language Knowledge-Based System** *MKB* for a given structure *MKB-ST* is a 3-tuple *MKB* = (*MKB-ST*, M_Σ, M_Ω) where:

1. *MKB-ST* = (*MLA*, U, M_L, M_Δ, B) is a Multi-Language Knowledge-Based Structure, and

2. M_Σ assigns a concrete signature $M_\Sigma(u) = (Oper, Sort, Func)$ for the language $M_L(u)$ of each unit identifier u, such that $Func : Oper \to Sort$, gives a type in $Sort$, for each element in the alphabet $Oper$.
3. M_Ω assigns a set of formulas (initial local theory) built upon M_Σ to each unit identifier, i.e. $M_\Omega : U \to \bigcup_{i \in J} 2^{L_j}$ such that if $M_L(u) = L_k$ then $M_\Omega(u) \subseteq L_k$.

As a consequence of the execution of a MKB, the local theory associated to u_i will change due to the intra and inter-unit deductions. A bridge rule $r \in B(u_i, u_j)$ is an inference rule between unit u_i and unit u_j whose premises are in language $L_i = M_L(u_i)$ and the consequent is in a (possibly different) language $L_j = M_L(u_j)$, that is $r \subseteq 2^{L_i} \times L_j$. Notice that the sets of bridge rules can be empty. An empty set $B(u_i, u_j) = \emptyset$ means that no communication between units u_i and u_j is defined. An empty set $M_\Delta(u_i) = \emptyset$ means that the local theory attached to unit u_i will evolve only if inter-unit bridge rules leading to that unit are defined, otherwise the unit will remain static along the execution process of the MKB.

The distinction between architecture and knowledge base structure is in some cases not as sharp as implied by the above definitions. Often some inference rules are already fixed in the architecture (e.g. DESIRE, MILORD II), and the number and/or structure and/or role of units that can be built is also fixed (e.g. DESIRE, MILORD II, ML2). So, these definitions have to be considered as a very general approach to *MLA* and MKB descriptions.

The next example illustrates the way this notation can be used to formalize the BMS architecture [Tan, 1992] as a Multi-Language Knowledge-Based Structure.

EXAMPLE 4. Let be MLA_{BMS} [2] $= (L, \Delta, S, T)$, where each component is described as follows.

- $L = \{PL^*, FOL^*\}$, where PL^* is a propositional language with negation ($\sim$), conjunction (&) and implication ($\subset$) and sentences are pairs of classical propositional formulas and truth values of the set $\{1, 0, u\}$, meaning *true*, *false* and *unknown* respectively. FOL^* is a first order language with predicate signature (T, PA) where $T(s)$ means that s is true and $PA(s)$ means that s is a plausible assumption. Constants of FOL^* are quoting of propositions in PL^*.

- $\Delta = \{\Delta_{PL^*,PL^*}, \Delta_{PL^*,FOL^*}, \Delta_{FOL^*,FOL^*}, \Delta_{FOL^*,PL^*}\}$, where Δ_{PL^*,PL^*} and Δ_{FOL^*,FOL^*} contain the instances of *modus ponens*, *and-introduction*, and possibly others inference rules. The bridge rules are:

$$\Delta_{PL^*,FOL^*} = \{\frac{(\varphi,1)}{T(\varphi)}, \frac{(\varphi,u)}{\neg T(\varphi)}, \frac{(\varphi,0)}{T(\varphi)}\}$$

[2] This *MLA* is a generalization of the BMS architecture [Tan, 1992]

$$\Delta_{FOL^*,PL^*} = \{\frac{PA(\varphi)}{(\varphi,1)}\}$$

- S is the set of symbol identifiers used in BMS and T the possible topologies, consisting of all possible pairs of unit interconnections.

The corresponding *MKB-ST* of the BMS system, which is a particular structure over the above architecture, contains just a couple of units. That is, *MKB-ST*$_{BMS} = (MLA, U, M_L, M_\Delta, B)$, where

- $U = \{a, b\}$,
- $M_L = \{a \mapsto PL^*, b \mapsto FOL^*\}$,
- $M_\Delta = \{a \mapsto \Delta_{PL^*,PL^*}, b \mapsto \Delta_{FOL^*,FOL^*}\}$, and
- $B = \{(a,b) \mapsto \Delta_{PL^*,FOL^*}, (b,a) \mapsto \Delta_{FOL^*,PL^*}\}$.

4 DDL = QUOTING MKB-STS IN PDL.

In this section our aim is to present our logical tools to represent and reason about the computational dynamics of Multi-Language Knowledge-Bases. According to the hierarchy of concepts introduced in the previous section, the modeling of such knowledge bases is done at two different levels.

1. First we define an extension of *PDL*, called *Descriptive Dynamic Logic*, *DDL* for short, to describe knowledge-base structures. This extension consists of:
   - Defining a language to represent the basic components of the structure of a particular *MKB-ST*, that is units, languages, inference rules and topology, and
   - Fixing a set of axioms, and the corresponding class of models, to describe the common behaviour of Multi-Language Architectures.
2. The second step to model a particular knowledge base is then to build a particular theory, in the *DDL* describing its structure, containing formulas representing both the formulas of its units and the inference rules used in each unit and between different units.

Our final goal is to faithfully describe the computational behaviour of the multi-language knowledge base by performing logical deduction in *DDL* theories, or in other words, to be able to check some properties of the multi-language system by means of proofs in *DDL*.

To define *DDL* we need first of all to fix the set of atomic formulas and the set of programs. Given a *MKB-ST* = (MLA, U, M_L, M_Δ, B), the set of atomic formulas of *DDL* will be defined as the set of "quoted" formulas built upon the languages L in *MLA* together with concrete signatures for each unit, and indexed by the unit identifiers in U. The set of *DDL* atomic programs will be restricted to represent deduction steps in those languages. For notation simplicity, in the following definitions the set of formulas [3] of a unit u will be denoted by $M_L(u)$, that is, without explicit reference to the corresponding signature.

DEFINITION 5. Given a *MKB-ST*, the set of atomic formulas of *DDL* is defined as the following *finite* set:

$$\Phi_0 = \{u{:}\lceil\varphi\rceil \mid u \in U, \varphi \in M_L(u)\}$$

that is, formulas are indexed by unit names, using the notation unit – identifier: ⌈formula⌉.

In the following definition, we build up the set of atomic programs Π_0 from both intra-unit inference rules and inter-unit inference rules, or bridge rules.

DEFINITION 6. Given a *MKB-ST*, the set Π_0 of atomic programs of *DDL* is defined as the union of the intra-unit inference rules Π_0^{Intra} and the *MKB-ST* bridge rules Π_0^{Inter}, i.e. $\Pi_0 = \Pi_0^{Intra} \cup \Pi_0^{Inter}$, where:

$$\Pi_0^{Intra} = \bigcup_{k \in K} \Pi_{0_{kk}}^{Intra}, \quad \Pi_0^{Inter} = \bigcup_{k,l \in K, k \neq l} \Pi_{0_{kl}}^{Inter}$$

being

$$\Pi_{0_{kk}}^{Intra} = \{\lceil\Gamma \vdash_{kk} \varphi\rceil \mid \Gamma \cup \{\varphi\} \subseteq M_L(u_k), (\Gamma, \varphi) \in M_\Delta(u_k)\}$$

and

$$\Pi_{0_{kl}}^{Inter} = \{\lceil\Gamma \vdash_{kl} \varphi\rceil \mid \Gamma \subseteq M_L(u_k), \varphi \in M_L(u_l), (\Gamma, \varphi) \in B(u_k, u_l)\}$$

where $\lceil\Gamma \vdash_{kl} \varphi\rceil$ is an abbreviation for the quoting function applied to a deduction step.

Notice that atomic programs contain not only the inference rules applied but also the formulas involved in the deduction steps. Notice also that the quoting function of *DDL* is similar to that of OMEGA [Attardi and Simi, 1995]. So, having defined the quoting function for formulas, we extend it to sets of formulas and deduction as follows, where $\Gamma = \{\gamma_1, \ldots, \gamma_n\}$:

$$\lceil\Gamma\rceil = set(\lceil\gamma_1\rceil, \ldots, \lceil\gamma_n\rceil)$$

$$\lceil\Gamma \vdash_{kl} \varphi\rceil = proof(\lceil\Gamma\rceil, \lceil\varphi\rceil, \lceil k\rceil, \lceil l\rceil)$$

[3] We will denote the formulas of the languages of the architectures with $\varphi, \psi, \ldots$.

It is clear then that the access to components of quoted proofs is possible by means of the appropriate accessor functions.

Compound programs representing arbitrary applications of intra-unit and inter-unit inference rules can be built by means of indeterministic unions of atomic programs. Therefore, no commitments are done about any particular control strategy. For instance, we define below the compound programs corresponding to the intra and inter-unit deduction.

DEFINITION 7. Given the set Π_0 of atomic programs, we define the next two sets of indeterministic compound programs standing for inter and intra-unit deductions:

- Intra-unit deduction is $\vdash_{kk}= \bigcup\{\alpha \mid \alpha \in \Pi_{0_{kk}}^{Intra}\}$
- Inter-unit deduction is $\vdash_{kl}= \bigcup\{\alpha \mid \alpha \in \Pi_{0_{kl}}^{Inter}\}$

Notice again that given the finiteness of the languages in the architectures, the set Π_0 of atomic programs is finite, and thus the above compound programs representing intra and inter deductions are well defined.

5 SYNTAX AND SEMANTICS OF DDL

5.1 *The intended models of DDL*

The semantics of *DDL* is defined, as in *PDL*, relative to a structure M of the form $M = (W, \tau, \rho)$, as a particularization of the general semantics for *PDL* by adding a particular interpretation of both the *DDL* atomic formulas and the *DDL* atomic programs. The computational systems one could model in *DDL* is very wide, showing very different behaviours from a logical point of view. Some of these systems are non-monotonic, some are conservative, etc.

The class of models we define for *DDL* restrict the possible transitions between states to those satisfying a set of requirements that capture the basic properties of the deductive steps performed in multi-language architectures. The idea in mind is that performing a deductive step $\Gamma \vdash_{kl} \varphi$ in any unit means two things: first, the formulas in Γ of unit u_k were previously proved, and second, the formula φ of unit u_l becomes proved. Then, the intuition behind the semantics of *DDL* is that states provide which formulas are proved in a particular moment of the execution of the system and therefore transitions corresponding to atomic programs, which represent deductive steps of the kind $\Gamma \vdash_{kl} \varphi$, have to satisfy the following natural requirements:

1. an atomic program $\lceil\Gamma \vdash_{kl} \varphi\rceil$ is executable in all states that satisfy the quoted formulas from Γ,
2. an atomic program $\lceil\Gamma \vdash_{kl} \varphi\rceil$ leads to states that satisfy $l{:}\lceil\varphi\rceil$,

3. an atomic formula is true in the target state whenever is true in the source state, (that is, we have the property of monotonicity),

4. an atomic formula different from the conclusion in the program and true in the target state must also be true in the source state, (that is, there are no side-effects),

5. if an atomic program is executable in a source state of another atomic program, it is also executable in the target state of that program, (that is, we have a persistence property), and finally

6. atomic programs are partial functions, that is, if an atomic program is executable in a given state, there exists a unique target state (atomic programs are meant to represent a single execution of a rule of inference).

Notice that condition five can be extended to compound programs that do not contain test programs as components, since a test program could be applicable before the execution of an atomic program but not after. This is so because the meaning of the falsity of the quoted formula p in a given state s is that the formula has not been proved in the logic of the unit. Therefore, $\langle \neg p? \rangle true$ is true in s. But nothing prevents $\langle \neg p? \rangle true$ to become false after the application of an inference rule proving p. Henceforth we shall denote by $\Pi_{?-free}$ the set of programs that does not contain test programs as components.

The above requirements impose a set of constraints in the possible transitions allowed for a ρ in the class of models. In what follows, for any state s, $f(s)$ will stand for the atomic truth set of s, composed of the atomic DDL-formulas which are true in s, i.e. $f(s) = \{p \mid r \in \tau(p), p \in \Phi_0\}$.

DEFINITION 8 (Class of models $\mathcal{C}^{DDL}$). . The class of structures $\mathcal{C}^{DDL}$ consists of structures (W, τ, ρ) satisfying the following property: given a set of sentences $\Gamma = \{\psi_i \mid i \in I\}$ from a unit u_k and a sentence φ from a unit u_l, the following conditions hold for any program $\lceil \Gamma \vdash_{kl} \varphi \rceil$:

C1 if $s \in \bigcap_{i \in I} \tau(k{:}\lceil \psi_i \rceil)$ then there exists t such that $(s,t) \in \rho(\lceil \Gamma \vdash_{kl} \varphi \rceil)$

C2 if $(s,t) \in \rho(\lceil \Gamma \vdash_{kl} \varphi \rceil)$ then $s \in \bigcap_{i \in I} \tau(k{:}\lceil \psi_i \rceil)$ and $t \in \tau(l{:}\lceil \varphi \rceil)$

C3 if $(s,t) \in \rho(\lceil \Gamma \vdash_{kl} \varphi \rceil)$ then if $s \in \tau(m{:}\lceil \psi \rceil)$ then $t \in \tau(m{:}\lceil \psi \rceil)$, for all $\psi \in M_L(u_m)$

C4 if $(s,t) \in \rho(\lceil \Gamma \vdash_{kl} \varphi \rceil)$ then if $t \in \tau(p)$ then $s \in \tau(p)$, for all $p \in \Phi_0$ and $p \not\equiv l{:}\lceil \varphi \rceil$

C5 if $(s,t) \in \rho(\lceil \Gamma \vdash_{kl} \varphi \rceil)$ then if $(s,t') \in \rho(\beta)$ then there exists r such that $(t,r) \in \rho(\beta)$, for all $\beta \in \Pi_{?-free}$.

C6 if $(s,t) \in \rho(\lceil \Gamma \vdash_{kl} \varphi \rceil)$ then if $(s,t') \in \rho(\lceil \Gamma \vdash_{kl} \varphi \rceil)$ then $t = t'$.

We will use the notations $\models_{\mathcal{DDL}} A$ and $\Delta \models_{\mathcal{DDL}} A$ to say that A is valid formula in the class of models $\mathcal{C}^{DDL}$, and that A is logical consequence of a set of formulas Δ in $\mathcal{C}^{DDL}$, respectively.

To have an idea about what kind of formulas can be proved to be valid in $\mathcal{C}^{DDL}$, let us consider the case where a unit u_k of a multi-language logical system is endowed with the modus ponens inference rule. The behaviour of this inference rule can be described by the following *DDL* formula:

$$MP: \langle\beta\rangle(k{:}\lceil\varphi\rceil \wedge k{:}\lceil\varphi \to \psi\rceil) \to \langle\beta; \vdash_{kk}^{mp}\rangle k{:}\lceil\psi\rceil$$

where the modus ponens program $\vdash_k k^{mp}$is defined as the following indeterministic finite union:

$$\vdash_{kk}^{mp} = \bigcup_{\gamma,\delta \in M_L(u_k)} \lceil\{\gamma, \gamma \to \delta\} \vdash_{kk} \delta\rceil$$

The above formula MP says that if it is possible to prove φ and $\varphi \to \psi$, after some previous deduction (represented by the program β), then it is also possible to prove ψ after β followed by an application of the modus ponens inference rule. Notice that, since we can only have atomic programs corresponding to *instances* of deduction steps, the program standing for the modus ponens inference rule is actually defined as the finite union of those atomic programs corresponding to all possible particular instantiations of the rule in the unit u_k. Then, it is easy to show that the formula MP is valid in $\mathcal{C}^{DDL}$.

The following formula schemes will become the axioms of *DDL*.

1. DED-1: $\langle\lceil\Gamma \vdash_{ij} \psi\rceil\rangle true \leftrightarrow \bigwedge_{\varphi_k \in \Gamma} i{:}\lceil\varphi_k\rceil$

2. DED-2: $[\lceil\Gamma \vdash_{ij} \psi\rceil] j{:}\lceil\psi\rceil$

3. MON: $p \to [\alpha]p$, for $p \in \Phi_0$

4. SEA: $\langle\lceil\Gamma \vdash_{ij} \psi\rceil\rangle p \to p$ for $p \in \Phi_0$ and $p \neq j{:}\lceil\psi\rceil$

5. PER: $\langle\alpha\rangle true \wedge \langle\beta\rangle true \to \langle\alpha;\beta\rangle true$ for $\beta \in \Pi_{?-free}$

6. PFUN: $\langle\alpha\rangle A \to [\alpha]A$, for $\alpha \in \Pi_0$

We will see now, in Theorem 9, that they exactly correspond to the previous conditions C_1 to C_6 imposed over the class of models $\mathcal{C}^{\mathcal{DDL}}$, except for the $PFUN$ schema for which we can only prove its validity.

THEOREM 9. *Let* $M = (W, \tau, \rho)$ *belong to the class* $\mathcal{C}^{\mathcal{PDL}}$ *of standard* PDL *models. Then the following conditions are verified:*

1. *M satisfies C1 and C2 iff DED-1 and DED-2 are valid in M*

2. *M satisfies C3 iff MON is valid in M*
3. *M satisfies C4 iff SEA is valid in M*
4. *if $\rho(\alpha)$ is partially functional for any atomic program α, then M satisfies C5 iff PER is valid in M*
5. *$PFUN$ is valid in M if M satisfies C6*

Proof.

1. M satisfies C1 and C2 iff DED-1 and DED-2 are valid in M:
   Let $\beta \equiv \lceil \Gamma \vdash_{ij} \varphi \rceil$.
   DED-1 is valid in M iff for all s in M, $(M, s) \models \langle\beta\rangle true \leftrightarrow \bigwedge_{\gamma\in\Gamma} i{:}\lceil\gamma\rceil$ iff

   $$\forall s(\exists t((s,t) \in \rho(\beta)) \text{ iff } s \in \bigcap_{\gamma\in\Gamma} \tau(i{:}\lceil\gamma\rceil)) \qquad \text{(i)}$$

   It is clear that condition $C1$ corresponds to the right to left implication in (i). Now DED-2 is valid in M iff for all s in M, $(M, s) \models [\beta]j{:}\lceil\varphi\rceil$ iff

   $$\forall s(\forall t(\text{ if } (s,t) \in \rho(\beta) \text{ then } t \in \tau(j{:}\lceil\varphi\rceil))) \qquad \text{(ii)}$$

   It is easy to check that condition $C2$ corresponds to both (ii) and the left to right implication in (i). Altogether we finally get that both conditions $C1$ and $C2$ hold iff both (i) and (ii) hold.

2. M satisfies C3 iff MON is valid in M:
   MON is valid in M iff for every s in M, $(M, s) \models p \to [\alpha]p$ (for $p \in \Phi_0$) iff $\forall s(s \in \tau(p) \Rightarrow \forall t((s,t) \in \rho(\alpha) \Rightarrow t \in \tau(p)$ iff C3.

3. M satisfies C4 iff SEA is valid in M:
   Let $\beta \equiv \lceil \Gamma \vdash_{ij} \varphi \rceil$. SEA is valid in M iff for every s in M, $(M, s) \models [\langle\beta\rangle p \to p$ (where $p \neq j{:}\lceil\varphi\rceil$) iff $\forall s((\exists t((s,t) \in \rho(\beta) \& t \in \tau(p)) \Rightarrow s \in \tau(p))$ iff C4.

4. Suppose $\rho(\alpha)$ is partially functional, $\alpha \in \Pi_0$, i.e., if (s,t) and (s,t') belong to $\rho(\alpha)$ then $t = t'$ and $\beta \in \Pi_{?-free}$. In this case, we have:

PER is valid in M	iff	$(M,s) \models \langle\alpha\rangle true \wedge \langle\beta\rangle true \to \langle\alpha;\beta\rangle true$
	iff	$(\exists t)((s,t) \in \rho(\alpha))$ and $(\exists t')((s,t') \in \rho(\beta))$
		implies
		$(\exists t_1, t_2)((s,t_1) \in \rho(\alpha)$ and $(t_1,t_2) \in \rho(\beta))$

   but, since $\rho(\alpha)$ is partially functional, $t_1 = t$ and therefore M satisfies C5. On the other hand, if M satisfies C5, it is easy to show that PER is valid.

5. Straightforward. ■

This theorem gives the validity of the above schemes in the class $\mathcal{C}^{DDL}$ since any model $M = (W, \tau, \rho) \in \mathcal{C}^{\mathcal{DDL}}$ satisfies C6, and therefore, $\rho(\alpha)$ is partially functional for all atomic program α, and thus PER is valid too in M.

COROLLARY 10. *Let $M = (W, \tau, \rho)$ be a model in $\mathcal{C}^{PDL}$ for which $\rho(\alpha)$ is partially functional for any atomic program α. Then, it follows that DED-1, DED-2, MON, SEA and PER are valid in M iff $M \in \mathcal{C}^{DDL}$. As a consequence, the schemes DED-1, DED-2, MON, SEA, PER and PFUN are valid in the class of models $\mathcal{C}^{\mathcal{DDL}}$.*

5.2 *Axiomatic of* DDL

In this section we provide an axiom system for *DDL*, and prove its soundness and completeness with respect to the previously defined class of models $\mathcal{C}^{\mathcal{DDL}}$.

DEFINITION 11. Given a Multi-Language Knowledge-Base Structure, the corresponding *DDL* logic is defined as the following extension of *PDL*:

- The language of *DDL* is defined upon the sets of atomic formulas Φ_0 and atomic programs Π_0 as given in definitions 5 and 6 respectively.
- The additional axioms of *DDL* are the schemes DED-1, DED-2, MON, SEA, PER and PFUN.

THEOREM 12 (Soundness and completeness). DDL *is sound and complete with respect to the class of models $\mathcal{C}^{\mathcal{DDL}}$.*

Proof.

1. *Soundness.* Let $\vdash_{DDL} \varphi$ (i.e. φ is a theorem of the logic *DDL*). Thus, $DDL \vdash_{PDL} \varphi$ (where *DDL* is the set of instances of the scheme axioms of the logic *DDL*). Therefore $\vdash_{PDL} \delta_1 \wedge \cdots \wedge \delta_n \rightarrow \varphi$, where the formulas $\delta_1, \ldots, \delta_n$ are the axioms in *DDL* used in the deduction. The number is finite because of the finiteness of the deduction, and we use the deduction theorem: $\psi \vdash \pi \Rightarrow \vdash \psi \rightarrow \pi$). Then $\models_{\mathcal{PDL}} \delta_1 \wedge \cdots \wedge \delta_n \rightarrow \varphi$, because *PDL* is sound (see [Goldblatt, 1992]). Then, $\models_{\mathcal{DDL}} \delta_1 \wedge \cdots \wedge \delta_n \rightarrow \varphi$, because $\mathcal{C}^{\mathcal{DDL}} \subseteq \mathcal{C}^{\mathcal{PDL}}$. Finally, $\models_{\mathcal{DDL}} \varphi$, because $\delta_1, \ldots, \delta_n$ are valid in *DDL* by corollary 10.

2. *Completeness* Let $\models_{\mathcal{DDL}} \varphi$. Thus, $DDL \models_{\mathcal{PDL}} \varphi$ (by corollary 10). Therefore, $\models_{\mathcal{PDL}} \bigwedge_{\delta \in DDL'} \delta \rightarrow \varphi$, where DDL' is a finite set of axioms obtained from the scheme axioms in *DDL*. We put in DDL' all the instances of $DED-1$, $DED-2$ and SEA. To reduce the instances of MON, PER and $PFUN$ to a finite number we use the Fischer-Ladner closure. Then, $\vdash_{PDL} \bigwedge_{\delta \in DDL'} \delta \rightarrow \varphi$, by completeness of *PDL* (see [Goldblatt, 1992]). Thus, $DDL' \vdash_{PDL} \varphi$, using the propositional rules of the calculus. Then,

$DDL \vdash_{PDL} \varphi$, by monotonicity ($DDL' \subseteq DDL$). Finally, $\vdash_{DDL} \varphi$, since the logic *DDL* is obtained when adding to *PDL* the axioms in *DDL*. ∎

5.3 *Representing Deductive Closures in* DDL

Since, in our framework, programs represent deductions, it is interesting to see how to define, for any program α, the program representing the deductive closure of α. These programs are easily defined in *PDL* as

$$\alpha^c = \alpha^*; (\bigwedge_{\varphi \in \Phi_0} (\langle \alpha^* \rangle \varphi \rightarrow \varphi))?$$

where α^c represents the closure of program α, meaning that this program will lead to a state in which no different state is reachable by another application of program α. Some intuitive properties that α^c should verify can be actually proved. But before doing that, we need some previous results.

PROPOSITION 13. *Let* $(W, \tau, \rho) \in \mathcal{C}^{\mathcal{DDL}}$. *Then the following condition holds for any* $\alpha, \beta \in \Pi_{?-free}$: *if* $(s, t_1) \in \rho(\alpha)$ *and* $(s, t_2) \in \rho(\beta)$ *then there exists* r *such that* $(t_1, r) \in \rho(\beta)$ *and* $(t_2, r) \in \rho(\alpha)$

Proof. Let $\alpha, \beta \in \Pi_{?-free}$. The proof will by induction on the structure of the programs α and β.

1. Atomic case. Let $\alpha = \lceil \Gamma \vdash_{kl} \varphi \rceil$ and $\beta = \lceil \Gamma' \vdash_{ij} \varphi' \rceil$. Then by condition C5, there exists r_1 such that $(t_1, r_1) \in \rho(\beta)$ and $f(r_1) = f(s) \cup \{l{:}\lceil\varphi\rceil, j{:}\lceil\varphi'\rceil\}$. Analogously, there exists r_2 such that $(t_2, r_2) \in \rho(\alpha)$ and $f(r_2) = f(s) \cup \{l{:}\lceil\varphi\rceil, j{:}\lceil\varphi'\rceil\}$. So, r_1 and r_2 denote the same state.

2. Let $\alpha = \delta_1; \delta_2$, being $\delta_1, \delta_2 \in \Pi_{?-free}$, and suppose that δ_1, δ_2 and β verify the condition. Since $(s, t_1) \in \rho(\delta_1; \delta_2)$, there exists u_1 such that $(s, u_1) \in \rho(\delta_1)$ and $(u_1, t_1) \in \rho(\delta_2)$. By induction hypothesis, it is easy to show that exist u_2 and r such that:

   $(u_1, u_2) \in \rho(\beta)$ and $(t_2, u_2) \in \rho(\delta_1)$, and

   $(t_1, r) \in \rho(\beta)$ and $(u_2, r) \in \rho(\delta_2)$,
   Therefore, we have proved that $(t_1, r) \in \rho(\beta)$ and $(t_2, r) \in \rho(\delta_1; \delta_2)$.

3. Let $\alpha = \delta_1 \cup \delta_2$, being $\delta_1, \delta_2 \in \Pi_{?-free}$, and suppose that δ_1, δ_2 and β verify the condition. Let $(s, t_1) \in \rho(\delta_1 \cup \delta_2)$ and $(s, t_2) \in \rho(\beta)$. Assume $(s, t_1) \in \rho(\delta_1)$ then by the induction hypothesis there exists r such that $(t_1, r) \in \rho(\beta)$ and $(t_2, r) \in \rho(\delta_1)$ and therefore $(t_2, r) \in \rho(\delta_1 \cup \delta_2)$. The case $(s, t_1) \in \rho(\delta_2)$ is analogous.

4. Let $\alpha = \delta^*$, being $\delta \in \Pi_{?-free}$ and suppose δ, and β verify the condition. Let $(s, t_1) \in \rho(\alpha*)$ and $(s, t_2) \in \rho(\beta)$. There is a finite sequence of applications of α leading from s to t_1. In each intermediate step we can apply the

induction hypothesis getting a new connecting chain by α of states leading from t_2 to r, and a step by β from t_1 to r.

The rest of the cases are proved in a similar way. ■

As a particular case of this proposition, by making $\alpha = \beta$ we get that the transition relations of test-free programs are weakly directed. A binary relation R is weakly directed if when $(s, t_1), (s, t_2) \in R$ there exists r such that $(t_1, r), (t_2, r) \in R$. It is known that in Modal Logics the axiom corresponding to this kind of relations is $\Diamond\Box\varphi \to \Box\Diamond\varphi$, and is called axiom G. The validity of this axiom in *DDL* is expressed also in the next corollary.

COROLLARY 14. *If* $(W, \tau, \rho) \in \mathcal{C}^{\mathcal{DDL}}$ *then* $\rho(\alpha)$ *is weakly directed for all* $\alpha \in \Pi_{?-free}$*; therefore, the formula*

$$G: \langle\alpha\rangle[\alpha]A \to [\alpha]\langle\alpha\rangle A$$

is valid in $\mathcal{C}^{\mathcal{DDL}}$ *for all* $\alpha \in \Pi_{?-free}$.

Proof. The first part is straightforward. Let $(W, \tau, \rho) \in \mathcal{C}^{\mathcal{DDL}}$. We want to prove that $\tau(\langle\alpha\rangle[\alpha]A) \subseteq \tau([\alpha]\langle\alpha\rangle A)$. Let $s \in \tau(\langle\alpha\rangle[\alpha]A)$, then there exists t such that $(s, t) \in \rho(\alpha)$ and $t \in \tau([\alpha]A)$. Let $(s, t') \in \rho(\alpha)$. We want to find r such that $(t', r) \in \rho(\alpha)$ and $r \in \tau(A)$. By the first part of the corollary $\rho(\alpha)$ is weakly directed so, there exists r such that $(t, r), (t', r) \in \rho(\alpha)$. But $t \in \tau([\alpha]A)$, so $r \in \tau(A)$. ■

THEOREM 15. *The following formulas are valid in* $\mathcal{C}^{\mathcal{DDL}}$*:*

1. $\langle\alpha^*\rangle p \to [\alpha^c]p, \ \forall\alpha \in \Pi_{?-free}, \forall p \in \Phi_0$
2. $\langle\alpha\rangle p \to [\alpha^c]p, \ \forall\alpha \in \Pi_{?-free}, \forall p \in \Phi_0$

Proof.

1. Let $s \models \langle\alpha^*\rangle p$, i.e. there exists t_0 such that $(s, t_0) \in \rho(\alpha^*)$ and $t_0 \models p$. We have to prove that $s \models [\alpha^c]p$, that is, for all $(s, t) \in \rho(\alpha^c)$ it holds $t \models p$. Notice that $(s, t) \in \rho(\alpha^c)$ if $(s, t) \in \rho(\alpha^*)$ and $t \models \langle\alpha^*\rangle\psi \to \psi$, for any $\psi \in \Phi_0$. Now, take $(s, t) \in \rho(\alpha^c)$. In particular, $(s, t) \in \rho(\alpha^*)$. Then, since $(s, t_0) \in \rho(\alpha^*)$ and $\rho(\alpha^*)$ is weakly directed due to corollary 14, there must exist r such that $(t, r), (t_0, r) \in \rho(\alpha^*)$. Therefore $r \models p$ since we already had $t_0 \models p$. But now, since $(t, r) \in \rho(\alpha^*)$, it is the case that $t \models \langle\alpha^*\rangle p$. But by hypothesis, $t \models \langle\alpha^*\rangle p \to p$, hence it is the case that $t \models p$.

2. Trivial by noticing that $\langle\alpha\rangle A \to \langle\alpha^*\rangle A$ is a valid formula in *PDL*. ■

LEMMA 16. *The following are provable formulas in* PDL.

1. $\langle\alpha\rangle true \leftrightarrow ([\alpha]A \rightarrow \langle\alpha\rangle A)$,

2. $\langle\alpha^c\rangle A \rightarrow \langle\alpha^*\rangle A$.

Proof. (1) It is a standard result in modal logic. For (2) we prove the following successive implications, where B is a shorthand for $\bigwedge_{p\in\Phi_0}(\langle\alpha^*\rangle p \rightarrow p)$:
$\langle\alpha^c\rangle p \rightarrow \langle\alpha^*; B?\rangle p$, by definition
$\langle\alpha^*; B?\rangle p \rightarrow \langle\alpha^*\rangle\langle B?\rangle p$, by axiom (A2)
$\langle\alpha^*\rangle\langle B?\rangle p \rightarrow \langle\alpha^*\rangle(B \wedge p)$, by axiom (A5)
$\langle\alpha^*\rangle(B \wedge p) \rightarrow \langle\alpha^*\rangle B \wedge \langle\alpha^*\rangle p$, by standard modal calculus
$\langle\alpha^*\rangle A \wedge \langle\alpha^*\rangle p \rightarrow \langle\alpha^*\rangle p$, by propositional calculus. ■

THEOREM 17. DDL *proves*

$$\langle\alpha\rangle true \rightarrow (\langle\alpha^c\rangle p \leftrightarrow [\alpha^c]p),$$

for $\alpha \in \Pi_{?-free}$ *and for* $p \in \Phi_0$.

Proof. Easy from (1) of Theorem 15, and using the previous lemma 16. ■

6 DESCRIBING REFLECTIVE KNOWLEDGE-BASED SYSTEMS WITH *DDL*

In this section we describe, by means of *DDL*, the reasoning dynamics of the reflective architecture *BMS* [Treur, 1992], and a reflective component of the *GET* language [Benerecetti and Spalazzi, 1996]. In the *BMS* system the reflection mechanism is declarative, in the sense that formulas in a unit are *reified* by bridge rules into another unit, which plays the role of a meta-level reasoner. The formulas deduced at this meta-level unit are then *reflected* back. In the case of *GET* we describe its program tactics mechanization, which is based on a set of primitive tactics that *GET* represent as *names* denoting code, plus a set of tacticals that combine the primitive ones to generate complex proof strategies. The reflection consists in the semantic attachment between names (and terms) of tactics and the code associated to them.

6.1 *Example 1:* BMS reasoning dynamics

The inference system of the structure of the *BMS* system, presented in the example of Section 3, can be described in *DDL* as a set of formulas. Some of them, that will be needed later, are the following ones (remember that a is the unit identifier for the $object$ unit and b for the $meta$ unit:

$$\langle \alpha \rangle (b{:}\lceil \varphi \rceil \wedge b{:}\lceil \psi \rceil) \to \langle \alpha; \vdash_{bb} \rangle b{:}\lceil \varphi \& \psi \rceil \quad (1)$$
$$\langle \alpha \rangle (b{:}\lceil \varphi \supset \psi \rceil \wedge b{:}\lceil \varphi \rceil) \to \langle \alpha; \vdash_{bb} \rangle b{:}\lceil \psi \rceil \quad (2)$$
$$\langle \alpha \rangle a{:}\lceil (\varphi, 1) \rceil \to \langle \alpha; \vdash_{ab} \rangle b{:}\lceil T(\varphi) \rceil \quad (3)$$
$$\langle \alpha \rangle a{:}\lceil (\varphi, u) \rceil \to \langle \alpha; \vdash_{ab} \rangle b{:}\lceil \sim T(\varphi) \rceil \quad (4)$$
$$\langle \alpha \rangle a{:}\lceil (\varphi, 0) \rceil \to \langle \alpha; \vdash_{ab} \rangle b{:}\lceil \sim T(\varphi) \rceil \quad (5)$$
$$\langle \alpha \rangle b{:}\lceil PA(\varphi) \rceil \to \langle \alpha; \vdash_{ba} \rangle a{:}\lceil (\varphi, 1) \rceil \quad (6)$$

Now consider the well-known *tweety* example as the particular knowledge base MKB_{TWEETY}, over the above structure, defined as the 3-tuple $(MKB - ST_{BMS}, M_\Sigma, M_\Omega)$, where:

1. $M_\Sigma(a) = (\{penguin, bird, flies\}, \{Prop\}, \{penguin \mapsto Prop, bird \mapsto Prop, flies \mapsto Prop\})$,
2. $M_\Sigma(b) = (\{\mathtt{penguin}, \mathtt{bird}, \mathtt{flies}\}, \{Constant\}, \{\mathtt{penguin} \mapsto Constant, \mathtt{bird} \mapsto Constant, \mathtt{flies} \mapsto Constant\})$,
   where $\{\mathtt{penguin}, \mathtt{bird}, \mathtt{flies}\}$ are a set of constants representing the propositions in $M_{\Sigma(a)}$,
3. $M_\Omega(a) = \{(penguin \supset \sim flies, 1), (penguin \supset bird, 1)\}$, and
4. $M_\Omega(b) = \{\sim T(\mathtt{penguin}) \& T(\mathtt{bird}) \supset PA(\mathtt{flies})\}$

This knowledge base is mapped into the following *DDL* theory, which we call it R_{Tweety}:

$$a{:}\lceil (penguin \supset \sim flies, 1) \rceil \quad (7)$$
$$a{:}\lceil (penguin \supset bird, 1) \rceil \quad (8)$$
$$b{:}\lceil \sim T(\mathtt{penguin}) \& T(\mathtt{bird}) \supset PA(\mathtt{flies}) \rceil \quad (9)$$

To prove in R_{Tweety} that from knowing *bird* and not knowing *penguin* we get *flies* is equivalent to prove the next formula in *DDL*:

$$R_{Tweety} \cup \{a{:}\lceil (bird, 1) \rceil, a{:}\lceil (penguin, u) \rceil\} \vdash_{DDL} [BMS_{Control}] a{:}\lceil (flies, 1) \rceil$$

which means that after the execution of the *BMS* system we get to a state in which the formula *flies* is true. The implicit control in *BMS* system, represented by the compound program $BMS_{Control} = (\vdash_a^c; \vdash_{ab}^c; \vdash_b^c; \vdash_{ba}^c)^c$ means that the control of *BMS* makes all possible deductions at unit a, reflects up to unit b all formulas in a, makes all possible deductions in unit b and finally reflects down to unit a, which can be proved, as the next proof tree shows. In this proof we repeatedly use (2) of theorem 15. In the proof, only the modus ponens in *DDL* is used.

$$\cfrac{\cfrac{\cfrac{\cfrac{\cfrac{\cfrac{a{:}\lceil(bird,1)\rceil}{[\vdash^c_{ab}]b{:}\lceil T(\mathtt{bird})\rceil}\,(3) \qquad \cfrac{a{:}\lceil(penguin,u)\rceil}{[\vdash^c_{ab}]b{:}\lceil\sim T(\mathtt{penguin})\rceil}\,(4)}{[\vdash^c_{ab};\vdash^c_{bb}]b{:}\lceil T(\mathtt{bird})\ \&\ {\sim}T(\mathtt{penguin})\rceil}\,(1)}{[\vdash^c_{ab};\vdash^c_{bb}]b{:}\lceil PA(\mathtt{flies})\rceil}\,(2{,}9)}{[\vdash^c_{ab};\vdash^c_{bb};\vdash^c_{ba}]a{:}\lceil(flies,1)\rceil}\,(6)}{[\vdash^c_{aa};\vdash^c_{ab};\vdash^c_{bb};\vdash^c_{ba}]a{:}\lceil(flies,1)\rceil}\,(\mathrm{MON})}{[(\vdash^c_{aa};\vdash^c_{ab};\vdash^c_{bb};\vdash^c_{ba})^c]a{:}\lceil(flies,1)\rceil}\,(\text{Ths. 15, 17; Lem. 16})$$

Similarly, it can be proved that:

$$R_{Tweety} \cup \{a{:}(bird,1)\ a{:}(penguin,1)\} \vdash_{DDL} [BMS_{Control}]\ a{:}({\sim}flies,1)$$

6.2 *Example 2:* Tacticals in *GET*

We model here tacticals provided in *GET* [Benerecetti and Spalazzi, 1996] as compounds programs in *DDL*. For each tactical T in *GET* we define a compound program T^{DDL} with an *equivalent* behaviour. To make more precise what we want to say by equivalent behaviour, we need first to fix the meaning of three different concepts related to tacticals, as they are understood in *GET*.

1. *Tactical failure*. In *GET* a tactic (or tactical) F_T is a function that is said to *fail* for a given set of arguments $args$ if the function is not defined for $args$; where $args$ are a set of formulas. In such cases, the function defining a tactical in *GET*, T, is constructed by extending the original one returning the symbol "fail" for the undefined cases.

$$T(args) = \begin{cases} \text{fail} & F_T(args) \text{ is undefined} \\ F_T(args) & \text{otherwise} \end{cases}$$

In *DDL* the meaning of *program failure* in a state s can be understood as $s \models_{DDL} \neg\langle T^{DDL}\rangle true$. That is, the program is not applicable in that state. So we can set the next relation between *GET* and *DDL*:

$$T(args) = \text{fail} \Leftrightarrow \lceil args\rceil \models_{DDL} \bot_{T^{DDL}}$$

where $\bot_{T^{DDL}}$ is a shortcut for $[T^{DDL}]false$.

2. *Non-effectiveness*. Another aspect that needs to be clarified is that *GET* tacticals return their first argument as result when the program execution of T *does not change* the *working memory*. In our case we will make it equivalent to the situation of the program T^{DDL} being *not effective*, that is the application of the program T^{DDL} in a state s_{args} will lead us to the same state s_{args}. This condition about the non-effectiveness of program T^{DDL} can be expressed by the following relation:

$$T(args) = first(args) \Leftrightarrow \lceil args \rceil \models_{DDL} \langle T^{DDL} \rangle \varphi \rightarrow \varphi, \forall \varphi \in \Phi_0$$

Notice that the axiom MON together with the condition above allow us to see the relation as:

$$T(args) = first(args) \Leftrightarrow \forall s_{args} \text{ such that} f(s_{args}) \supseteq args, \exists ! t :$$

$$(s_{args}, t) \in \rho(T^{DDL}) \text{ being } t = s_{args}.$$

3. *Effectiveness.* When the computation of an inference rule over an appropriate set of arguments produces a formula that was not present in the arguments, tacticals return that formula as result. It means, in terms of DDL, that the program is effective, that is, there is a transition from a state satisfying the arguments to a state satisfying the inferred formula. This fact can be represented by the next relation:

$$T(args) = result \Leftrightarrow \lceil args \rceil \models_{DDL} \langle T^{DDL} \rangle true \wedge [T^{DDL}] \lceil result \rceil$$

So with these considerations in mind and looking at the definitions of tacticals in [Benerecetti and Spalazzi, 1996] we will make equivalent the computation of a tactical over a set of arguments to the deduction of a particular formula in a suitable theory in DDL. The set of formulas, in a particular unit u, associated to the tacticals in GET is defined as the right part of the next equivalences:

$$T(args) = \text{fail} \Leftrightarrow u{:}\lceil args \rceil \rightarrow \perp_{T^{DDL}}$$

$$T(args) = first(args) \Leftrightarrow u{:}\lceil args \rceil \rightarrow (\langle T^{DDL} \rangle \varphi \rightarrow \varphi)$$

$$T(args) = result \Leftrightarrow u{:}\lceil args \rceil \rightarrow [T^{DDL}] \lceil u{:}\lceil result \rceil \rceil$$

The programs associated to tactics and tacticals are the next ones:

1. Names of *GET* primitive tactics T representing natural deduction inference rules, $\Delta \vdash \phi$, are represented as sets of atomic programs representing all their possible instances [4], that is $T^{DDL} = \bigcup \{ \lceil \Gamma \vdash \varphi \rceil | \Gamma \text{ matches } \Delta, \varphi$ matches $\phi\}$. So, for example, when $T(\Gamma) = \varphi$ then clearly $\Gamma \models_{DDL} [\lceil \Gamma \vdash \varphi \rceil] \varphi$

2. Tacticals are the next compound programs in DDL:

   - $T = (T_1 \; THEN \; T_2)$ is represented as:
     $T^{DDL} = \neg \perp_{T_1^{DDL}}?; T_1^{DDL}; T_2^{DDL}$

[4]The finiteness of the languages ensures that the set of instances is finite. However, the inference rule of implication introduction needs to be modeled by special axioms. Its detailed explanation is omitted here.

- $T = (T_1\ ORELSE\ T_2)$ is represented as:
$T^{DDL} = T_1^{DDL} \cup (\bot_{T_1^{DDL}}?; T_2^{DDL})$
- $T = (TRY\ T_1)$ is represented as:
$T^{DDL} = T_1^{DDL} \cup \bot_{T_1^{DDL}}?$
- $T = (PROGRESS\ T_1)$ is represented as:
$T^{DDL} = ((\bigwedge_{\varphi \in \Phi_0} \langle T_1^{DDL} \rangle \varphi \rightarrow \varphi)?; false?) \cup (\bigvee_{\varphi \in \Phi_0} \langle T_1^{DDL} \rangle \varphi \wedge \neg\varphi)?$
- $T = (REPEAT\ T_1)$ is represented as:
$T^{DDL} = (\bot_{T_1^{DDL}}? \cup \neg \bot_{T_1^{DDL}}?; T_1^{DDL})^*$

To check whether the representation of tacticals in *DDL* corresponds to the intended meaning of failure and effectiveness of tacticals, we will analyze in detail three examples in the following proposition.

PROPOSITION 18. *Given* $T = (TRY\ T_1)$ *and* $T' = (T_1\ THEN\ T_2)$ *we have that for all* s_{args} *such that* $f(s_{args}) \supseteq args$:

1. *If* $T(args) = first(args)$ *Then* $(s_{args}, s_{args}) \in \rho(T^{DDL})$
2. *If* $T(args) = result$*,* $result \neq$ fail*,* $first(args)$*,*
Then $(s_{args}, s_{args} \cup \{result\}) \in \rho(T^{DDL})$
3. *If* $T'(args) =$ fail *Then* $s_{args} \models_{DDL} [T'^{DDL}]false$

Proof. The proof is made as an induction step assuming that T_1 and T_2 have the desired behaviour. The test for the initial case, that is, for primitive tactics, is straightforward. The proof is a semantical one taking into account the model for GET tacticals. Given the completeness of DDL, there is also a syntactic one.

1. By the semantics of tacticals in *GET* we have that $(TRY\ T_1)(args) = first(args)$ iff $T_1(args) =$ fail. By the induction hypothesis $T_1(args) =$ fail if and only if $s_{args} \models_{DDL} [T_1^{DDL}]false$.
But if $s_{args} \models_{DDL} [T_1^{DDL}]false$ then $(s_{args}, s_{args}) \in \rho([T_1^{DDL}]false?)$ and then by the properties of union of programs $(s_{args}, s_{args}) \in \rho(T_1^{DDL} \cup [T_1^{DDL}]false?)$, that is $(s_{args}, s_{args}) \in \rho(T_1^{DDL} \cup \bot_{T_1^{DDL}}?)$, or in other words $(s_{args}, s_{args}) \in \rho(T^{DDL})$.

2. By the semantics of tacticals in *GET* we have that $(TRY\ T_1)(args) = result$ iff $T_1(args) = result$. By the induction hypothesis we have that $T_1(args) = result$ entails $(s_{args}, s_{args} \cup \{result\}) \in \rho(T_1^{DDL})$. So, by the properties of ρ clearly $(s_{args}, s_{args} \cup \{result\}) \in \rho(T_1^{DDL} \cup \bot_{T_1^{DDL}}?)$, that is, $(s_{args}, s_{args} \cup \{result\}) \in \rho(T^{DDL})$.

3. By the semantics of *GET* we have that $(T_1\ THEN\ T_2)(args) =$ fail if one of the next two cases holds:

(a) $T_1(args)$ = fail. In this case we have by induction hypothesis that if $T_1(args)$ = fail then $s_{args} \models_{DDL} \perp_{T_1^{DDL}}$. In such case we have $s_{args} \not\models_{DDL} \neg \perp_{T_1^{DDL}}$ and by the properties of test $s_{args} \not\models_{DDL} \langle \neg \perp_{T_1^{DDL}}? \rangle true$ and also $s_{args} \not\models_{DDL} \langle \neg \perp_{T_1^{DDL}}?; \beta \rangle true$ for any β. Then, obviously, $s_{args} \models_{DDL} [\neg \perp_{T_1^{DDL}}?; T_1^{DDL}; T_2^{DDL}] false$.

(b) $T_1(args) \neq$ fail and $T_2(args \cup T_1(args))$ = fail. In this situation we have that $s_{args} \models_{DDL} \neg \perp_{T_1^{DDL}}$, $s_{args} \models_{DDL} [T_1^{DDL}] true$ and also $s_{args \bigcup \{result\}} \models_{DDL} [t_2^{DDL}] false$. Given that $(s_{args}, s_{args \bigcup \{result\}})$ is a transition in $\rho(\neg \perp_{T_1^{DDL}}?; T_1^{DDL})$ it is clear that $s_{args} \models_{DDL} [\neg \perp_{T_1^{DDL}}?; T_1^{DDL}; T_2^{DDL}] false$. ■

7 CONCLUSION

In this work we have focused on using dynamic logics to describe a particular case of computational dynamic programs called Multi-Language Knowledge-Bases Systems. These systems are built upon several languages, connected through bridge rules that in some cases implement Reification/Reflection mechanisms. Many Artificial Intelligence architectures embed such formalisms as the basic reasoning system as it is the case in MC [Giunchiglia and Serefini, 1994; Giunchiglia and Traverso, 1991; Giunchiglia *et al.*, 1993], *BMS* [Tan, 1992], ML^2 [Balder *et al.*, 1993], $DESIRE$ [Treur, 1992], $MILORDII$ [Agusti *et al.*, 1991; Agusti *et al.*, 1994; Sierra and Godo, 1993], $OMEGA$ [Attardi and Simi, 1995], FOL [Weyhrauch, 1980], and in the Object-centered language for knowledge modeling called NOOS [Arcos and Plaza, 1996].

We have presented a family of Dynamic Logics called Descriptive Dynamic Logics that provides a general framework to describe and compare such multi-language architectures, because *DDL* allows to neatly model their operational semantics. This is the first attempt to formalize executions in multi-language knowledge-based systems and in particular reflective architectures as proofs in Dynamic Logic. Other attempts are based on temporal semantics [Treur, 1994] or on General Logics [Clavel and Meseguer, 1996]. As a next research step, we will show how the most relevant multi-language and multi-agent architectures can be described and compared in DDL.

ACKNOWLEDGMENTS

We are gratefully indebted to Wiebe van der Hoek, John-Jules Meyer, Yao-Hua Tan, Fausto Giunchiglia, Paolo Traverso and Luciano Serafini for many useful comments made on a previous draft of this paper. This Research has been supported by the European TMR number PL93-0186 *VIM*, the Esprit III Basic Research Action number 6156 *DRUMS II*, and spanish projects DGCYT PS93-0212 and *ANALOG* TIC122-93.

This chapter is a reprint of a previously published paper in *Future Generation Computer Systems Journal*, pp. 157–172, 1996; and is reproduced here with the kind permission from the publishers, Elsevier.

Carles Sierra, Lluís Godo, Ramon López de Màntaras
Artificial Intelligence Research Institute, Bellaterra, Spain.

Mario Manzano
Universitat de Barcelona, Spain.

BIBLIOGRAPHY

[Agusti *et al.*, 1991] J. Agusti, F. Esteva, P. Garcia, L. Godo and C. Sierra. Combining multiple-valued logics in modular expert systems. In *Proc. 7th Conf. on Uncertainty in AI*, Morgan Kaufmann, Los Angeles, 17–25, 1991.

[Agusti *et al.*, 1994] J. Agusti, F. Esteva, P. Garcia, L. Godo, R. Lopez de Mantaras and C. Sierra. Local multi-valued logics in modular expert systems. *J. Exper. Theoret. Artificial Intelligence*, **6**, 303–321, 1994.

[Arcos and Plaza, 1996] J.L. Arcos and E. Plaza. Inference and reflection in the object-centered representation language NOOS. *Future Generation Computer System Journal,*, **12**, 119–121, 1996.

[Attardi and Simi, 1995] G. Attardi and M. Simi. A formalisation of viewpoints. *Fundamenta Informaticae*, **23**, 149–174, 1995.

[Balder *et al.*, 1993] J. Balder, F. Van Harmelen and M. Aben. A KADS/(ML)^2 model of a scheduling task. In *Formal Specification of Complex Reasoning Systems*, Jan Treur and Thomas Wetter (eds.), Ellis Horwood, Chichester, UK, 1993.

[Benerecetti and Spalazzi, 1996] M. Benerecetti and L. Spalazzi. *Metafol: Program tactics and logic tactics plus reflection. Future Generation Computer System Journal,*, **12**, 139–156, 1996.

[Clavel and Meseguer, 1996] M. Clavel and J. Meseguer. Axiomatizing reflective logis and languages. In *Proceedngs of Reflction '96*, G. Kiczales, ed. pp. 263–288, Xerox PARC, 1996.

[Giunchiglia and Serefini, 1994] F. Giunchiglia and L. Serafini. Multilanguage hierarchical logics (or: How we can do without modal logics). *Artificial Intelligence*, **65**, 29–70, 1994.

[Giunchiglia and Traverso, 1991] F. Giunchiglia and P. Traverso. Reflective reasoning with and between a declarative metatheory. *IJCAI-91*, 111–117, 1991.

[Giunchiglia *et al.*, 1993] F. Giunchiglia, P. Traverso and E. Giunchiglia. Multicontext systems as a specification framework for complex reasoning systems. In *Formal Specification of Complex Reasoning Systems*, Jan Treur and Thomas Wetter (eds.), Ellis Horwood, Chichester, UK, 1993.

[Goldblatt, 1992] R. Goldblatt. Logics of time and computation. *Lecture Notes*, **7**, CSLI, 1992.

[Harel, 1984] D. Harel. Dynamic logic. In *Handbook of Phylosophical Logic*, D.M. Gabbay and F. Guenthner (eds.), Reidel, Dordrecht, 497–604, 1984.

[F. Van Harmelen *et al.*, 1993] F. Van Harmelen, R. Lopez de Mantaras, J. Malec and J. Treur. Comparing formal specification languages for complex reasoning systems. In *Formal Specification of Complex Reasoning Systems*, Jan Treur and Thomas Wetter (eds.), Ellis Horwood, Chichester, UK, 258–282, 1993.

[Manzano, 1996] M. Manzano. *Extensions of first order logic*, Cambridge Tracts in Theoretical Computer Science. Cambridge University Press, Cambridge, 1996.

[Pratt, 1976] V.R. Pratt. Semantical cosnsiderations on floyd-hoare logic. *Proc. 17th IEEE Symp. Found. Comput. Sci.*, 109–121, 1976.

[Sierra and Godo, 1993] C. Sierra and L. Godo. Specifying simple scheduling tasks in a reflective and modular architecture. In *Formal Specification of Complex Reasoning Systems*, Jan Treur and Thomas Wetter (eds.), Ellis Horwood, Chichester, UK, 199–232, 1993.

[Sierra *et al.*, 1995] C. Sierra, L. Godo and R. Lopez de Mantaras. A dynamic logic framework for reflective architectures. In *IJCAI-95 Workshop on Reflection and Meta-Level Architectures and* their applications in AI, Mamdouh Ibrahim (ed.), 94–102, 1995.

[Tan, 1992] Y.H. Tan. *Non-monotonic Reasoning: Logical Architecture and philosophical applications*. Ph.D. Thesis, Vrije Universitaet Amsterdam, Amsterdam, 1992.

[Treur, 1992] J. Treur. On the use of reflection principles in modelling comples reasoning. *Internat. J. Intelligent Systems*, **6**, 277–294, 1992.

[Treur, 1994] J. Treur. Temporal semantics of meta-level architectures for dynamic control. *Proc. META'94*, Pisa, 1994, pp. 353–376, LNCS Vol. 883, Springer, 1994.

[Weyhrauch, 1980] R. Weyhrauch. Prolegomena to a theory of mechanized formal reasoning. *Artificial Intelligence*, **13**, 133–170, 1980.

JOERI ENGELFRIET, CATHOLIJN M. JONKER
AND JAN TREUR

COMPOSITIONAL VERIFICATION OF MULTI-AGENT SYSTEMS IN TEMPORAL MULTI-EPISTEMIC LOGIC

1 INTRODUCTION

It is a recent trend in the literature on verification to study the use of compositionality and abstraction to structure the process of verification; for example, see [Abadi and Lamport, 1993; Dams *et al.*, 1996; Hooman, 1994]. In [Jonker and Treur, 1998] a compositional verification method was introduced for (formal specifications of) multi-agent systems. In that paper, properties to be verified were formalized semantically in terms of temporal epistemic models, and proofs were constructed by hand. The current paper focuses on the requirements for the choice and use of a suitable logic within which both the properties to be verified and their proofs can be adequately formalized. For the particular application of the logic within a compositional multi-agent system development process in practice, the following requirements for the logic itself and for the use of the logic are of importance:

- compositional structure: properties and proofs can be structured in a compositional manner, in accordance with the compositional structure of the system design.
- dynamics and time: dynamic properties can be expressed, reasoning and induction over time is possible.
- incomplete information states can be expressed.
- it is possible to use some form of default persistence to avoid the necessity to specify within the system design for each time point everything that has to persist over the next time step(s)
- transparency: the proof system and the semantics are transparent and not unnecessarily complicated.

In the following sections, *Temporal Multi-Epistemic Logic* (TMEL) is introduced and shown to be a suitable logic; this logic is a generalization of the Temporal Epistemic Logic TEL introduced in [Engelfriet and Treur, 1996; Engelfriet and Treur, 1996a]; see also [Engelfriet, 1996; Engelfriet and Treur, 1997]. The generalization is made by adding multiple epistemic operators according to the hierarchical compositional structure of the system to be verified. This generalization was

J.J.Ch. Meyer and J. Treur (eds.),
Handbook of Defeasible Reasoning and Uncertainty Management Systems, Vol. 7, 221–250.

inspired by [Fisher and Wooldridge, 1997], were multiple modal operators were introduced (in their case without hierarchical compositional structure) to verify multi-agent systems specified in Concurrent METATEM. By choosing temporal epistemic logic as a point of departure, a choice was made for a discrete and linear time structure and for time to be global.

The structure of the chapter is as follows. In Section 2 the compositional verification method for multi-agent systems is briefly described and an example is given. In Section 3 the temporal multi-epistemic logic is defined. Section 4 discusses compositional temporal theories, Section 5 compositional proof structures, and Section 6 focuses on how to treat non- classical semantics related to default persistence of information.

2 COMPOSITIONAL VERIFICATION

The purpose of verification is to prove that, under a certain set of assumptions, a system satisfies a certain set of properties, for example, the design requirements. In the approach introduced in [Jonker and Treur, 1998], this is done by mathematical proof (i.e., a proof in the form mathematicians are accustomed to), which proves that the specification of the system together with the assumptions implies the properties that the system needs to fulfill. A compositional multi-agent system can be viewed and specified at different levels of abstraction. Viewed from the top level, denoted by L_0, the complete multi-agent system is one component S, where internal information and processes are left unspecified at this level of abstraction (information and process hiding). At the next level of abstraction, L_1, the internal structure of the system is given in terms of its components (as an example, see the agents A and B and the external world EW in Figure 1), but the details of the components are hidden. At the next lower level of abstraction, L_2, (for example) the agent A is specified as a composition of sub-components (see Figure 2). Some components may not be composed of sub-components; such components are called *primitive*. The example has been designed using the compositional development method DESIRE, see [Brazier *et al.*, 1995]. This is a method to develop multi-agent systems according to a compositional structure. The approach to compositional verification addressed in this paper can be used for multi-agent systems designed on the basis of DESIRE, but also for systems designed on the basis of any other method using compositionality as a design principle.

Compositional verification takes into account this compositional structure during the verification process. Properties of a component are only to be expressed using the language specified for the components interfaces (and not the languages specified for sub-components or super-components); this drastically restricts the space of the properties that can be formulated. Verification of a composed component is done using properties of the sub- components it embeds and the components specification (which specifies how it is composed of its sub-components). The assumptions on its sub-components under which the component functions properly,

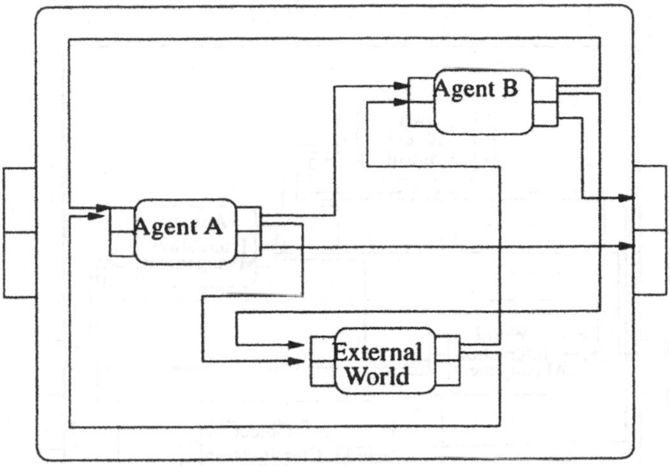

Figure 1. Example multi-agent system for cooperative information gathering

are properties to be proven for these sub-components. This implies that properties at different levels of abstraction are involved in the verification process. These properties have hierarchical logical relations in the sense that at each level, given the component's specification, a property is logically implied by (a conjunction of) the lower level properties that relate to it in the hierarchy (see Figure 3); of course, also logical relations between properties within one abstraction level may exist.

The example multi-agent model used in this paper is composed of two co-operative information gathering agents, A and B, and a component EW representing the external world (see Figure 1). Each of the agents is able to acquire partial information about the external world (by observation). Each agent's own observations are insufficient to draw conclusions of a desired type, but the combined information of both agents is sufficient. Therefore communication is required to be able to draw conclusions. The agents can communicate their own observation results and requests for observation information of the other agent. This quite common situation is simplified to the following materialized form. The world situation consists of an object that has to be classified. One agent can only observe the bottom view of the object (e.g., a circle), the other agent the side view (e.g., a square). By exchanging and combining observation information they are able to classify the object (e.g., a cylinder, expressed by the atom object_type(cylinder)).

Communication from the agent A to B takes place in the following manner:

- the agent A generates at its output interface a statement of the form:

$$\mathsf{to\_be\_communicated\_to}(\langle\mathsf{type}\rangle, \langle\mathsf{atom}\rangle, \langle\mathsf{sign}\rangle, \mathsf{B})$$

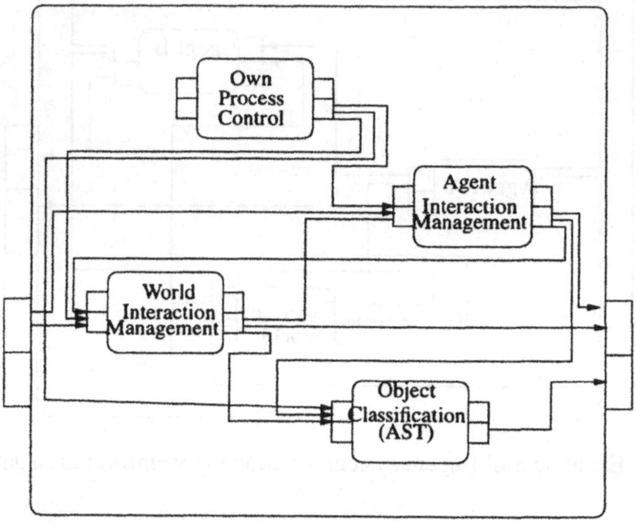

Figure 2. Composition of an agent

- the information is transferred to B; thereby it translated into

  communicated_by(⟨type⟩, ⟨atom⟩, ⟨sign⟩, A)

In the example ⟨ type⟩ can be filled with a label request or world_info, ⟨ atom⟩ is an atom expressing information on the world, and ⟨ sign⟩, is one of pos or neg, to indicate truth or falsity. Interaction between agent A and the world takes place as follows:

- the agent A generates at its output interface a statement of the form:

  to_be_observed(⟨atom⟩)

- the information is transferred to EW; thereby it is translated into

  to_be_observed_by(⟨atom⟩, A)

- the external world EW generates at its output interface a statement of the form:

  observation_result_for(⟨atom⟩, ⟨sign⟩, A)

- the information is transferred to A; thereby it is translated into

  observation_result(⟨atom⟩, ⟨sign⟩)

Part of the output of an agent are conclusions about the classification of the object of the form object_type(ot); these are transferred to the output of the system. As an illustration, an example trace of an agent is shown.

EXAMPLE 1. The input and output information states of agent A during an example process are shown. The information states are denoted by the set of ground literals true in the state. Where r is used, it indicates all instantiated literals with r taken from the set of shapes (circle, square, ...).

A	*input state*	*output state*
0		
1		to_be_observed(view(A, r)) to_be_communicated_to (request, view(B, r), pos, B)
2	observation_result(view(A, circle), pos)	to_be_observed(view(A, r)) to_be_communicated_to (request, view(B, r), pos, B)
3	observation_result(view(A, circle), pos)	to_be_observed(view(A, r)) to_be_communicated_to (request, view(B, r), pos, B) to_be_communicated_to (world_info, view(A, circle), pos, B)
4	observation_result(view(A, circle), pos) communicated_by (world_info, view(B, square), pos, B)	to_be_observed(view(A, r)) to_be_communicated_to (request, view(B, r), pos, B) to_be_communicated_to (world_info, view(A, circle), pos, B)
5	observation_result(view(A, circle), pos) communicated_by (world_info, view(B, square), pos, B)	to_be_observed(view(A, r)) to_be_communicated_to (request, view(B, r), pos, B) to_be_communicated_to (world_info, view(A, circle), pos, B) object_type(cylinder)

This trace shows that the agent A takes the initiative to observe his own view on the object, without having received any input. Moreover, she asks B for his view (apparently A is a pro-active agent for observation and for asking). She receives the result of her observation, namely that her view is a circle, and subsequently communicates this to agent B (apparently A is also a pro-active agent for communication of world information). Next she receives from B information about his view on the object, namely that his view is a square (B is assumed to be reactive for observation and communication of world information). Finally A has drawn the conclusion that the object is a cylinder. Note that for simplicity of the example the information is not retracted.

To be able to perform its tasks, each agent is composed of four components, see Figure 2: three for generic agent tasks (world interaction management, or WIM for short, which reasons about the interaction with the outside world, agent interaction management, or AIM, which reasons about the interaction with other agents, and own process control, or OPC, which reasons about the control of the agent itself; in this example it determines the agent characteristics, for example whether the agent is pro-active or reactive), and one for an agent specific task (object classification, or OC). Since the two agents have a similar architecture, the notation A.WIM is used, for example, to denote component WIM of agent A. As an example of how this agent model works, information describing communication by the agent B to the agent A is transferred to the (input interface of the) component AIM within A (in the form of an atom communicated_by(⟨type⟩, ⟨atom⟩, ⟨sign⟩, A)). In the component AIM the communicated information is identified (by a meta-reasoning process that interprets the communication) and at the output interface of AIM the atom new_world_info(⟨atom⟩, ⟨sign⟩) is generated. From this output interface the information is transferred to the component OC, where it is stored as object level information in the form ⟨atom⟩ or not ⟨atom⟩, depending on whether ⟨sign⟩ is pos or neg. A similar process takes place when observation information is received by the agent, this time through the component WIM.

This example multi-agent system has been verified for all 64 cases where each of the two agents may be pro-active or reactive with respect to observation, communication and/or reasoning in any combination (see [Jonker and Treur, 1998]); in Figure 3 a small part of the properties and logical relations found is depicted). The example used to illustrate the formalization in the current chapter is restricted to a pro-active agent A and a reactive agent. The *compositional verification method* can be formulated informally as follows (for a formalization, see Section 5 below):

A. Verifying one abstraction level against the other
For each abstraction level the following procedure is followed:

1. Determine which properties are of interest for the (higher level) component *D*; these properties can be expressed only in terms of the vocabulary defined for the interfaces of *D*.

2. Determine assumed properties for the lower level components (expressed in terms of their interface languages) that guarantee *D*'s properties.

3. Prove *D*'s properties on the basis of the properties of its sub-components, using the system specification that defines how *D* is composed.

B. Verifying a primitive component
For primitive knowledge-based components a number of verification techniques exist in the literature, for example, [Treur and Willems, 1994; Halpern and Vardi, 1986].

C. The overall verification process
To verify the complete system:

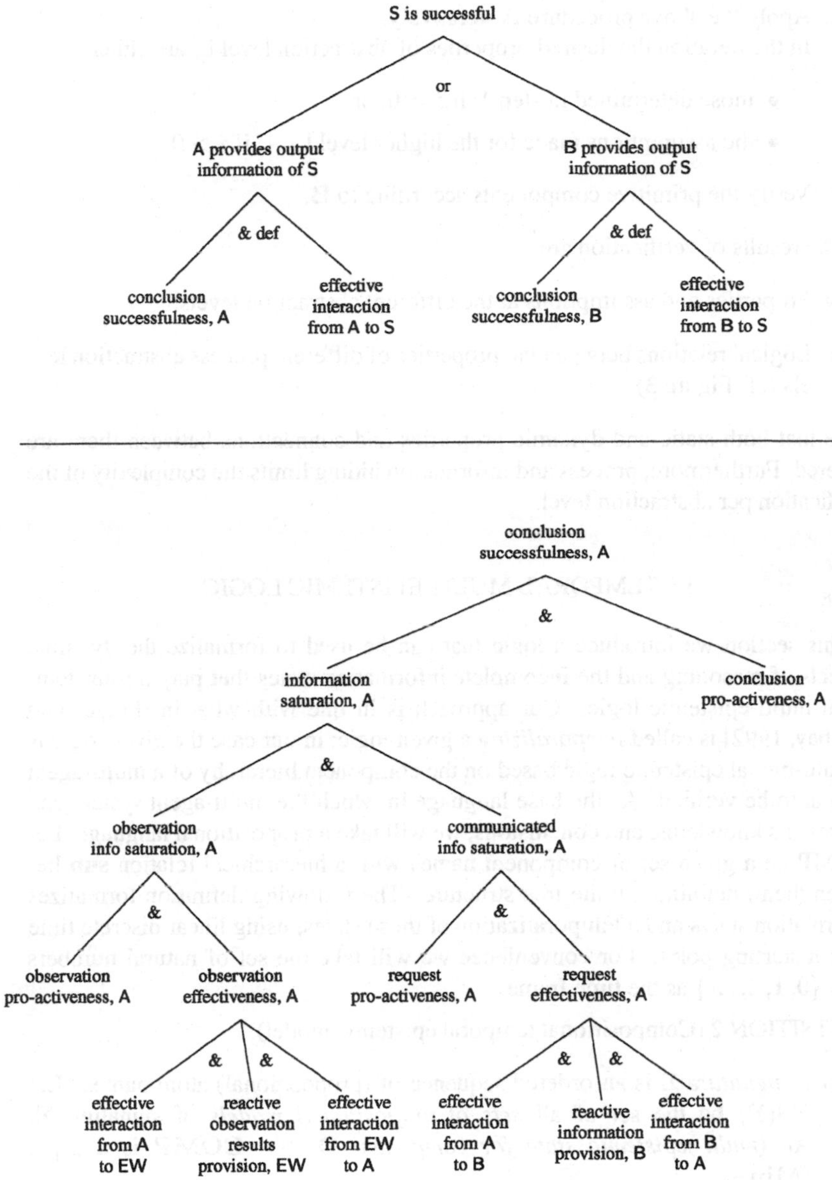

Figure 3. Logical relations between properties at different levels of abstraction for the example system

1. Determine the properties are that are desired for the whole system.
2. Apply the above procedure **A** iteratively.
   In the iteration the desired properties of abstraction level L_i are either:
   - those determined in step 1, if $i = 0$, or
   - the assumptions made for the higher level L_{i-1}, if $i > 0$
3. Verify the primitive components according to **B**.

The results of verification are:

- Properties and assumptions at the different abstraction levels.
- Logical relations between the properties of different process abstraction levels (cf. Figure 3).

Note that both static and dynamic properties and connections between them are covered. Furthermore, process and information hiding limits the complexity of the verification per abstraction level.

3 TEMPORAL MULTI-EPISTEMIC LOGIC

In this section we introduce a logic that can be used to formalize the dynamic aspects of reasoning and the incomplete information states that play a role: temporal multi-epistemic logic. Our approach is in line with what in [Finger and Gabbay, 1992] is called *temporalizing* a given logic; in our case the given logic is a multi-modal epistemic logic based on the component hierarchy of a multi-agent system to be verified. As the base language in which the multi-agent system can express its knowledge and conclusions, we will take a propositional language. Let **COMP** be a given set of component names with a hierarchical relation **sub** between them, defining a finite tree structure. The following definition formalizes information states and a temporalization of these states, using linear discrete time with a starting point. For convenience we will take the set of natural numbers $\mathrm{N} = \{0, 1, 2, \ldots\}$ as the time frame.

DEFINITION 2 (Compositional temporal epistemic model).

(a) A *signature* Σ is an ordered sequence of (propositional) atom names. Let $\mathbf{ES}(\Sigma)$ be the set of all *sets of propositional models* of signature Σ. An *(multi-)epistemic state for component* $\mathbf{D} \in$ **COMP** is a triple $(\mathbf{Min}_D,$
$\mathbf{Mint}_D, \mathbf{Mout}_D) \in \mathbf{ES}(\Sigma) \times \mathbf{ES}(\Sigma) \times \mathbf{ES}(\Sigma)$. A *compositional (multi-) epistemic state*, or compositional information state, based on Σ, is a collection

$$\mathbf{S} = (\mathbf{S}_D)_{\mathrm{D} \in \mathrm{COMP}}$$

of epistemic states $\mathbf{S}_D$ for each of the components **D** in **COMP**.

The set of all possible compositional information states based on Σ is denoted by **CIS**(Σ), or shortly **CIS**.

(b) Let Σ be a signature. A (propositional) *compositional temporal epistemic model* $\mathcal{M}$ of signature Σ is a mapping

$$\mathcal{M} : \mathbb{N} \to \mathbf{CIS}(\Sigma).$$

We will also use the notation $\mathcal{M}_t$ for $\mathcal{M}(t)$.

The definition is illustrated for a example trace.

EXAMPLE 3. The traces described in Example 1 can be defined formally as follows. The signature Σ is as defined for the example agents just above Example 1. For shortness we leave out the internal states. Again, r indicates all instantiated literals with r taken from the set of shapes).

The state $\mathcal{M}_0$ at time point 0 is described by the triple ($\mathbf{Min}_A$, $\mathbf{Mint}_A$, $\mathbf{Mout}_A$) with

$$\begin{array}{lcl} \mathbf{Min}_A & = & \{m \mid m \text{ propositional model}\} \\ \mathbf{Mout}_A & = & \{m \mid m \text{ propositional model}\} \end{array}$$

The state $\mathcal{M}_1$ at time point 1 is described by the triple ($\mathbf{Min}_A$, $\mathbf{Mint}_A$, $\mathbf{Mout}_A$) with

$$\begin{array}{ll} \mathbf{Min}_A & = \{m \mid m \text{ propositional model}\} \\ \mathbf{Mout}_A & = \{m \mid m \text{ propositional model with } m \models \text{to\_be\_observed(view(A, r))} \\ & \quad \text{and} \\ & \quad m \models \text{to\_be\_communicated\_to(request, view(B, r), pos, B)}\} \end{array}$$

The state $\mathcal{M}_2$ at time point 2 is described by the triple ($\mathbf{Min}_A$, $\mathbf{Mint}_A$, $\mathbf{Mout}_A$) with

$$\begin{array}{ll} \mathbf{Min}_A & = \{m \mid m \text{ propositional model with } m \models \text{observation\_result} \\ & \quad \text{(view(A, circle), pos)}\} \\ \mathbf{Mout}_A & = \{m \mid m \text{ propositional model with } m \models \text{to\_be\_observed(view(A, r))} \\ & \quad \text{and} \\ & \quad m \models \text{to\_be\_communicated\_to(request, view(B, r), pos, B)}\} \end{array}$$

The state $\mathcal{M}_3$ at time point 3 is described by the triple ($\mathbf{Min}_A$, $\mathbf{Mint}_A$, $\mathbf{Mout}_A$) with

$$\begin{array}{ll} \mathbf{Min}_A & = \{m \mid m \text{ propositional model with } m \models \text{observation\_result} \\ & \quad \text{(view(A, circle), pos)}\} \\ \mathbf{Mout}_A & = \{m \mid m \text{ propositional model with } m \models \text{to\_be\_observed(view(A, r))}, \\ & \quad m \models \text{to\_be\_communicated\_to(request, view(B, r), pos, B) and} \\ & \quad m \models \text{to\_be\_communicated\_to(world\_info, view(A, circle), pos, B)}\} \end{array}$$

The state $\mathcal{M}_4$ at time point 4 is described by the triple $(\mathbf{Min}_A, \mathbf{Mint}_A, \mathbf{Mout}_A)$ with

$$\begin{aligned}\mathbf{Min}_A &= \{m \mid m \text{ propositional model with } m \models \text{observation\_result}\\ &\quad (\text{view(A, circle), pos) and}\\ &\quad m \models \text{communicated\_by(world\_info, view(B, square), pos, B)}\}\\ \mathbf{Mout}_A &= \{m \mid m \text{ propositional model with } m \models \text{to\_be\_observed(view(A, r))},\\ &\quad m \models \text{to\_be\_communicated\_to(request, view(B, r), pos, B) and}\\ &\quad m \models \text{to\_be\_communicated\_to(world\_info, view(A, circle), pos, B)}\}\end{aligned}$$

The state $\mathcal{M}_5$ at time point 5 is described by the triple $(\mathbf{Min}_A, \mathbf{Mint}_A, \mathbf{Mout}_A)$ with

$$\begin{aligned}\mathbf{Min}_A &= \{m \mid m \text{ propositional model with } m \models \text{observation\_result}\\ &\quad (\text{view(A, circle), pos) and}\\ &\quad m \models \text{communicated\_by(world\_info, view(B, square), pos, B)}\}\\ \mathbf{Mout}_A &= \{m \mid m \text{ propositional model with } m \models \text{to\_be\_observed(view(A, r))},\\ &\quad m \models \text{to\_be\_communicated\_to(request, view(B, r), pos, B)},\\ &\quad m \models \text{to\_be\_communicated\_to(world\_info, view(A, circle), pos, B)}\\ &\quad \text{and}\\ &\quad m \models \text{object\_type(cylinder)}\}\end{aligned}$$

In the language we introduce modal operators $\mathbf{Cin}_D$, $\mathbf{Cint}_D$, $\mathbf{Cout}_D$for each component **D** in **COMP**, expressing the input, internal, and output knowledge of the component. We call these operators the *epistemic operators*. Modal formulae can be evaluated in compositional epistemic states at any point in time: a modal formula $\mathbf{Cout}_D\alpha$ (where α is propositional) is true in a compositional epistemic state **M**, denoted $\mathbf{M} \models \mathbf{Cout}_D\alpha$, if $\mathbf{m} \models \alpha$ for all $\mathbf{m} \in \mathbf{Mout}_D$ (and similarly for $\mathbf{Cin}_D$ and $\mathbf{Cint}_D$). The operators $\mathbf{Cin}_D$, $\mathbf{Cint}_D$, $\mathbf{Cout}_D$ are very similar to the modal **K** operator, so for instance the formula $\neg\mathbf{Cout}_D\alpha \wedge \neg\mathbf{Cout}_D\neg\alpha$ denotes that α is unknown in the output state of component **D** (i.e., neither known to be true nor known to be false). More precisely, the semantics of the epistemic formule are defined by:

DEFINITION 4 (Semantics of the epistemic formulae). Let a compositional epistemic state $\mathbf{S} = (\mathbf{S}_D)_{D \in \mathbf{COMP}}$ be given with for each $\mathbf{D} \in \mathbf{COMP}$ the (multi-)epistemic component state $\mathbf{S}_D = (\mathbf{Min}_D, \mathbf{Mint}_D, \mathbf{Mout}_D) \in ES(\Sigma) \times ES(\Sigma) \times ES(\Sigma)$.

(a) The semantics of the epistemic operators are defined by (where α is a propositional formula):

$$\begin{aligned}\mathbf{S} \models \mathbf{Cin}_D\alpha &\quad\Leftrightarrow\quad m \models \alpha \text{ for all } m \in \mathbf{Min}_D\\ \mathbf{S} \models \mathbf{Cint}_D\alpha &\quad\Leftrightarrow\quad m \models \alpha \text{ for all } m \in \mathbf{Mint}_D\\ \mathbf{S} \models \mathbf{Cout}_D\alpha &\quad\Leftrightarrow\quad m \models \alpha \text{ for all } m \in \mathbf{Mout}_D\end{aligned}$$

(b) For an epistemic formula α:

$$\mathbf{S} \models \neg\alpha \Leftrightarrow \text{ it is not the case that } \mathbf{S} \models \alpha$$

(c) For a set A of epistemic formulae:

$$\mathbf{S} \models \wedge A \Leftrightarrow \text{ for all } \varphi \in A : \mathbf{S} \models \varphi.$$

Furthermore the connectives $\vee$ and $\rightarrow$ are introduced as the usual abbreviations

EXAMPLE 5. For the states occurring in the trace described in Example 3 the following epistemic formulae are true (again r indicates all instantiated literals with r taken from the set of shapes):

$\mathcal{M}_t \models$ **Cout**$_\mathsf{A}$to_be_observed(view(A, r))$\wedge$
to_be_communicated_to(request, view(B, r), pos, B)
for all t $\geq$ 1.

$\mathcal{M}_t \models$ **Cin**$_\mathsf{A}$observation_result(view(A, circle), pos)
for all $t \geq 2$.

$\mathcal{M}_t \models$ **Cout**$_\mathsf{A}$to_be_communicated_to(world_info, view(A, circle), pos, B)
for all $t \geq 3$.

$\mathcal{M}_t \models$ **Cin**$_\mathsf{A}$communicated_by(world_info, view(B, square), pos, B)
for all $t \geq 4$.

$\mathcal{M}_t \models$ **Cout**$_\mathsf{A}$object_type(cylinder)
for all $t \geq 5$.

We need a language to express changes over time. To this end in [Engelfriet and Treur, 1996; Engelfriet and Treur, 1996a] the temporal (uni-modal) epistemic language TEL and its semantics were introduced. To obtain a compositional temporal logic, this logic TEL is generalized in the following manner (the result is called Temporal Multi-Epistemic Logic, or TMEL). Formulae of the form $\mathbf{Cin}_\mathcal{D}\alpha$, $\mathbf{Cint}_\mathcal{D}\alpha$ and $\mathbf{Cout}_\mathcal{D}\alpha$ play the role of atomic propositions. The temporal operators **X, Y, F** and **G** are used. Intuitively, the temporal formula $\mathbf{F}\alpha$ is true at time t means that viewed from time point t, the formula α will be true at *some* time in the future (in some future information state), $\mathbf{G}\alpha$ is true at time t means that viewed from time point t, the formula α will be true at *all* time points in the future, and $\mathbf{X}\alpha$ is true at time t means that α will be true in the next information state. The operator **Y** means 'true at the previous time point'. The semantics of the temporal operators are defined more precisely as follows:

DEFINITION 6 (Semantics of the temporal formulae).

(a) Semantics of the temporal operators ($s \in \mathbb{N}$):

$$\begin{array}{lcl}
(\mathcal{M}, s) \models \mathbf{F}\varphi & \Leftrightarrow & \text{there exists } t \in \mathbb{N}, t > s \text{ such that } (\mathcal{M}, t) \models \varphi \\
(\mathcal{M}, s) \models \mathbf{G}\varphi & \Leftrightarrow & \text{for all } t \in \mathbb{N} \text{ with } t > s : (\mathcal{M}, t) \models \varphi \\
(\mathcal{M}, s) \models \mathbf{X}\varphi & \Leftrightarrow & (\mathcal{M}, s+1) \models \varphi \\
(\mathcal{M}, s) \models \mathbf{Y}\varphi & \Leftrightarrow & (\mathcal{M}, s-1) \models \varphi \text{ if } s > 0 \\
 & & \text{false if } s = 0
\end{array}$$

(b) For a non-temporal formula φ:

$$(\mathcal{M}, t) \models \varphi \quad \Leftrightarrow \quad \mathcal{M}_t \models \varphi$$

(c) For a temporal formula α:

$$(\mathcal{M}, s) \models \neg\alpha \quad \Leftrightarrow \quad \text{it is not the case that } (\mathcal{M}, s) \models \alpha$$

(d) For a set A of temporal formula:

$$(\mathcal{M}, s) \models \wedge A \quad \Leftrightarrow \quad \text{for all } \varphi \in A : (\mathcal{M}, s) \models \varphi$$

(e) A formula φ is true in a model $\mathcal{M}$, denoted $\mathcal{M} \models \varphi$, if for all $s \in \mathbb{N}$: $(\mathcal{M}, s) \models \varphi$.

(f) A set of formulae $\mathbf{T}$ is true in a model $\mathcal{M}$, denoted $\mathcal{M} \models \mathbf{T}$, if for all $\varphi \in \mathbf{T}, \mathcal{M} \models \varphi$. We call $\mathcal{M}$ a model of $\mathbf{T}$.

Furthermore the connectives $\vee$ and $\rightarrow$ are introduced as the usual abbreviations.

Temporal formulae can be used to express aspects of agent behaviour. Specific types of pro-activeness and reactiveness (cf. [Wooldridge and Jennings, 1995] for a general informal explanation) can be defined formally as follows.

Pro-activeness properties can be formalized by expressing that in all traces a specific output will occur. By instantiating this specific type of output, a specific variant of pro-activeness can be defined, for example to perform a specific action, observation or communication. We address some variants.

An agent A is called *observation pro-active* for the set of propositional formulae Φ if for all $\varphi \in \Phi$ it holds at the initial time point of each trace

$$\mathbf{FCout}_{\mathrm{A}}\mathsf{to\_be\_observed}(\varphi)$$

The agent A is called *request pro-active* for the set of propositional formulae Φ if for all $\varphi \in \Phi$ it holds at the initial time point of each trace

$$\mathbf{F}(\mathbf{Cout}_{\mathrm{A}}\mathsf{to\_be\_communicated\_to}(\mathsf{request}, \varphi, \mathsf{pos}, \mathsf{B})$$

The agent A is called *inform pro-active* for the set of propositional formulae Φ if for all $\varphi \in \Phi$ it holds at the initial time point of each trace

$$\mathbf{F}(\mathbf{Cout}_{\mathrm{A}}\mathsf{to\_be\_communicated_to}(\mathsf{world_info}, \varphi, \mathsf{pos}, \mathsf{B}) \vee$$
$$\mathbf{Cout}_{\mathrm{A}}\mathsf{to\_be\_communicated_to}(\mathsf{world_info}, \varphi, \mathsf{neg}, \mathsf{B}))$$

An agent can also be able to initiate actions by generating at its output to_be_performed(a) where a is an action (this does not occur in the example information gathering agents in this chapter). Such an agent A is called *action pro-active* for the set of actions $\aleph$ if for all $a \in \aleph$ it holds at the initial time point of each trace

$$\mathbf{FCout}_{\mathrm{A}}\mathsf{to\_be\_performed}(\mathsf{a})$$

Reactiveness properties relate the occurrence of a specific output to a specific input that has occurred. Also reactiveness properties can be defined for specific types of action, observation or communication. Some variants are as follows.

An agent A is called *observation reactive* with respect to an agent B for the set of propositional formulae Φ if for all $\varphi \in \Phi$ it holds for each trace (i.e., for each time point in the trace)

$$(\mathbf{Cin}_\mathrm{A}\,\mathsf{communicated\_by}(\mathsf{request}, \varphi, \mathsf{pos}, \mathsf{B}) \wedge \\ \neg\mathbf{YCin}_\mathrm{A}\,\mathsf{communicated\_by}(\mathsf{request}, \varphi, \mathsf{pos}, \mathsf{B})) \\ \rightarrow \mathbf{FCout}_\mathrm{A}\,\mathsf{to\_be\_observed}(\varphi)$$

The agent A is called *inform reactive* with respect to agent B for the set of propositional formulae Φ if for all $\varphi \in \Phi$ it holds for each trace

$$(\mathbf{Cin}_\mathrm{A}\,\mathsf{communicated\_by}(\mathsf{request}, \varphi, \mathsf{pos}, \mathsf{B}) \wedge \\ \neg\mathbf{YCin}_\mathrm{A}\,\mathsf{communicated\_by}(\mathsf{request}, \varphi, \mathsf{pos}, \mathsf{B})) \\ \rightarrow \mathbf{F}(\mathbf{Cout}_\mathrm{A}\,\mathsf{to\_be\_communicated\_to}(\mathsf{world\_info}, \varphi, \mathsf{pos}, \mathsf{B}) \vee \\ \mathbf{Cout}_\mathrm{A}\,\mathsf{to\_be\_communicated\_to}(\mathsf{world\_info}, \varphi, \mathsf{neg}, \mathsf{B}))$$

An agent A which is also able to perform actions is called *action reactive* with respect to agent B for the set of actions $\aleph$ if for all $a \in \aleph$ it holds for each trace

$$(\mathbf{Cin}_\mathrm{A}\,\mathsf{communicated\_by}(\mathsf{request}, a, \mathsf{pos}, \mathsf{B}) \wedge \\ \neg\mathbf{YCin}_\mathrm{A}\,\mathsf{communicated\_by}(\mathsf{request}, a, \mathsf{pos}, \mathsf{B})) \\ \rightarrow \mathbf{FCout}_\mathrm{A}\,\mathsf{to\_be\_performed}(a)$$

Note that, a pro-activeness property is stronger than a reactive propertry of the same type: the latter is a conditional variant of the former, and hence implied by it.

EXAMPLE 7. The trace $\mathcal{M}$ of agent A described in Examples 3 and 5 satisfies the temporal formulae for all pro-activeness and reactiveness properties introduced above, for a suitable choice of the set of formulae Φ.

Taking $\Phi = \{\mathsf{view(A, r)} \mid \mathsf{r\ shape}\}$ in the observation pro-activeness definition, indeed for all shapes r at the initial time point of the trace $\mathcal{M}$ it holds

$$\mathbf{FCout}_\mathrm{A}\,\mathsf{to\_be\_observed}(\mathsf{view}(\mathsf{A}, \mathsf{r}))$$

Taking $\Phi = \{\mathsf{view(B, r)} \mid \mathsf{r\ shape}\}$ in the request pro-activeness definition, indeed for all shapes r at the initial time point of the trace $\mathcal{M}$ it holds

$$\mathbf{FCout}_\mathrm{A}\,\mathsf{to\_be\_communicated\_to}(\mathsf{request}, \mathsf{view}(\mathsf{B}, \mathsf{r}), \mathsf{pos}, \mathsf{B})$$

Taking $\Phi = \{\mathsf{view(A, r)} \mid \mathsf{r\ shape}\}$ in the request pro-activeness definition, indeed for all shapes r at the initial time point of the trace $\mathcal{M}$ it holds

$$\mathbf{F}(\mathbf{Cout}_\mathrm{A}\,\mathsf{to\_be\_communicated\_to}(\mathsf{world\_info}, \mathsf{view}(\mathsf{A}, \mathsf{r}), \mathsf{pos}, \mathsf{B}) \vee \\ \mathbf{Cout}_\mathrm{A}\,\mathsf{to\_be\_communicated\_to}(\mathsf{world\_info}, \mathsf{view}(\mathsf{A}, \mathsf{r}), \mathsf{neg}, \mathsf{B}))$$

For more details of TEL, see [Engelfriet and Treur, 1996; Engelfriet and Treur, 1996a]. For temporal epistemic logic different entailment relations can be used, both classical and non-classical; see e.g., [Engelfriet, 1996; Engelfriet and Treur, 1997]. A discussion about this choice can be found in Section 6.

4 COMPOSITIONAL TEMPORAL THEORIES

In order to embed the compositional verification proofs in temporal multi-epistemic logic, a multi-agent system specification is translated into a temporal theory. We require of this translation that the compositional structure is preserved. This means that instead of one global temporal theory, each component **C** in the hierarchy is translated into a separate, local temporal theory $\mathbf{T}_C$ for this component. Each of these local temporal theories is expressed in its own local language $\mathbf{L}_C$. Therefore, we introduce collections of sub-languages and collections of temporal theories that are labelled by the set of components **COMP**. A language for a component defines the (restricted) terms in which its internal information, as well as the information in its input and output interface can be expressed. Each language $\mathbf{L}_C$ defines a subset of the epistemic formulae; it only contains symbols from a (local) subset of the overall signature Σ and it contains the epistemic operators $\mathbf{Cin}_C$, $\mathbf{Cint}_C$ and $\mathbf{Cout}_C$ (and no epistemic operators for other components). All temporal operators can be applied.

DEFINITION 8 (language composition). Let **COMP** be a set of component names with a binary sub-component relation **sub**. *Primitive* components are elements $\mathbf{D} \in \mathbf{COMP}$ for which no $\mathbf{C} \in \mathbf{COMP}$ exists with **C sub D**. The other components are called *composed*.

a) A *language composition* is a collection of sub-languages

$$(\mathbf{L}_C)_{C \in \mathbf{COMP}}$$

where in each language $\mathbf{L}_C$ only the epistemic operators $\mathbf{Cin}_C$, $\mathbf{Cint}_C$ and $\mathbf{Cout}_C$ are used (and no epistemic operators for other components).

b) The collection of *interface languages* for the language composition $(\mathbf{L}_C)_{C \in \mathbf{COMP}}$ is the collection

$$(\mathbf{L}^{\mathrm{if}}_C)_{C \in \mathbf{COMP}}$$

where for any component **D**, the language $\mathbf{L}^{\mathrm{if}}_\mathbf{D}$ is the restriction of $\mathbf{L}_D$ to formulae in which the epistemic operator $\mathbf{Cint}_D$ does not occur.

c) The collection of *bridge languages* for the language composition $(\mathbf{L}_C)_{C \in \mathbf{COMP}}$ is the collection

$$(\mathbf{L}^{+}_C)_{C \in \mathbf{COMP}}$$

defined for any component **C** by

$$\mathbf{L}_C^+ = \mathbf{L}_C \cup \bigcup_{C' \mathbf{sub} C} \mathbf{L}_{C'}^{\mathrm{if}}$$

d) The *cumulative language composition* for the language composition $(\mathbf{L}_C)_{C \in \mathbf{COMP}}$ is the collection

$$(\mathbf{L}_C^*)_{C \in \mathbf{COMP}}$$

defined for any component **C** by

$$\mathbf{L}_C^* = \mathbf{L}_C \cup \bigcup_{C' \mathbf{sub} C} \mathbf{L}_C^* \quad \text{if C is a composed component}$$
$$\mathbf{L}_C^* = \mathbf{L}_C \quad \text{if C is a primitive component}$$

EXAMPLE 9 (language composition). We give part of the languages of some of the components of the example multi-agent system (for ot varying over the object types, r over shapes, X is the agent A or B, sign is pos or neg):

$\mathbf{L}_S$	$\mathbf{Cout}_S$	object_type(ot)
$\mathbf{L}_A$	$\mathbf{Cout}_A$	to_be_observed(view(A, r)),
	$\mathbf{Cout}_A$	to_be_communicated_to(request, view(B, r), pos, B)
	$\mathbf{Cin}_A$	observation_result(view(A, r), pos)
	$\mathbf{Cin}_A$	communicated_by(world_info, view(B, r), pos, B)
$\mathbf{L}_{EW}$	$\mathbf{Cin}_{EW}$	to_be_observed_by(view(X, r), X),
	$\mathbf{Cout}_{EW}$	observation_result_for(view(X, r), sign, X)

A theory composition consists of a collection of (local) theories $\mathbf{T}_C$ for the different components C; for example, see Figure 4. A more precise definition is as follows:

DEFINITION 10 (Theory composition). Let $(\mathbf{L}_C)_{C \in \mathbf{COMP}}$ be a language composition.

a) A *compositional temporal theory* for $(\mathbf{L}_C)_{C \in \mathbf{COMP}}$ is a collection $(\mathbf{T}_C)_{C \in \mathbf{COMP}}$ where each temporal theory $\mathbf{T}_C$ is a theory in the language $\mathbf{L}_C^+$.

b) Let $(\mathbf{T}_C)_{C \in \mathbf{COMP}}$ be a compositional temporal theory. The *collection of cumulative theories* $(\mathbf{T}_C^*)_{C \in \mathbf{COMP}}$ is defined for any component **C** as:

$$\mathbf{T}_C^* = \mathbf{T}_C \cup \bigcup_{C' \mathbf{sub} C} T_{C'}^* \quad \text{if } C \text{ is a composed component}$$
$$\mathbf{T}_C^* = \mathbf{T}_C \quad \text{if } C \text{ is a primitive component}$$

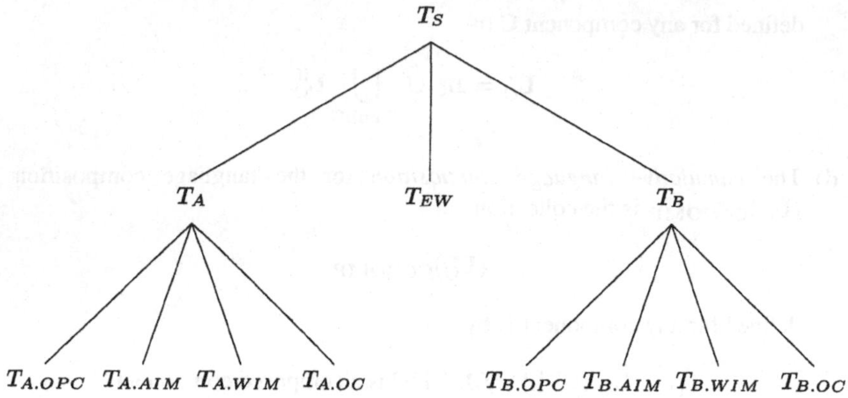

Figure 4. Structure of a theory composition for the example system

EXAMPLE 11 (partial compositional theory; a composed component). For each of the components of the multi-agent system its specification can be translated into a temporal theory (see Figure 4). The part of the theory for the top level component that is relevant to prove success of the system is the following (again, ot ranges over the object types, r over shapes, X is the agent A or B, sign is pos or neg):

$\mathbf{T}_{:}$ $\mathbf{Y}\ \mathbf{Cout}_{X}\text{to\_be\_observed(view(X, r))}$
$\quad \rightarrow \mathbf{Cin}_{EW}\text{to\_be\_observed\_by(view(X, r), X)}$
$\mathbf{Y}\ \mathbf{Cout}_{A}\text{to\_be\_communicated\_to(request, view(B, r), pos, B)}$
$\quad \rightarrow \mathbf{Cin}_{B}\text{communicated\_by(request, view(B, r), pos, A)}$
$\mathbf{Cout}_{B}\text{to\_be\_communicated\_to(world\_info, view(B, r), sign, A)}$
$\quad \rightarrow \mathbf{Cin}_{A}\text{communicated\_by(world\_info, view(B, r), sign, B)}$
$\mathbf{Y}\ \mathbf{Cout}_{X}\text{object\_type(ot)} \rightarrow \mathbf{Cout}_{S}\text{object\_type(ot)}$
$\mathbf{Y}\ \mathbf{Cout}_{X}\text{object\_type(ot)} \rightarrow \mathbf{Cout}_{S}\text{object\_type(ot)}$
$\mathbf{Y}\ \mathbf{Cout}_{EW}\text{observation\_result\_for(view(X, r), sign, X)}$
$\quad \rightarrow \mathbf{Cin}_{X}\text{observation\_result(view(X, r), sign, X)}$

For example, the last formula is part of the description of the information links from EW to A and from EW to B. This formula expresses that the information previously in the output of EW is currently contained in the input interface of the agent A (under a simple translation). The part of the theory for agent A that is relevant to prove successfulness of the system is the following:

$$
\begin{aligned}
\mathbf{T}_A: \quad & \mathbf{Y}\ \mathbf{Cin}_{\mathsf{A}}\mathsf{observation\_result}(\mathsf{view}(\mathsf{A},\mathsf{r}),\mathsf{sign}) \\
& \quad \rightarrow \mathbf{Cin}_{\mathsf{A.WIM}}\mathsf{observation\_result}(\mathsf{view}(\mathsf{A},\mathsf{r}),\mathsf{sign}) \\
& \mathbf{Y}\ \mathbf{Cin}_{\mathsf{A}}\mathsf{communicated\_by}(\mathsf{world\_info},\mathsf{view}(\mathsf{B},\mathsf{r}),\mathsf{sign},\mathsf{B}) \\
& \quad \rightarrow \mathbf{Cin}_{\mathsf{A.AIM}}\mathsf{communicated\_by}(\mathsf{world\_info},\mathsf{view}(\mathsf{B},\mathsf{r}),\mathsf{sign},\mathsf{B}) \\
& \mathbf{Y}\ \mathbf{Cout}_{\mathsf{A.WIM}}\mathsf{to\_be\_observed}(\mathsf{view}(\mathsf{A},\mathsf{r}),\mathsf{sign}) \\
& \quad \rightarrow \mathbf{Cout}_{\mathsf{A}}\mathsf{to\_be\_observed}(\mathsf{view}(\mathsf{A},\mathsf{r}),\mathsf{sign}) \\
& \mathbf{Y}\ \mathbf{Cout}_{\mathsf{A.AIM}}\mathsf{to\_be\_communicated\_to}(\mathsf{request},\mathsf{view}(\mathsf{B},\mathsf{r}),\mathsf{pos},\mathsf{B}) \\
& \quad \rightarrow \mathbf{Cout}_{\mathsf{A}}\mathsf{to\_be\_communicated\_to}(\mathsf{request},\mathsf{view}(\mathsf{B},\mathsf{r}),\mathsf{pos},\mathsf{B}) \\
& \mathbf{Y}\ \mathbf{Cout}_{\mathsf{A.OC}}\mathsf{object\_type}(\mathsf{ot}) \rightarrow \mathbf{Cout}_{\mathsf{A}}\mathsf{object\_type}(\mathsf{ot}) \\
& \mathbf{Y}\ \mathbf{Cout}_{\mathsf{A.OC}}\mathsf{object\_type}(\mathsf{ot}) \rightarrow \mathbf{Cout}_{\mathsf{A}}\mathsf{object\_type}(\mathsf{ot}) \\
& \mathbf{Y}\ \mathbf{Cout}_{\mathsf{A.AIM}}\mathsf{communicated\_by}(\mathsf{request},\mathsf{view}(\mathsf{A},\mathsf{r}),\mathsf{sign},\mathsf{B}) \\
& \quad \rightarrow \mathbf{Cin}_{\mathsf{A.WIM}}\mathsf{requested}(\mathsf{view}(\mathsf{A},\mathsf{r}))
\end{aligned}
$$

$$
\begin{aligned}
& \mathbf{Y}\ \mathbf{Cout}_{\mathsf{A.AIM}}\mathsf{new\_world\_info}(\mathsf{view}(\mathsf{B},\mathsf{r}),\mathsf{pos}) \rightarrow \mathbf{Cin}_{\mathsf{A.OC}}\mathsf{view}(\mathsf{B},\mathsf{r}) \\
& \mathbf{Y}\ \mathbf{Cout}_{\mathsf{A.AIM}}\mathsf{new\_world\_info}(\mathsf{view}(\mathsf{B},\mathsf{r}),\mathsf{neg}) \rightarrow \mathbf{Cin}_{\mathsf{A.OC}}\neg\mathsf{view}(\mathsf{B},\mathsf{r}) \\
& \mathbf{Y}\ \mathbf{Cout}_{\mathsf{A.WIM}}\mathsf{new\_world\_info}(\mathsf{view}(\mathsf{A},\mathsf{r}),\mathsf{pos}) \rightarrow \mathbf{Cin}_{\mathsf{A.OC}}\mathsf{view}(\mathsf{B},\mathsf{r}) \\
& \mathbf{Y}\ \mathbf{Cout}_{\mathsf{A.WIM}}\mathsf{new\_world\_info}(\mathsf{view}(\mathsf{A},\mathsf{r}),\mathsf{neg}) \rightarrow \mathbf{Cin}_{\mathsf{A.OC}}\neg\mathsf{view}(\mathsf{B},\mathsf{r})
\end{aligned}
$$

EXAMPLE 12 (partial compositional theory; a primitive component). Primitive components can, for example, be specified by logical rules of the form 'conjunction of literals' implies 'literal'), as is the case in DESIRE. Consider the following rule of the knowledge base of the primitive component *object classification*:

if view(A, circle) and view(B, square) then object_type(cylinder)

This rule can be formalized in TMEL by:

$$
\begin{aligned}
& \phi \wedge \mathbf{Y}\mathbf{Cin}_{\mathsf{X.OC}}\mathsf{view}(\mathsf{A},\mathsf{circle}) \wedge \mathbf{Y}\mathbf{Cin}_{\mathsf{X.OC}}\mathsf{view}(\mathsf{B},\mathsf{square}) \rightarrow \\
& \quad \mathbf{Cout}_{\mathsf{X.OC}}\mathsf{object\_type}(\mathsf{cylinder})
\end{aligned}
$$

where ϕ is a formula expressing control information that allows the rule to be used (for example, the component should be active).

5 COMPOSITIONAL PROOF STRUCTURES

Verification proofs are composed of proofs at different levels of abstraction (see Figure 3). These proofs involve properties of the components at these abstraction levels.

DEFINITION 13 (composition of properties). A *composition of properties* for a language composition $(\mathbf{L}_C)_{C\in\mathbf{COMP}}$ is a collection

$$(\mathbf{P}_C)_{C\in\mathbf{COMP}}$$

where for each **C** the set $\mathbf{P}_C$ is a set of temporal statements in the language $\mathbf{L}^{\mathrm{if}}_C$.

Note that in our approach it is not allowed to phrase properties of a component in terms other than those of its interface language.

EXAMPLE 14. In the proof of the success property of S (a small part of which is depicted in Figure 3) the following composition of properties is used (see also Example 9):

In the proof of the properties shown in Example 5.2, the theories shown in Example 11 and 12 are used.

To formalize the notion of proof, some distinctions have to be made. In the first place, a proof can either be local or global. A *local proof* relates properties of one component to properties of its immediate sub-components and (domain-specific) assumptions. A *global proof* relates top level properties to assumptions. In the second place, proofs can be related to different types of semantics. A straightforward choice would be to take the *standard semantics* as defined in Section 3. However, as already mentioned, to avoid the necessity to specify every aspect of the system that persists over a time step, also non-classical forms of semantics can be considered. For example, semantics defined by a subset of *preferred* temporal epistemic models, that cover *default persistence*: aspects of states only change if explicitly specified. The issue of default persistence is addressed in more detail in Section 6. To be able to express all these possibilities, a given but arbitrary entailment relation $\mid\!\sim$ is taken as a parameter in the next definitions. One of the possible instances for this entailment relation is classical provability, but other possible instances can be found in relation to non-classical types of semantics.

DEFINITION 15 (compositional and global provability). For the language composition $(\mathbf{L}_C)_{C\in\mathbf{COMP}}$, let a composition of properties $(\mathbf{P}_C)_{C\in\mathbf{COMP}}$ and a compositional temporal theory $(\mathbf{T}_C)_{C\in\mathbf{COMP}}$ be given. Let $\mid\!\sim$ be an entailment relation for temporal multi-epistemic logic.

a) The composition of properties $(\mathbf{P}_C)_{C\in\mathbf{COMP}}$ is *compositionally provable* with respect to $\mid\!\sim$ from the compositional temporal theory $(\mathbf{T}_C)_{C\in\mathbf{COMP}}$ if for each component C the following holds:

$$\mathbf{T}_C \mid\!\sim \{\wedge \bigcup_{C' \mathrm{sub} C} \mathbf{P}_{C'} \rightarrow \wedge \mathbf{P}_C\} \quad \text{if C is composed}$$

$$\mathbf{T}_C \mid\!\sim \mathbf{P}_C \quad \text{if C is primitive}$$

b) The composition of properties is *globally provable* with respect to $\mid\!\sim$ from the compositional temporal theory $(\mathbf{T}_C)_{C\in\mathbf{COMP}}$ if for each component C the following holds:

$$\mathbf{T}^*_C \mid\!\sim \mathbf{P}_C.$$

For example, the collection of success properties of Example 14 turns out to be globally provable from the compositional temporal theory $(\mathbf{T}_C)_{C\in\mathbf{COMP}}$, with

System S *as a whole*

$\mathbf{P}_S$: $\wedge_{ot}$(**F Cout**$_S$object_type(ot) ∨ **F Cout**$_S$¬object_type(ot))

Agent A (the pro-active agent)

$\mathbf{P}_A$: [$\wedge_r$(**Cin**$_A$observation_result(view(A, r), pos)∨
Cin$_A$observation_result(view(A, r), neg))]
∧[$\wedge_r$(**Cin**$_A$communicated_by(world_info, view(B, r), pos, B)∨
Cin$_A$communicated_by(world_info, view(B, r), neg, B))]
→ $\wedge_{ot}$(*F***Cout**$_A$object_type(ot) ∨ **F Cout**$_A$¬object_type(ot))
(conclusion pro-activeness, A)
$\wedge_r$**F Cout**$_A$to_be_observed(view(A, r))
(observation pro-activeness, A)
$\wedge_r$**F Cout**$_A$to_be_communicated_to(request, view(B, r), pos, B),
(request pro-activeness, A)

Agent B (the reactive agent)

$\mathbf{P}_B$: $\wedge_r$[**Cin**$_B$communicated_by(request, view(B, r), pos, A)
→ (**F Cout**$_B$to_be_communicated_to(world_info, view(B, r), pos)∨
F Cout$_B$to_be_communicated_to(world_info, view(B, r), neg))]
(reactive information provision, B)

External World EW

$\mathbf{P}_{EW}$: $\wedge_r$[**Cin**$_{EW}$to_be_observed_by(view(X, r), X)
→ (**F Cout**$_{EW}$observation_result_for(view(X, r), pos)∨
F Cout$_{EW}$observation_result_for(view(X, r), neg))]
(reactive observation results provision, EW)

Components within A

$\mathbf{P}_{A.OPC}$: F**Cout**$_{A.OPC}$pro-active
(pro-activeness, OPC)

$\mathbf{P}_{A.AIM}$: [**Cin**$_{A.AIM}$pro-active
→ $\wedge_r$F**Cout**$_{A.AIM}$to_be_communicated_to(request, view(B, r), pos, B)],
(conditional request pro-activeness, AIM)
[$\wedge_r$**Cin**$_{A.AIM}$communicated_by(world_info, view(B, r), sign, B)
→ F**Cout**$_{A.AIM}$new_world_info(view(B, r), sign)]

$\mathbf{P}_{A.WIM}$: [**Cin**$_{A.WIM}$pro-active → $\wedge_r$F**Cout**$_{A.WIM}$to_be_observed (view(A, r))],
(conditional observation pro-activeness, WIM)
[$\wedge_r$**Cin**$_{A.WIM}$observation_result(view(A, r), sign)
→ F**Cout**$_{A.WIM}$new_world_info(view(A, r), sign)]

$\mathbf{P}_{A.OC}$: $\wedge_r$X(**Cin**$_{A.OC}$view(X, r) ∨ **Cin**$_{A.OC}$view(X, r))
→ $\wedge_{ot}$(F**Cout**$_{A.OC}$object_type(ot) ∨ F**Cout**$_{A.OC}$ ¬object_type(ot))

respect to the provability relation of classical entailment in TMEL, augmented with a default persistence assumption (see the next section).

Compositional provability does not necessarily imply global provability. However, the implication holds if the entailment relation satisfies, apart from reflexivity (if $V \subseteq W$, then $W|\!\!\sim V$), the property of transitivity:

$$\mathbf{T}|\!\!\sim U \quad \& \quad U|\!\!\sim W \Rightarrow \mathbf{T}|\!\!\sim W \qquad \text{(Transitivity)}$$

for all sets of formulae T, U, W. It is well-known that transitivity and reflexivity imply monotonicity. Moreover, in the proof below the following straighforward properties are used:

$$\begin{array}{lll} T|\!\!\sim V & \& & U|\!\!\sim W \Rightarrow T \cup U|\!\!\sim V \cup W \\ A|\!\!\sim C & \Rightarrow & A \cup B|\!\!\sim C \cup B \end{array}$$

Finally, the following relation to classical entailment $\vdash$ is assumed:

$$T|\!\!\sim V \quad \& \quad V \vdash U \Rightarrow T|\!\!\sim U.$$

This means that on the right hand side classical deduction can be done.

PROPOSITION 16. *If the entailment relation $|\!\!\sim$ satisfies, in addition to reflexivity, transitivity, then compositional provability with respect to $|\!\!\sim$ implies global provability with respect to $|\!\!\sim$. In particular, if $\vdash$ is a classical provability relation for temporal multi-epistemic logic, then compositional provability with respect to $\vdash$ implies global provability with respect to $\vdash$.*

Proof. The proof is by induction on the depth of the tree structure of the component hierarchy.

a) For trees with depth 0, consisting of a single component C, both compositional provability and global provability are by definition equivalent to

$$\mathbf{T}_C|\!\!\sim \mathbf{P}_C$$

b) For a tree with depth at least one with top level component C, suppose compositional provability holds for a composition of properties, and by induction hypothesis the global provability already holds for all proper subtrees. Global provability for the top level component is defined as

$$\mathbf{T}^*_C|\!\!\sim \mathbf{P}_C$$

where

$$\mathbf{T}^*_C = \mathbf{T}_C \cup \bigcup_{C' \mathbf{sub} C} \mathbf{T}^*_{C'}$$

So, to be proven:

$$\mathbf{T}_C \cup \bigcup_{C' \mathbf{sub} C} \mathbf{T}^*_{C'} \mathbf{P}_C \qquad (0)$$

The induction hypothesis implies that for all subcomponents C' of C:

$$\mathbf{T}^*_{C'} \mid\!\sim \mathbf{P}_{C'}$$

Therefore, by taking unions left and right

$$\bigcup_{C' \mathbf{sub} C} \mathbf{T}^*_{C'} \mid\!\sim \bigcup_{C' \mathbf{sub} C} \mathbf{P}_{C'}$$

By taking left and right the union with $\mathbf{T}_C$ it follows

$$\mathbf{T}_C \cup \bigcup_{C' \mathbf{sub} C} \mathbf{T}^*_{C'} \mid\!\sim \mathbf{T}_C \cup \bigcup_{C' \mathbf{sub} C} \mathbf{P}_{C'} \qquad (1)$$

As compositional provability holds for the whole tree, for the composed component C we have

$$\mathbf{T}_C \mid\!\sim \{\wedge \bigcup_{C' \mathbf{sub} C} \mathbf{P}_{C'} \rightarrow \wedge \mathbf{P}_C\} \qquad (2)$$

¿From (1) and (2) it follows

$$\mathbf{T}_C \cup \bigcup_{C' \mathbf{sub} C} \mathbf{T}^*_{C'} \mid\!\sim \{\wedge \bigcup_{C' \mathbf{sub} C} \mathbf{P}_{C'} \rightarrow \wedge \mathbf{P}_C\} \cup \bigcup_{C' \mathbf{sub}\, C} \mathbf{P}_{C'}$$

and from this it follows by classical deduction at the right hand side:

$$\mathbf{T}_C \cup \bigcup_{C' \mathbf{sub} C} \mathbf{T}^*_{C'} \mid\!\sim \mathbf{P}_C$$

which had to be proven: (0). ■

This proposition shows that for classical entailment the implication holds. But, for example, for an entailment relation taking into account minimal change, the implication does not hold. In the next section it is discussed what kind of provability relation, in the light of these results, to formalize compositional verification processes is an adequate choice.

6 DEFAULT PERSISTENCE

The conditions under which a classical inference relation can be used depend on the specific form of semantics. For example, in DESIRE a default persistence assumption has been made: it is only specified what has to be changed; all other information is meant to persist in time. An exception is made for information that has to be retracted because it was derived from information that does not hold any more. We now discuss a manner in which default persistence and revision can be treated within temporal multi-epistemic logic. In principle, a compositional specification can be formalized using executable temporal formulae. Roughly, executable temporal formulae are temporal formulae of the form

declarative past $\Rightarrow$ imperative future

For more details on this paradigm, and the different variants within, see [Barringer *et al.*, 1991; Barringer *et al.*, 1996]. For our purposes the following definition is chosen. *Simplified executable temporal formulae* are formulae of the form

past and present $\Rightarrow$ present

The right hand side of these formulae φ are called *heads*, denoted by **head**(φ); they are taken from the set

$$\begin{aligned}\mathbf{HEADS} \;=\; & \{\, CL \mid L \text{ propositional literal, } C \text{ epistemic operator} \,\} \cup \\ & \{\neg CA \wedge \neg C\neg A \mid A \text{ propositional atom, } C \text{ epistemic operator} \,\}\end{aligned}$$

The left hand side of φ is called *body*, denoted by **body**(φ). Within the body, the 'past' part is a formula that refers strictly to the past. The 'present' part is a conjunction of temporal literals that are either of the form CL or $\neg CL$.

These formulae only specify what has to be changed. All other information is meant to persist (default persistence) in time, with an exception for information that has to be revised because it was derived from information that does not hold anymore. In principle this entails non-classical semantics. However, a translation is possible into temporal theories with classical semantics if a form of temporal completion (similar to Clark's completion in logic programming) is applied (here $\sim L$ denotes the complementary literal of L):

DEFINITION 17. Let **T** be a temporal theory consisting of simplified executable temporal formulae. For each $H \in \mathbf{HEADS}$ define

$$\mathbf{T}_H = \{\varphi \in T \mid \mathbf{head}(\varphi) = H\}$$

Let **L** be a literal and C an epistemic operator; define

$$\begin{aligned} tc(\mathbf{T}_{CL}) \;=\; & [\bigvee\{\text{body}(\varphi) \mid \varphi \in \mathbf{T}_{CL}\} \vee \\ & (\neg \bigvee\{\text{body}(\varphi) \mid \\ & \quad \varphi \in T_{C\sim L}\} \wedge \neg \bigvee\{\text{body}(\varphi) \mid \varphi \in T_{\neg CL \wedge \neg C \sim L}\} \wedge \mathbf{Y}CL)] \\ & \leftrightarrow CL \end{aligned}$$

and

$$\begin{aligned} tc(T_{\neg CL \wedge \neg C \sim L} \;=\; & [\bigvee\{\text{body}(\varphi) \mid \varphi \in T_{\neg CL \wedge \neg C \sim L}\} \vee \\ & (\neg \bigvee\{\text{body}(\varphi) \mid \varphi \in T_{C\sim L}\} \wedge \\ & \neg \bigvee\{\text{body}(\varphi) \mid \varphi \in T_{CL}\} \wedge \\ & \neg \mathbf{Y}CL \wedge \neg \mathbf{Y}C \sim L)] \\ & \leftrightarrow \neg CL \wedge \neg C \sim L \end{aligned}$$

The intuition behind these formulae is the following: a literal is (known to be) true in a component exactly when either there is an applicable rule making it true, or it was true before, and no rule making the literal false or unknown, is applicable.

DEFINITION 18 (temporal completion).

Let **T** be a temporal theory consisting of simplified executable temporal formulae.

a) The *temporal completion* of **T** is defined by

$$\begin{aligned} tc(\mathbf{T}) \;=\; & \{tc(\mathbf{T}_{CL}) \mid \; L \text{ literal}, C \text{ epistemic operator}\} \cup \\ & \{tc(\mathbf{T}_{\neg CL \wedge \neg C{\sim}L}) \mid \; L \text{ literal}, C \text{ epistemic operator}\} \end{aligned}$$

b) The non-classical entailment relation $\mid\!\sim_{tc}$ based on temporal completion is defined by

$$T \mid\!\sim_{tc} U \Leftrightarrow tc(T) \vdash U$$

Under a consistency assumption the second part

$$\{tc(\mathbf{T}_{\neg CL \wedge \neg C{\sim}L}) \mid L \text{ literal}, C \text{ epistemic operator}\}$$

of the above union is already implied by the first part

$$\{tc(\mathbf{T}_{CL}) \mid L \text{ literal}, C \text{ epistemic operator}\}.$$

EXAMPLE 19 (temporal completion of a link formalization). Let **T** be the temporal theory (a subset of $\mathbf{T}_S$) that formalizes the information link from EW to the agent X; see Example 11. The temporal completion of T contains the set of formulae:

$$\begin{aligned} & [\mathbf{YCout}_{EW}\text{observation\_result\_for(view(X, r), sign, X))} \vee \\ & \quad (\neg\mathbf{YCout}_{EW}\neg\text{observation\_result\_for(view(X, r), sign, X))} \wedge \\ & \qquad \mathbf{YCin}_X\text{observation\_result(view(X, r), sign, X))]} \\ & \leftrightarrow \mathbf{Cin}_X\text{observation\_result(view(X, r), sign, X)} \\ & [\mathbf{YCout}_{EW}\neg\text{observation\_result\_for(view(X, r), sign, X)} \vee \\ & \quad (\neg\mathbf{YCout}_{EW}\text{observation\_result\_for(view(X, r), sign, X))} \vee \\ & \qquad \mathbf{YCin}_X\neg\text{observation\_result(view(X, r), sign, X))]} \\ & \leftrightarrow \mathbf{Cin}_X\neg\text{observation\_result(view(X, r), sign, X)} \end{aligned}$$

Note that the result of temporal completion is a temporal theory that is not any more in executable format.

The temporal completion allows to formalize proofs in a classical proof system. This means that, given a compositional theory $(\mathbf{T}_C)_{C\in\mathbf{COMP}}$, we should consider the completion of the union of these theories, i.e., $tc(T_S^*)$ where S is the component of the entire system, for global provability. On the other hand, for compositional provability, we have to consider $(tc(\mathbf{T}_C))_{C\in\mathbf{COMP}}$. In general, however, $tc(\mathbf{T}_S^*)$

need not be identical to the union of $(\boldsymbol{tc}(\mathbf{T}_C))_{C \in \mathbf{COMP}}$. This may occur when a literal occurs in the head of two rules belonging to different components. Then there will be one formula $\boldsymbol{tc}(\mathbf{T}_{CL})$ in $\boldsymbol{tc}(\mathbf{T}^*_S)$, combining the two rules (and this is intended), but there will be two in the union of $(\boldsymbol{tc}(\mathbf{T}_C))_{C \in \mathbf{COMP}}$, one for each component (and this is not intended). In the case of simplified executable temporal formulae we can give a simple criterion which ensures that $\boldsymbol{tc}(\mathbf{T}^*_S)$ is equal to the union of $(\boldsymbol{tc}(\mathbf{T}_C))_{C \in \mathbf{COMP}}$. The only thing that is required is that for each formula CL, the temporal formulae defining it, are all in one component, i.e., $\mathbf{T}_{CL} \subseteq \mathbf{T}_C$ for some component **C**. It is easy to see that this requirement is sufficient, and it is a requirement satisfied at least by all theories describing components in DESIRE.

Given that this requirement is satisfied, we can prove:

PROPOSITION 20. *For the language composition* $(\mathbf{L}_C)_{C \in \mathbf{COMP}}$, *let a composition of properties* $(\mathbf{P}_C)_{C \in \mathbf{COMP}}$ *and a compositional temporal theory* $(\mathbf{T}_C)_{C \in \mathbf{COMP}}$ *be given. If* $(\mathbf{P}_C)_{C \in \mathbf{COMP}}$ *is compositionally provable with respect to* $\mid\!\sim_{tc}$ *from the compositional temporal theory* $(\mathbf{T}_C)_{C \in \mathbf{COMP}}$ *then* $(\mathbf{P}_C)_{C \in \mathbf{COMP}}$ *is globally provable with respect to* $\mid\!\sim_{tc}$ *from the compositional theory* $(\mathbf{T}_C)_{C \in \mathbf{COMP}}$.

Proof. Suppose $(\mathbf{P}_C)_{C \in \mathbf{COMP}}$ is compositionally provable with respect to tc from the compositional temporal theory $(\mathbf{T}_C)_{C \in \mathbf{COMP}}$. Then for each component **C** the following holds:

$$\begin{array}{ll} \mathbf{T}_C \cup \mid\!\sim_{tc} \bigwedge \bigcup_{C' \mathbf{sub} C} \mathbf{P}_{C'} \to \mathbf{P}_C & \text{if } \mathbf{C} \text{ is composed} \\ \mathbf{T}_C \mid\!\sim_{tc} \mathbf{P}_C & \text{if } \mathbf{C} \text{ is primitive} \end{array}$$

Let $\vdash$ be a classical provability relation for temporal multi-epistemic logic. Then by definition of $\mid\!\sim_{tc}$,

$$\begin{array}{ll} \boldsymbol{tc}(\mathbf{T}_C) \vdash \bigwedge \bigcup_{C' \mathbf{sub} D} \mathbf{P}_{C'} \to \mathbf{P}_C & \text{if } \mathbf{C} \text{ is composed} \\ \boldsymbol{tc}(\mathbf{T}_C) \vdash \mathbf{P}_C & \text{if } \mathbf{C} \text{ is primitive} \end{array}$$

This means that the collection of properties $(\mathbf{P}_C)_{C \in \mathbf{COMP}}$ is compositionally provable with respect to $\vdash$ from the compositional temporal theory $(\boldsymbol{tc}(\mathbf{T}_C))_{C \in \mathbf{COMP}}$.

By Proposition 16 it follows that the collection of properties $(\mathbf{P}_C)_{C \in \mathbf{COMP}}$ is globally provable with respect to from the compositional temporal theory $(tc(\mathbf{T}_C))_{C \in \mathbf{COMP}}$, or, in other words:

$$\boldsymbol{tc}(\mathbf{T}_C) \cup \bigcup_{C' \mathbf{sub} C} \boldsymbol{tc}(\mathbf{T}_{C'})^* \vdash \mathbf{P}_C$$

By commutativity of taking unions and the ***tc*** operator, it follows:

$$\boldsymbol{tc}(\mathbf{T}_C \cup \bigcup_{C' \mathbf{sub} C} T^*_{C'}) \vdash \mathbf{P}_C$$

Therefore

$$\mathbf{T}_C \cup \bigcup_{C' \mathbf{sub} C} T^*_{C'} \vdash_{tc} \mathbf{P}_C$$

or in other words, the collection of properties $(\mathbf{P}_C)_{C \in \mathbf{COMP}}$ is globally provable with respect to *tc* from the compositional temporal theory $(\mathbf{T}_C)_{C \in \mathbf{COMP}}$. ■

The notion of temporal completion defined above expresses default persistence for all information in the system. This implies that in all cases where no default persistence is intended, explicit temporal rules are required that prohibit the persistence. For example, to describe retraction of information that deductively depends on other information that was revised (such as occurs, for example, in the truth maintenance process of primitive components in DESIRE), it is needed in addition to explicitly express a temporal rule, e.g., (for the Example 12) of the form:

$$\phi \wedge \neg(\mathbf{YCin}_{X.OC}\mathsf{view(A, circle)} \wedge Y\mathbf{Cin}_{X.OC}\mathsf{view(B, circle)}) \rightarrow$$
$$\neg\mathbf{Cout}_{X.OC}\mathsf{object\_type(sphere)} \wedge \neg\mathbf{Cout}_{X.OC}\neg\mathsf{object\_type(sphere)}$$

where ϕ is again a formula expressing control information that allows the rule to be used (for example, the component should be active). Another approach is to define a more sensitive form of temporal completion already taking this into account, in which case these separate rules for retraction are not needed.

7 CONCLUSIONS

The compositional verification method formalized in this paper can be applied to a broad class of multi-agent systems. Compositional verification for one process abstraction level deep is based on the following very general assumptions:

- a multi-agent system consists of a number of agents and external world components.
- agents and components have explicitly defined input and output interface languages; all other information is hidden; information exchange between components can only take place via the interfaces (*information hiding*).
- a formal description exists of the manner in which agents and world components are composed to form the whole multi-agent system (*composition relation*).
- the semantics of the system can be described by the evolution of states of the agents and components at the different levels of abstraction (*state-based semantics*).

This non-iterative form of compositional verification can be applied to many existing approaches, for example, to systems designed using Concurrent METATEM [Fisher, 1994; Fisher and Wooldridge, 1997]. Compositional verification involving more abstraction levels assumes, in addition:

- some of the agents and components are composed of sub-components.
- a formal description exists of the manner in which agents or components are composed of sub-components (*composition relation*).
- information exchange between components is only possible between two components at the same or adjacent levels (*information hiding*).

Currently not many approaches to multi-agent system design exist that exploit iterative compositionality. One approach that does is the compositional development method DESIRE. The compositional verification method formalized in this paper fits well to DESIRE, but not exclusively.

Two main advantages of a compositional approach to modelling are the transparent structure of the design and support for reuse of components and generic models. The compositional verification method extends these main advantages to (1) a well-structured verification process, and (2) the reusability of proofs for properties of components that are reused.

The first advantage entails that both conceptually and computationally the complexity of the verification process can be handled by compositionality at different levels of abstraction. Apart from the work reported in [Jonker and Treur, 1998], a generic model for diagnosis has been verified [Cornelissen *et al.*, 1997] and a multi-agent system with agents negotiating about load-balancing of electricity use [Brazier *et al.*, 1998]. The second advantage entails: if a modified component satisfies the same properties as the previous one, the proof of the properties at the higher levels of abstraction can be reused to show that the new system has the same properties as the original. This has high value for a library of reusable generic models and components. The verification of generic models forces one to find the assumptions under which the generic model is applicable for the considered domain, as is also discussed in [Fensel and Benjamins, 1996]. A library of reusable components and generic models may consist of both specifications of the components and models, and their design rationale. As part of the design rationale, at least the properties of the components and their logical relations can be documented.

The usefulness of a temporal multi-epistemic logic, TMEL, a generalization of temporal epistemic logic was investigated to formalize verification proofs. As a test, the properties and proofs for verification of an example multi-agent system for co-operative information gathering [Jonker and Treur, 1998] were successfully formalized within the logic TMEL. Our study shows that TMEL provides expressivity for dynamics and reasoning about time, and formalizes incomplete information states in an adequate manner. To obtain the right structure in accordance with the compositional system design, the logic is equipped with a number of compositional structures: compositions of sub-languages, compositional theories, and compositional provability. It was established that under the assumption that the provability relation is reflexive and transitive, compositional provability implies global provability. Therefore this logic is adequate if the executable temporal theories formalizing a specification are temporally completed, a temporal variant of Clark's completion for logic programs. In this case classical provability can be used, which is much more transparent than the more complicated non-classical provability relations that are possible.

In [Fisher and Wooldridge, 1997] a temporal belief logic, TBL, was introduced to define semantics and verify properties for systems specified in Concurrent METATEM [Fisher, 1994]. A similarity with our approach as introduced above is that in both cases modal operators are used to distinguish knowledge of different agents, and a discrete linear time temporal logic is built on top of the multimodal logic. A main difference in comparison to [Fisher and Wooldridge, 1997] is that our approach exploits compositionality. In Concurrent METATEM no iterated compositional structures can be defined, as is the case in DESIRE. Therefore verification in TBL always takes place at the global level, instead of the iterated compositional approach to verification in TMEL. Another difference is that in our approach the states in the base logic are in principle three-valued, whereas the states in Concurrent METATEM are two-valued: an atom in a state that is not true is assumed false in this state.

A similarity between our approach and the approach introduced in [Benerecetti *et al.*, 1998] is that both employ a combination of temporal logic and modal logic, and exploit modular system structures to limit complexity. A difference is that in [Benerecetti *et al.*, 1998] branching time logic is used and the agent architecture is restricted to the BDI-architecture. Another difference is that they use model checking as the verification approach, whereas our approach is based on proofs. In our case it is easier to include a description of a verification proof as part of the design rationale documentation.

The notion of multi-language systems to formally specify complex reasoning patterns has a longer tradition in the literature. In [Konolige, 1986; Langevelde *et al.*, 1992; Giunchiglia *et al.*, 1993; Giunchiglia and Serafini, 1994] some of the main earely contributions can be found. In [Treur and Wetter, 1993] eight of such approaches are discussed, including the DESIRE approach in this paper. In this book [Treur and Wetter, 1993], these approaches are applied to a common reasoning problem and compared in detail; see [Hermelen *et al.*, 1993]. Some of the differences between the approach of [Giunchiglia *et al.*, 1993; Giunchiglia and Serafini, 1994] and DESIRE is that in DESIRE the dynamics of the reasoning pattern is explitly modelled, and dynamic meta-level and control constructs are available. The semantics of DESIRE is explicitly predefined as temporal and formalised on the basis of a form of temporal logic, whereas in Giunchiglia's approach the notion of contexts which have no explicit predefined temporal meaning plays a main role in the semantics; meta-level constructs are only available for static aspects.

The combination of epistemic operators and temporal operators in one logic has been studied in different variants. Examples of work in this area are [Halpern and Vardi, 1986; van der Meyden, 1994]; for an extensive overview, see [Halpern *et al.*, 1999]. A main difference is that our logic TMEL does not allow arbitrary combinations of epistemic and temporal operators. Epistemic operators cannot be nested, and cannot be not applied to temporal formulae. Therefore, our language is much more restricted, tailored to the intended use. In [21], four properties of logics of knowledge and time are considered:

- unique initial state
- perfect recall
- the agent does not learn
- synchronous

Our logic only satisfies the first of these properties, not the other three. Moreover, our logic uses constructs for interaction with duration, and considers non-classical semantics to incorporate default persistence. Another difference is that in our logic a hierarchy of process abstraction levels is explitly formalised.

A future continuation of this work will consider the development of tools for compositional verification, and the integration of verification and requirements specification.

ACKNOWLEDGEMENTS

An extended abstract of this work was presented at the Fifth International Workshop on Agent Theories, Architectures and Languages, ATAL'98.

Joeri Engelfriet, Catholijn M. Jonker and Jan Treur
Department of Artificial Intelligence, Vrije Universiteit Amsterdam, The Netherlands.

BIBLIOGRAPHY

[Abadi and Lamport, 1993] M. Abadi and L. Lamport. Composing specifications, *ACM Transactions on Programming Languages and Systems*, **15**, 73–132, 1993.

[Barringer *et al.*, 1991] H. Barringer, M. Fisher, D. Gabbay and A. Hunter. Meta-reasoning in executable temporal logic. In J. Allen, R. Fikes, E. Sandewall, eds., *Proceedings of the 2nd International Conference on Principles of Knowledge Representation and Reasoning, KR'91*, 1991.

[Barringer *et al.*, 1996] H. Barringer, M. Fisher, D. Gabbay, R. Owens and M. Reynolds. *The Imperative Future: Principles of Executable Temporal Logic*, Research Studies Press Ltd. and John Wiley & Sons, 1996.

[Benerecetti *et al.*, 1998] M. Benerecetti, F. Giunchiglia and L. Serafini (1998). A model-checking algorithm for multiagent systems. In Proc. of the International Workshop on Agent theories, Architectures and Languages, ATAL'98, 1998.

[van Benthem, 1983] J. F. A. K. van Benthem. *The Logic of Time: a Model-theoretic Investigation into the Varieties of Temporal Ontology and Temporal Discourse*, Reidel, Dordrecht, 1983.

[Brazier *et al.*, 1998] F. M. T. Brazier, F. Cornelissen, R. Gustavsson, C. M. Jonker, O. Lindeberg, B. Polak and J. Treur. Compositional design and verification of a multi-agent system for one-to-many negotiation. In: Y. Demazeau (ed.), *Proceedings of the Third International Conference on Multi-Agent Systems*, pp. 49–56, IEEE Computer Society Press, 1998.

[Brazier *et al.*, 1995] F. M. T. Brazier, B. M. Dunin-Keplicz, N. R. Jennings and J. Treur. Formal specification of multi-agent systems: a real world case. In: Lesser, V. (ed.), *Proceedings of the First International Conference on Multi-Agent Systems*, pp. 25–32. MIT Press, 1995. Extended version in: Huhns, M. and Singh, M. (eds.), *International Journal of Co-operative Information Systems, IJCIS*, **6**, 67–94, special issue on Formal Methods in Co-operative Information Systems: Multi-Agent Systems.

[Cornelissen *et al.*, 1997] F. Cornelissen, C. M. Jonker and J. Treur. Compositional verification of knowledge-based systems: a case study for diagnostic reasoning. In: E. Plaza, R. Benjamins (eds.), *Knowledge Acquisition, Modelling and Management, Proceedings of the 10th EKAW*, pp. 65–80. Lecture Notes in AI, vol. 1319, Springer Verlag, 1997.

[Dams *et al.*, 1996] D. Dams, R. Gerth, and P. Kelb. Practical symbolic model checking of the full μ-calculus using compositional abstractions. Report, Eindhoven University of Technology, Department of Mathematics and Computer Science, 1996.

[Engelfriet, 1996] J. Engelfriet. Minimal temporal epistemic logic, *Notre Dame Journal of Formal Logic*, **37**, 233–259, 1996. (special issue on Combining Logics).

[Engelfriet and Treur, 1996] J. Engelfriet and J. Treur. Specification of nonmonotonic reasoning. In *Proceedings International Conference on Formal and Applied Practical Reasoning*, pp. 111–125. Springer-Verlag, Lecture Notes in Artificial Intelligence, vol. 1085, 1996.

[Engelfriet and Treur, 1996a] J. Engelfriet and J. Treur. Executable temporal logic for nonmonotonic reasoning, *Journal of Symbolic Computation*, **22**, 615–625, 1996.

[Engelfriet and Treur, 1997] J. Engelfriet and J. Treur. An interpretation of default logic in temporal epistemic logic. *Journal of Logic, Language and Information*, **7**, 369–388, 1997.

[Fensel and Benjamins, 1996] D. Fensel and R. Benjamins. Assumptions in model-based diagnosis. In B. R. Gaines and M. A. Musen (eds.), *Proceedings of the 10th Banff Knowledge Acquisition for Knowledge-based Systems workshop, Calgary*, pp. 5/1–5/18. SRDG Publications, Department of Computer Science, University of Calgary, 1996.

[Fensel *et al.*, 1996] D. Fensel, A. Schonegge, R. Groenboom and B. Wielinga. Specification and verification of knowledge-based systems. In *Proceedings of the 10th Banff Knowledge Acquisition for Knowledge-based Systems workshop*, B. R. Gaines and M. A. Musen, eds. pp. 4/1–4/20. SRDG Publications, Calgary. Department of Computer Science, University of Calgary, 1996.

[Finger and Gabbay, 1992] M. Finger and D. Gabbay. Adding a temporal dimension to a logic system. *Journal of Logic, Language and Information*, **1**, 203–233, 1992.

[Fisher, 1994] M. Fisher. A survey of Concurrent METATEM — the language and its applications. In Temporal Logic. In *Proceedings of the First International Conference*, D. M. Gabbay and H. J. Ohlbach, eds. pp. 480–505. Lecture Notes in AI, vol. 827, Springer-Verlag, 1994.

[Fisher and Wooldridge, 1997] M. Fisher and M. Wooldridge. On the formal specification and verification of multi-agent systems. In *International Journal of Co-operative Information Systems, IJCIS*, **6**, Special issue on Formal Methods in Co-operative Information Systems: Multi-Agent Systems, M. Huhns and M. Singh, eds. 37–65, 1997.

[Giunchiglia *et al.*, 1993] E. Giunchiglia, P. Traverso and F. Giunchiglia. Multi-Context Systems as a Specification Framework for Complex Reasoning Systems. In [Treur and Wetter, 1993], pp. 45–72, 1993

[Giunchiglia and Serafini, 1994] F. Giunchiglia and L. Serafini. Multilanguage Hierarchical Logics (or: How we can do without modal logics), *Artificial Intelligence*, **65**, 29–70, 1994.

[Halpern *et al.*, 1999] J. Y. Halpern, R. van der Meyden and M. Y. Vardi. Complete Axiomatizations for Reasoning about Knowledge and Time. Report, 1999.

[Halpern and Vardi, 1986] J. Y. Halpern and M. Y. Vardi. The complexity of reasoning about knowledge and time. In *Proc. of the 18th ACM Symposium on the Theory of Computing*, pp. 319–315, 1986.

[Hermelen *et al.*, 1993] F. van Hermelen, J. Malec, R. Lopez de Manataras and J. Treur. Comparing Formal Specification Languages for complex reasoning Systems. In [Treur and Wetter, 1993], pp. 257–282, 1993.

[Hooman, 1994] J. Hooman. Compositional verification of a distributed real-time arbitration protocol. *Real-Time Systems*, **6**, 173–206, 1994.

[Jonker and Treur, 1998] C. M. Jonker and J. Treur. Compositional verification of multi-agent Systems: a formal analysis of pro-activeness and reactiveness. In *Proceedings of the International Symposium on Compositionality*, W. P. De Roever, H. Langmaack and A. Pnueli, eds. Lecture Notes in Computer Science, vol. 1536, 350–380, Springer Verlag, 1998. Extended version is to appear in the *International Journal of Cooperative Information Systems*, 2002.

[Konolige, 1986] K. Konolige. *A Deduction Model of Belief.* Pitman, 1986.

[Langevelde *et al.*, 1992] I. A. Langevelde, A. W. Philipsen and J. Treur. Formal Specification of Compositional Architectures. In *Proc. 10th European Conference on Artificial Intelligence, ECAI '92*, B. Neumann, ed. pp. 272–276. Wiley and Sons, 1992.

[van der Meyden, 1994] R. van der Meyden. Axioms for Knowledge and Time in Distributed Systems with Perfect Recall. In *Proc. IEEE Symposium on Logic in Computer Science*, Paris, pp. 448–457, 1994.

[Treur and Wetter, 1993] J. Treur and T. Wetter. Formal Specification of Complex Reasoning Systems. Ellis Horwood, 1993.

[Treur and Willems, 1994] J. Treur and M. Willems. A logical foundation for verification. In *Proceedings of the Eleventh European Conference on Artificial Intelligence, ECAI '94*, A. G. Cohn, ed. pp. 745–749. John Wiley & Sons, Ltd., 1994.

[Wooldridge and Jennings, 1995] M. J. Wooldridge and N. R. Jennings. Intelligent Agents: Theory and practice. *Knowledge Engineering Review*, **10**, 115–152, 1995.

PART IIIB

FORMAL ANALYSIS:

LOGICS FOR AGENTS

B. VAN LINDER, W. VAN DER HOEK AND
J.-J. CH. MEYER

FORMALISING ABILITIES AND OPPORTUNITIES OF AGENTS

1 INTRODUCTION

The last ten years have witnessed an intense flowering of interest in artificial agents, both on a theoretical and on a practical level. The ACM devoted a special issue of its 'Communications' to intelligent agents [CACM, 1994], and Scientific American ranked intelligent software agents among the key technologies for the 21st century [Maes, 1995]. Also various conferences and workshops were initiated that specifically address agents, their theories, languages, architectures and applications [Fiadeiro and Schobbens, 1996; Lesser, 1995; Wooldridge and Jennings, 1995b; Wooldridge *et al.*, 1996]. Consequently, terms like agent-based computing, agent-based software engineering and agent-oriented programming have become widely used in research on AI. Despite its wide use, there is no agreement on what the term 'agent' means. Riecken remarks that 'at best, there appears to be a rich set of emerging views' and that 'the terminology is a bit messy' [Riecken, 1994]. Existing definitions range from 'any entity whose state is viewed as consisting of mental objects ' [Shoham, 1993] and 'autonomous objects with the capacity to learn, memorize and communicate' [Foner, 1993], to 'systems whose behavior is neither casual nor strictly causal, but teleonomic, goal-oriented toward a certain state of the world' [Castelfranchi, 1995]. Other authors, and truly not the least, use the term 'robot' instead of agent [Lespérance *et al.*, 1996], or take the common-sense definition of agents for granted [Rao and Georgeff, 1991b]. In practical applications agents are 'personal assistant[s] who [are] collaborating with the user in the same work environment' [Maes, 1994], or 'computer programs that simulate a human relationship, by doing something that another person could otherwise do for you' [Selker, 1994].

The informal description of an (artificial) agent in its most primitive form, which we distill from the definitions given above and which the reader is advised to keep at the back of his/her mind throughout reading this paper, is that of an entity which has the possibility to execute certain *actions*, and is in the possession of certain *information*, which allows it to *reason* about its own and other agents' actions. In general, these agents will also have motives that explain why they act the way they do. The treatment of these motives is however not the subject of this paper (but see [Meyer *et al.*, 1999]). Moreover, although we borrow a lot of terminology and notions

J.J.Ch. Meyer and J. Treur (eds.),
Handbook of Defeasible Reasoning and Uncertainty Management Systems, Vol. 7, 253–307.

from philosophy, the reader should keep in mind that it is our main goal to describe artificial agents, rather than humans.

Currently several applications of agent-technology are in use. Among those listed by Wooldridge & Jennings [1995a] are air-traffic control systems, spacecraft control, telecommunications network management and particle acceleration control. Furthermore, interface agents are used that for instance take care of email administration, as well as information agents that deal with information management and retrieval. In all probability, these implemented agents will be rather complex. In addition, life-critical implementations like air-traffic control systems and spacecraft control systems need to be highly reliable. To guarantee reliability it is probably necessary to use formal methods in the development of these agent systems, since such a guarantee can never be given by just performing tests on the systems. Besides this general reason for using formal techniques in any branch of AI and computer science, there is another reason when dealing with agents. These agents will in general be equipped with features representing common-sense concepts as knowledge, belief and ability. Since most people do have their own conception of these concepts, it is very important to unambiguously establish what is meant by these concepts when ascribed to some specific implemented agent. Formal specifications allow for such an unambiguous definition.

The formal tool that we propose to model agency is *modal logic* [Chellas, 1980; Hughes and Cresswell, 1968; Hughes and Cresswell, 1984]. Using modal logics offers a number of advantages. First, using an intensional logic like modal logic allows one to come up with an intuitively acceptable formalisation of intensional notions with much less effort than it would take to do something similar using fully-fledged first-order logic. Secondly, the reducibility of modal logic to (fragments of) first-order logic ensures that methods and techniques developed for first-order logic are still applicable to modal logic. Lastly, using possible worlds models as originally proposed by Kripke [1963], provides for a uniform, clear, intelligible, and intuitively acceptable means to give mathematical meaning to a variety of modal operators. The modal systems that we propose to formalise agents belong to what we call the *KARO-framework*. In this framework, the name of which is inspired by the well-known BDI-architecture [Rao and Georgeff, 1991b], special attention is paid to the agents' *knowledge* and *abilities*, and to the *results* of and *opportunities* for their actions. We present two different systems, both belonging to the KARO-framework, that differ in their treatment of abilities for certain actions. To show the expressive power of the framework we formalise various notions that are interesting maybe from a philosophical point of view, but that above all should help to understand and model artificially intelligent agents—we like to stress that our aim is to describe artificial agents like softbots and robots by means of these notions, rather than human agents, which are far more complex and for which one probably needs more complicated descriptions.

The agent attitudes in this paper are limited to knowledge, and abilities, results and opportunities with respect to his/its actions, i.e. the KARO framework. We stress that the purpose of this paper is to give a thorough treatment of this KARO framework, which has been used by us as a basis for a much more extensive description of agents, incorporating such notions as observations [van Linder *et al.*, 1994b], communication [van Linder *et al.*, 1994a], default reasoning [van Linder *et al.*, 1997], belief revision [van Linder *et al.*, 1995]), and goals [Meyer *et al.*, 1999].

The philosophy adhered to in this endeavour is that the primary attitude of agents is to *act*, by the very meaning of the word 'agent', so that a specification logic for the behaviour of agents should start out from a logic of action (for which we have chosen an extension of dynamic logic in which knowledge and ability is expressible as well). In this enterprise we owe to Bob Moore's work combining a version of dynamic logic and epistemic logic for the first time [Moore, 1985] So here we deviate from the philosophy of other foundational work on agents, in particular that of Rao & Georgeff [1991b; 1991a; 1993], who take belief, desire, intentions as well as time as primitive notions for agents. (One could argue that our approach is more in line with that of Cohen & Levesque [1990]. They, too, take actions as basic building blocks for their theory of agents. However, they consider only *models* of their framework and provide no formal proof system. Furthermore, they are mainly concerned with the formalisation of motivational attitudes such as goals and intentions of agents, and employ actions merely as a basis to obtain this, while here we are interested in actions and aspects of these, such as opportunities and abilities, in their own right.)

Therefore, the main contribution of this paper is to investigate the KARO logic and provide meta-results such as completeness, which, particularly by the addition of abilities, will turn out to be a non-trivial extension of that for basic dynamic logic.

Organisation of the paper The rest of the paper is organised as follows. In Section 2 we look at the philosophical foundations of the KARO-framework. In Section 3 we present the formal definitions constituting the two systems belonging to the KARO-framework. We start by defining the language common to the two systems, where-after a common class of models and two different interpretations for the formulas from the language in the models are presented. In Section 4 various properties of knowledge and action in the KARO-framework are considered. In Section 5 we consider the notion of practical possibility, and formalise part of the reasoning of agents on the correctness and feasibility of their plans. In Section 6 we present two slightly different proof systems that are sound and complete with respect to the notions of validity associated with the two interpretations. Section 7 concludes this paper with a brief summary, an overview of related work, and some suggestions for future research. In the appendix we present the proofs of soundness and completeness in considerable detail.

2 THE KARO-FRAMEWORK FROM A PHILOSOPHICAL PERSPECTIVE

As mentioned in the previous section, in its simplest form an agent is an entity that performs actions and possesses information. The informational attitude that we equip our agents with is termed *knowledge*. Our use of the term knowledge agrees with the common one in AI and computer science [Halpern and Moses, 1992; Meyer and van der Hoek, 1995], i.e. knowledge is veridical information with respect to which the agent satisfies conditions of both positive and negative introspection. Veridicality implies that only true formulas are known by agents, positive introspection states that agents know that they know something whenever they know it, and negative introspection states that agents know that they do not know something as soon as they do not know it.

To explain the concept of action, we first have to spend some words on the ontology of states of affairs that we presuppose. By a *state of affairs* we mean the way the world is, or one way it might be, at a moment. The (currently) actual state of affairs is composed of the facts about the world as it actually is at this very moment. But there are, presumably, various other states of affairs which could have applied to the world as this moment instead, or that would apply as the result of a given action. An agent, with limited knowledge of the facts, might consider various merely hypothetical states of affairs consistent with the agent's knowledge *Actions* are now considered to be descriptions of causal processes, which upon execution by an agent may turn one state of affairs into another one. Thus, our intuitive idea of actions corresponds to what Von Wright calls the *generic* view on actions [Wright, 1967]. An *event* consists of the performance of a particular action by a particular agent, and is as such related to Von Wright's *individual* view on actions [Wright, 1967]. We will use the plain term events, although perhaps the term *agent-driven event* would be more appropriate, here. Given the ontology of actions and events as somehow causing transitions between states of affairs, we deem two aspects of these notions to be crucial: when is it possible for an agent to perform an action, and what are the effects of the event consisting of the performance by a particular agent of a particular action in a particular state of affairs? To investigate these questions we focus on three aspects of actions and events that are in our opinion essential, viz. *result*, *opportunity* and *ability*. Slightly simplifying ideas of Von Wright [1963], we consider any aspect of the state of affairs brought about by the occurrence of an event in some state of affairs to be among the results of that particular event in that particular state of affairs. In adopting this description of results we abstract from all kinds of aspects of results that would probably have to be dealt with in order to come up with an account that is completely acceptable from a philosophical point of view, such as for instance the question whether all changes in a state of affairs have to be ascribed to the occurrence of some event, thereby excluding the possibility of external factors influencing these changes. However, it is not our aim to provide a complete theory of results incorporating all these aspects, but instead combine results with other notions that are important for agency. From this point of view it seems that our definition of results is adequate to investigate the effects of actions, and, given the complexity already associated with this simple definition, it does not make much sense to pursue even more complex ones.

Along with the notion of the result of events, the notions of ability and opportunity are among the most discussed and investigated in analytical philosophy. Ability plays an important part in various philosophical theories, as for instance the theory of free will and determinism, the theory of refraining and seeing-to-it, and deontic theories. Following Kenny [1975], we consider ability to be the complex of physical, mental and moral capacities, internal to an agent, and being a positive explanatory factor in accounting for the agent's performing an action. Opportunity on the other hand is best described as circumstantial possibility, i.e. possibility by virtue of the circumstances. The opportunity to perform some action is external to the agent and is often no more than the absence of circumstances that would prevent or interfere with the performance. Although essentially different, abilities and opportunities are interconnected in that abilities can be exercised only when opportunities for their exercise present themselves, and opportunities can be taken only by those who have the appropriate abilities. From this point of view it is important to remark that abilities are understood to be *reliable* (cf. [Brown, 1988]), i.e. having the ability to perform a certain action suffices to take the opportunity to perform the action every time it presents itself. The combination of ability and opportunity determines whether or not an agent has the (practical) possibility to perform an action.

For our artificial agents, we will in Section 5 study 'correctness' of α for i to bring about ϕ in terms of having the opportunity: for the artificial agents that we have in mind and the actions ('programs') they perform correctness has to do with intrinsic (rather than external) features of the action: its halting and the intended outcome. Such features are still beyond the scope of the agent's abilities, of course.

3 THE KARO-FRAMEWORK FROM A FORMAL PERSPECTIVE

For the reasons already given in Section 1, we propose the use of a propositional multi-modal language to formalise the knowledge and abilities of agents, and the results of and opportunities for their actions. In contrast with most philosophical accounts, but firmly in the tradition of theoretical computer science, this language is an *exogenous* one, i.e. actions are represented explicitly. Although it is certainly possible to come up with accounts of action without representing actions (see for instance [Pörn, 1977; Segerberg, 1989]), we are convinced that many problems that plague these endogenous formalisations can be avoided in exogenous ones.

The language contains modal operators to represent the knowledge of agents as well as to represent the result and opportunity of events. The ability of agents is formalised by a factually non-modal operator. Following the representation of Hintikka [1962] we use the operator $\mathbf{K}_{\_}\_$ to refer to the agents' knowledge: $\mathbf{K}_i\varphi$ denotes the fact that agent i knows φ to hold. To formalise results and opportunities we borrow constructs from dynamic logic: $\langle \mathrm{do}_i(\alpha)\rangle\varphi$ denotes that agent i has the opportunity to perform the action α and that φ will result from this performance. The abilities of agents are formalised through the $\mathbf{A}_{\_}\_$ operator: $\mathbf{A}_i\alpha$ states that agent i has the ability to perform the action α. The class of actions that we consider here is built up from a set of atomic actions using a variety of constructors. These constructors deviate somewhat from the standard actions from dynamic logic [Goldblatt, 1992; Harel, 1984], but are both well-known from high-level programming languages and somewhat closer to philosophical views on actions than the standard constructors. When defining the models we will ensure that atomic actions are deterministic, i.e. the event consisting of an agent performing an action in some state of affairs has a unique outcome. As we will see later on this ensures that all actions are deterministic.

DEFINITION 1. The language $\mathrm{L}(\Pi, \mathrm{A}, \mathrm{At})$ is founded on three denumerable, non-empty sets, each of which is disjoint of the others: Π is the set of propositional variables, $\mathrm{A} \subseteq \mathbb{N}$ is the set of (names of) agents, and At is the set of atomic actions. The alphabet contains the well-known connectives $\neg$ and $\wedge$, the epistemic operator $\mathbf{K}_{\_}\_$, the dynamic operator $\langle \mathrm{do}_{\_}(\_)\rangle\_$, the ability operator $\mathbf{A}_{\_}\_$, the action constructors `confirm_` (confirmations), $\_;\_$ (sequential composition), `if_then_else_fi` (conditional composition) and `while_do_od` (repetitive composition).

DEFINITION 2. The language $\mathrm{L}(\Pi, \mathrm{A}, \mathrm{At})$ is the smallest superset of Π such that

- if $\varphi \in \mathrm{L}(\Pi, \mathrm{A}, \mathrm{At})$ and $\psi \in \mathrm{L}(\Pi, \mathrm{A}, \mathrm{At})$ then $\neg\varphi \in \mathrm{L}(\Pi, \mathrm{A}, \mathrm{At})$ and $\varphi \wedge \psi \in \mathrm{L}(\Pi, \mathrm{A}, \mathrm{At})$
- if $\varphi \in \mathrm{L}(\Pi, \mathrm{A}, \mathrm{At})$, $i \in \mathrm{A}$, $\alpha \in \mathrm{Ac}(\mathrm{At})$ then $\mathbf{K}_i\varphi \in \mathrm{L}(\Pi, \mathrm{A}, \mathrm{At})$,

$$\langle \mathrm{do}_i(\alpha)\rangle\varphi \in \mathrm{L}(\Pi, \mathrm{A}, \mathrm{At}) \text{ and } \mathbf{A}_i\alpha \in \mathrm{L}(\Pi, \mathrm{A}, \mathrm{At})$$

where the class Ac(At) of actions is the smallest superset of At such that

- if $\varphi \in \mathrm{L}(\Pi, \mathrm{A}, \mathrm{At})$ then $\mathtt{confirm}\,\varphi \in \mathrm{Ac}(\mathrm{At})$
- if $\alpha_1 \in \mathrm{Ac}(\mathrm{At}), \alpha_2 \in \mathrm{Ac}(\mathrm{At})$ then $\alpha_1; \alpha_2 \in \mathrm{Ac}(\mathrm{At})$
- if $\varphi \in \mathrm{L}(\Pi, \mathrm{A}, \mathrm{At}), \alpha_1 \in \mathrm{Ac}(\mathrm{At}), \alpha_2 \in \mathrm{Ac}(\mathrm{At})$ then $\mathtt{if}\,\varphi\,\mathtt{then}\,\alpha_1\,\mathtt{else}\,\alpha_2\,\mathtt{fi} \in \mathrm{Ac}(\mathrm{At})$
- if $\varphi \in \mathrm{L}(\Pi, \mathrm{A}, \mathrm{At}), \alpha \in \mathrm{Ac}(\mathrm{At})$ then $\mathtt{while}\,\varphi\,\mathtt{do}\,\alpha\,\mathtt{od} \in \mathrm{Ac}(\mathrm{At})$

The constructs $\vee$, $\rightarrow$, $\leftrightarrow$, $\top$, denoting the canonical tautology and $\bot$, denoting the canonical contradiction, are defined in the usual way. Other constructs are introduced by definitional abbreviation:

$$\begin{array}{lll}
\mathbf{M}_i\varphi & =^{\mathrm{def}} & \neg\mathbf{K}_i\neg\varphi \\
[\mathrm{do}_i(\alpha)]\varphi & =^{\mathrm{def}} & \neg\langle \mathrm{do}_i(\alpha)\rangle\neg\varphi \\
\mathtt{skip} & =^{\mathrm{def}} & \mathtt{confirm}\,\top \\
\mathtt{fail} & =^{\mathrm{def}} & \mathtt{confirm}\,\bot \\
\alpha^0 & =^{\mathrm{def}} & \mathtt{skip} \\
\alpha^{n+1} & =^{\mathrm{def}} & \alpha; \alpha^n
\end{array}$$

The following letters, possibly marked, are used as typical elements:

- p, q, r for the elements of Π
- i, j for the elements of A
- a, b, c for the elements of At
- φ, ψ, ρ for the elements of $\mathrm{L}(\Pi, \mathrm{A}, \mathrm{At})$
- α, β, γ for the elements of Ac(At)

Whenever the sets Π, A, At are understood, which we assume to be the case unless explicitly stated otherwise, we write L and Ac rather than $\mathrm{L}(\Pi, \mathrm{A}, \mathrm{At})$ and Ac(At).

The intuitive interpretation of formulas $\mathbf{K}_i\varphi$, $\langle \mathrm{do}_i(\alpha)\rangle\varphi$ and $\mathbf{A}_i\alpha$ is discussed above. The formula $\mathbf{M}_i\varphi$ is the dual of $\mathbf{K}_i\varphi$ and represents the epistemic possibility of φ for agent i, i.e. on the basis of its knowledge, i considers φ to be possible. The formula $[\mathrm{do}_i(\alpha)]\varphi$ is the dual of $\langle \mathrm{do}_i(\alpha)\rangle\varphi$; this formula is noncommittal about the opportunity of agent i to perform the action α but states that if the opportunity to do α is present, then φ would be among the results of $\mathrm{do}_i(\alpha)$. The action constructors presented in Definition 2 constitute the class of so-called ***strict programs*** (cf. [Halpern and Reif, 1983; Harel, 1984]). Their intuitive interpretation is as follows:

$\texttt{confirm}\,\varphi$	verify φ
$\alpha_1;\alpha_2$	α_1 followed by α_2
$\texttt{if}\,\varphi\,\texttt{then}\,\alpha_1\,\texttt{else}\,\alpha_2\,\texttt{fi}$	α_1 if φ holds and α_2 otherwise
$\texttt{while}\,\varphi\,\texttt{do}\,\alpha\,\texttt{od}$	α as long as φ holds

The action **skip** represents the void action, and **fail** denotes the abort action. The action α^n consists of sequentially doing α n times.

3.1 The KARO-framework: semantics

The vast majority of all interpretations proposed for modal languages is based on the use of Kripke-style possible worlds models. The models that we use to interpret formulas from L contain a set S of possible worlds, representing actual and hypothetical states of affairs, a valuation π on the elements of Π, indicating which atomic propositions are true in which possible world, a relation R denoting epistemic accessibility, and two functions $\mathbf{r}_0$ and $\mathbf{c}_0$ dealing with (the result, opportunity and ability for) atomic actions. In the sequel $S^\cdot$ denotes the lift of S, i.e. $S^\cdot =^{\text{def}} S\cup\{\emptyset\}$, and bool $=^{\text{def}} \{\mathbf{1},\mathbf{0}\}$ is a set of truth values.

DEFINITION 3. A model M for L is a tuple consisting of the following five elements:

- a non-empty set S of possible worlds or states.
- a valuation $\pi : \Pi \times S \to$ bool on propositional symbols.
- a function $R : A \to \wp(S \times S)$ indicating the epistemic alternatives of agents. This function is required to be such that $R(i)$ is an equivalence relation for all $i \in A$.
- a function $\mathbf{r}_0 : A \times At \to S \to S^\cdot$ indicating the state-transitions caused by the execution of atomic actions.
- a function $\mathbf{c}_0 : A \times At \to S \to$ bool determining the abilities of agents with regard to atomic actions.

Note that models in principle depend on the language: valuations π depend on Π, there are epistemic alternatives for each agent and the state transitions and abilities are agent- and atomic action-dependent. However, we will in the semantics often omit reference to the sets Π, A and At. The class containing all models for L is denoted by **M**. The letter M, possibly marked, denotes a typical model, and s, t, u, possibly marked, are used as typical elements of the set of states.

The relation $R(i)$ indicates which pairs of worlds are indistinguishable for agent i on the basis of its knowledge: if $(s, s') \in R(i)$ then whenever s is the description of the actual world, s' might as well be for all agent i knows. To ensure that knowledge indeed has the properties sketched in Section 2, it is demanded that $R(i)$ is an equivalence relation for all i. That this demand ensures that knowledge behaves as desired is stated in Proposition 5 and explained in Proposition 11. The function $\mathbf{r}_0$ characterises occurrences of atomic events, i.e. events consisting of an agent performing an atomic action: whenever s is some possible world, then $\mathbf{r}_0(i,a)(s)$ represents the state of affairs following execution of the atomic action a in the possible world s by the agent i. Since atomic actions are inherently deterministic, $\mathbf{r}_0(i,a)(s)$ yields at most one state of affairs as the one resulting from the occurrence of the event $\mathrm{do}_i(a)$ in s. If $\mathbf{r}_0(i,a)(s) = \emptyset$, we will sometimes say that execution of a by i in s leads to the (unique) *counterfactual state of affairs*, i.e. a state of affairs which is neither actual nor hypothetical, but counterfactual. One may think of $\mathbf{r}_0(i,a)(s) = \emptyset$ as indicating a serious failure, rather than just a disappointment: from $\emptyset$, no further actions can be taken. The function $\mathbf{c}_0$ acts as a kind of valuation on atomic actions, i.e. $\mathbf{c}_0(i,a)(s)$ indicates whether agent i has the ability to perform the action a in the possible world s.

Formulas from the language L are interpreted on the possible worlds in the models from $\mathbf{M}$. Propositional symbols are directly interpreted using the valuation π: a propositional symbol p is true in a state s iff $\pi(p,s)$ yields the value $\mathbf{1}$. Negations and conjunctions are interpreted as in classical logic: a formula $\neg\varphi$ is true in a state s iff φ is not true in s and $\varphi \wedge \psi$ is true in s iff both φ and ψ are true in s. The knowledge formulas $\mathbf{K}_i\varphi$ are interpreted using the epistemic accessibility relation $R(i)$: agent i knows that φ in s iff φ is true in all the possible worlds that are epistemically equivalent to s, for that agent. The dynamic formulas $\langle \mathrm{do}_i(\alpha)\rangle\varphi$ and the ability formulas $\mathbf{A}_i\alpha$ are interpreted through the extensions $\mathbf{r}$ and $\mathbf{c}$ of the functions $\mathbf{r}_0$ and $\mathbf{c}_0$, respectively. These extensions $\mathbf{r}$ and $\mathbf{c}$ will be defined in Definition 4. Informally, a formula $\langle \mathrm{do}_i(\alpha)\rangle\varphi$ is true in some possible world s, if the extension $\mathbf{r}$ of $\mathbf{r}_0$ applied to i, α and s yields some successor state s' in which the formula φ holds. A formula $\mathbf{A}_i\alpha$ is true in a state s if the extension $\mathbf{c}$ of $\mathbf{c}_0$ yields the value $\mathbf{1}$ when applied to i, α and s. Before defining the extended versions of $\mathbf{r}_0$ and $\mathbf{c}_0$, we first motivate the choices underlying these extensions.

Results and opportunities for composite actions

Recall from the introduction to this section that $\langle \mathrm{do}_i(\alpha)\rangle\varphi$ denotes that agent i has the opportunity to perform action α in such a way that φ will result from this performance. Thus, we can define the opportunity *sec* to do α as $\langle \mathrm{do}_i(\alpha)\rangle\top$. Note that under our assumption about determinism of actions, the formula $\langle \mathrm{do}_i(\alpha)\rangle\varphi$ is in fact stronger than $[\mathrm{do}_i(\alpha)]\varphi$: whereas the diamond-formula expresses that i has the opportunity to do α and φ will be among the results of i's doing α, the box-formula conditions φ being a result of i performing α upon i's opportunity to do α.

The extension $\mathbf{r}$ of the function $\mathbf{r}_0$ as we will present it is originally due to Halpern & Reif [Halpern and Reif, 1983]. Although Halpern & Reif's logic is meant to reason about computer programs and not about agents performing actions, we argue that their definition is also adequate for our purposes. Using this definition, actions $\mathtt{confirm}\varphi$ are interpreted as genuine confirmations: whenever the formula φ is true in a state s, s is its own $\mathrm{do}_i(\mathtt{confirm}\varphi)$-successor. If φ does not hold in a possible world s, then the $\mathtt{confirm}\varphi$ action fails, and no successor state results. In practice this implies that (all) agents have the opportunity to confirm the truth of a certain formula iff the formula holds. Execution of such an action does not have any effects in the case that the formula that is confirmed holds, and leads to the counterfactual state of affairs if the formula does not hold.[1]

Since the action $\alpha_1;\alpha_2$ is intuitively interpreted as 'α_1 followed by α_2', the transition caused by execution of an action $\alpha_1;\alpha_2$ equals the 'sum' of the transition caused by α_1 and the one caused by α_2 in the state brought about by execution of α_1. In the case that execution of α_1 leads to an empty set of states, execution of the action $\alpha_1;\alpha_2$ also leads to an empty set: there is no escape from the counterfactual state of affairs. In practice this implies that an agent has the opportunity to perform a sequential composition $\alpha_1;\alpha_2$ iff it has the opportunity to do α_1 (now), and doing α_1 results in the agent having the opportunity to do α_2. The results of performing $\alpha_1;\alpha_2$ equal the results of doing α_2, having done α_1.

Given its intuitive meaning, it is obvious that the transition caused by a conditional composition $\mathtt{if}\,\varphi\,\mathtt{then}\,\alpha_1\,\mathtt{else}\,\alpha_2\,\mathtt{fi}$ equals the one associated with α_1 in the case that φ holds and the one caused by execution of α_2 in the case that $\neg\varphi$ holds. This implies that an agent has the opportunity to perform an action $\mathtt{if}\,\varphi\,\mathtt{then}\,\alpha_1\,\mathtt{else}\,\alpha_2\,\mathtt{fi}$ if (it has the opportunity to confirm that) φ holds and it has the opportunity to do α_1, or (it has the opportunity to confirm that) $\neg\varphi$ holds and the agent has the opportunity to do α_2. The result of performing $\mathtt{if}\,\varphi\,\mathtt{then}\,\alpha_1\,\mathtt{else}\,\alpha_2\,\mathtt{fi}$ equals the result of α_1 in the case that φ holds and that of α_2 otherwise.

[1]Originally in dynamic logic [Harel, 1984; Kozen and Tiuryn, 1990] these actions were referred to as tests instead of confirmations. As long as one deals with the behaviour of computer programs, the term 'test' is quite acceptable. However, as soon as formalisations of (human) agents are concerned, one should be careful with using this term. The common-sense notion of test is that of an action, execution of which provides some kind of information (in our terminology of [van Linder *et al.*, 1995], a test is a 'knowledge producing action' and 'informative'). For example dope-tests and eye-tests are performed in order to acquire information on whether some athlete has been taking drugs, or whether someone's eyesight is adequate. The nature of this kind of tests is not captured by the action which just checks for the truth of some proposition, without yielding any information whatsoever. To avoid confusion we have chosen to refer to these latter kinds of actions as confirmations. Thus, in terms of our models, we think of a confirmation as an action that does not change the state, whereas an agent testing for φ might end up in a different (epistemic) state.

The definition of the extension $\mathbf{r}$ of $\mathbf{r}_0$ for the repetitive composition is based on the idea that execution of the action $\texttt{while}\,\varphi\,\texttt{do}\,\alpha\,\texttt{od}$ comes down to sequentially testing for the truth of φ and executing α until a state is reached in which $\neg\varphi$ holds. For deterministic while-loops $\texttt{while}\,\varphi\,\texttt{do}\,\alpha\,\texttt{od}$, at most one of the actions $\beta_k = ((\texttt{confirm}\,\varphi;\alpha)^k;\texttt{confirm}\,\neg\varphi)$, with $k \in \mathbb{N}$, has an execution which does not lead to the counterfactual state of affairs. Now if such an action β_k exists, the resulting state of execution of the while-loop is defined to be the state resulting from execution of β_k, and otherwise execution of the loop is taken to lead to the counterfactual state of affairs.

Abilities for composite actions

Whereas the extension $\mathbf{r}$ of $\mathbf{r}_0$ for composite actions is more or less standard, the extension $\mathbf{c}$ of $\mathbf{c}_0$ as determining the abilities of agents for composite actions, is not. Since we are (among) the first to give a formal, exogenous account of ability, extending the function $\mathbf{c}_0$ to the class of all actions involves a couple of personal choices.

We start with motivating our definitions of ability for confirmations and conditional compositions since neither of these is really controversial: the definition of ability for confirmations is indisputable since it represents a highly personal choice (and there is no accounting for tastes), and that of the ability for the conditional composition is too obvious and natural to be questioned.

We have decided to let an agent have the ability to confirm any formula that is actually true. Since confirmations do not correspond to any actions usually performed by humans, this definition seems to be perfectly acceptable, or at least it is hard to come up with any convincing counter-arguments to it. Note that this definition implies that in a situation where some proposition is true, (all) agents have both the opportunity and the ability to confirm this proposition.

Let us continue with defining abilities for conditionally composed actions. For these actions, ability is defined analogously to opportunity: an agent is able to perform the action $\texttt{if}\,\varphi\,\texttt{then}\,\alpha_1\,\texttt{else}\,\alpha_2\,\texttt{fi}$ iff either it is able to confirm the condition φ and perform α_1 afterwards, or it is able to confirm the negation of the condition and perform α_2. In practice this implies that having the ability to perform an action $\texttt{if}\,\varphi\,\texttt{then}\,\alpha_1\,\texttt{else}\,\alpha_2\,\texttt{fi}$ boils down to being able to do α_1 whenever φ holds and being able to do α_2 whenever φ does not hold. In our opinion this is *the* natural way to define the ability for conditionally composed actions, thereby accepting the import of some oddities of conditionals like the following. With our definition, an agent may claim on Tuesday that it has the ability to jump over the moon if it is Wednesday and scratch its nose otherwise'.[2] This may be undesirable,

[2]These oddities, and, in particular, this example, was suggested by a referee of a preliminary version of this paper

but also note that it is not the same as (the even worse) claim that on Wednesday it is able to jump over the moon.

Whereas the definitions of the ability for confirmations and conditional compositions are easily explained and motivated, this is not the case for those describing the ability for sequential and repetitive compositions, even though the basic ideas underlying these definitions are perfectly clear.

Informally, having the ability to perform a sequentially composed action $\alpha_1;\alpha_2$ is defined as having the ability to do α_1 now, while being able to do α_2 as a result of having done α_1. If the opportunity to perform α_1 exists, i.e. performing α_1 does not result in the counterfactual state of affairs, there is no question concerning the intuitive correctness of this definition, but things are different when this opportunity is absent. It is not clear how the abilities of agents are to be determined in the counterfactual state of affairs. Probably the most acceptable approach would be to declare the question of whether the agent is able to perform an action in the counterfactual state of affairs to be meaningless, which could be formalised by extending the set of truth-values to contain an element representing undefinedness of a proposition. Since this would necessitate a considerable complication of our classical, two-valued approach, we have chosen not to explore this avenue, which leaves us with the task of assigning a classical truth-value to the agents' abilities in the counterfactual state of affairs. In general we see two ways of doing this, the first of which would be to treat all actions equally and come up with a uniform truth value for the abilities of all agents to perform any action in the counterfactual state of affairs. This approach is relatively simply to formalise, and is in fact the one that we will pursue. The second approach would be to treat each action individually, and determine the agents' abilities through other means, such as by assuming an agent to be in the possession of certain default, or typical, abilities. This approach is further discussed in Section 7. Coming back to the first approach, it is obvious that — given that there are exactly two truth-values — two ways exist to treat all actions equally with respect to the agents' abilities in the counterfactual state of affairs. The first of these could be called an *optimistic*, or bold, approach, and states that agents are omnipotent in the counterfactual state of affairs. According to this approach, in situations where an agent does have the ability but not the opportunity to perform an action α_1 it is concluded that the agent has the ability to perform the sequential composition $\alpha_1;\alpha_2$ for arbitrary actions α_2. The second approach is a *pessimistic*, or careful one. In this approach agents are assumed to be nilpotent in counterfactual situations. Thus, in situations in which an agent does have the ability but not the opportunity to perform an action α_1 it is concluded that the agent is unable to perform the sequential composition $\alpha_1;\alpha_2$ for all α_2. Note that in the case that the agent has the opportunity to do α_1, optimistic and pessimistic approaches towards the agent's ability to do $\alpha_1;\alpha_2$ coincide. Although there is a case for both definitions, neither is completely acceptable. Consider the example of a lion in a cage, which is perfectly well capable of eating a zebra, but ideally never has the opportunity to do so. Using the first definition we would have to conclude that the lion is capable of performing the sequential composition 'eat zebra; fly to the moon', which hardly seems intuitive. Using the second definition it follows that the lion is unable to perform the action 'eat zebra; do nothing', which seems equally counterintuitive. Fortunately, the problems associated with these definitions are not really serious. They occur only in situations where an agent has the ability but not the opportunity to perform some action. And since it is exactly the combination of opportunity and ability that is important, no unwarranted conclusions can be drawn in these situations. Henceforth, we pursue both the optimistic and the pessimistic approach; in Section 7 we suggest alternative approaches in which the aforementioned counterintuitive situations do not occur.

Defining abilities for while-loops is even more hazardous than for sequential compositions. Intuitively it seems a good point of departure to let an agent be able to perform a while-loop only if it is at any point during execution capable of performing the next step. However, using this intuitive definition one has to be careful not to jump to undesired conclusions in the case of an action for which execution does not terminate. It seems highly counterintuitive to declare an agent, be it artificial or not, to have the reliable ability to perform an action that goes on indefinitely. For no agent is eternal: human agents die, artificial agents break down, and after all even the lifespan of the earth and the universe is bounded. Hence agents should not be able to perform actions that take infinite time. Therefore it seems reasonable to equate the ability to perform a while-loop with the ability to perform some finite-length sequence of confirmations and actions constituting the body of the while-loop, which ends in a confirmation for the negation of the condition of the loop, analogously to the equation used in extending the function $\mathbf{r}_0$ to while-loops. Accepting this equation, it is obvious that the discussion concerning the ability of agents for sequentially composed actions also becomes relevant for the repetitive composition, i.e. also with respect to abilities for while-loops a distinction between optimistic and pessimistic agents can be made. In the case that the while-loop terminates, optimistic and pessimistic approaches coincide, but in the case that execution of the action leads to the counterfactual state of affairs, they differ. Consider the situation of an agent that up to a certain point during the execution of an action $\mathtt{while}\,\varphi\,\mathtt{do}\,\alpha\,\mathtt{od}$ has been able to perform the confirmation for φ followed by α, and now finds itself in a state where φ holds, it is able to do α but does not have the opportunity for α. An optimistic agent concludes that it would have been able to finish the finite-length sequence constituting the while-loop after the (counterfactual) execution of α, and therefore considers itself to be capable of performing the while-loop. A pessimistic agent considers itself unable to finish the sequence, and thus is unable to perform the while-loop. The demand for finiteness of execution of the while-loop and the pessimistic view on abilities provide for a very interesting combination. For in order for an agent to be able to perform an action $\texttt{while}\,\varphi\,\texttt{do}\,\alpha\,\texttt{od}$ it has to have the opportunity to perform all the steps in the execution of $\texttt{while}\,\varphi\,\texttt{do}\,\alpha\,\texttt{od}$, possibly except for the last one. Furthermore, as as result of performing the last but one step in the execution the agent should obtain the ability to perform the last one, which is a confirmation for $\neg\varphi$. Since ability and opportunity coincide for confirmations this implies that the agent has the opportunity to confirm $\neg\varphi$, i.e. the agent has the opportunity to perform the last step in the execution of $\texttt{while}\,\varphi\,\texttt{do}\,\alpha\,\texttt{od}$. But then the agent has the opportunity to perform all the steps in the execution of the while-loop, and thus has the opportunity to perform the while-loop. Hence in the pessimistic approach the ability to perform a while-loop implies the opportunity!

Formally interpreting knowledge, abilities, results and opportunities

To interpret dynamic and ability formulas from L in a model M for L, the functions $\mathbf{r}_0$ and $\mathbf{c}_0$ from M are extended to deal with composite, i.e. non-atomic actions. To account for the difference between the optimistic and the pessimistic outlook on the agents' abilities, we define two different extensions of $\mathbf{c}_0$, and thereby also two different interpretations. The optimistic and the pessimistic approach coincide in their extension of $\mathbf{r}_0$, but differ in the extension of $\mathbf{c}_0$ for sequentially composed actions, and hence also in their treatment of ability for repetitive compositions. The following definition presents the extensions of $\mathbf{r}_0$ and $\mathbf{c}_0$. Here functions with the superscript **1** correspond to the optimistic view, and those with the superscript **0** to the pessimistic view on the agents' abilities in the counterfactual state of affairs.

DEFINITION 4. For $\mathbf{b} \in$ bool we inductively define the binary relation $\models^{\mathbf{b}}$ between a formula from L and a pair M, s consisting of a model M for L and a state s in M for the dynamic and ability formulas as follows:

$$
\begin{array}{lcl}
\mathrm{M}, s \models^{\mathbf{b}} p & \Leftrightarrow & \pi(p,s) = \mathbf{1} \text{ for } p \in \Pi \\
\mathrm{M}, s \models^{\mathbf{b}} \neg\varphi & \Leftrightarrow & \text{not } (\mathrm{M}, s \models^{\mathbf{b}} \varphi) \\
\mathrm{M}, s \models^{\mathbf{b}} \varphi \wedge \psi & \Leftrightarrow & \mathrm{M}, s \models^{\mathbf{b}} \varphi \text{ and } \mathrm{M}, s \models^{\mathbf{b}} \psi \\
\mathrm{M}, s \models^{\mathbf{b}} \mathbf{K}_i\varphi & \Leftrightarrow & \forall s' \in \mathrm{S}((s,s') \in \mathrm{R}(i) \Rightarrow \mathrm{M}, s' \models^{\mathbf{b}} \varphi) \\
\mathrm{M}, s \models^{\mathbf{b}} \langle \mathrm{do}_i(\alpha)\rangle\varphi & \Leftrightarrow & \exists s' \in \mathrm{S}(s' = \mathbf{r}^{\mathbf{b}}(i,\alpha)(s) \,\&\, \mathrm{M}, s' \models^{\mathbf{b}} \varphi) \\
\mathrm{M}, s \models^{\mathbf{b}} \mathbf{A}_i\alpha & \Leftrightarrow & \mathbf{c}^{\mathbf{b}}(i,\alpha)(s) = \mathbf{1}
\end{array}
$$

where $\mathbf{r}^{\mathbf{b}}$ and $\mathbf{c}^{\mathbf{b}}$ are defined by:

$$
\begin{array}{lcl}
\mathbf{r}^{\mathbf{b}} & : & \mathrm{A} \times \mathrm{Ac} \to \mathrm{S}^{\cdot} \to \mathrm{S}^{\cdot} \\
\mathbf{r}^{\mathbf{b}}(i,a)(s) & = & \mathbf{r}_0(i,a)(s) \\
\mathbf{r}^{\mathbf{b}}(i,\texttt{confirm}\,\varphi)(s) & = & s \text{ if } \mathrm{M}, s \models^{\mathbf{b}} \varphi
\end{array}
$$

$$
\begin{array}{lll}
 & = & \emptyset \text{ otherwise} \\
\mathbf{r^b}(i, \alpha_1; \alpha_2)(s) & = & \mathbf{r^b}(i, \alpha_2)(\mathbf{r^b}(i, \alpha_1)(s)) \\
\mathbf{r^b}(i, \mathtt{if}\,\varphi\,\mathtt{then}\,\alpha_1\,\mathtt{else}\,\alpha_2\,\mathtt{fi})(s) & = & \mathbf{r^b}(i, \alpha_1)(s) \text{ if } \mathrm{M}, s \models^{\mathbf{b}} \varphi \\
 & = & \mathbf{r^b}(i, \alpha_2)(s) \text{ otherwise} \\
\mathbf{r^b}(i, \mathtt{while}\,\varphi\,\mathtt{do}\,\alpha\,\mathtt{od})(s) & = & s' \text{ if } s' = \mathbf{r^b}(i, (\mathtt{confirm}\,\varphi; \alpha)^k; \\
 & & \quad \mathtt{confirm}\,\neg\varphi)(s) \text{ for some } k \in \mathbb{N} \\
 & = & \emptyset \text{ otherwise} \\
\mathbf{r^b}(i, \alpha)(\emptyset) & = & \emptyset \\
 & & \\
\mathbf{c^b} & : & \mathrm{A} \times \mathrm{Ac} \to \mathrm{S}^{\cdot} \to \mathrm{bool} \\
\mathbf{c^b}(i, a)(s) & = & \mathrm{c}_0(i, a)(s) \\
\mathbf{c^b}(i, \mathtt{confirm}\,\varphi)(s) & = & \mathbf{1} \text{ iff } \mathrm{M}, s \models^{\mathbf{b}} \varphi \\
\mathbf{c^b}(i, \alpha_1; \alpha_2)(s) & = & \mathbf{1} \text{ iff } \mathbf{c^b}(i, \alpha_1)(s) = \mathbf{1} \,\&\, \mathbf{c^b}(i, \alpha_2) \\
 & & \quad (\mathbf{r^b}(i, \alpha_1)(s)) = \mathbf{1} \\
\mathbf{c^b}(i, \mathtt{if}\,\varphi\,\mathtt{then}\,\alpha_1\,\mathtt{else}\,\alpha_2\,\mathtt{fi})(s) & = & \mathbf{1} \text{ iff } \mathbf{c^b}(i, \mathtt{confirm}\,\varphi; \alpha_1)(s) = \mathbf{1} \text{ or} \\
 & & \quad \mathbf{c^b}(i, \mathtt{confirm}\,\neg\varphi; \alpha_2)(s) = \mathbf{1} \\
\mathbf{c^b}(i, \mathtt{while}\,\varphi\,\mathtt{do}\,\alpha\,\mathtt{od})(s) & = & \mathbf{1} \text{ iff } \mathbf{c^b}(i, (\mathtt{confirm}\,\varphi; \alpha)^k; \mathtt{confirm} \\
 & & \quad \neg\varphi)(s) = \mathbf{1} \text{ for some } k \in \mathbb{N} \\
\mathbf{c^b}(i, \alpha)(\emptyset) & = & \mathbf{b}
\end{array}
$$

The formula φ is $\models^{\mathbf{b}}$-satisfiable in the model M iff $\mathrm{M}, s \models^{\mathbf{b}} \varphi$ for some s in M; φ is $\models^{\mathbf{b}}$-valid in M, denoted by $\mathrm{M} \models^{\mathbf{b}} \varphi$, iff $\mathrm{M}, s \models^{\mathbf{b}} \varphi$ for all s in M. The formula φ is $\models^{\mathbf{b}}$-satisfiable in **M** iff φ is $\models^{\mathbf{b}}$-satisfiable in some $\mathrm{M} \in \mathbf{M}$; φ is $\models^{\mathbf{b}}$-valid in **M**, denoted by $\models^{\mathbf{b}} \varphi$, iff φ is $\models^{\mathbf{b}}$-valid in all $\mathrm{M} \in \mathbf{M}$. Whenever $\models^{\mathbf{b}}$ is clear from the context, we drop it as a prefix and simply speak of a formula φ being satisfiable or valid in a (class of) model(s). For a given model M, we define $[s]_{\mathrm{R}(i)} =^{\mathrm{def}} \{s' \in \mathrm{S} \mid (s, s') \in \mathrm{R}(i)\}$ and $[\![\varphi]\!]_{\mathrm{M}} =^{\mathrm{def}} \{s \in \mathrm{S} \mid \mathrm{M}, s \models^{\mathbf{b}} \varphi\}$. Whenever the model M is clear from the context, the latter notion is usually simplified to $[\![\varphi]\!]$.

4 PROPERTIES OF KNOWLEDGE AND ACTIONS IN THE KARO-FRAMEWORK

In this section we look at the properties that knowledge and actions have in the KARO-framework. We furthermore consider additional properties, and show how some of these additional properties can be brought about by imposing constraints on the interpretation of atomic actions. We start with the properties of knowledge. When demanding the agents' epistemic accessibility relations to be equivalence relations, the modal operator **K** indeed formalises the notion of knowledge discussed in Section 2.

PROPOSITION 5. For all $i \in \mathrm{A}$ and $\varphi, \psi \in \mathrm{L}$ we have:

1. $\models^{\mathbf{b}} \mathbf{K}_i(\varphi \to \psi) \to (\mathbf{K}_i\varphi \to \mathbf{K}_i\psi)$ K

2. $\models^{\mathbf{b}} \varphi \Rightarrow \models^{\mathbf{b}} \mathbf{K}_i\varphi$ N

3. $\models^{\mathbf{b}} \mathbf{K}_i\varphi \rightarrow \varphi$ T

4. $\models^{\mathbf{b}} \mathbf{K}_i\varphi \rightarrow \mathbf{K}_i\mathbf{K}_i\varphi$ 4

5. $\models^{\mathbf{b}} \neg\mathbf{K}_i\varphi \rightarrow \mathbf{K}_i\neg\mathbf{K}_i\varphi$ 5

The first two items of Proposition 5 formalise that $\mathbf{K}_i$ is a normal modal operator: $\mathbf{K}_i$ satisfies both the K-axiom and the necessitation rule N (the names of these and other modal axioms are according to the Chellas classification [Chellas, 1980]). Furthermore, $\mathbf{K}_i$ satisfies the axioms of veridicality (the T-axiom), positive introspection (axiom 4) and negative introspection (axiom 5).

Although $\models^{\mathbf{1}}$ differs from $\models^{\mathbf{0}}$, the compositional behaviour of actions with respect to opportunities and results is identical in the two interpretations.

PROPOSITION 6. For $\mathbf{b} \in$ bool, $i \in$ A, $\alpha, \alpha_1, \alpha_2 \in$ Ac and $\varphi, \psi \in$ L we have:

1. $\models^{\mathbf{b}} \langle \mathrm{do}_i(\mathtt{confirm}\,\varphi)\rangle\psi \leftrightarrow (\varphi \wedge \psi)$
2. $\models^{\mathbf{b}} \langle \mathrm{do}_i(\alpha_1;\alpha_2)\rangle\psi \leftrightarrow \langle \mathrm{do}_i(\alpha_1)\rangle\langle \mathrm{do}_i(\alpha_2)\rangle\psi$
3. $\models^{\mathbf{b}} \langle \mathrm{do}_i(\mathtt{if}\,\varphi\,\mathtt{then}\,\alpha_1\,\mathtt{else}\,\alpha_2\,\mathtt{fi})\rangle\psi \leftrightarrow ((\varphi \wedge \langle \mathrm{do}_i(\alpha_1)\rangle\psi) \vee$ $(\neg\varphi \wedge \langle \mathrm{do}_i(\alpha_2)\rangle\psi))$
4. $\models^{\mathbf{b}} \langle \mathrm{do}_i(\mathtt{while}\,\varphi\,\mathtt{do}\,\alpha\,\mathtt{od})\rangle\psi \leftrightarrow ((\neg\varphi \wedge \psi) \vee (\varphi \wedge \langle \mathrm{do}_i(\alpha)\rangle$ $\langle \mathrm{do}_i(\mathtt{while}\,\varphi\,\mathtt{do}\,\alpha\,\mathtt{od})\rangle\psi))$
5. $\models^{\mathbf{b}} [\mathrm{do}_i(\alpha)](\varphi \rightarrow \psi) \rightarrow ([\mathrm{do}_i(\alpha)]\varphi \rightarrow [\mathrm{do}_i(\alpha)]\psi)$
6. $\models^{\mathbf{b}} \psi \Rightarrow \models^{\mathbf{b}} [\mathrm{do}_i(\alpha)]\psi$

Proposition 6 is in fact nothing but a formalisation of the intuitive ideas on results and opportunities for composite actions as expressed above. The first item states that agents have the opportunity to confirm exactly the formulas that are true, and that no state-transition takes place as the result of such a confirmation. The second item deals with the separation of the sequential composition into its elements: an agent has the opportunity to do $\alpha_1;\alpha_2$ with result ψ iff it has the opportunity to do α_1 (now) and doing so will result in having the opportunity to do α_2 with result ψ. The third item states that a conditionally composed action equals its 'then'-part in the case that the condition holds, and its 'else'-part if the condition does not hold. The fourth item formalises a sort of fixed-point equation for execution of while-loops: if an agent has the opportunity to perform a while-loop then it keeps this opportunity under execution of the body of the loop as long as the condition holds. The result of performing a while-loop is also fixed under executions of the body of the loop in states where φ holds, and is determined by the propositions that are true in the first state where $\neg\varphi$ holds. Note that a validity like this one does not suffice to axiomatise the repetitive composition: although it captures the idea of while-loops representing fixed-points, it fails to force termination, i.e. this formula on its own does not guarantee that agents do not have the opportunity to bring an infinitely non-terminating while-loop to its end. In the proof systems that we present in Section 6 this problem is solved by including suitable proof rules guiding the repetitive composition. The last two items state the normality of $[\mathrm{do}_i(\alpha)]$.

As soon as the abilities of agents come into play, the differences between $\models^1$ and $\models^0$ become visible, in particular for sequential and repetitive compositions.

PROPOSITION 7. For $\mathbf{b} \in$ bool, $i \in$ A, $\alpha, \alpha_1, \alpha_2 \in$ Ac and $\varphi \in$ L we have:

1. $\models^{\mathbf{b}} \mathbf{A}_i\mathtt{confirm}\,\varphi \leftrightarrow \varphi$
2. $\models^{\mathbf{1}} \mathbf{A}_i\alpha_1;\alpha_2 \leftrightarrow \mathbf{A}_i\alpha_1 \wedge [\mathrm{do}_i(\alpha_1)]\mathbf{A}_i\alpha_2$
3. $\models^{\mathbf{0}} \mathbf{A}_i\alpha_1;\alpha_2 \leftrightarrow \mathbf{A}_i\alpha_1 \wedge \langle\mathrm{do}_i(\alpha_1)\rangle\mathbf{A}_i\alpha_2$
4. $\models^{\mathbf{b}} \mathbf{A}_i\mathtt{if}\,\varphi\,\mathtt{then}\,\alpha_1\,\mathtt{else}\,\alpha_2\,\mathtt{fi} \leftrightarrow ((\varphi \wedge \mathbf{A}_i\alpha_1) \vee (\neg\varphi \wedge \mathbf{A}_i\alpha_2))$
5. $\models^{\mathbf{1}} \mathbf{A}_i\mathtt{while}\,\varphi\,\mathtt{do}\,\alpha\,\mathtt{od} \leftrightarrow (\neg\varphi \vee (\varphi \wedge \mathbf{A}_i\alpha \wedge [\mathrm{do}_i(\alpha)]\mathbf{A}_i\mathtt{while}\,\varphi\,\mathtt{do}\,\alpha\,\mathtt{od}))$
6. $\models^{\mathbf{0}} \mathbf{A}_i\mathtt{while}\,\varphi\,\mathtt{do}\,\alpha\,\mathtt{od} \leftrightarrow (\neg\varphi \vee (\varphi \wedge \mathbf{A}_i\alpha \wedge \langle\mathrm{do}_i(\alpha)\rangle\mathbf{A}_i\mathtt{while}\,\varphi\,\mathtt{do}\,\alpha\,\mathtt{od}))$

The first and the fourth items of Proposition 7 deal with the actions for which abilities are defined in a straightforward manner: agents are able to confirm exactly the true formulas, and having the ability to perform a conditional composition comes down to having the 'right' ability, dependent on the truth or falsity of the condition. The differences between the optimistic and the pessimistic outlook on abilities in the counterfactual state of affairs are clearly visible in the other items of Proposition 7. Optimistic agents are assumed to be omnipotent in counterfactual situations, and therefore it suffices for the agent to be able to do α_2 as a conditional result of doing α_1. A pessimistic agent needs certainty, and therefore demands to have the opportunity to do α_1 before concluding anything on its abilities following execution of α_1. This behaviour of optimistic and pessimistic agents is formalised in the second and the third item, respectively. The fifth and sixth item formalise an analogous behaviour for repetitive compositions: optimistic agents are satisfied with conditional results (item 5) whereas pessimistic agents demand certainty (item 6).

The compositional behaviour of sequential and repetitive compositions differs for the two interpretations only in situations where an agent lacks opportunities. If all appropriate opportunities are present, there is no difference for the two interpretations, a property which is formalised in the following corollary.

COROLLARY 8. For $i \in \mathrm{A}$, $\alpha, \alpha_1, \alpha_2 \in \mathrm{Ac}$ and $\varphi \in \mathrm{L}$ we have:

- $\models^1 \langle \mathrm{do}_i(\alpha_1)\rangle\top \rightarrow (\mathbf{A}_i\alpha_1;\alpha_2 \leftrightarrow \mathbf{A}_i\alpha_1 \wedge \langle \mathrm{do}_i(\alpha_1)\rangle \mathbf{A}_i\alpha_2)$
- $\models^1 \langle \mathrm{do}_i(\mathtt{while}\,\varphi\,\mathbf{do}\,\alpha\,\mathbf{od})\rangle\top \rightarrow$
  $(\mathbf{A}_i\mathtt{while}\,\varphi\,\mathbf{do}\,\alpha\,\mathbf{od} \leftrightarrow$
  $(\neg\varphi \vee (\varphi \wedge \mathbf{A}_i\alpha \wedge \langle \mathrm{do}_i(\alpha)\rangle\ \mathbf{A}_i\mathtt{while}\,\varphi\,\mathbf{do}\,\alpha\,\mathbf{od})))$

4.1 Frames and correspondences

To investigate properties of knowledge and actions, it will often prove useful to refer to *schemas*, which are sets of formulas, usually of a particular form. Using schemas one may abstract from particular agents and particular formulas, thereby having the possibility to formulate certain qualities of knowledge and action in a very general way. For instance, the axiom 4, $\mathbf{K}_i\varphi \rightarrow \mathbf{K}_i\mathbf{K}_i\varphi$, considered as a schema *in* φ expresses positive introspection of the agent i, and, as a schema in φ and i, it denotes positive introspection of all of the agents. If context allows it, we will remain implicit about what exactly are the varying elements in a schema.

Where schemas are used to express general properties of knowledge and actions on the syntactic level, *frames* can be used to do so on the semantic level. Informally speaking, a frame can be seen as a model without a valuation. By leaving out the valuation one may abstract from particular properties of knowledge and actions that are due to the valuation rather than inherently due to the nature of knowledge and/or action itself. Truth in a frame is defined in terms of truth in all models that can be constructed by adding a valuation to the frame.

DEFINITION 9. A frame F for a model $\mathrm{M} \in \mathbf{M}$ is a tuple consisting of the elements of M except for the valuation π. The class of all frames for models from $\mathbf{M}$ is denoted by $\mathbf{F}$. If F is some frame then (F, π) denotes the model generated by the elements of F and the valuation π. For F a frame, s one of its states and $\varphi \in \mathrm{L}$ we define

- $\mathrm{F}, s \models \varphi \Leftrightarrow (\mathrm{F}, \pi), s \models \varphi$ for all valuations π
- $\mathrm{F} \models \varphi \Leftrightarrow \mathrm{F}, s \models \varphi$ for all states s of F.
- $\mathbf{F} \models \varphi \Leftrightarrow \mathrm{F} \models \varphi$ for all $\mathrm{F} \in \mathbf{F}$

Since schemas are used to express general properties of knowledge and action syntactically, and frames can be used to do this semantically, the question arises as to how these notions relate. In particular, it is both interesting and important to try to single out first-order constraints on frames that exactly correspond to certain properties of knowledge and/or action, expressed in the form of schemas. The area of research called *correspondence theory* deals with finding relations — *correspondences* — between schemas and (first-order expressible) constraints on frames. A modal schema is said to correspond to a first-order constraint on frames if the schema is satisfied in exactly those frames that obey the constraint. A good introduction into correspondence theory is given in [Benthem, 1984].

DEFINITION 10. If φ is a schema and F is some frame then $\mathrm{F} \models \varphi$ iff $\mathrm{F} \models \chi$ for all formulas χ that are an instantiation of φ. If P is a formula in the first-order language subsuming the functions $\mathrm{R}, \mathrm{r}_0, \mathrm{c}_0$ and equality, then $\mathrm{F} \models^{\mathrm{fo}} P$ iff F satisfies P. The schema φ corresponds to the first-order formula P, notation $\varphi \sim P$ iff $\forall \mathrm{F}, (\mathrm{F} \models \varphi \Leftrightarrow \mathrm{F} \models^{\mathrm{fo}} P)$.

As already hinted at above, the properties that we require knowledge to obey correspond to constraints on the epistemic accessibility relations $\mathrm{R}(i)$. In Definition 3 we required these relations to be equivalence relations, and this demand indeed corresponds to knowledge being veridical and satisfying the properties of positive and negative introspection. The proof of the following proposition is standard and well-known from the literature [Hughes and Cresswell, 1984; Meyer and van der Hoek, 1995].

PROPOSITION 11. The following correspondences hold.

1. $\mathrm{T} \sim \forall s((s,s) \in \mathrm{R}(i))$, i.e. $\mathrm{R}(i)$ is reflexive
2. $4 \sim \forall s, s', s''((s,s') \in \mathrm{R}(i) \,\&\, (s',s'') \in \mathrm{R}(i) \Rightarrow (s,s'') \in \mathrm{R}(i))$, i.e. $\mathrm{R}(i)$ is transitive
3. $5 \sim \forall s, s', s''((s,s') \in \mathrm{R}(i) \,\&\, (s,s'') \in \mathrm{R}(i) \Rightarrow (s',s'') \in \mathrm{R}(i))$, i.e. $\mathrm{R}(i)$ is Euclidean

4.2 Additional properties of actions

The language L is sufficiently expressive to formalise various properties of knowledge, actions and their interplay, that are interesting both from a philosophical point of view as from the point of view of AI. The first of the properties that we consider here is *accordance.* Informally speaking, accordant actions are known to behave according to plan, i.e. for an accordant action it will be the case that things that an agent expects — on the basis of its knowledge — to hold in the future state of affairs that will result from it executing the action, are indeed known to be true by the agent when that future state of affairs has been brought about. Accordance of actions may be an important property in the context of agents planning to achieve certain goals. For if the agent knows (now) that performing some accordant action will bring about some goal, then it will be satisfied after it has executed the action: the agent knows that the goal is brought about. From a formal point of view, i-accordance of an action α corresponds to the schema $\mathbf{K}_i[\mathrm{do}_i(\alpha)]\varphi \rightarrow [\mathrm{do}_i(\alpha)]\mathbf{K}_i\varphi$.

The notion of ***determinism*** was already touched upon in the explanation of Definition 3 where it was stated that atomic actions are inherently deterministic. As we will see later on, viz. in Proposition 15, the determinism of atomic actions implies that of all actions. The notion of i-determinism of an action α is formalised through the schema $\langle \mathrm{do}_i(\alpha)\rangle\varphi \rightarrow [\mathrm{do}_i(\alpha)]\varphi$.

Whenever an action is ***idempotent***, consecutively executing the action twice — or in general an arbitrary number of times — will have exactly the same results as performing the action just once. In a sense, the state of affairs reached after the first performance of the action can be seen as a kind of fixed-point of execution of the action. The simplest idempotent action in our framework is the void action `skip`: performing it once, twice or an arbitrary number of times will not affect the state of affairs in any way whatsoever. More interesting idempotent actions were determined in our paper on actions that change the agent's epistemic state [van Linder *et al.*, 1995]); there, we claimed that such actions (we distinguished *retracting*, *expanding* and *revising*) have idempotency as a characterising property. Formally, i-idempotence of an action α corresponds to the schema $[\mathrm{do}_i(\alpha;\alpha)]\varphi \leftrightarrow [\mathrm{do}_i(\alpha)]\varphi$, or equivalently $\langle \mathrm{do}_i(\alpha;\alpha)\rangle\varphi \leftrightarrow \langle \mathrm{do}_i(\alpha)\rangle\varphi$.

Agents always have the opportunity to perform ***realisable*** actions, regardless of the circumstances, i.e. there never is an external factor that may prevent the performance of such an action. Typical realisable actions are, again, those in which the agent changes its information; an agent always has the opportunity to change its mind. The property of ***A-realisability*** relates ability and opportunity. For actions that are A-realisable, ability implies opportunity, i.e. whenever an agent is able to perform the action it automatically has the opportunity to perform it. Realisable actions are trivially A-realisable, and so are actions that no agent is ever capable of performing, but it seems hard to think of non-trivial examples of regular, mundane actions that an agent is able to execute and therefore automatically has the opportunity to do so. Adopting the A-realisability schema as an axiom schema would not be desirable for a general-purpose account of actions, opportunities and abilities, but might be appropriate for some specialized investigations. In any case it is far more reasonable to assume that ability implies opportunity than the reverse, given the fact that 'abilities are states that are acquired with effort [whereas] opportunities are there for the taking until they pass' [Kenny, 1975, p. 133]. Realisability of an action α for agent i is formalised through the schema $\langle \mathrm{do}_i(\alpha)\rangle\top$ in $i \in \mathbf{A}$; A-realisability of α for agent i corresponds to $\mathbf{A}_i\alpha \rightarrow \langle \mathrm{do}_i(\alpha)\rangle\top$.

The following definition summarises the properties discussed above in a formal way.

DEFINITION 12. Let $\alpha \in \text{Ac}$ be some action, i an agent and let F be a frame. The right-hand side of the following definitions is to be understood as a schema in φ.

- α is i-accordant in F iff $\text{F} \models \mathbf{K}_i[\text{do}_i(\alpha)]\varphi \rightarrow [\text{do}_i(\alpha)]\mathbf{K}_i\varphi$
- α is i-deterministic in F iff $\text{F} \models \langle\text{do}_i(\alpha)\rangle\varphi \rightarrow [\text{do}_i(\alpha)]\varphi$
- α is i-idempotent in F iff $\text{F} \models [\text{do}_i(\alpha;\alpha)]\varphi \leftrightarrow [\text{do}_i(\alpha)]\varphi$
- α is realisable for i in F iff $\text{F} \models \langle\text{do}_i(\alpha)\rangle\top$
- α is A-realisable for i in F iff $\text{F} \models \mathbf{A}_i\alpha \rightarrow \langle\text{do}_i(\alpha)\rangle\top$

We often omit explicit reference to the agent i in the above properties. Then, for instance, naming α accordant may either mean that is i-accordant for all agents i, or that mentioning the particular agent is clear from context, or not important. If Prop is any of the properties defined above, we say that α has the property Prop in **F** iff α has the property Prop in every $\text{F} \in \mathbf{F}$.

Here we show how these properties can be brought about to hold for all actions by imposing constraints on the functions R, $\mathbf{r}_0$ and $\mathbf{c}_0$. On the level of atomic actions, these properties correspond to first-order expressible constraints on R, $\mathbf{r}_0$ and $\mathbf{c}_0$. In Proposition 14 we present the correspondences for the properties of accordance, determinism, idempotence, realisability and A-realisability, respectively. Since we have defined two possible interpretations, viz. $\models^{\mathbf{1}}$ and $\models^{\mathbf{0}}$, for schemas from L in frames from **F** we have to be precise on the meaning of these correspondences.

DEFINITION 13. For $\mathbf{b} \in \text{bool}$ we define the schema φ to correspond to the first-order formula P given thc interpretation $\models^{\mathbf{b}}$ iff $\forall \text{F}(\text{F} \models^{\mathbf{b}} \varphi \Leftrightarrow \text{F} \models^{\text{fo}} P)$. In such a case, we write $\varphi \sim^{\mathbf{b}} P$.

PROPOSITION 14. For atomic actions $a \in \text{At}$, the following correspondences hold in the class **F** of frames for **M** both for $\mathbf{b} = \mathbf{1}$ and $\mathbf{b} = \mathbf{0}$. The left-hand side of these correspondences is to be understood as a schema in φ.

1. $\mathbf{K}_i[\text{do}_i(a)]\varphi \rightarrow [\text{do}_i(a)]\mathbf{K}_i\varphi \sim^{\mathbf{b}}$
   $\forall s_0 \in \text{S} \forall s_1 \in \text{S}(\exists s_2 \in \text{S}(s_2 = \mathbf{r}_0(i,a)(s_0) \,\&\, (s_2, s_1) \in \text{R}(i)) \Rightarrow$
   $\exists s_3 \in \text{S}((s_0, s_3) \in \text{R}(i) \,\&\, s_1 = \mathbf{r}_0(i,a)(s_3)))$
2. $\langle\text{do}_i(a)\rangle\varphi \rightarrow [\text{do}_i(a)]\varphi \sim^{\mathbf{b}}$
   $\forall s \in \text{S} \forall s' \in \text{S} \forall s'' \in \text{S}(\mathbf{r}_0(i,a)(s) = s' \,\&\, \mathbf{r}_0(i,a)(s) = s'' \Rightarrow s' = s'')$
3. $[\text{do}_i(a;a)]\varphi \leftrightarrow [\text{do}_i(a)]\varphi \sim^{\mathbf{b}} \forall s \in \text{S}(\mathbf{r}_0(i,a)(\mathbf{r}_0(i,a)(s)) = \mathbf{r}_0(i,a)(s))$

4. $\langle \mathrm{do}_i(a)\rangle\top \sim^{\mathbf{b}} \forall s \in \mathrm{S}(\mathbf{r}_0(i,a)(s) \neq \emptyset)$

5. $\mathbf{A}_i a \rightarrow \langle \mathrm{do}_i(a)\rangle\top \sim^{\mathbf{b}} \forall s \in \mathrm{S}(\mathbf{c}_0(i,a)(s) = \mathbf{1} \Rightarrow \mathbf{r}_0(i,a)(s) \neq \emptyset)$

Since the functions $\mathbf{r}_0$ and $\mathbf{c}_0$ are defined for atomic actions only, and the functions $\mathbf{r}^{\mathbf{b}}$ and $\mathbf{c}^{\mathbf{b}}$ — which are the extensions of $\mathbf{r}_0$ and $\mathbf{c}_0$ for arbitrary actions — are constructed out of $\mathbf{r}_0$ and $\mathbf{c}_0$ and have no existence on their own, it is not possible to prove correspondences like those of Proposition 14 for non-atomic actions. There simply is no semantic entity to correspond the syntactic schemas with. This implies that it is in general not possible to ensure that arbitrary actions satisfy a certain property. However, it turns out that some of the properties considered above straightforwardly extend from the atomic level to the level of arbitrary actions, regardless of the interpretation that is used. This is in particular the case for the properties of A-realisability and determinism.

PROPOSITION 15. The following lifting results hold for all F and $i \in \mathrm{A}$, in the case of $\models^{\mathbf{1}}$ as well as that of $\models^{\mathbf{0}}$:

- $\forall a \in \mathrm{At}(a$ is A-realisable for i in F$) \Rightarrow \forall \alpha \in \mathrm{Ac}(\alpha$ is A-realisable for i in F$)$
- $\forall a \in \mathrm{At}(a$ is i-deterministic in F$) \Rightarrow \forall \alpha \in \mathrm{Ac}(\alpha$ is i-deterministic in F$)$

Since the range of the function $\mathbf{r}_0$ is the set $\mathrm{S}^{\cdot}$, it follows directly that atomic actions are deterministic in $\mathbf{F}$: for if $a \in \mathrm{At}$, $i \in \mathrm{A}$ and s a state in some model, then $\mathbf{r}_0(i,a)(s)$ is either the empty set, or a single state from S, and hence the frame condition for determinism as given in Proposition 14 is satisfied. Using the lifting result obtained in Proposition 15 one then concludes that all actions are deterministic in $\mathbf{F}$.

COROLLARY 16. All actions $\alpha \in \mathrm{Ac}$ are deterministic in $\mathbf{F}$, both for $\models^{\mathbf{1}}$ and $\models^{\mathbf{0}}$.

Thus two of the properties formalised in Definition 12 can be ensured to hold for arbitrary actions by imposing suitable constraints on the frames for $\mathbf{M}$. For the other three properties, viz. accordance, idempotence and realisability, constraining the function $\mathbf{r}_0$ for atomic actions does not suffice, since this does not conservatively extend to the class of all actions. That realisability may not be lifted is easily seen by considering the action `fail`. Independent of the realisability of atomic actions, `fail` will never be realisable: the formula $\neg\langle \mathrm{do}_i(\mathtt{fail})\rangle\top$ is valid, both for $\models^{\mathbf{1}}$ and for $\models^{\mathbf{0}}$. The following examples show why accordance and idempotence are in general not to be lifted.

EXAMPLE 17. Consider the language $\mathrm{L}(\Pi, \mathrm{A}, \mathrm{At})$ with $\Pi = \{p, q\}$, $i \in \mathrm{A}$ and At arbitrary. Let $\mathrm{F} \in \mathbf{F}$ be a frame such that the set S of states in F contains at least two elements, say s and t, on which the relation R(i) is defined to be universal, and the first-order property corresponding with accordance of atomic actions is met. Let π be a valuation such that $\pi(p, s) = \pi(q, s) = \mathbf{1}, \pi(p, t) = \pi(q, t) = \mathbf{0}$. Then we have that $(\mathrm{F}, \pi), s \models^{\mathbf{b}} \mathbf{K}_i[\mathrm{do}_i(\mathtt{confirm}\,p)]q$, and furthermore that $(\mathrm{F}, \pi), s \not\models^{\mathbf{b}} [\mathrm{do}_i(\mathtt{confirm}\,p)]\mathbf{K}_i q$. Hence $\mathrm{F} \not\models^{\mathbf{b}} \mathbf{K}_i[\mathrm{do}_i(\mathtt{confirm}\,p)]q \rightarrow [\mathrm{do}_i(\mathtt{confirm}\,p)]\mathbf{K}_i q$, which provides a counterexample to the lifting of accordance.

EXAMPLE 18. Consider the language L(Π, A, At) with $\Pi = \{p\}$, $i \in \mathrm{A}$ $\langle \mathrm{S}, \mathrm{R}, \mathrm{r_0}, \mathrm{c_0} \rangle$, where

- $\mathrm{S} = \{s_1, s_2, s_3, s_4\}$
- R(i) is an arbitrary equivalence relation on S
- $\mathrm{r_0}(i, a_1)(s_1) = s_1$ $\mathrm{r_0}(i, a_1)(s_2) = s_3$ $\mathrm{r_0}(i, a_1)(s_3) = s_3$ $\mathrm{r_0}(i, a_1)(s_4) = \emptyset$
  $\mathrm{r_0}(i, a_2)(s_1) = s_2$ $\mathrm{r_0}(i, a_2)(s_2) = s_2$ $\mathrm{r_0}(i, a_2)(s_3) = s_4$ $\mathrm{r_0}(i, a_2)(s_4) = s_4$
- $\mathrm{c_0} : \mathrm{A} \times \mathrm{S} \rightarrow$ bool is arbitrary

It is easily checked that both a_1 and a_2 are idempotent in F. However, it is not the case that all actions that can be built on At are idempotent in F. For it holds for arbitrary $\mathbf{b} \in$ bool that $\mathrm{F} \not\models^{\mathbf{b}} [\mathrm{do}_i((a_1; a_2); (a_1; a_2))]p \leftrightarrow [\mathrm{do}_i(a_1; a_2)]p$. To see this take $\mathrm{M} = (\mathrm{F}, \pi)$ where $\pi(p, s_2) \neq \pi(p, s_4)$. In this model it holds that $\mathrm{M}, s_1 \models^{\mathbf{b}} [\mathrm{do}_i((a_1; a_2); (a_1; a_2))]p \leftrightarrow [\mathrm{do}_i(a_1; a_2)]\neg p$. Hence $\mathrm{M} \not\models^{\mathbf{b}} [\mathrm{do}_i((a_1; a_2); (a_1; a_2))]p \leftrightarrow [\mathrm{do}_i(a_1; a_2)]p$, and therefore also $\mathrm{F} \not\models^{\mathbf{b}} [\mathrm{do}_i((a_1; a_2); (a_1; a_2))]p \leftrightarrow [\mathrm{do}_i(a_1; a_2)]p$. Thus neither for $\models^{\mathbf{1}}$ nor for $\models^{\mathbf{0}}$ is $a_1; a_2$ idempotent in F.

Although we showed in Example 17 that accordance is not to be lifted from atomic actions to general ones, we can prove a restricted form of lifting for accordance. That is, if we leave confirmations out of consideration, we can prove that accordance is lifted.

PROPOSITION 19. Let Ac^- be the confirmation-free fragment of Ac, i.e. the fragment built from atomic actions through sequential, conditional or repetitive composition. Then we have for all $\mathrm{F} \in \mathbf{F}$ and for all $\mathbf{b} \in$ bool:

- $\forall a \in \mathrm{At}(a$ is accordant for $\models^{\mathbf{b}}$ in F$) \Rightarrow \forall \alpha \in \mathrm{Ac}^-(\alpha$ is accordant for $\models^{\mathbf{b}}$ in F$)$

The properties of idempotence and (A-)realisability are in general undesirable ones. If all actions were idempotent, it would be impossible to walk the roads by taking one step at a time. Realisability would render the notion of opportunity meaningless and A-realisability would tie ability and opportunity in a way that we feel is unacceptable. Therefore we consider neither the lifting result for A-realisability to be very important, nor the absence of such a result for idempotence and realisability. And even though the property of accordance is, or may be, important, it is not one that typically holds in the lively world of human agents. Therefore we consider this property to be an exceptional one, that holds for selected actions only. Hence also for accordance the absence of a lifting result is not taken too seriously.

5 CORRECTNESS AND FEASIBILITY OF ACTIONS: PRACTICAL POSSIBILITY

Within the KARO-framework, several notions concerning agency may be formalised that are interesting not only from a philosophical point of view, but also when analysing agents in planning systems. The most important one of these notions formalises the knowledge that agents have about their practical possibilities. We consider the notion of practical possibility as relating an agent, an action, and a proposition: agents may have the practical possibility to bring about (truth of) the proposition by performing the action. We think of practical possibility as consisting of two parts, viz. correctness and feasibility. Correctness implies that no external factors will prevent the agent from performing the action and thereby making the proposition true. As such, correctness is defined in terms of opportunity and result: an action is correct for some agent to bring about some proposition iff the agent has the opportunity to perform the action in such a way that its performance results in the proposition being true. Feasibility captures the internal aspect of practical possibility. It states that it is within the agent's capacities to perform the action, and as such is nothing but a reformulation of ability. Together, correctness and feasibility constitute practical possibility.

DEFINITION 20. For $\alpha \in \mathrm{Ac}$, $i \in \mathrm{A}$ and $\varphi \in \mathrm{L}$ we define:

- $\mathbf{Correct}_i(\alpha, \varphi) =^{\mathrm{def}} \langle \mathrm{do}_i(\alpha) \rangle \varphi$
- $\mathbf{Feasible}_i \alpha =^{\mathrm{def}} \mathbf{A}_i \alpha$
- $\mathbf{PracPoss}_i(\alpha, \varphi) =^{\mathrm{def}} \mathbf{Correct}_i(\alpha, \varphi) \wedge \mathbf{Feasible}_i \alpha$

The counterintuitive situations that occurred with respect to the ability of agents as described previously do not take root for practical possibility. That is, a lion that has the ability but not the opportunity to eat a zebra will neither have the practical possibility to eat a zebra first and thereafter fly to the moon nor have the practical possibility to eat a zebra and rest on its laurels afterwards. Thus even though the notion of ability suffers from problems like these, the more important notion of practical possibility does not. The importance of practical possibility manifests itself particularly when ascribing — from the outside — certain qualities to an agent.

It seems that for the agent itself practical possibilities are relevant in so far as the agent has knowledge of these possibilities. For one may not expect an agent to act on its practical possibilities if the agent does not know of this possibilities. To formalise this kind of knowledge, we introduce the Can-predicate and the Cannot-predicate. The first of these predicates concerns the knowledge of agents about their practical possibilities, the latter predicate does the same for their practical impossibilities.

DEFINITION 21. For $\alpha \in$ Ac, $i \in$ A and $\varphi \in$ L we define:

- $\mathbf{Can}_i(\alpha, \varphi) =^{\text{def}} \mathbf{K}_i\mathbf{PracPoss}_i(\alpha, \varphi)$
- $\mathbf{Cannot}_i(\alpha, \varphi) =^{\text{def}} \mathbf{K}_i\neg\mathbf{PracPoss}_i(\alpha, \varphi)$

The Can-predicate and the Cannot-predicate integrate knowledge, ability, opportunity and result, and seem to formalise one of the most important notions of agency. In fact it is probably not too bold to say that knowledge like that formalised through the Can-predicate, although perhaps in a weaker form by taking aspects of uncertainty into account, underlies all acts performed by rational agents. For rational agents act only if they have some information on both the possibility to perform the act, and its possible outcome; at least in this paper we restrict ourselves to such actions, leaving mere *experiments* out of our scope. It therefore seems worthwhile to take a closer look at both the Can-predicate and the Cannot-predicate. The following proposition focuses on the behaviour of the *means*-part of the predicates, which is the α in $\mathbf{Can}_i(\alpha, \varphi)$ and $\mathbf{Cannot}_i(\alpha, \varphi)$.

PROPOSITION 22. For all $\mathbf{b} \in$ bool, $i \in$ A, $\alpha, \alpha_1, \alpha_2 \in$ Ac and $\varphi, \psi \in$ L we have:

1. $\models^{\mathbf{b}} \mathbf{Can}_i(\mathtt{confirm}\,\varphi, \psi) \leftrightarrow \mathbf{K}_i(\varphi \wedge \psi)$
2. $\models^{\mathbf{b}} \mathbf{Cannot}_i(\mathtt{confirm}\,\varphi, \psi) \leftrightarrow \mathbf{K}_i(\neg\varphi \vee \neg\psi)$
3. $\models^{\mathbf{b}} \mathbf{Can}_i(\alpha_1; \alpha_2, \varphi) \leftrightarrow \mathbf{Can}_i(\alpha_1, \mathbf{PracPoss}_i(\alpha_2, \varphi))$
4. $\models^{\mathbf{b}} \mathbf{Can}_i(\alpha_1; \alpha_2, \varphi) \rightarrow \langle \mathrm{do}_i(\alpha_1)\rangle \mathbf{Can}_i(\alpha_2, \varphi)$ for α_1 accordant in **F**
5. $\models^{\mathbf{b}} \mathbf{Cannot}_i(\alpha_1; \alpha_2, \varphi) \leftrightarrow \mathbf{Cannot}_i(\alpha_1, \mathbf{PracPoss}_i(\alpha_2, \varphi))$
6. $\models^{\mathbf{b}} \mathbf{Can}_i(\mathtt{if}\,\varphi\,\mathtt{then}\,\alpha_1\,\mathtt{else}\,\alpha_2\,\mathtt{fi}, \psi) \wedge \mathbf{K}_i\varphi \leftrightarrow \mathbf{Can}_i(\alpha_1, \psi) \wedge \mathbf{K}_i\varphi$
7. $\models^{\mathbf{b}} \mathbf{Can}_i(\mathtt{if}\,\varphi\,\mathtt{then}\,\alpha_1\,\mathtt{else}\,\alpha_2\,\mathtt{fi}, \psi) \wedge \mathbf{K}_i\neg\varphi \leftrightarrow \mathbf{Can}_i(\alpha_2, \psi) \wedge \mathbf{K}_i\neg\varphi$
8. $\models^{\mathbf{b}} \mathbf{Cannot}_i(\mathtt{if}\,\varphi\,\mathtt{then}\,\alpha_1\,\mathtt{else}\,\alpha_2\,\mathtt{fi}, \psi) \wedge \mathbf{K}_i\varphi \leftrightarrow$
   $\mathbf{Cannot}_i(\alpha_1, \psi) \wedge \mathbf{K}_i\varphi$
9. $\models^{\mathbf{b}} \mathbf{Cannot}_i(\mathtt{if}\,\varphi\,\mathtt{then}\,\alpha_1\,\mathtt{else}\,\alpha_2\,\mathtt{fi}, \psi) \wedge \mathbf{K}_i\neg\varphi \leftrightarrow$
   $\mathbf{Cannot}_i(\alpha_2, \psi) \wedge \mathbf{K}_i\neg\varphi$

10. $\models^{\mathbf{b}} \mathbf{Can}_i(\mathtt{while}\,\varphi\,\mathbf{do}\,\alpha\,\mathbf{od}, \psi) \wedge \mathbf{K}_i\varphi \leftrightarrow$
$\mathbf{Can}_i(\alpha, \mathbf{PracPoss}_i(\mathtt{while}\,\varphi\,\mathbf{do}\,\alpha\,\mathbf{od}, \psi)) \wedge \mathbf{K}_i\varphi$

11. $\models^{\mathbf{b}} \mathbf{Can}_i(\mathtt{while}\,\varphi\,\mathbf{do}\,\alpha\,\mathbf{od}, \psi) \wedge \mathbf{K}_i\varphi \rightarrow$
$\langle \mathrm{do}_i(\alpha)\rangle \mathbf{Can}_i(\mathtt{while}\,\varphi\,\mathbf{do}\,\alpha\,\mathbf{od}, \psi)$ for α accordant in $\mathbf{F}$

12. $\models^{\mathbf{b}} \mathbf{Can}_i(\mathtt{while}\,\varphi\,\mathbf{do}\,\alpha\,\mathbf{od}, \psi) \rightarrow \mathbf{K}_i(\varphi \vee \psi)$

13. $\models^{\mathbf{b}} \mathbf{Cannot}_i(\mathtt{while}\,\varphi\,\mathbf{do}\,\alpha\,\mathbf{od}, \psi) \wedge \mathbf{K}_i\neg\varphi \leftrightarrow \mathbf{K}_i(\neg\varphi \wedge \neg\psi)$

14. $\models^{\mathbf{b}} \mathbf{Cannot}_i(\mathtt{while}\,\varphi\,\mathbf{do}\,\alpha\,\mathbf{od}, \psi) \wedge \mathbf{K}_i\varphi \leftrightarrow$
$\mathbf{Cannot}_i(\alpha; \mathtt{while}\,\varphi\,\mathbf{do}\,\alpha\,\mathbf{od}, \psi) \wedge \mathbf{K}_i\varphi$

Proposition 22 supports the claim about appropriateness of the Can-predicate and Cannot-predicate as formalising knowledge of practical possibilities of actions performed by rational agents. In particular items 6 through 9 and item 14 are genuine indications of the rationality of the agents that we formalised. Consider for example item 7. This item states that whenever an agent knows both that it has the practical possibility to bring about ψ by performing $\mathtt{if}\,\varphi\,\mathtt{then}\,\alpha_1\,\mathtt{else}\,\alpha_2\,\mathtt{fi}$ and that the negation of the condition of $\mathtt{if}\,\varphi\,\mathtt{then}\,\alpha_1\,\mathtt{else}\,\alpha_2\,\mathtt{fi}$ holds, it also knows that performing the else-part of the conditional composition provides the practical possibility to achieve ψ. Conversely, if agent i knows that it has the practical possibility to bring about ψ by performing α_2 while at the same time knowing that the proposition φ is false, then the agent knows that performing a conditional composition $\mathtt{if}\,\varphi\,\mathtt{then}\,\alpha_1\,\mathtt{else}\,\alpha_2\,\mathtt{fi}$ would also bring about ψ, regardless of α_1. For since it knows that $\neg\varphi$ holds, it knows that this compositional composition comes down to the else-part α_2. Items 4 and 11 explicitly use the accordance of actions. For it is exactly this property of accordance that causes the agent's knowledge of its practical possibilities to persist under execution of the first part of the sequential composition in item 4 and the body of the while-loop in item 11.

In the following proposition we characterise the relation between the Can-predicate and the Cannot-predicate. Furthermore some properties are presented that concern the *end*-part of these predicates, i.e. the φ in $\mathbf{Can}_i(\alpha, \varphi)$ and $\mathbf{Cannot}_i(\alpha, \varphi)$.

PROPOSITION 23. For all $\mathbf{b} \in$ bool, $i \in$ A, $\alpha \in$ Ac and $\varphi, \psi \in$ L we have:

1. $\models^{\mathbf{b}} \mathbf{Can}_i(\alpha, \varphi) \rightarrow \neg\mathbf{Can}_i(\alpha, \neg\varphi)$
2. $\models^{\mathbf{b}} \mathbf{Can}_i(\alpha, \varphi) \rightarrow \neg\mathbf{Cannot}_i(\alpha, \varphi)$
3. $\models^{\mathbf{b}} \mathbf{Can}_i(\alpha, \varphi) \rightarrow \mathbf{Cannot}_i(\alpha, \neg\varphi)$
4. $\models^{\mathbf{b}} \mathbf{Can}_i(\alpha, \varphi \wedge \psi) \leftrightarrow \mathbf{Can}_i(\alpha, \varphi) \wedge \mathbf{Can}_i(\alpha, \psi)$
5. $\models^{\mathbf{b}} \mathbf{Cannot}_i(\alpha, \varphi) \vee \mathbf{Cannot}_i(\alpha, \psi) \rightarrow \mathbf{Cannot}_i(\alpha, \varphi \wedge \psi)$

6. $\models^{\mathbf{b}} \mathbf{Can}_i(\alpha, \varphi) \vee \mathbf{Can}_i(\alpha, \psi) \rightarrow \mathbf{Can}_i(\alpha, \varphi \vee \psi)$

7. $\models^{\mathbf{b}} \mathbf{Cannot}_i(\alpha, \varphi \vee \psi) \leftrightarrow \mathbf{Cannot}_i(\alpha, \varphi) \wedge \mathbf{Cannot}_i(\alpha, \psi)$

8. $\models^{\mathbf{b}} \mathbf{Can}_i(\alpha, \varphi) \wedge \mathbf{K}_i[\mathrm{do}_i(\alpha)](\varphi \rightarrow \psi) \rightarrow \mathbf{Can}_i(\alpha, \psi)$

9. $\models^{\mathbf{b}} \mathbf{Cannot}_i(\alpha, \varphi) \wedge \mathbf{K}_i[\mathrm{do}_i(\alpha)](\psi \rightarrow \varphi) \rightarrow \mathbf{Cannot}_i(\alpha, \psi)$

Even more than Proposition 22 does Proposition 23 make out a case for the rationality of agents. Take for example item 3, which states that whenever an agent knows that it has the practical possibility to achieve φ by performing α it also knows that α does not provide for a means to achieve $\neg\varphi$. Items 4 through 7 deal with the decomposition of the end-part of the Can-predicate and the Cannot-predicate, which behaves as desired. Note that the reverse implication of item 5 is not valid: it is quite possible that even though an agent knows that α is not correct to bring about $(p \wedge \neg p)$ it might still be that it knows that α is correct for either p or $\neg p$. An analogous line of reasoning shows the invalidity of the reverse implication of item 6. Items 8 and 9 formalise that agents can extend their knowledge about their practical (im-)possibilities by combining it with their knowledge of the (conditional) results of actions.

6 PROOF THEORY

Here we present a proof theory for the semantic framework defined in the previous section. In general the purpose of a proof theory is to provide a syntactic counterpart of the semantic notion of validity for a given interpretation and a given class of models. The idea is to define a predicate denoting deducibility, which holds for a given formula iff the formula is valid. This predicate is to be defined purely syntactically, i.e. it should depend only on the syntactic structure of formulas, without making any reference to semantic notions such as truth, validity, satisfiability etc. We present two such predicates, viz. $\vdash^{\mathbf{1}}$ and $\vdash^{\mathbf{0}}$, which characterise the notions of validity associated with $\models^{\mathbf{1}}$ and $\models^{\mathbf{0}}$, respectively. The definition of these predicates is based on a set of axioms and proof rules, which together constitute a proof system. The proof systems that we define deviate somewhat from the ones that are common in (modal) logics, the most notable difference being the use of infinitary proof rules. Given the relative rarity of this kind of rules, we feel that some explanation is justified.

6.1 Infinitary proof rules

The proof rules that are commonly employed in proof systems, are inference schemes of the form $\mathrm{P}_1, \ldots, \mathrm{P}_m \,/\, \mathrm{C}$, where the premises $\mathrm{P}_1, \ldots, \mathrm{P}_m$ and the conclusion C are elements of the language under consideration. Informally, a rule like this denotes that one may deduce C as soon as $P_1, \ldots, P_m$ have been deduced. An infinitary[3] proof rule is a rule containing an infinite number of premises. Although not very common, infinitary proof rules have been used in a number of proof systems: Hilbert used an infinitary proof rule in axiomatising number theory [Hilbert, 1931], Schütte uses infinitary proof rules in a number of systems [Schütte, 1960], and both Kröger [1980] and Goldblatt [1982a; 1982b] use infinitary proof rules in logics of action.

In finitary proof systems proofs can be carried out completely within the formal system. A proof is usually taken to be a finite sequence of formulas that are either axioms of the proof system or conclusions of proof rules applied to formulas that appear earlier in the sequence. Since finitary proof rules can be applied as soon as all of their finitely many premises have been deduced, there is no need to step outside of the formal system. In order to apply an infinitary rule, a meta-logical investigation on the deducibility of the (infinitely many) premises needs to be carried out, which makes it in general impossible to carry out proofs completely within the proof system. As such, proofs are no longer 'schematically' constructed, and theorems are not recursively enumerable. However, there are also advantages associated with the use of infinitary proof rules. One such advantage is that for some systems *strong completeness* can be achieved using infinitary proof rules, whereas this is not possible using finitary proof rules (cf. [Goldblatt, 1982a; Schütte, 1960]). The notion of strong completeness implies that fewer sets of formulas are consistent, and in particular that sets of formulas that are seen to be inconsistent can also be proved to be so. After the presentation of the proof systems, we will return to the property of strong completeness in the presence of infinitary rules. Besides the possibility to achieve strong completeness when using infinitary proof rules, there are two other arguments that influenced our decision to use this kind of rule. The first of these is its intuitive acceptability. In particular when dealing with notions with an infinitary character, like for instance while-loops, infinitary proof rules provide a much better formalisation of human intuition on the nature of these notions than do finitary proof rules. The second, perhaps less convincing but certainly more compelling, argument is given by the fact that our attempts to come up with finitary axiomatisations remained unavailing.

6.2 Logics of capabilities

Before presenting the actual axiomatisations, we first make some notions precise that were already informally discussed above. An axiom is a schema in L. A proof rule is a schema of the form $\varphi_1, \varphi_2, \ldots \,/\, \psi$ where $\varphi_1, \varphi_2, \ldots, \psi$

[3] We decided to follow the terminology of Goldblatt [Goldblatt, 1982a; Goldblatt, 1982b] and refer to these rules as being infinitary. Other authors call these rules infinite [Kröger, 1980; Schütte, 1960].

are schemas in L. A proof system is a pair consisting of a set of axioms and a set of proof rules. As mentioned above, the presence of infinitary proof rules forces us to adopt a more abstract approach to the notions of deducibility and theoremhood than the one commonly employed in finitary proof systems. Usually, a formula φ is defined to be a theorem of some proof system if there exists a finite-length sequence of formulas of which φ is the last element and such that each formula in the sequence is either an instance of an axiom or the conclusion of a proof rule applied to earlier members of the sequence. An alternative formulation, which is equally usable in finitary and in infinitary proof systems, is to define φ to be a theorem of a proof system iff it belongs to the smallest subset of L containing all (instances of all) axioms and being closed under the proof rules. This latter notion of deducibility is actually the one that we will employ here. We define a logic for a given proof system to be a subset of L containing all instances of the axioms of the proof system and being closed under its proof rules. A formula is a theorem for a given proof system iff it is an element of the smallest logic for the proof system. These notions are formalised in Definitions 27 through 29.

To axiomatise the behaviour of while-loops we propose two infinitary rules. Both these rules are based on the idea to equate a repetitive composition $\texttt{while}\,\varphi\,\texttt{do}\,\alpha\,\texttt{od}$ with the infinite set $\{(\texttt{confirm}\,\varphi;\alpha)^k;\texttt{confirm}\,\neg\varphi \mid k \in \mathbb{N}\}$. The two proof rules take as their premises an infinite set of formulas built around this infinite set and have as their conclusion a formula built around $\texttt{while}\,\varphi\,\texttt{do}\,\alpha\,\texttt{od}$. To make this idea of 'building formulas around actions' explicit, we introduce the concept of *admissible forms.* The notion of admissible forms as given in Definition 24 is an extension of that used by Goldblatt in his language of program schemata [Goldblatt, 1982a]. In his investigation of infinitary proof rules, Kröger found that, in order to prove completeness, he needed rules in which the context of the while-loop and of the set $\{(\texttt{confirm}\,\varphi;\alpha)^k;\texttt{confirm}\,\neg\varphi \mid k \in \mathbb{N}\}$ is taken into account [Kröger, 1980]. The concept of admissible forms provides an abstract generalisation of this idea of taking contexts into account.

DEFINITION 24. The set of admissible forms for L, denoted by Afm(L), is defined by the following BNF.

$$\phi ::= \# \mid [\mathrm{do}_i(\alpha)]\phi \mid \mathbf{K}_i\phi \mid \psi \to \phi$$

where $i \in$ A, $\alpha \in$ Ac and $\psi \in$ L. We use ϕ as a typical element of Afm(L).

Usually 'admissible form' is abbreviated to 'afm'. By definition, each afm has a unique occurrence of the special symbol #. By instantiating this symbol with a formula from L, afms are turned into genuine formulas. If ϕ is an afm and $\psi \in$ L is some formula we denote by $\phi(\psi)$ the formula that is obtained by replacing (the unique occurrence of) # in ϕ by ψ.

The following definition introduces two abbreviations that will be used in formulating the infinitary rules.

DEFINITION 25. For all $\psi, \varphi \in \mathrm{L}$, $i \in \mathrm{A}$, $\alpha \in \mathrm{Ac}$ and $l \in \mathbb{N}$ we define:

- $\psi_l(i, \varphi, \alpha) =^{\mathrm{def}} [\mathrm{do}_i((\mathtt{confirm}\,\varphi; \alpha)^l; \mathtt{confirm}\,\neg\varphi)]\psi$
- $\varphi_l(i, \alpha) =^{\mathrm{def}} \mathbf{A}_i((\mathtt{confirm}\,\varphi; \alpha)^l; \mathtt{confirm}\,\neg\varphi)$

The formulas introduced in Definition 25 are used to define the premises of the infinitary rules. The rule formalising the behaviour of while-loops with respect to results and opportunities has as premises all sentences in the infinite set $\{\phi(\psi_l(i, \varphi, \alpha)) \mid l \in \mathbb{N}\}$ for some $\phi \in \mathrm{Afm(L)}\}$. The conclusion of this rule is the formula $\phi([\mathrm{do}_i(\mathtt{while}\,\varphi\,\mathtt{do}\,\alpha\,\mathtt{od})]\psi)$. Leaving the context provided by ϕ out of consideration, this rules intuitively states that if it is deducible that ψ holds after executing the actions $(\mathtt{confirm}\,\varphi; \alpha)^k; \mathtt{confirm}\,\neg\varphi$, for every $k \in \mathbb{N}$, then it is also deducible that ψ holds after executing $\mathtt{while}\,\varphi\,\mathtt{do}\,\alpha\,\mathtt{od}$. The rule used in formalising the ability of agents for while-loops has as its premises the set $\phi(\neg(\varphi_l(i, \alpha)))$ for $l \in \mathbb{N}, \phi \in \mathrm{Afm(L)}$, and a conclusion $\phi(\neg\mathbf{A}_i\mathtt{while}\,\varphi\,\mathtt{do}\,\alpha\,\mathtt{od})$. This rule states that whenever it is deducible that an agent i is not capable of performing any of the actions $(\mathtt{confirm}\,\varphi; \alpha)^k; \mathtt{confirm}\,\neg\varphi$, where $k \in \mathbb{N}$, then it is also deducible that the agent is incapable of performing the while-loop itself. Or read in its contrapositive form, that an agent is able to perform a while-loop only if it is able to perform some finite-length sequence of confirmations and actions constituting the while-loop. As such, this rule is easily seen to be the proof-theoretic counterpart of the negated version of the (semantic) definition of $\mathrm{c^b}$ for while-loops. For read in its negative form this semantic definition states that $\mathrm{c^b}(i, \mathtt{while}\,\varphi\,\mathtt{do}\,\alpha\,\mathtt{od})(s) = \mathbf{0}$ iff $\mathrm{c^b}(i, \varphi_l(i, \alpha))(s) = \mathbf{0}$ for all $l \in \mathbb{N}$.

The axioms that are used to build the two proof systems are formulated using the necessity operator for actions, i.e. $[\mathrm{do}_-(-)]_-$, rather than its dual $\langle \mathrm{do}_-(-)\rangle_-$. The reason for this is essentially one of convenience: in proving completeness of the axiomatisations it turns out to be useful to deal with two necessity operators, viz. $\mathbf{K}_-$ and $[\mathrm{do}_-(-)]_-$, to allow proofs by analogy. Since $[\mathrm{do}_-(-)]_-$ and $\langle \mathrm{do}_-(-)\rangle_-$ are inter-definable this does not create any essential differences.

DEFINITION 26. The following axioms and proof rules are used to constitute the two proof systems that we consider here. Both the axioms as well as the premises and conclusions of the proof rules are to be taken as schemas in $i \in \mathrm{A}, \varphi, \psi \in \mathrm{L}$ and $\alpha, \alpha_1, \alpha_2 \in \mathrm{Ac}$. The ϕ occurring in the two infinitary rules ΩI and ΩIA is taken to be a meta-variable ranging over $\mathrm{Afm(L)}$.

A1. All propositional tautologies and their epistemic and dynamic instances

A2. $\mathbf{K}_i(\varphi \to \psi) \to (\mathbf{K}_i\varphi \to \mathbf{K}_i\psi)$

A3. $\mathbf{K}_i\varphi \to \varphi$

A4. $\mathbf{K}_i\varphi \to \mathbf{K}_i\mathbf{K}_i\varphi$

A5. $\neg\mathbf{K}_i\varphi \to \mathbf{K}_i\neg\mathbf{K}_i\varphi$

A6. $[\mathrm{do}_i(\alpha)](\varphi \to \psi) \to ([\mathrm{do}_i(\alpha)]\varphi \to [\mathrm{do}_i(\alpha)]\psi)$

A7. $[\mathrm{do}_i(\mathtt{confirm}\,\varphi)]\psi \leftrightarrow (\neg\varphi \vee \psi)$

A8. $[\mathrm{do}_i(\alpha_1;\alpha_2)]\varphi \leftrightarrow [\mathrm{do}_i(\alpha_1)][\mathrm{do}_i(\alpha_2)]\varphi$

A9. $[\mathrm{do}_i(\mathbf{if}\,\varphi\,\mathbf{then}\,\alpha_1\,\mathbf{else}\,\alpha_2\,\mathbf{fi})]\psi \leftrightarrow$
$([\mathrm{do}_i(\mathtt{confirm}\,\varphi;\alpha_1)]\psi \wedge [\mathrm{do}_i(\mathtt{confirm}\,\neg\varphi;\alpha_2)]\psi)$

A10. $[\mathrm{do}_i(\mathtt{while}\,\varphi\,\mathbf{do}\,\alpha\,\mathbf{od})]\psi \leftrightarrow ([\mathrm{do}_i(\mathtt{confirm}\,\neg\varphi)]\psi \wedge$
$[\mathrm{do}_i(\mathtt{confirm}\,\varphi;\alpha)][\mathrm{do}_i(\mathtt{while}\,\varphi\,\mathbf{do}\,\alpha\,\mathbf{od})]\psi)$

A11. $[\mathrm{do}_i(\alpha)]\varphi \vee [\mathrm{do}_i(\alpha)]\neg\varphi$

A12. $\mathbf{A}_i\mathtt{confirm}\,\varphi \leftrightarrow \varphi$

A13₁. $\mathbf{A}_i(\alpha_1;\alpha_2) \leftrightarrow \mathbf{A}_i\alpha_1 \wedge [\mathrm{do}_i(\alpha_1)]\mathbf{A}_i\alpha_2$

A13₀. $\mathbf{A}_i(\alpha_1;\alpha_2) \leftrightarrow \mathbf{A}_i\alpha_1 \wedge \langle \mathrm{do}_i(\alpha_1)\rangle\mathbf{A}_i\alpha_2$

A14. $\mathbf{A}_i\mathbf{if}\,\varphi\,\mathbf{then}\,\alpha_1\,\mathbf{else}\,\alpha_2\,\mathbf{fi} \leftrightarrow$
$(\mathbf{A}_i\mathtt{confirm}\,\varphi;\alpha_1 \vee \mathbf{A}_i\mathtt{confirm}\,\neg\varphi;\alpha_2)$

A15₁. $\mathbf{A}_i\mathtt{while}\,\varphi\,\mathbf{do}\,\alpha\,\mathbf{od} \leftrightarrow (\mathbf{A}_i(\mathtt{confirm}\,\neg\varphi)\vee$
$(\mathbf{A}_i\mathtt{confirm}\,\varphi;\alpha \wedge [\mathrm{do}_i(\mathtt{confirm}\,\varphi;\alpha)]\mathbf{A}_i\mathtt{while}\,\varphi\,\mathbf{do}\,\alpha\,\mathbf{od}))$

A15₀. $\mathbf{A}_i\mathtt{while}\,\varphi\,\mathbf{do}\,\alpha\,\mathbf{od} \leftrightarrow (\mathbf{A}_i(\mathtt{confirm}\,\neg\varphi)\vee$
$(\mathbf{A}_i\mathtt{confirm}\,\varphi;\alpha \wedge \langle\mathrm{do}_i(\mathtt{confirm}\,\varphi;\alpha)\rangle\mathbf{A}_i\mathtt{while}\,\varphi\,\mathbf{do}\,\alpha\,\mathbf{od}))$

R1. $\phi(\psi_l(i,\varphi,\alpha))$ all $l \in \mathbb{N}$ / $\phi([\mathrm{do}_i(\mathtt{while}\,\varphi\,\mathbf{do}\,\alpha\,\mathbf{od})]\psi)$ ΩI

R2. $\phi(\neg(\varphi_l(i,\alpha)))$ all $l \in \mathbb{N}$ / $\phi(\neg\mathbf{A}_i\mathtt{while}\,\varphi\,\mathbf{do}\,\alpha\,\mathbf{od})$ ΩIA

R3. $\varphi,\ \varphi \to \psi \,/\, \psi$ MP

R4. $\varphi \,/\, \mathbf{K}_i\varphi$ KN

R5. $\varphi \,/\, [\mathrm{do}_i(\alpha)]\varphi$ AN

Most of the axioms are fairly obvious, in particular given the discussion on the validities presented in Section 4. Rule R1, the Omega Iteration rule, is adopted from the axiomatisations given by Goldblatt [1982a; 1982b]. Both ΩI and rule R2, which is the Omega Iteration rule for Ability, were already discussed above. Rule R3 is the rule of Modus Ponens, well known from, and used in, both classical and modal logics. R4 and R5 are both instances of the rule of necessitation, which is known to hold for necessity operators. These rules state that whenever some formula is deducible, it is also deducible that an arbitrary agent knows the formula, and that all events have this formula among their conditional results, respectively. Axioms A2 and A6, and the rules R4 and R5 indicate that both knowledge and conditional results are formalised through normal modal operators.

The axioms and proof rules given above are used to define two different proof systems. One of these proof systems embodies the optimistic view on abilities in the counterfactual state of affairs, the other employs a pessimistic view.

DEFINITION 27. The proof system Σ_1 contains the axioms A1 through A12, A13$_1$, A14, A15$_1$ and the proof rules R1 through R5. The proof system Σ_0 contains the axioms A1 through A12, A13$_0$, A14, A15$_0$ and the proof rules R1 through R5.

As mentioned above, a logic for a given proof system is a set encompassing the proof system.

DEFINITION 28. A **b**-logic is a set Λ that contains all the instances of the axioms of $\Sigma_\mathbf{b}$ and is closed under the proof rules of $\Sigma_\mathbf{b}$. The intersection of all **b**-logics, which is itself a **b**-logic, viz. the smallest one, is denoted by LCap$_\mathbf{b}$. Whenever the underlying proof system is either irrelevant or clear from the context, we refer to a **b**-logic simply as a logic.

Deducibility in a given proof system is now defined as being an element of the smallest logic for the proof system.

DEFINITION 29. For Λ some logic, the unary predicate $\vdash^\Lambda \subseteq \mathrm{L}$ is defined by: $\vdash^\Lambda \varphi \Leftrightarrow \varphi \in \Lambda$. As an abbreviation we occasionally write $\vdash^\mathbf{b} \varphi$ for $\vdash^{\mathrm{LCap}_\mathbf{b}}$. Whenever $\vdash^\Lambda \varphi$ holds we say that φ is deducible in Λ or alternatively that φ is a theorem of Λ.

The proof systems Σ_1 and Σ_0 provide sound and complete axiomatisations of validity for $\models^1$ and $\models^0$ respectively. This is summarized in the following theorem, of which the proof is provided in the appendix.

THEOREM 30. For $\mathbf{b} \in$ bool and all $\varphi \in \mathrm{L}$ we have: $\vdash^\mathbf{b} \varphi \Leftrightarrow\ \models^\mathbf{b} \varphi$.

Besides the notion of deducibility *per se*, it is also interesting to look at deducibility from a set of premises. In modal logics one may distinguish two notions of deducibility from premises. In the first of these, the premises are considered to be additional axioms, on which also rules of necessitation may be applied. The second notion of deducibility allows necessitation only on the axioms of the proof system, and not on the premises. This latter notion of deducibility is perhaps the more natural one, and is in fact the one that we will concentrate on.

To account for deducibility from premises with respect to the alternative notion of deducibility as being an element of some set of formulas, we introduce the notion of a theory of a logic. Corresponding to the idea that the rules of necessitation are not to be applied on premises, we do not demand that a theory be closed under these rules. A formula is now defined to be deducible from some set of premises iff it is contained in every theory that encompasses the set of premises.

DEFINITION 31. For Λ some logic, we define a Λ-theory to be any subset Θ of L that contains Λ and is closed under the rules ΩI, ΩIA, and MP.

DEFINITION 32. Let Λ be some logic and $\Phi \cup \{\varphi\} \subseteq L$. The binary relation $\vdash^{\Lambda} \subseteq \wp(L) \times L$ is defined by:

$$\Phi \vdash^{\Lambda} \varphi \Leftrightarrow \varphi \in \bigcap\{\Gamma \subseteq L \mid \Phi \subseteq \Gamma \text{ and } \Gamma \text{ is a } \Lambda\text{-theory}\}$$

Whenever $\Phi \vdash^{\Lambda} \varphi$ we say that φ is deducible from Φ in Λ. A set $\Phi \subseteq L$ is called Λ-inconsistent iff $\Phi \vdash^{\Lambda} \bot$, and Λ-consistent iff it is not Λ-inconsistent.

Given the 'overloading' of the symbol $\vdash^{\Lambda}$ as representing both deducibility *per se* and deducibility from premises, it is highly desirable that the two uses of this symbol coincide in the case that the set of premises is empty: deducibility from an empty set of premises should not differ from deducibility *per se*.

PROPOSITION 33. For Λ some logic and $\varphi \in L$ we have: $\vdash^{\Lambda} \varphi \Leftrightarrow \emptyset \vdash^{\Lambda} \varphi$.

As already mentioned before, using infinitary rules to describe the behaviour of while-loops allows one to achieve strong completeness, the notion which states that every consistent set of formulas is simultaneously satisfiable. Achieving strong completeness is in general not possible when just finitary rules are used. To see this consider the set $\Omega = \{[\text{do}_i(a^k)]p \mid k \in \mathbb{N}\} \cup \{\langle \text{do}_i(\texttt{while}\, p\, \texttt{do}\, a\, \texttt{od})\rangle \top\}$. It is obvious that Ω is not satisfiable. For whenever $M, s \models^{\mathbf{b}} [\text{do}_i(a^k)]p$ for all $k \in \mathbb{N}$ then execution of $\texttt{while}\, p\, \texttt{do}\, a\, \texttt{od}$ does not terminate, and hence $M, s \not\models \langle \text{do}_i(\texttt{while}\, p\, \texttt{do}\, a\, \texttt{od})\rangle \top$. However, when using just finitary rules to describe while-loops (like for instance the well-known Hoare rule [Hoare, 1969]), the set Ω will be consistent. For when restricting oneself to finitary rules, consistency of an infinite set of formulas corresponds to consistency of each of its finite subsets. And in every axiomatisation that is to be sound, all finite subsets of Ω should be consistent, and therefore Ω itself is consistent. In the infinitary proof systems Σ_1 and Σ_0, $\bot$ is deducible from Ω, i.e. Ω is inconsistent. More generally, the property of strong completeness holds for both Σ_1 and Σ_0.

PROPOSITION 34. The proof systems Σ_1 and Σ_0 are strongly complete, i.e. every set $\Phi \subseteq L$ that is $\text{LCap}_{\mathbf{b}}$-consistent is $\models^{\mathbf{b}}$-satisfiable.

Just as deducibility *per se* is the proof theoretic counterpart of the semantic notion of validity, there is also a semantic counterpart to the notion of deducibility from premises.

PROPOSITION 35. For $\mathbf{b} \in \text{bool}$, $\Phi \subseteq L$ and $\varphi \in L$ we have:

- $\Phi \vdash^{\mathbf{b}} \varphi \Leftrightarrow \Phi \models^{\mathbf{b}} \varphi$

where $\Phi \models^{\mathbf{b}} \varphi$ iff $M, s \models^{\mathbf{b}} \Phi$ implies $M, s \models^{\mathbf{b}} \varphi$ for all $M \in \mathbf{M}$ with state s.

In the light of the strong completeness property, Proposition 35 is not very surprising. In fact, the right-to-left implication is a direct consequence of the strong completeness property. The left-to-right implication follows from the observation that the set of formulas that is satisfied in some world forms a theory.

7 SUMMARY AND CONCLUSIONS

In this paper we introduced the KARO-framework, a formal framework based on a combination of various modal logics that can be used to formalise agents. After a somewhat philosophical exposition on knowledge, actions and events, we presented two formal systems, both belonging to the KARO-framework, that share a common language and a common class of models but that differ in the interpretation of dynamic and ability formulas. The language common to the two systems is a propositional, multi-modal, exogenous language, containing modalities representing knowledge, opportunity and result, and an operator formalising ability. The models that are used to interpret formulas from the language L are Kripke-style possible worlds models. These models interpret knowledge by means of an accessibility relation on worlds; opportunity, result and ability are interpreted using designated functions. We explained our intuition on the composite behaviour of results, opportunities and abilities, and presented two formal interpretations that comply with this intuition. These interpretations differ in their treatment of abilities of agents for sequentially composed actions. We considered various properties of knowledge and action in the KARO-framework. In defining some of these properties we used the notions of schemas, frames and correspondences. Using the various modalities present in the framework, we proposed a formalisation of the knowledge of agents about their practical possibilities, a notion which captures an important aspect of agency, particularly in the context of planning agents. We presented two proof systems that syntactically characterise the notion of validity in the two interpretations that we defined. The most remarkable aspect of these proof systems is the use of infinitary proof rules, which on the one hand allows for a better correspondence between the semantic notion of validity and its syntactic counterpart, and on the other hand forces one to generalise the usual notions of proof and theorem.

In the KARO-framework we proposed two definitions for the ability of agents to execute a sequentially composed action $\alpha_1;\alpha_2$ in cases where execution of α_1 leads to the counterfactual state of affairs. The simplicity of these definitions, both at a conceptual and at a technical level, may lead to counterintuitive situations. Recall that using the so-called optimistic approach it is possible that an agent is considered to be capable of performing $\alpha;\mathtt{fail}$, whereas in the pessimistic approaches agents may be declared unable to perform $\alpha;\mathtt{skip}$, for $\alpha \in \mathrm{Ac}$. A more realistic approach would be not to treat all actions equally, but instead to determine for each action individually whether it makes sense to declare an agent (un)able to perform the action in the counterfactual state of affairs. One way to formalise this consists of extending the models from $\mathbf{M}$ with an additional function $\mathtt{t} : \mathrm{A} \times \mathrm{Ac} \to \mathrm{S} \to \mathrm{S}$ which is such that $\mathtt{t}(i,\alpha)(s) = \mathtt{r}(i,\alpha)(s)$ whenever $\mathtt{r}(i,\alpha)(s) \neq \emptyset$. Hence in the case that $\mathtt{r}(i,\alpha)(s) \neq \emptyset$, $\mathtt{t}(i,\alpha)(s)$ equals

$\mathtt{r}(i,\alpha)(s)$ and in other cases $\mathtt{t}(i,\alpha)(s)$ is definitely not empty. The function $\mathtt{t}$ denotes the outcome of actions when 'abstracting away' from opportunities, so to speak. The ability for the sequential composition is then defined by

$$\mathtt{c}(i,\alpha_1;\alpha_2)(s) = \mathbf{1} \Leftrightarrow \mathtt{c}(i,\alpha_1)(s) = \mathbf{1} \,\&\, \mathtt{c}(i,\alpha_2)(\mathtt{t}(i,\alpha_1)(s)) = \mathbf{1}$$

Applying this definition implies that $\mathbf{A}_i\alpha_1;\mathtt{fail}$ is no longer satisfiable, and that $\mathbf{A}_i\alpha_1;\mathtt{skip}$ holds in cases where $\mathbf{A}_i\alpha_1$ is true, regardless of the truth of $\langle \mathrm{do}_i(\alpha_1)\rangle\top$. A special instantiation of this approach corresponds to the idea that abilities of agents do not tend to change. Therefore it could seem reasonable to assume that agents retain their abilities when ending up in the counterfactual state of affairs. Formally this can be brought about by demanding $\mathtt{t}(i,\alpha)(s)$ to equate s in cases where $\mathtt{r}(i,\alpha)(s) = \emptyset$. Since this is but a special case of the general idea discussed above, it also avoids the counterintuitive situations where agents are declared to be able to do $\alpha;\mathtt{fail}$ or unable to do $\alpha;\mathtt{skip}$.

A A PROOF OF SOUNDNESS AND COMPLETENESS

Below we prove the soundness and completeness of deducibility in $\mathrm{LCap}_\mathbf{b}$ for $\models^\mathbf{b}$-validity in $\mathbf{M}$. As far as we know, this is one of the very few proofs of completeness that concerns a proof system in which both knowledge and actions are dealt with, and it is probably the very first in which abilities are also taken into consideration.

Rather than restricting ourselves to $\mathrm{LCap}_\mathbf{b}$ we will for the greater part consider general logics, culminating in a very general and rather powerful result from which the soundness and completeness proof for $\mathrm{LCap}_\mathbf{b}$ can be derived as a corollary. Globally, the proof given below can be split into three parts. In the first part of the proof, canonical models are constructed for the logics induced by the proof systems $\Sigma_\mathbf{1}$ and $\Sigma_\mathbf{0}$. The possible worlds of these canonical models are given by so-called maximal theories. In the second part, the truth-theorem is proved, which states that truth in a possible world of a canonical model corresponds to being an element of the maximal theory that constitutes the possible world. In the last, and almost trivial, part of the proof it is shown how the general truth-theorem implies soundness and completeness of $\mathrm{LCap}_\mathbf{b}$ for $\mathbf{b}$-validity in $\mathbf{M}$.

The definition of canonical models as we give it is, as far as actions and dynamic constructs are concerned, based on the construction given by Goldblatt [1982a]. The proof of the truth-theorem is inspired by the one given by Spruit [1993] to show completeness of the Segerberg axiomatisation for propositional dynamic logic. Due to the fact that formulas and actions are strongly related, the subformula or subaction relation does not provide an adequate support for induction in the proof of the truth-theorem. Instead a fairly complex ordering is used, well-foundedness of which is proved using some very powerful (and partly automated) techniques that are well-known from the theory of Term Rewriting Systems [Dershowitz and Jouannaud, 1990; Klop, 1992].

Some preliminary definitions, propositions and lemmas are needed before the canonical models can be constructed.

PROPOSITION 36. For all $\mathrm{M} \in \mathbf{M}$ with state s and all $i \in \mathrm{A}$, $\alpha \in \mathrm{Ac}$, $\varphi \in \mathrm{L}$ and $\phi \in \mathrm{Afm(L)}$ we have:

- $\mathrm{M}, s \models^{\mathbf{b}} \phi([\mathrm{do}_i(\mathtt{while}\,\varphi\,\mathbf{do}\,\alpha\,\mathbf{od})]\psi)$ iff for all $l \in \mathbb{N}$, $\mathrm{M}, s \models^{\mathbf{b}} \phi(\psi_l(i, \alpha, \varphi))$
- $\mathrm{M}, s \models^{\mathbf{b}} \phi(\neg\mathbf{A}_i\mathtt{while}\,\varphi\,\mathbf{do}\,\alpha\,\mathbf{od})$ iff for all $l \in \mathbb{N}$, $\mathrm{M}, s \models^{\mathbf{b}} \phi(\neg(\varphi_l(i, \alpha)))$

Proof. We prove both items by induction on the structure of ϕ.

- Let $\mathrm{M} \in \mathbf{M}$ with state s, and $i \in \mathrm{A}$, $\varphi, \psi \in \mathrm{L}$ and $\alpha \in \mathrm{Ac}$ be arbitrary.

  1. $\phi = \#$:

  $$\begin{aligned}
  &\mathrm{M}, s \models^{\mathbf{b}} [\mathrm{do}_i(\mathtt{while}\,\varphi\,\mathbf{do}\,\alpha\,\mathbf{od})]\psi \\
  \Leftrightarrow\ &\mathrm{M}, t \models^{\mathbf{b}} \psi \text{ for all } t \in \mathrm{S} \text{ such that } t = \mathbf{r^b}(i, \mathtt{while}\,\varphi\,\mathbf{do}\,\alpha\,\mathbf{od})(s) \\
  \Leftrightarrow\ &\mathrm{M}, t \models^{\mathbf{b}} \psi \text{ for all } t \in \mathrm{S} \text{ such that } t = \mathbf{r^b}(i, (\mathtt{confirm}\,\varphi; \alpha)^l; \\
  &\quad \mathtt{confirm}\,\neg\varphi)(s) \text{ for all } l \in \mathbb{N} \\
  \Leftrightarrow\ &\mathrm{M}, s \models^{\mathbf{b}} [\mathrm{do}_i(\mathtt{confirm}\,\varphi; \alpha)^l; \mathtt{confirm}\,\neg\varphi)]\psi \text{ for all } l \in \mathbb{N} \\
  \Leftrightarrow\ &\mathrm{M}, s \models^{\mathbf{b}} \psi_l(i, \varphi, \alpha) \text{ for all } l \in \mathbb{N}
  \end{aligned}$$

  2. $\phi = \mathbf{K}_i\phi'$:

  $$\begin{aligned}
  &\mathrm{M}, s \models^{\mathbf{b}} (\mathbf{K}_i\phi')([\mathrm{do}_i(\mathtt{while}\,\varphi\,\mathbf{do}\,\alpha\,\mathbf{od})]\psi) \\
  \Leftrightarrow\ &\mathrm{M}, s \models^{\mathbf{b}} \mathbf{K}_i(\phi'([\mathrm{do}_i(\mathtt{while}\,\varphi\,\mathbf{do}\,\alpha\,\mathbf{od})]\psi)) \\
  \Leftrightarrow\ &\mathrm{M}, t \models^{\mathbf{b}} \phi'([\mathrm{do}_i(\mathtt{while}\,\varphi\,\mathbf{do}\,\alpha\,\mathbf{od})]\psi) \text{ for all } t \in \mathrm{S} \text{ such that} \\
  &\quad (s, t) \in \mathrm{R}(i) \\
  \Leftrightarrow\ &\mathrm{M}, t \models^{\mathbf{b}} \phi'(\psi_l(i, \alpha, \varphi)) \text{ for all } l \in \mathbb{N}, \\
  &\quad \text{for all } t \in \mathrm{S} \text{ such that } (s, t) \in \mathrm{R}(i) \\
  &\quad \text{(by induction hypothesis)} \\
  \Leftrightarrow\ &\mathrm{M}, s \models^{\mathbf{b}} \mathbf{K}_i\phi'(\psi_l(i, \alpha, \varphi)) \text{ for all } l \in \mathbb{N} \\
  \Leftrightarrow\ &\mathrm{M}, s \models^{\mathbf{b}} (\mathbf{K}_i\phi')(\psi_l(i, \alpha, \varphi)) \text{ for all } l \in \mathbb{N}
  \end{aligned}$$

  3. The cases where $\phi = [\mathrm{do}_i(\beta)]\phi'$ and $\phi = \psi' \to \phi'$ are analogous to the case where $\phi = \mathbf{K}_i\phi'$.

- Let again $\mathrm{M} \in \mathbf{M}$ with state s, $i \in \mathrm{A}$, $\varphi \in \mathrm{L}{=}$ and $\alpha \in \mathrm{Ac}$ be arbitrary.

  1. $\phi = \#$:

  $$\begin{aligned}
  &\mathrm{M}, s \models^{\mathbf{b}} \neg\mathbf{A}_i\mathtt{while}\,\varphi\,\mathbf{do}\,\alpha\,\mathbf{od} \\
  \Leftrightarrow\ &\mathrm{not}(\mathrm{M}, s \models^{\mathbf{b}} \mathbf{A}_i\mathtt{while}\,\varphi\,\mathbf{do}\,\alpha\,\mathbf{od}) \\
  \Leftrightarrow\ &\mathrm{not}(\exists k \in \mathbb{N}(\mathbf{c^b}(i, (\mathtt{confirm}\,\varphi; \alpha)^k; \mathtt{confirm}\,\neg\varphi)(s) = \mathbf{1}))
  \end{aligned}$$

$$\Leftrightarrow \forall k \in \mathbb{N}(\text{not}(\mathbf{c^b}(i, (\texttt{confirm}\,\varphi; \alpha)^k; \texttt{confirm}\,\neg\varphi)(s) = \mathbf{1}))$$
$$\Leftrightarrow \forall k \in \mathbb{N}(\text{not}(\mathrm{M}, s \models^{\mathbf{b}} \varphi_k(i, \alpha)))$$
$$\Leftrightarrow \forall k \in \mathbb{N}(\mathrm{M}, s \models^{\mathbf{b}} \neg(\varphi_k(i, \alpha_1)))$$

2. $\phi = [\mathrm{do}_i(\beta)]\phi'$:

$$\mathrm{M}, s \models^{\mathbf{b}} ([\mathrm{do}_i(\beta)]\phi')(\neg\mathbf{A}_i\texttt{while}\,\varphi\,\mathbf{do}\,\alpha\,\mathbf{od})$$
$$\Leftrightarrow \mathrm{M}, s \models^{\mathbf{b}} [\mathrm{do}_i(\beta)](\phi'(\neg\mathbf{A}_i\texttt{while}\,\varphi\,\mathbf{do}\,\alpha\,\mathbf{od}))$$
$$\Leftrightarrow \mathrm{M}, t \models^{\mathbf{b}} \phi'(\neg\mathbf{A}_i\texttt{while}\,\varphi\,\mathbf{do}\,\alpha\,\mathbf{od}) \text{ for all } t \in \mathrm{S} \text{ such that } t = \mathbf{r^b}(i, \beta)(s)$$
$$\Leftrightarrow \mathrm{M}, t \models^{\mathbf{b}} \phi'(\neg(\varphi_l(i, \alpha))) \text{ for all } l \in \mathbb{N}, \text{ for all } t \in \mathrm{S} \text{ such that } t = \mathbf{r^b}(i, \beta)(s) \text{ (by induction hypothesis)}$$
$$\Leftrightarrow \mathrm{M}, s \models^{\mathbf{b}} [\mathrm{do}_i(\beta)](\phi'(\neg(\varphi_l(i, \alpha)))) \text{ for all } l \in \mathbb{N}$$
$$\Leftrightarrow \mathrm{M}, s \models^{\mathbf{b}} ([\mathrm{do}_i(\beta)]\phi')(\neg(\varphi_l(i, \alpha))) \text{ for all } l \in \mathbb{N}$$

3. The cases where $\phi = \mathbf{K}_i\phi'$ and $\phi = (\psi' \rightarrow \phi')$ are analogous to the case where $\phi = [\mathrm{do}_i(\beta)]\phi'$. ■

PROPOSITION 37. If $\mathrm{M} \in \mathbf{M}$ is a well-defined model from $\mathbf{M}$, then $\Lambda^{\mathrm{M}}_{\mathbf{b}} =^{\mathrm{def}} \{\varphi \in \mathrm{L} \mid \mathrm{M} \models^{\mathbf{b}} \varphi\}$ is a **b**-logic.

Proof. We need to check for a given model $\mathrm{M} \in \mathbf{M}$ that the axioms of $\Sigma_{\mathbf{b}}$ are valid in M and that M is validity-preserving for the proof rules of $\Sigma_{\mathbf{b}}$. The validity of the axioms A1–A9 and A12–A14 is easily checked. Axiom A10 follows from the determinism of all actions as stated in Corollary 16. Axiom $\mathrm{A15}_{\mathbf{1}}$ is shown in Proposition 7, and $\mathrm{A15}_{\mathbf{0}}$ is shown analogously. The validity-preservingness of M for the rules R1 and R2 follows from Proposition 36; M is easily seen to be validity-preserving for the other rules. As an example we show here the validity of axiom A10.

$$\mathrm{M}, s \models^{\mathbf{b}} [\mathrm{do}_i(\texttt{while}\,\varphi\,\mathbf{do}\,\alpha\,\mathbf{od})]\psi$$
$$\Leftrightarrow \mathrm{M}, t \models^{\mathbf{b}} \psi \text{ for all } t \in \mathrm{S} \text{ such that } t = \mathbf{r^b}(i, \texttt{while}\,\varphi\,\mathbf{do}\,\alpha\,\mathbf{od})(s)$$
$$\Leftrightarrow \mathrm{M}, t \models^{\mathbf{b}} \psi \text{ for all } t \in \mathrm{S} \text{ such that } t = \mathbf{r^b}(i, \texttt{confirm}\,\neg\varphi)(s) \text{ and } \mathrm{M}, t \models^{\mathbf{b}} \psi \text{ for all } t \in \mathrm{S} \text{ such that } t = \mathbf{r^b}(i, (\texttt{confirm}\,\varphi; \alpha); \texttt{while}\,\varphi\,\mathbf{do}\,\alpha\,\mathbf{od})(s)$$
$$\Leftrightarrow \mathrm{M}, s \models^{\mathbf{b}} [\mathrm{do}_i(\texttt{confirm}\,\neg\varphi)]\psi \text{ and } \mathrm{M}, s \models^{\mathbf{b}} [\mathrm{do}_i((\texttt{confirm}\,\varphi; \alpha); \texttt{while}\,\varphi\,\mathbf{do}\,\alpha\,\mathbf{od})]\psi$$
$$\Leftrightarrow \mathrm{M}, s \models^{\mathbf{b}} [\mathrm{do}_i(\texttt{confirm}\,\neg\varphi)]\psi \wedge [\mathrm{do}_i(\texttt{confirm}\,\varphi; \alpha)][\mathrm{do}_i(\texttt{while}\,\varphi\,\mathbf{do}\,\alpha\,\mathbf{od})]\psi$$

■

PROPOSITION 38. *Let Λ be a logic. The following properties are shared by all Λ-theories Γ, for all $\varphi, \psi \in \mathrm{L}$, $i \in \mathrm{A}$, $\alpha \in \mathrm{Ac}$ and all $\phi \in \mathrm{Afm(L)}$:*

1. $\top \in \Gamma$

2. *if* $\Gamma \vdash^{\Lambda} \varphi$ *then* $\varphi \in \Gamma$

3. *if* $\vdash^{\Lambda} (\varphi \to \psi)$ *and* $\varphi \in \Gamma$ *then* $\psi \in \Gamma$

4. Γ *is* Λ*-consistent iff* $\bot \notin \Gamma$ *iff* $\Gamma \neq \mathrm{L}$

5. $(\varphi \wedge \psi) \in \Gamma$ *iff* $\varphi \in \Gamma$ *and* $\psi \in \Gamma$

6. *if* $\varphi \in \Gamma$ *or* $\psi \in \Gamma$ *then* $(\varphi \vee \psi) \in \Gamma$

7. $\phi([\mathrm{do}_i(\mathtt{while}\,\varphi\,\mathbf{do}\,\alpha\,\mathbf{od})]\psi) \in \Gamma$ *iff* $\{\phi(\psi_l(i,\varphi,\alpha)) \mid l \in \mathbb{N}\} \subseteq \Gamma$

8. $\phi(\neg \mathbf{A}_i\mathtt{while}\,\varphi\,\mathbf{do}\,\alpha\,\mathbf{od}) \in \Gamma$ *iff* $\{\phi(\neg(\varphi_l(i,\alpha))) \mid l \in \mathbb{N}\} \subseteq \Gamma$

Proof. The items 1 to 6 are fairly standard, and are proved by Goldblatt [1982a]. The cases 7 and 8 follow from the fact that theories contain the axioms A10 and A15$_b$ and are closed under ΩI and ΩIA. ■

DEFINITION 39. Let Λ be a logic. A maximal Λ-theory is a consistent Λ-theory Γ such that $\varphi \in \Gamma$ or $\neg\varphi \in \Gamma$ for all $\varphi \in \mathrm{L}$.

PROPOSITION 40. *The following properties are shared by all maximal* Λ*-theories* Γ, *for* Λ *some logic, and* $\varphi, \psi \in \mathrm{L}$.

1. $\bot \notin \Gamma$

2. *exactly one of* φ *and* $\neg\varphi$ *belongs to* Γ, *for all* $\varphi \in \mathrm{L}$

3. $(\varphi \vee \psi) \in \Gamma$ *iff* $\varphi \in \Gamma$ *or* $\psi \in \Gamma$

PROPOSITION 41. *For* Λ *a logic and all* $\varphi, \psi \in \mathrm{L}$, $\Phi, \Psi \subseteq \mathrm{L}$, $i \in \mathrm{A}$, $\alpha \in \mathrm{Ac}$ *and* $\phi \in \mathrm{Afm(L)}$ *we have:*

1. *if* $\varphi \in \Phi$ *then* $\Phi \vdash^{\Lambda} \varphi$

2. *if* $\Phi \vdash^{\Lambda} \varphi$ *and* $\Phi \subseteq \Psi$ *then* $\Psi \vdash^{\Lambda} \varphi$

3. $\vdash^{\Lambda} \varphi$ *iff* $\emptyset \vdash^{\Lambda} \varphi$

4. *if* $\Phi \vdash^{\Lambda} (\varphi \to \psi)$ *and* $\Phi \vdash^{\Lambda} \varphi$ *then* $\Phi \vdash^{\Lambda} \psi$

5. *if* $\Phi \vdash^{\Lambda} \phi(\psi_l(i,\varphi,\alpha))$ *for all* $l \in \mathbb{N}$ *then*
   $\Phi \vdash^{\Lambda} \phi([\mathrm{do}_i(\mathtt{while}\,\varphi\,\mathbf{do}\,\alpha\,\mathbf{od})]\psi)$

6. *if* $\Phi \vdash^{\Lambda} \phi(\neg(\varphi_l(i,\alpha)))$ *for all* $l \in \mathbb{N}$ *then* $\Phi \vdash^{\Lambda} \phi(\neg \mathbf{A}_i\mathtt{while}\,\varphi\,\mathbf{do}\,\alpha\,\mathbf{od})$

THEOREM 42 (The deduction theorem). *For* Λ *some logic and all* $\varphi, \psi \in \mathrm{L}$ *and* $\Phi \subseteq \mathrm{L}$ *we have that* $\Phi \cup \{\varphi\} \vdash^{\Lambda} \psi$ *iff* $\Phi \vdash^{\Lambda} (\varphi \to \psi)$.

Proof. We will prove the 'iff' by proving two implications:

'$\Leftarrow$' This case follows directly from items 1, 2, and 4 of Proposition 41.

'⇒' Assume that $\Phi \cup \{\varphi\} \vdash^{\Lambda} \psi$. Let $\Gamma =^{\text{def}} \{\rho \in \mathrm{L} \mid \Phi \vdash^{\Lambda} (\varphi \to \rho)\}$. We have to show that $\psi \in \Gamma$. For this it suffices to show that Γ is a Λ-theory containing $\Phi \cup \{\varphi\}$. We show here that Γ is closed under ΩIA; the proof of the other properties is easy and left to the reader. Assume that $\{\phi(\neg(\varphi'_l(i,\alpha))) \mid l \in \mathbb{N}\} \subseteq \Gamma$. Then $\Phi \vdash^{\Lambda} (\varphi \to \phi(\neg(\varphi'_l(i,\alpha))))$ for all $l \in \mathbb{N}$. Applying case 6 of Proposition 41 to the set $\{(\varphi \to \phi(\neg(\varphi'_l(i,\alpha)))) \mid l \in \mathbb{N}\}$ yields $\Phi \vdash^{\Lambda} (\varphi \to \phi(\neg \mathbf{A}_i \mathtt{while}\, \varphi' \,\mathbf{do}\, \alpha \,\mathbf{od}))$, hence $\phi(\neg \mathbf{A}_i \mathtt{while}\, \varphi' \,\mathbf{do}\, \alpha \,\mathbf{od}) \in \Gamma$. Thus Γ is closed under ΩIA. ■

COROLLARY 43. *For Λ some logic and all $\varphi \in \mathrm{L}$ and $\Phi \subseteq \mathrm{L}$ we have:*

- $\Phi \cup \{\varphi\}$ *is Λ-consistent iff* $\Phi \nvdash^{\Lambda} \neg\varphi$
- $\Phi \cup \{\neg\varphi\}$ *is Λ-consistent iff* $\Phi \nvdash^{\Lambda} \varphi$

DEFINITION 44. For $\Phi \subseteq \mathrm{L}$, $i \in \mathrm{A}$ and $\alpha \in \mathrm{Ac}$ we define:

- $\Phi/\mathbf{K}_i =^{\text{def}} \{\varphi \in \mathrm{L} \mid \mathbf{K}_i\varphi \in \Phi\}$
- $\mathbf{K}_i\Phi =^{\text{def}} \{\mathbf{K}_i\varphi \in \mathrm{L} \mid \varphi \in \Phi\}$
- $\Phi/[\mathrm{do}_i(\alpha)] =^{\text{def}} \{\varphi \in \mathrm{L} \mid [\mathrm{do}_i(\alpha)]\varphi \in \Phi\}$
- $[\mathrm{do}_i(\alpha)]\Phi =^{\text{def}} \{[\mathrm{do}_i(\alpha)]\varphi \in \mathrm{L} \mid \varphi \in \Phi\}$

PROPOSITION 45. *For Λ some logic and all $\varphi \in \mathrm{L}$, $i \in \mathrm{A}$ and $\Phi \subseteq \mathrm{L}$ we have:*

- *if* $\Phi \vdash^{\Lambda} \varphi$ *then* $\mathbf{K}_i\Phi \vdash^{\Lambda} \mathbf{K}_i\varphi$
- *if* $\Phi \vdash^{\Lambda} \varphi$ *then* $[\mathrm{do}_i(\alpha)]\Phi \vdash^{\Lambda} [\mathrm{do}_i(\alpha)]\varphi$

Proof. We show the first case; the second case is completely analogous. So let Γ be a Λ-theory such that $\mathbf{K}_i\Phi \subseteq \Gamma$. We need to show that $\mathbf{K}_i\varphi \in \Gamma$. Let $\Delta =^{\text{def}} \Gamma/\mathbf{K}_i$. Since $\Phi \vdash^{\Lambda} \varphi$, it suffices to show that Δ is a Λ-theory containing Φ. Then $\varphi \in \Delta$ and hence $\mathbf{K}_i\varphi \in \Gamma$.

1. $\Phi \subseteq \Delta$: If $\psi \in \Phi$, then $\mathbf{K}_i\psi \in \Gamma$ and hence $\psi \in \Delta$.
2. Δ contains Λ: If $\vdash^{\Lambda} \psi$, then by NK, $\vdash^{\Lambda} \mathbf{K}_i\psi$ and, since Γ is a Λ-theory, then $\mathbf{K}_i\psi \in \Gamma$, which implies $\psi \in \Delta$.
3. Δ is closed under MP, ΩI and ΩIA.
   - MP: If $\psi \in \Delta$ and $(\psi \to \psi_1) \in \Delta$, then $\mathbf{K}_i\psi \in \Gamma$ and $\mathbf{K}_i(\psi \to \psi_1) \in \Gamma$. Since Γ contains axiom A2, this implies $\mathbf{K}_i\psi_1 \in \Gamma$ and hence $\psi_1 \in \Delta$.

- ΩI: If $\{\phi(\psi_l(j,\varphi',\alpha)) \mid l \in \mathbb{N}\} \subseteq \Delta$, then $\{\mathbf{K}_i\phi(\psi_l(j,\varphi',\alpha)) \mid l \in \mathbb{N}\} \subseteq \Gamma$. Applying ΩI to the set $\{\mathbf{K}_i\phi(\psi_l(j,\varphi',\alpha)) \mid l \in \mathbb{N}\}$ yields $\mathbf{K}_i\phi([\mathrm{do}_j(\mathtt{while}\,\varphi'\,\mathbf{do}\,\alpha\,\mathbf{od})]\psi) \in \Gamma$, and hence $\phi([\mathrm{do}_j(\mathtt{while}\,\varphi'\,\mathbf{do}\,\alpha\,\mathbf{od})]\psi) \in \Delta$.
- ΩIA: If $\{\phi(\neg(\varphi_l'(j,\alpha))) \mid l \in \mathbb{N}\} \subseteq \Delta$, then $\{\mathbf{K}_i\phi(\neg(\varphi_l'(j,\alpha))) \mid l \in \mathbb{N}\} \subseteq \Gamma$. Applying ΩIA to $\{\mathbf{K}_i\phi(\neg(\varphi_l'(j,\alpha))) \mid l \in \mathbb{N}\}$ yields $\mathbf{K}_i\phi(\neg\mathbf{A}_j\mathtt{while}\,\varphi'\,\mathbf{do}\,\alpha\,\mathbf{od}) \in \Gamma$, and hence $\phi(\neg\mathbf{A}_j\mathtt{while}\,\varphi'\,\mathbf{do}\,\alpha\,\mathbf{od}) \in \Delta$.

It follows that Δ is closed under MP, ΩI and ΩIA.

Since Δ contains Λ and is closed under MP, ΩI and ΩIA it follows that Δ is a Λ-theory. ■

COROLLARY 46. *Let* Λ *be some logic. For all* Λ*-theories* Γ*, and for* $i \in \mathrm{A}$, $\alpha \in \mathrm{Ac}$ *and* $\varphi \in \mathrm{L}$ *we have:*

- $\mathbf{K}_i\varphi \in \Gamma$ *iff* $\Gamma/\mathbf{K}_i \vdash^{\Lambda} \varphi$
- $[\mathrm{do}_i(\alpha)]\varphi \in \Gamma$ *iff* $\Gamma/[\mathrm{do}_i(\alpha)] \vdash^{\Lambda} \varphi$

PROPOSITION 47. *Let* Λ *be some logic. For all maximal* Λ*-theories* Γ *we have that if* $\Gamma/[\mathrm{do}_i(\alpha)]$ *is* Λ*-consistent then* $\Gamma/[\mathrm{do}_i(\alpha)]$ *is a maximal* Λ*-theory.*

Proof. Suppose that $\Gamma/[\mathrm{do}_i(\alpha)]$ is Λ-consistent. We show that $\Gamma/[\mathrm{do}_i(\alpha)]$ is a Λ-theory and that for all $\varphi \in \mathrm{L}$, either $\varphi \in \Gamma/[\mathrm{do}_i(\alpha)]$ or $\neg\varphi \in \Gamma/[\mathrm{do}_i(\alpha)]$. Since by assumption $\Gamma/[\mathrm{do}_i(\alpha)]$ is consistent, this suffices to conclude that $\Gamma/[\mathrm{do}_i(\alpha)]$ is a maximal Λ-theory.

1. $\Gamma/[\mathrm{do}_i(\alpha)]$ contains Λ: If $\vdash^{\Lambda} \varphi$ then by NA, $\vdash^{\Lambda} [\mathrm{do}_i(\alpha)]\varphi$, and, since Γ is a Λ-theory, $[\mathrm{do}_i(\alpha)]\varphi \in \Gamma$. This implies that $\varphi \in \Gamma/[\mathrm{do}_i(\alpha)]$.
2. $\Gamma/[\mathrm{do}_i(\alpha)]$ is closed under MP, ΩI and ΩIA:
   - MP: Assume that $(\varphi \to \psi) \in \Gamma/[\mathrm{do}_i(\alpha)]$ and $\varphi \in \Gamma/[\mathrm{do}_i(\alpha)]$. Then $[\mathrm{do}_i(\alpha)](\varphi \to \psi) \in \Gamma$ and $[\mathrm{do}_i(\alpha)]\varphi \in \Gamma$, which implies, since Γ contains A6 and is closed under MP, that $[\mathrm{do}_i(\alpha)]\psi \in \Gamma$. This implies that $\psi \in \Gamma/[\mathrm{do}_i(\alpha)]$.
   - ΩI: If $\{\phi(\psi_l(j,\varphi,\beta)) \mid l \in \mathbb{N}\} \subseteq \Gamma/[\mathrm{do}_i(\alpha)]$, then $\{[\mathrm{do}_i(\alpha)]\phi(\psi_l(j,\varphi,\beta)) \mid l \in \mathbb{N}\} \subseteq \Gamma$. Applying ΩI to the set of afms $\{[\mathrm{do}_i(\alpha)]\phi(\psi_l(j,\varphi,\beta)) \mid l \in \mathbb{N}\}$, yields that $[\mathrm{do}_i(\alpha)]\phi([\mathrm{do}_j(\mathtt{while}\,\varphi\,\mathbf{do}\,\beta\,\mathbf{od})]\psi) \in \Gamma$, and $\phi([\mathrm{do}_j(\mathtt{while}\,\varphi\,\mathbf{do}\,\beta\,\mathbf{od})]\psi) \in \Gamma/[\mathrm{do}_i(\alpha)]$.

- ΩIA: Let $\{\phi(\neg(\varphi_l(j,\beta))) \mid l \in \mathbb{N}\} \subseteq \Gamma/[\text{do}_i(\alpha)]$.
  Then $\{[\text{do}_i(\alpha)]\phi(\neg(\varphi_l(j,\beta))) \mid l \in \mathbb{N}\} \subseteq \Gamma$.
  By applying ΩI to the set $\{[\text{do}_i(\alpha)]\phi(\neg(\varphi_l(j,\beta))) \mid l \in \mathbb{N}\}$ it follows that
  $[\text{do}_i(\alpha)]\phi(\neg\mathbf{A}_j\texttt{while}\,\varphi\,\textbf{do}\,\beta\,\textbf{od}) \in \Gamma$.
  Hence $\phi(\neg\mathbf{A}_j\texttt{while}\,\varphi\,\textbf{do}\,\beta\,\textbf{od}) \in \Gamma/[\text{do}_i(\alpha)]$.

3. Since Γ is a theory, Γ contains axiom A11: $[\text{do}_i(\alpha)]\varphi \vee [\text{do}_i(\alpha)]\neg\varphi$ for all i, α and φ. Since Γ is maximal, $[\text{do}_i(\alpha)]\varphi \in \Gamma$ or $[\text{do}_i(\alpha)]\neg\varphi \in \Gamma$ for all i, α and φ. But this implies that $\varphi \in \Gamma/[\text{do}_i(\alpha)]$ or $\neg\varphi \in \Gamma/[\text{do}_i(\alpha)]$, for all $\varphi \in \text{L}$.

By items 1, 2, and 3 it follows that $\Gamma/[\text{do}_i(\alpha)]$ is a maximal Λ-theory if $\Gamma/[\text{do}_i(\alpha)]$ is Λ-consistent. ∎

DEFINITION 48. For Λ some logic, the set S_Λ is defined by $\text{S}_\Lambda =^{\text{def}} \{\Gamma \subseteq \text{L} \mid \Gamma \text{ is a maximal } \Lambda\text{-theory}\}$.

PROPOSITION 49. *For Λ some logic and all $\Phi \subseteq \text{L}$ and $\varphi \in \text{L}$ we have:*

- $\Phi \vdash^\Lambda \varphi$ *iff for all* $\Gamma \in \text{S}_\Lambda$ *such that* $\Phi \subseteq \Gamma$ *holds that* $\varphi \in \Gamma$
- $\vdash^\Lambda \varphi$ *iff for all* $\Gamma \in \text{S}_\Lambda$ *holds that* $\varphi \in \Gamma$

Proof. The second item follows by instantiating the first item with $\Phi = \emptyset$ and using item 3 of Proposition 41. We show the first item by proving two implications.

'⇒' By definition of $\Phi \vdash^\Lambda \varphi$.

'⇐' We show: if $\Phi \not\vdash^\Lambda \varphi$ then some $\Gamma \in \text{S}_\Lambda$ exists such that $\Phi \subseteq \Gamma$ and $\varphi \notin \Gamma$. We construct a Γ that satisfies this demand. To this end, we start by making an enumeration $\rho_0, \rho_1, \ldots$ of the formulas of L. Using this enumeration, the increasing sequence of sets $\Gamma_l \subseteq \text{L}$ is for $l \in \mathbb{N}$ inductively defined as follows:

1. $\Gamma_0 = \Phi \cup \{\neg\varphi\}$
2. Assume that Γ_k has been defined. The set Γ_{k+1} is defined by the following algorithm, written in a high-level programming language pseudocode:

   if $\Gamma_k \vdash^\Lambda \rho_k$ then $\Gamma_{k+1} = \Gamma_k \cup \{\rho_k\}$
   elsif ρ_k is of the form $\phi([\text{do}_i(\texttt{while}\,\varphi\,\textbf{do}\,\alpha\,\textbf{od})]\psi)$
   then $\Gamma_{k+1} = \Gamma_k \cup \{\neg\phi(\psi_j(i,\varphi,\alpha))\} \cup \{\neg\rho_k\}$,
   where j is the least number such that $\Gamma_k \not\vdash^\Lambda \phi(\psi_j(i,\varphi,\alpha))$
   (this j exists since otherwise application of ΩI would yield $\Gamma_k \vdash^\Lambda \rho_k$)

elsif ρ_k is of the form $\phi(\neg\mathbf{A}_i \mathtt{while}\,\varphi\,\mathbf{do}\,\alpha\,\mathbf{od})$
then $\Gamma_{k+1} = \Gamma_k \cup \{\neg\phi(\neg(\varphi_j(i,\alpha)))\} \cup \{\neg\rho_k\}$, where
j is the least number such that $\Gamma_k \not\vdash^\Lambda \phi(\neg(\varphi_j(i,\alpha)))$
(this j exists since otherwise application of ΩIA would yield $\Gamma_k \vdash^\Lambda \rho_k$)
else $\Gamma_{k+1} = \Gamma_k \cup \{\neg\rho_k\}$
fi

Now Γ is defined by $\Gamma =^{\text{def}} \cup_{l\in\mathbb{N}}\Gamma_l$. We show that Γ is a maximal Λ-theory. ■

LEMMA 50. *The set Γ_l is Λ-consistent for all $l \in \mathbb{N}$.*

Proof. We prove the lemma by induction on l. Since $\Phi \not\vdash^\Lambda \varphi$, we have that $\Phi \cup \{\neg\varphi\} = \Gamma_0$ is Λ-consistent by Corollary 43. Now assume that Γ_k is consistent. Consider the four possibilities for the definition of Γ_{k+1}:

1. If $\Gamma_k \vdash^\Lambda \rho_k$, then, since Γ_k is assumed to be Λ-consistent, $\Gamma_k \not\vdash^\Lambda \neg\rho_k$, and hence, by Corollary 43, $\Gamma_{k+1} = \Gamma_k \cup \{\rho_k\}$ is Λ-consistent.

2. If $\Gamma_k \cup \{\neg\phi(\psi_j(i,\varphi,\alpha))\} \cup \{\neg\rho_k\}$ were to be Λ-inconsistent, we would have $\Gamma_k \cup \{\neg\phi(\psi_j(i,\varphi,\alpha))\} \vdash^\Lambda \rho_k$, where $\rho_k = \phi([\mathrm{do}_i(\mathtt{while}\,\varphi\,\mathrm{do}\,\alpha\,\mathrm{od})]\psi)$. Since we have $\vdash^\Lambda \rho_k \to \phi(\psi_l(i,\varphi,\alpha))$ for all $l \in \mathbb{N}$, we also have $\Gamma_k \cup \{\neg\phi(\psi_j(i,\varphi,\alpha))\} \vdash^\Lambda \phi(\psi_j(i,\varphi,\alpha))$, which implies that $\Gamma_k \cup \{\neg\phi(\psi_j(i,\varphi,\alpha))\}$ is Λ-inconsistent. But then, by Corollary 43, $\Gamma_k \vdash^\Lambda \phi(\psi_j(i,\varphi,\alpha))$ which contradicts the fact that $\Gamma_k \not\vdash^\Lambda \phi(\psi_j(i,\varphi,\alpha))$. Hence Γ_{k+1} is Λ-consistent.

3. If $\Gamma_k \cup \{\neg(\phi(\neg\varphi_j(i,\alpha)))\} \cup \{\neg\rho_k\}$ were to be Λ-inconsistent, we would have $\Gamma_k \cup \{\neg\phi(\neg(\varphi_j(i,\alpha)))\} \vdash^\Lambda \rho_k$, where $\rho_k = \phi(\neg\mathbf{A}_i \mathtt{while}\,\varphi\,\mathbf{do}\,\alpha\,\mathbf{od})$. Since we have $\vdash^\Lambda \rho_k \to \phi(\neg(\varphi_l(i,\alpha)))$ for all $l \in \mathbb{N}$, we also have $\Gamma_k \cup \{\neg\phi(\neg(\varphi_j(i,\alpha)))\} \vdash^\Lambda \phi(\neg(\varphi_j(i,\alpha)))$, which implies that $\Gamma_k \cup \{\neg\phi(\neg(\varphi_j(i,\alpha)))\}$ is Λ-inconsistent. Then $\Gamma_k \vdash^\Lambda \phi(\neg(\varphi_j(i,\alpha)))$ which contradicts the fact that $\Gamma_k \not\vdash^\Lambda \phi(\neg(\varphi_j(i,\alpha)))$. Hence Γ_{k+1} is Λ-consistent.

4. If $\Gamma_k \not\vdash^\Lambda \rho_k$ then $\Gamma_k \cup \{\neg\rho_k\}$ is Λ-consistent by Corollary 43. ■

LEMMA 51. *The set Γ as constructed above is maximal, i.e. for all $\varphi \in$ L, exactly one of φ and $\neg\varphi$ is an element of Γ.*

Proof. Let $\psi \in$ L be arbitrary, then $\psi = \rho_k$ for some $k \in \mathbb{N}$. By construction, now either $\rho_k \in \Gamma_{k+1}$ or $\neg\rho_k \in \Gamma_{k+1}$, hence either $\psi \in \Gamma$ or $\neg\psi \in \Gamma$. Suppose both ψ and $\neg\psi$ in Γ. Then for some $k \in \mathbb{N}$, $\{\psi,\neg\psi\} \subseteq \Gamma_k$, which would make Γ_k inconsistent. Since this contradicts the result of Lemma 50 given above, it follows that ψ and $\neg\psi$ are not both in Γ. ■

LEMMA 52. *The set* Γ *as constructed above is a* Λ*-theory.*

Proof. We need to show that Γ contains Λ and is closed under MP, ΩI, and ΩIA. So let $\varphi, \psi \in \mathrm{L}$, $i \in \mathrm{A}$ and $\alpha \in \mathrm{Ac}$ be arbitrary.

1. Γ contains Λ: If $\vdash^{\Lambda} \varphi$, where $\varphi = \rho_k$ for some $k \in \mathbb{N}$, then $\Gamma_k \vdash^{\Lambda} \rho_k$ and hence $\varphi = \rho_k \in \Gamma_{k+1} \subseteq \Gamma$.
2. Closure under MP, ΩI, and ΩIA:
   - MP: Suppose that $\varphi, \varphi \to \psi \in \Gamma$. If $\psi \notin \Gamma$, then $\neg\psi \in \Gamma$, since Γ is maximal by Lemma 51. Hence $\{\varphi, \varphi \to \psi, \neg\psi\} \in \Gamma_k$ for some $k \in \mathbb{N}$, which would make Γ_k Λ-inconsistent. This leads to a contradiction with Lemma 50, hence $\psi \in \Gamma$.
   - ΩI: Suppose $\{\phi(\psi_l(i,\varphi,\alpha)) \mid l \in \mathbb{N}\} \subseteq \Gamma$.
     Let $\phi([\mathrm{do}_i(\mathtt{while}\,\varphi\,\mathtt{do}\,\alpha\,\mathtt{od})]\psi) = \rho_k$, for some $k \in \mathbb{N}$. If $\rho_k \notin \Gamma$, then $\Gamma_k \nvdash^{\Lambda} \rho_k$, and, by case 2 of the construction of Γ_{k+1}, this implies that $\neg\phi(\psi_j(i,\varphi,\alpha)) \in \Gamma_{k+1}$, where $j \in \mathbb{N}$ is the least number such that $\Gamma_k \nvdash^{\Lambda} \phi(\psi_j(i,\varphi,\alpha))$. Hence $\neg\phi(\psi_j(i,\varphi,\alpha)) \in \Gamma$, and by Lemma 51, $\phi(\psi_j(i,\varphi,\alpha)) \notin \Gamma$, which contradicts the assumption that $\{\phi(\psi_l(i,\varphi,\alpha)) \mid l \in \mathbb{N}\} \subseteq \Gamma$.
     Hence $\phi([\mathrm{do}_i(\mathtt{while}\,\varphi\,\mathtt{do}\,\alpha\,\mathtt{od})]\psi) \in \Gamma$.
   - ΩIA: Suppose $\{\phi(\neg(\varphi_l(i,\alpha))) \mid l \in \mathbb{N}\} \subseteq \Gamma$.
     Let $\phi(\neg\mathbf{A}_i\mathtt{while}\,\varphi\,\mathtt{do}\,\alpha\,\mathtt{od}) = \rho_k$, for some $k \in \mathbb{N}$. If $\rho_k \notin \Gamma$, then $\Gamma_k \nvdash^{\Lambda} \rho_k$, and by case 2 of the construction of Γ_{k+1}, this implies that $\neg(\phi(\neg\varphi_j(i,\alpha))) \in \Gamma_{k+1}$, for $j \in \mathbb{N}$ the least number such that $\Gamma_k \nvdash^{\Lambda} \phi(\neg(\varphi_j(i,\alpha)))$. Hence $\neg\phi(\neg(\varphi_j(i,\alpha))) \in \Gamma$, and $\phi(\neg(\varphi_j(i,\alpha))) \notin \Gamma$, which contradicts the assumption that $\{\phi(\neg(\varphi_l(i,\alpha))) \mid l \in \mathbb{N}\} \subseteq \Gamma$. Hence $\phi(\neg\mathbf{A}_i\mathtt{while}\,\varphi\,\mathtt{do}\,\alpha\,\mathtt{od}) \in \Gamma$.

   We conclude that Γ is closed under MP, ΩI and ΩIA.

Since Γ contains Λ and is closed under MP, ΩI and ΩIA, we conclude that Γ is a Λ-theory. ■

Now if Γ is Λ-inconsistent, then $\Gamma \vdash^{\Lambda} \bot$. Since, by Lemma 52, Γ is a Λ-theory, it follows by Proposition 38(2) that $\bot \in \Gamma$. Then $\bot \in \Gamma_k$ for some $k \in \mathbb{N}$, which contradicts the Λ-consistency of Γ_k which was shown in Lemma 50. Hence Γ is a Λ-theory (Lemma 52) which is maximal (Lemma 51) and Λ-consistent, thus Γ is a maximal Λ-theory. Note that by construction of Γ, $\Phi \subseteq \Gamma$ and $\neg\varphi \in \Gamma$, which suffices to conclude the right-to-left implication.

DEFINITION 53. Let Λ be some logic. The canonical model M_Λ for Λ is defined by $\mathrm{M}_\Lambda =^{\mathrm{def}} \langle \mathrm{S}_\Lambda, \pi_\Lambda, \mathrm{R}_\Lambda, \mathrm{r}_\Lambda, \mathrm{c}_\Lambda \rangle$ where

1. S_Λ is the set of maximal Λ-theories

2. $\pi_\Lambda(p,s) = \mathbf{1}$ iff $p \in s$, for $p \in \Pi$ and $s \in \mathrm{S}_\Lambda$
3. $(s,t) \in \mathrm{R}_\Lambda(i)$ iff $s/\mathbf{K}_i \subseteq t$, for $s,t \in \mathrm{S}_\Lambda$ and $i \in \mathrm{A}$
4. $t = \mathbf{r}_\Lambda(i,a)(s)$ iff $s/\mathrm{do}_i(a) \subseteq t$, for $i \in \mathrm{A}$, $a \in \mathrm{At}$ and $s,t \in \mathrm{S}_\Lambda$
5. $\mathbf{c}_\Lambda(i,a)(s) = \mathbf{1}$ if $\mathbf{A}_i a \in s$ and $\mathbf{c}_\Lambda(i,a)(s) = \mathbf{0}$ if $\mathbf{A}_i a \notin s$ for $i \in \mathrm{A}$, $a \in \mathrm{At}$ and $s \in \mathrm{S}_\Lambda$

PROPOSITION 54. *Let Λ be some logic. The canonical model M_Λ for Λ as defined above is a well-defined model from* $\mathbf{M}$.

Proof. Let Λ be some logic. In order to show that M_Λ is a well-defined model from $\mathbf{M}$ we have to show that the demands determining well-definedness of models are met by M_Λ. It is easily seen that S_Λ, π_Λ, and $\mathbf{c}_\Lambda$ are well-defined, which leaves to show that R_Λ and $\mathbf{r}_\Lambda$ are. To prove that $\mathrm{R}(i)$ is an equivalence relation, assume that $i \in \mathrm{A}$ and that $\{s,t,u\} \subseteq \mathrm{S}_\Lambda$. We show:

1. $(s,s) \in \mathrm{R}(i)$, i.e. $\mathrm{R}(i)$ is reflexive.
2. if $(s,t) \in \mathrm{R}(i)$ and $(s,u) \in \mathrm{R}(i)$ then $(t,u) \in \mathrm{R}(i)$, i.e. $\mathrm{R}(i)$ is Euclidean.

To show the reflexivity of $\mathrm{R}(i)$, note that $(s,t) \in \mathrm{R}(i)$ iff $s/\mathbf{K}_i \subseteq t$. Now since s contains axiom A3: $\mathbf{K}_i\varphi \to \varphi$, we have for $\varphi \in s/\mathbf{K}_i$ that $\mathbf{K}_i\varphi \in s$ and hence $\varphi \in s$ by MP. Thus $s/\mathbf{K}_i \subseteq s$, hence $(s,s) \in \mathrm{R}(i)$. To show that $\mathrm{R}(i)$ is Euclidean assume that $\varphi \in t/\mathbf{K}_i$, i.e., $\mathbf{K}_i\varphi \in t$. To prove: $\varphi \in u$. Suppose $\varphi \notin u$. Then since $(s,u) \in \mathrm{R}(i)$, $\varphi \notin s/\mathbf{K}_i$, i.e., $\mathbf{K}_i\varphi \notin s$. Since s is a maximal Λ-theory this implies that $\neg\mathbf{K}_i\varphi \in s$, and since s contains axiom A5: $\neg\mathbf{K}_i\varphi \to \mathbf{K}_i\neg\mathbf{K}_i\varphi$, also $\mathbf{K}_i\neg\mathbf{K}_i\varphi \in s$. Since $(s,t) \in \mathrm{R}(i)$, $s/\mathbf{K}_i \subseteq t$ and thus $\neg\mathbf{K}_i\varphi \in t$. But then $\mathbf{K}_i\varphi \in t$ and $\neg\mathbf{K}_i\varphi \in t$ which contradicts the consistency of t. Thus $\mathrm{R}(i)$ is Euclidean, and, combined with the reflexivity, this ensures that $\mathrm{R}(i)$ is an equivalence relation.

To show that $\mathbf{r}_\Lambda$ is well-defined, it needs to be shown that for all $i \in \mathrm{A}$, $a \in \mathrm{At}$ and $s \in \mathrm{S}_\Lambda$ it holds that $\mathbf{r}_\Lambda(i,a)(s) \in \mathrm{S}_\Lambda$ or $\mathbf{r}_\Lambda(i,a)(s) = \emptyset$. To this end it suffices to show for arbitrary $i \in \mathrm{A}$, $a \in \mathrm{At}$ and $s,t,u \in \mathrm{S}_\Lambda$ that if $t = \mathbf{r}_\Lambda(i,a)(s)$ and $u = \mathbf{r}_\Lambda(i,a)(s)$ then $t = u$. By definition it follows that $s/[\mathrm{do}_i(a)] \subseteq t$ and $s/[\mathrm{do}_i(a)] \subseteq u$ if both $t = \mathbf{r}_\Lambda(i,a)(s)$ and $u = \mathbf{r}_\Lambda(i,a)(s)$. Since both t and u are maximal Λ-theories, both t and u are Λ-consistent, and hence $s/[\mathrm{do}_i(a)]$ is Λ-consistent. But then, by Proposition 47, $s/[\mathrm{do}_i(a)]$ is a maximal Λ-theory, which is properly contained only in L. Hence $s/[\mathrm{do}_i(a)] = t$ and $s/[\mathrm{do}_i(a)] = u$, which suffices to conclude that $\mathbf{r}_\Lambda$ is well-defined. ■

Up till now, the two proof systems Σ_0 and Σ_1 were dealt with identically, i.e. in none of the definitions or propositions given above one needs to distinguish the proof systems or the logics based on these proof systems. From this point on, however, we need to treat the two systems, and thereby the logics, differently. We start with finishing the proof of soundness and completeness for **1**-logics, and indicate thereafter how this proof needs to be modified to end up with one for **0**-logics.

The presence of the confirmation action, which tightly links actions and formulas, prevents the subformula- or subaction-relation from being an adequate parameter for induction in the proof of the truth-theorem, the theorem which links satisfiability in a state of the canonical model to being an element of the maximal theory which constitutes the state. Instead we need a more elaborate relation, which is defined below.

DEFINITION 55. The relation $\prec$ is the smallest relation on $\{0,1\} \times \mathrm{L}$ that satisfies for all $\varphi, \psi \in \mathrm{L}$, $i \in \mathrm{A}$, and $\alpha, \alpha_1, \alpha_2 \in \mathrm{Ac}$ the following constraints:

1. $(0, \varphi) \prec (0, \varphi \vee \psi)$
2. $(0, \psi) \prec (0, \varphi \vee \psi)$
3. $(0, \varphi) \prec (0, \neg\varphi)$
4. $(0, \varphi) \prec (0, \mathbf{K}_i\varphi)$
5. $(0, \varphi) \prec (0, [\mathrm{do}_i(\alpha)]\varphi)$
6. $(1, [\mathrm{do}_i(\alpha)]\varphi) \prec (0, [\mathrm{do}_i(\alpha)]\varphi)$
7. $(1, [\mathrm{do}_i(\alpha_1)][\mathrm{do}_i(\alpha_2)]\varphi) \prec (1, [\mathrm{do}_i(\alpha_1; \alpha_2)]\varphi)$
8. $(1, [\mathrm{do}_i(\alpha_2)]\varphi) \prec (1, [\mathrm{do}_i(\alpha_1; \alpha_2)]\varphi)$
9. $(1, [\mathrm{do}_i(\texttt{confirm}\,\varphi; \alpha_1)]\psi) \prec (1, [\mathrm{do}_i(\texttt{if}\,\varphi\,\mathbf{then}\,\alpha_1\,\mathbf{else}\,\alpha_2\,\texttt{fi})]\psi)$
10. $(1, [\mathrm{do}_i(\texttt{confirm}\,\neg\varphi; \alpha_2)]\psi) \prec (1, [\mathrm{do}_i(\texttt{if}\,\varphi\,\mathbf{then}\,\alpha_1\,\mathbf{else}\,\alpha_2\,\texttt{fi})]\psi)$
11. $(1, \psi_l(i, \varphi, \alpha)) \prec (1, [\mathrm{do}_i(\texttt{while}\,\varphi\,\mathbf{do}\,\alpha\,\mathbf{od})]\psi)$ for all $l \in \mathbb{N}$
12. $(0, \neg\varphi) \prec (1, [\mathrm{do}_i(\texttt{confirm}\,\varphi)]\psi)$
13. $(1, \mathbf{A}_i\alpha) \prec (0, \mathbf{A}_i\alpha)$
14. $(1, \mathbf{A}_i\alpha_1) \prec (1, \mathbf{A}_i\alpha_1; \alpha_2)$
15. $(1, \mathbf{A}_i\alpha_2) \prec (1, \mathbf{A}_i\alpha_1; \alpha_2)$
16. $(1, [\mathrm{do}_i(\alpha_1)]\mathbf{A}_i\alpha_2) \prec (1, \mathbf{A}_i\alpha_1; \alpha_2)$
17. $(1, \mathbf{A}_i\texttt{confirm}\,\varphi; \alpha_1) \prec (1, \mathbf{A}_i\texttt{if}\,\varphi\,\mathbf{then}\,\alpha_1\,\mathbf{else}\,\alpha_2\,\texttt{fi})$
18. $(1, \mathbf{A}_i\texttt{confirm}\,\neg\varphi; \alpha_2) \prec (1, \mathbf{A}_i\texttt{if}\,\varphi\,\mathbf{then}\,\alpha_1\,\mathbf{else}\,\alpha_2\,\texttt{fi})$

19. $(1, \varphi_l(i, \alpha)) \prec (1, \mathbf{A}_i\mathtt{while}\,\varphi\,\mathbf{do}\,\alpha\,\mathbf{od})$ for all $l \in \mathbb{N}$
20. $(0, \varphi) \prec (1, \mathbf{A}_i\mathtt{confirm}\,\varphi)$

DEFINITION 56. The ordering $<$ is defined as the transitive closure of $\prec$, and $\leq$ is defined as the reflexive closure of $<$.

PROPOSITION 57. *The ordering $<$ is well-founded.*

Proof. The proof of this proposition is quite elaborate; it can be found in [van der Hoek *et al.*, 1998] where it takes over three pages. Basically, the idea is to use a powerful technique well-known from the theory of Term Rewriting Systems, viz. the lexicographic path ordering. Using this technique it suffices to select an appropriate well-founded precedence on the function symbols of the language in order to conclude that the ordering $\prec$ is well-founded. Since the actual proof is not only rather elaborate but also contains many details that are completely outside the scope of this paper, it is omitted here; those who are interested can find all details in [van der Hoek *et al.*, 1998]. ■

Having proved that the ordering $<$ is well-founded, we can use it in the proof of the truth-theorem.

THEOREM 58 (The truth-theorem). *Let Λ be some* **1**-*logic. For any $\varphi \in \mathrm{L}$, and any $s \in \mathrm{S}_\Lambda$ we have: $\mathrm{M}_\Lambda, s \models^1 \varphi$ iff $\varphi \in s$.*

Proof. We prove the theorem by proving the following (stronger) properties for all $\varphi, \psi \in \mathrm{L}$, $i \in \mathrm{A}$, $\alpha \in \mathrm{Ac}$, and $s \in \mathrm{S}_\Lambda$:

1. For all $(0, \psi) \leq (0, \varphi)$ we have: $\mathrm{M}_\Lambda, s \models^1 \psi$ iff $\psi \in s$
2. For all $(1, [\mathrm{do}_i(\alpha)]\psi) < (0, \varphi)$ we have:
   (a) $\psi \in t$ for $t = \mathbf{r}^1(i, \alpha)(s) \Rightarrow [\mathrm{do}_i(\alpha)]\psi \in s$
   (b) if $t = \mathbf{r}^1(i, \alpha)(s)$ and $[\mathrm{do}_i(\alpha)]\psi \in s$ then $\psi \in t$
3. For all $(1, \mathbf{A}_i\alpha) < (0, \varphi)$ we have: $\mathbf{c}^1(i, \alpha)(s) = \mathbf{1}$ iff $\mathbf{A}_i\alpha \in s$

where $\mathbf{r}^1$ and $\mathbf{c}^1$ are the functions induced by $\mathbf{r}_\Lambda$ and $\mathbf{c}_\Lambda$ in the way described in Definition 4. The theorem then follows from the first item, since $(0, \varphi) \leq (0, \varphi)$. So let $\varphi \in \mathrm{L}$ be some fixed formula. We start by proving the first property. Let $\psi \in \mathrm{L}$ be such that $(0, \psi) \leq (0, \varphi)$. Consider the various cases for ψ:

- $\psi = p$, for $p \in \Pi$. By definition of π_Λ we have that $\pi_\Lambda(p, s) = \mathbf{1}$ iff $p \in s$.

- $\psi = \psi_1 \wedge \psi_2$. Since $(0, \psi_1) < (0, \psi_1 \wedge \psi_2)$ and $(0, \psi_2) < (0, \psi_1 \wedge \psi_2)$, we have that $\mathrm{M}_\Lambda, s \models^1 \psi_1 \wedge \psi_2$ iff ($\mathrm{M}_\Lambda, s \models^1 \psi_1$ and $\mathrm{M}_\Lambda, s \models^1 \psi_2$) iff $\psi_1 \in s$ and $\psi_2 \in s$ (by induction on (1)) iff $\psi_1 \wedge \psi_2 \in s$ (since s is a (maximal) theory).

- $\psi = \neg\psi_1$. Since $(0, \psi_1) < (0, \neg\psi_1)$ we have that $\mathrm{M}_\Lambda, s \models^1 \neg\psi_1$ iff not($\mathrm{M}_\Lambda, s \models^1 \psi_1$) iff not($\psi_1 \in s$) (by induction on (1)) iff $\neg\psi_1 \in s$ since s is (a) maximal (theory).

- $\psi = \mathbf{K}_i\psi_1$. We will prove two implications:

  '$\Leftarrow$' Suppose $\mathbf{K}_i\psi_1 \in s$. Then by definition of $\mathrm{R}_\Lambda(i)$, $\psi_1 \in t$ for all t such that $(s,t) \in \mathrm{R}_\Lambda(i)$. Since $(0, \psi_1) < (0, \mathbf{K}_i\psi_1)$, this implies that $\mathrm{M}_\Lambda, t \models^1 \psi_1$ for all $t \in \mathrm{S}_\Lambda$ with $(s,t) \in \mathrm{R}_\Lambda(i)$, hence $\mathrm{M}_\Lambda, s \models^1 \mathbf{K}_i\psi_1$.

  '$\Rightarrow$' Suppose $\mathrm{M}_\Lambda, s \models^1 \mathbf{K}_i\psi_1$. Now if for $t \in \mathrm{S}_\Lambda$, $(s,t) \in \mathrm{R}_\Lambda(i)$, then $\mathrm{M}_\Lambda, t \models^1 \psi_1$. Since $(0, \psi_1) < (0, \mathbf{K}_i\psi_1)$, we have by induction on (1) that $\psi_1 \in t$, for all $t \in \mathrm{S}_\Lambda$ with $(s,t) \in \mathrm{R}_\Lambda(i)$. This implies that ψ_1 belongs to every maximal theory containing $s/\mathbf{K}_i$, and by Proposition 49 we conclude that $s/\mathbf{K}_i \vdash^\Lambda \psi_1$. By Corollary 46 we conclude that $\mathbf{K}_i\psi_1 \in s$.

- $\psi = [\mathrm{do}_i(\alpha)]\psi_1$. We will prove two implications:

  '$\Leftarrow$' Let $[\mathrm{do}_i(\alpha)]\psi_1 \in s$. Let $t = \mathbf{r}^1(i, \alpha)(s)$. Since $(1, [\mathrm{do}_i(\alpha)]\psi_1) < (0, [\mathrm{do}_i(\alpha)]\psi_1)$, we find by induction on (2b) that $\psi_1 \in t$. Since $(0, \psi_1) < (0, [\mathrm{do}_i(\alpha)]\psi_1)$, we find by induction on (1) that $\mathrm{M}_\Lambda, t \models^1 \psi_1$, if $t = \mathbf{r}^1(i, \alpha)(s)$. But this implies that $\mathrm{M}_\Lambda, s \models^1 [\mathrm{do}_i(\alpha)]\psi_1$.

  '$\Rightarrow$' Suppose $\mathrm{M}_\Lambda, s \models^1 [\mathrm{do}_i(\alpha)]\psi_1$. This implies that $\mathrm{M}_\Lambda, t \models^1 \psi_1$ if $t = \mathbf{r}^1(i, \alpha)(s)$. Since $(0, \psi_1) < (0, [\mathrm{do}_i(\alpha)]\psi_1)$ we have by induction on (1) that $\psi_1 \in t$ if $t = \mathbf{r}^1(i, \alpha)(s)$. Now since $(1, [\mathrm{do}_i(\alpha)]\psi_1) < (0, [\mathrm{do}_i(\alpha)]\psi_1)$, we conclude by induction on (2a) that $[\mathrm{do}_i(\alpha)]\psi_1 \in s$.

- $\psi = \mathbf{A}_i\alpha$. Since $(1, \mathbf{A}_i\alpha) < (0, \mathbf{A}_i\alpha)$ we find by induction on (3) that $\mathrm{M}_\Lambda, s \models^1 \mathbf{A}_i\alpha$ iff $\mathbf{c}^1(i, \alpha)(s) = \mathbf{1}$ iff $\mathbf{A}_i\alpha \in s$.

Next we prove (2a). Let $(1, [\mathrm{do}_i(\alpha)]\psi) < (0, \varphi)$. Consider the various possibilities for α.

- $\alpha = a$, for $a \in \mathrm{At}$.
  Assume that $\psi \in t$ if $t = \mathbf{r}_\Lambda(i, a)(s)$. By definition of $\mathbf{r}_\Lambda$ this implies that ψ is in every maximal theory containing $s/[\mathrm{do}_i(a)]$, i.e. $s/[\mathrm{do}_i(a)] \vdash^\Lambda \psi$. By Corollary 46 we conclude that $[\mathrm{do}_i(a)]\psi \in s$.

- $\alpha = \texttt{confirm}\,\psi_1$.
Assume that $\psi \in t$ for $t = \mathbf{r}^1(i, \texttt{confirm}\,\psi_1)(s)$. If $\mathrm{M}_\Lambda, s \models^1 \psi_1$ we have that $s = t$, by definition of $\mathbf{r}^1$.
Then $\psi \in s$, and, since s is a theory, this implies $\neg\psi_1 \vee \psi \in s$, which in turn implies $[\mathrm{do}_i(\texttt{confirm}\,\psi_1)]\psi \in s$. If $\mathrm{M}_\Lambda, s \models^1 \neg\psi_1$ then, since $(0, \neg\psi_1) < (1, [\mathrm{do}_i(\texttt{confirm}\,\psi_1)]\psi)$, we have by induction on (1) that $\neg\psi_1 \in s$, hence $\neg\psi_1 \vee \psi \in s$, and thus $[\mathrm{do}_i(\texttt{confirm}\,\psi_1)]\psi \in s$.

- $\alpha = \alpha_1; \alpha_2$.
Assume that $\psi \in t$ for $t = \mathbf{r}^1(i, \alpha_1; \alpha_2)(s)$. By definition of $\mathbf{r}^1$ this implies that $\psi \in t$ for $t = \mathbf{r}^1(i, \alpha_2)(u)$ for $u = \mathbf{r}^1(i, \alpha_1)(s)$. Since $(1, [\mathrm{do}_i(\alpha_2)]\psi) < (1, [\mathrm{do}_i(\alpha_1; \alpha_2)]\psi)$ we have that $[\mathrm{do}_i(\alpha_2)]\psi \in u$ for $u = \mathbf{r}^1(i, \alpha_1)(s)$. Since furthermore $(1, [\mathrm{do}_i(\alpha_1)][\mathrm{do}_i(\alpha_2)]\psi) < (1, [\mathrm{do}_i(\alpha_1; \alpha_2)]\psi)$ we have that $[\mathrm{do}_i(\alpha_1)][\mathrm{do}_i(\alpha_2)]\psi \in s$. Since s is closed under the axioms of Σ_1 and MP, this implies that $[\mathrm{do}_i(\alpha_1; \alpha_2)]\psi \in s$.

- $\alpha = \texttt{if}\,\varphi\,\texttt{then}\,\alpha_1\,\texttt{else}\,\alpha_2\,\texttt{fi}$.
Let $\psi \in t$ for $t = \mathbf{r}^1(i, \texttt{if}\,\varphi\,\texttt{then}\,\alpha_1\,\texttt{else}\,\alpha_2\,\texttt{fi})(s)$. Then $\psi \in t$ for all $t = \mathbf{r}^1(i, \texttt{confirm}\,\varphi; \alpha_1)(s)$ and $\psi \in t$ for all $t = \mathbf{r}^1(i, \texttt{confirm}\,\neg\varphi; \alpha_2)(s)$.
Since we have both $(1, [\mathrm{do}_i(\texttt{confirm}\,\varphi; \alpha_1)]\psi) < (1, [\mathrm{do}_i(\texttt{if}\,\varphi\,\texttt{then}\,\alpha_1\,\texttt{else}\,\alpha_2\,\texttt{fi})]\psi)$ and $(1, [\mathrm{do}_i(\texttt{confirm}\,\neg\varphi; \alpha_2)]\psi) < (1, [\mathrm{do}_i(\texttt{if}\,\varphi\,\texttt{then}\,\alpha_1\,\texttt{else}\,\alpha_2\,\texttt{fi})]\psi)$ we have by induction that $[\mathrm{do}_i(\texttt{confirm}\,\varphi; \alpha_1)]\psi \in s$ and $[\mathrm{do}_i(\texttt{confirm}\,\neg\varphi; \alpha_2)]\psi \in s$, and, since s is a theory, $[\mathrm{do}_i(\texttt{if}\,\varphi\,\texttt{then}\,\alpha_1\,\texttt{else}\,\alpha_2\,\texttt{fi})]\psi \in s$.

- $\alpha = \texttt{while}\,\varphi\,\texttt{do}\,\alpha\,\texttt{od}$.
Assume that $\psi \in t$ for $t = \mathbf{r}^1(i, \texttt{while}\,\varphi\,\texttt{do}\,\alpha\,\texttt{od})(s)$.
Since $\mathbf{r}^1(i, \texttt{while}\,\varphi\,\texttt{do}\,\alpha\,\texttt{od})(s) = \cup_{k \in \mathbb{N}} \mathbf{r}^1(i, (\texttt{confirm}\,\varphi; \alpha)^k; \texttt{confirm}\,\neg\varphi)(s)$, we have that $\psi \in t$ for all $t = \mathbf{r}^1(i, (\texttt{confirm}\,\varphi; \alpha)^k; \texttt{confirm}\,\neg\varphi)(s)$, for all $k \in \mathbb{N}$.
Now since we have that $(1, [\mathrm{do}_i((\texttt{confirm}\,\varphi; \alpha)^k; \texttt{confirm}\,\neg\varphi)]\psi) < (1, [\mathrm{do}_i(\texttt{while}\,\varphi\,\texttt{do}\,\alpha\,\texttt{od})]\psi)$ for all $k \in \mathbb{N}$ we have by induction on (2a) that $\psi_k(i, \varphi, \alpha) \in s$ for all $k \in \mathbb{N}$, and since s is closed under ΩA this implies that $[\mathrm{do}_i(\texttt{while}\,\varphi\,\texttt{do}\,\alpha\,\texttt{od})]\psi \in s$.

We continue with proving (2b). So let again $(1, [\mathrm{do}_i(\alpha)]\psi) < (0, \varphi)$, and consider the various possibilities for α.

- $\alpha = a$, for $a \in \mathrm{At}$.
If $t = \mathbf{r}_\Lambda(i, a)(s)$ and $[\mathrm{do}_i(a)]\psi \in s$, then by definition of $\mathbf{r}_\Lambda$, $\psi \in t$.

- $\alpha = \texttt{confirm}\,\psi_1$.
Let $t = \mathbf{r}^1(i, \texttt{confirm}\,\psi_1)(s)$ and $[\mathrm{do}_i(\texttt{confirm}\,\psi_1)]\psi \in s$. By definition of $\mathbf{r}^1$, $\mathrm{M}_\Lambda, s \models^1 \psi_1$ and $s = t$. Since $(0, \psi_1) < (0, \neg\psi_1) <$

$(1, [\mathrm{do}_i(\mathtt{confirm}\,\psi_1)]\psi)$, we find by induction on (1) that $\psi_1 \in s$. Since s is a theory, $[\mathrm{do}_i(\mathtt{confirm}\,\psi_1)]\psi \in s$ implies that $\neg\psi_1 \vee \psi \in s$, and, since s is maximal, we conclude that $\psi \in s$.

- $\alpha = \alpha_1; \alpha_2$.
Let $t = \mathbf{r}^1(i, \alpha_1; \alpha_2)(s)$ and $[\mathrm{do}_i(\alpha_1; \alpha_2)]\psi \in s$. Then, by definition of $\mathbf{r}^1$, we have that $t = \mathbf{r}^1(i, \alpha_2)(u)$ for some $u \in \mathrm{S}_\Lambda$ such that $u = \mathbf{r}^1(i, \alpha_1)(s)$. Since s is closed under the axioms and proof rules of Σ_1 we have that $[\mathrm{do}_i(\alpha_1)][\mathrm{do}_i(\alpha_2)]\psi \in s$, and hence, since $(1, [\mathrm{do}_i(\alpha_1)][\mathrm{do}_i(\alpha_2)]\psi) < (1, [\mathrm{do}_i(\alpha_1; \alpha_2)]\psi)$, we have by induction on (2b) that $[\mathrm{do}_i(\alpha_2)]\psi \in u$. But this implies, since $(1, [\mathrm{do}_i(\alpha_2)]\psi) < (1, [\mathrm{do}_i(\alpha_1; \alpha_2)]\psi)$, that $\psi \in t$.

- $\alpha = \mathtt{if}\,\varphi\,\mathtt{then}\,\alpha_1\,\mathtt{else}\,\alpha_2\,\mathtt{fi}$.
Let $t = \mathbf{r}^1(i, \mathtt{if}\,\varphi\,\mathtt{then}\,\alpha_1\,\mathtt{else}\,\alpha_2\,\mathtt{fi})(s)$ and let furthermore $[\mathrm{do}_i(\mathtt{if}\,\varphi\,\mathtt{then}\,\alpha_1\,\mathtt{else}\,\alpha_2\,\mathtt{fi})]\psi \in s$.
Then either $t = \mathbf{r}^1(i, \mathtt{confirm}\,\varphi; \alpha_1)(s)$ or $t = \mathbf{r}^1(i, \mathtt{confirm}\,\neg\varphi; \alpha_2)(s)$.
If $[\mathrm{do}_i(\mathtt{if}\,\varphi\,\mathtt{then}\,\alpha_1\,\mathtt{else}\,\alpha_2\,\mathtt{fi})]\psi \in s$,then, since s is a theory, both $[\mathrm{do}_i(\mathtt{confirm}\,\varphi; \alpha_1)]\psi \in s$ and $[\mathrm{do}_i(\mathtt{confirm}\,\neg\varphi; \alpha_2)]\psi \in s$.
Since it holds that both $(1, [\mathrm{do}_i(\mathtt{confirm}\,\varphi; \alpha_1)]\psi) <$ $(1, [\mathrm{do}_i(\mathtt{if}\,\varphi\,\mathtt{then}\,\alpha_1\,\mathtt{else}\,\alpha_2\,\mathtt{fi})]\psi)$ and $(1, [\mathrm{do}_i(\mathtt{confirm}\,\neg\varphi; \alpha_2)]\psi) <$ $(1, [\mathrm{do}_i(\mathtt{if}\,\varphi\,\mathtt{then}\,\alpha_1\,\mathtt{else}\,\alpha_2\,\mathtt{fi})]\psi)$ we have by induction on (2b) that $\psi \in t$.

- $\alpha = \mathtt{while}\,\varphi\,\mathtt{do}\,\alpha\,\mathtt{od}$.
Let $t = \mathbf{r}^1(i, \mathtt{while}\,\varphi\,\mathtt{do}\,\alpha\,\mathtt{od})(s)$ and $[\mathrm{do}_i(\mathtt{while}\,\varphi\,\mathtt{do}\,\alpha\,\mathtt{od})]\psi \in s$. Since s is a theory, we have that $\psi_l(i, \varphi, \alpha) \in s$, for all $l \in \mathbb{N}$. By definition of $\mathbf{r}^1$, it holds that $t = \mathbf{r}^1(i, (\mathtt{confirm}\,\varphi; \alpha)^k; \mathtt{confirm}\,\neg\varphi)(s)$ for some $k \in \mathbb{N}$. Now since $(1, \psi_l(i, \varphi, \alpha)) < (1, [\mathrm{do}_i(\mathtt{while}\,\varphi\,\mathtt{do}\,\alpha\,\mathtt{od})]\psi)$ for all $l \in \mathbb{N}$, we conclude by induction on (2b) that $\psi \in t$.

Finally we come to the proof of item (3). Let $(1, \mathbf{A}_i\alpha) < (0, \varphi)$. Consider the various cases for α.

- $\alpha = a$, where $a \in \mathrm{At}$. Now $\mathrm{c}_\Lambda(i, a)(s) = \mathbf{1}$ iff $\mathbf{A}_i a \in s$, by definition of c_Λ.

- $\alpha = \mathtt{confirm}\,\psi_1$. By definition, $\mathrm{c}^1(i, \mathtt{confirm}\,\psi_1)(s) = \mathbf{1}$ iff $\mathrm{M}_\Lambda, s \models^1 \psi_1$ iff, since $(0, \psi_1) < (0, \mathbf{A}_i\mathtt{confirm}\,\psi_1)$, $\psi_1 \in s$ iff $\mathbf{A}_i\mathtt{confirm}\,\psi_1$ in s, since s is a theory.

- $\alpha = \alpha_1; \alpha_2$. We prove two implications:

  '$\Leftarrow$' Since s is a theory, $\mathbf{A}_i\alpha_1; \alpha_2 \in s$ iff $\mathbf{A}_i\alpha_1 \in s$ and $[\mathrm{do}_i(\alpha_1)]\mathbf{A}_i\alpha_2 \in s$. Since $(1, \mathbf{A}_i\alpha_1) < (1, \mathbf{A}_i\alpha_1; \alpha_2)$, we find by induction on (3) that $\mathrm{c}^1(i, \alpha_1)(s) = \mathbf{1}$. Now suppose $t = \mathbf{r}^1(i, \alpha_1)(s)$. Since $(1, [\mathrm{do}_i(\alpha_1)]\mathbf{A}_i\alpha_2) < (1, \mathbf{A}_i\alpha_1; \alpha_2)$, we find by induction on (2b) that $\mathbf{A}_i\alpha_2 \in t$. Furthermore, since $(1, \mathbf{A}_i\alpha_2) < (1, \mathbf{A}_i\alpha_1;\alpha_2)$, the latter implies that $\mathbf{c}^1(i,\alpha_2)(t) = \mathbf{1}$, for all $t = \mathbf{r}^1(i,\alpha_1)(s)$, which, together with $\mathbf{c}^1(i,\alpha_1)(s) = \mathbf{1}$, suffices to conclude that $\mathbf{c}^1(i,\alpha_1;\alpha_2)(s) = \mathbf{1}$.

'⇒' By definition, $\mathbf{c}^1(i,\alpha_1;\alpha_2)(s) = \mathbf{1}$ iff $\mathbf{c}^1(i,\alpha_1)(s) = \mathbf{1}$ and $\mathbf{c}^1(i,\alpha_2)(t) = \mathbf{1}$ for all $t = \mathbf{r}^1(i,\alpha_1)(s)$. Now since $(1, \mathbf{A}_i\alpha_1) < (1, \mathbf{A}_i\alpha_1;\alpha_2)$, we conclude by induction on (3) that $\mathbf{A}_i\alpha_1 \in s$. Furthermore, since $(1, \mathbf{A}_i\alpha_2) < (1, \mathbf{A}_i\alpha_1;\alpha_2)$, we have that $\mathbf{A}_i\alpha_2 \in t$, for all $t = \mathbf{r}^1(i,\alpha_1)(s)$. Now since $(1, [\mathrm{do}_i(\alpha_1)]\mathbf{A}_i\alpha_2) < (1, \mathbf{A}_i\alpha_1;\alpha_2)$, we find by induction on (2a) that $[\mathrm{do}_i(\alpha_1)]\mathbf{A}_i\alpha_2 \in s$. But then, since s is a theory, we conclude that $\mathbf{A}_i\alpha_1;\alpha_2 \in s$.

- $\alpha = \mathtt{if}\,\varphi\,\mathbf{then}\,\alpha_1\,\mathbf{else}\,\alpha_2\,\mathbf{fi}$.
  By definition of $<$ we have that $(1, \mathbf{A}_i\mathtt{confirm}\,\varphi;\alpha_1) < (1, \mathbf{A}_i\mathtt{if}\,\varphi\,\mathbf{then}\,\alpha_1\,\mathbf{else}\,\alpha_2\,\mathbf{fi})$ and furthermore that $(1, \mathbf{A}_i\mathtt{confirm}\,\neg\varphi;\alpha_2) < (1, \mathbf{A}_i\mathtt{if}\,\varphi\,\mathbf{then}\,\alpha_1\,\mathbf{else}\,\alpha_2\,\mathbf{fi})$. This implies that $\mathbf{c}^1(i, \mathtt{if}\,\varphi\,\mathbf{then}\,\alpha_1\,\mathbf{else}\,\alpha_2\,\mathbf{fi})(s) = \mathbf{1}$ iff $\mathbf{c}^1(i, \mathtt{confirm}\,\varphi;\alpha_1)(s) = \mathbf{1}$ or $\mathbf{c}^1(i, \mathtt{confirm}\,\neg\varphi;\alpha_2)(s) = \mathbf{1}$ iff — by induction on (3) — $\mathbf{A}_i\mathtt{confirm}\,\varphi;\alpha_1 \in s$ or $\mathbf{A}_i\mathtt{confirm}\,\neg\varphi;\alpha_2 \in s$ iff $\mathbf{A}_i\mathtt{if}\,\varphi\,\mathbf{then}\,\alpha_1\,\mathbf{else}\,\alpha_2\,\mathbf{fi} \in s$, since s is a theory.
- $\alpha = \mathtt{while}\,\varphi\,\mathbf{do}\,\alpha\,\mathbf{od}$. We prove two implications:

'⇐' Let $\mathbf{A}_i\mathtt{while}\,\varphi\,\mathbf{do}\,\alpha\,\mathbf{od} \in s$.
Then, since s is maximal, $\neg\mathbf{A}_i\mathtt{while}\,\varphi\,\mathbf{do}\,\alpha\,\mathbf{od} \notin s$, and, since s is closed under ΩIA, this implies that $\neg(\varphi_k(i,\alpha)) \notin s$, for some $k \in \mathbb{N}$, and, again since s is maximal, $\varphi_k(i,\alpha) \in s$. Since $(1, \varphi_l(i,\alpha)) < (1, \mathbf{A}_i\mathtt{while}\,\varphi\,\mathbf{do}\,\alpha\,\mathbf{od})$ for all $l \in \mathbb{N}$, we have by induction on (3) that $\mathbf{c}^1(i, (\mathtt{confirm}\,\varphi;\alpha)^k; \mathtt{confirm}\,\neg\varphi)(s) = \mathbf{1}$, and, by definition of $\mathbf{c}^1$, this implies $\mathbf{c}^1(i, \mathtt{while}\,\varphi\,\mathbf{do}\,\alpha\,\mathbf{od})(s) = \mathbf{1}$.

'⇒' If $\mathbf{c}^1(i, \mathtt{while}\,\varphi\,\mathbf{do}\,\alpha\,\mathbf{od})(s) = \mathbf{1}$, then $\mathbf{c}^1(i, (\mathtt{confirm}\,\varphi;\alpha_1)^k; \mathtt{confirm}\,\neg\varphi)(s) = \mathbf{1}$ for some $k \in \mathbb{N}$. Since $(1, \varphi_l(i,\alpha)) < (1, \mathbf{A}_i\mathtt{while}\,\varphi\,\mathbf{do}\,\alpha\,\mathbf{od})$ for all $l \in \mathbb{N}$, this implies by induction on (3) that $\varphi_k(i,\alpha) \in s$. Then, since s is a theory, $\neg(\varphi_k(i,\alpha)) \notin s$, and, by item 7 of Proposition 38, it follows that $\neg\mathbf{A}_i\mathtt{while}\,\varphi\,\mathbf{do}\,\alpha\,\mathbf{od} \notin s$. Now since s is maximal it follows that $\mathbf{A}_i\mathtt{while}\,\varphi\,\mathbf{do}\,\alpha\,\mathbf{od} \in s$.

Having proved the items (1), (2) and (3) suffices to prove that the truth-theorem holds. ■

The proof of the truth-theorem for **0**-logics is almost identical to the one given for Theorem 58. One just needs to change one clause in the definition of the $\prec$-relation, used to apply induction upon, and modify the proof of the truth-theorem accordingly.

DEFINITION 59. The ordering $<'$ is defined as the transitive closure of the smallest relation on $\{0,1\} \times \mathrm{L}$ satisfying the constraints 1 through 15 and 17 through 20 as given in Definition 55 and the constraint

16$'$. $(1, [\mathrm{do}_i(\alpha_1)]\neg \mathbf{A}_i\alpha_2) \prec (1, \mathbf{A}_i\alpha_1; \alpha_2)$

The ordering $\leq'$ is defined to be the reflexive closure of $<'$.

The only modification to the proof of the truth-theorem for **1**-logics that is required to end up with a proof of a truth-theorem for **0**-logics concerns the proof of property (3) for sequentially composed actions, i.e. the proof that $\mathbf{c^0}(i, \alpha_1; \alpha_2)(s) = \mathbf{1}$ iff $\mathbf{A}_i\alpha_1; \alpha_2 \in s$, whenever $(1, \mathbf{A}_i\alpha_1; \alpha_2) <' (0, \varphi)$. We will show this by proving two implications:

'$\Leftarrow$' Since s is a Λ-theory, $\mathbf{A}_i\alpha_1; \alpha_2$ in s iff $\mathbf{A}_i\alpha_1 \in s$ and $\neg[\mathrm{do}_i(\alpha_1)]\neg\mathbf{A}_i\alpha_2 \in s$. Since $(1, \mathbf{A}_i\alpha_1) <' (1, \mathbf{A}_i\alpha_1; \alpha_2)$ we find by induction on (3) that $\mathbf{c^0}(i, \alpha_1)(s) = \mathbf{1}$. Since $(1, [\mathrm{do}_i(\alpha_1)]\neg\mathbf{A}_i\alpha_2) <' (1, \mathbf{A}_i\alpha_1; \alpha_2)$, we find by induction on (2b), read in its contrapositive form, that for some $t \in \mathrm{S}_\Lambda$, $t = \mathbf{r^0}(i, \alpha_1)(s)$ with $\neg\mathbf{A}_i\alpha_2 \notin t$. Now since t is maximal, this implies that $\mathbf{A}_i\alpha_2 \in t$. Since $(1, \mathbf{A}_i\alpha_2) <' (1, \mathbf{A}_i\alpha_1; \alpha_2)$ the latter implies that $\mathbf{c^0}(i, \alpha_2)(t) = \mathbf{1}$. Together with $\mathbf{c^0}(i, \alpha_1)(s) = \mathbf{1}$ this suffices to conclude that $\mathbf{c^0}(i, \alpha_1; \alpha_2)(s) = \mathbf{1}$.

'$\Rightarrow$' By definition, $\mathbf{c^0}(i, \alpha_1; \alpha_2)(s) = \mathbf{1}$ iff $\mathbf{c^0}(i, \alpha_1)(s) = \mathbf{1}$ and for some $t \in \mathrm{S}_\Lambda$, $t = \mathbf{r^0}(i, \alpha_1)(s)$ and $\mathbf{c^0}(i, \alpha_2)(t) = \mathbf{1}$. Now since $(1, \mathbf{A}_i\alpha_1) <' (1, \mathbf{A}_i\alpha_1; \alpha_2)$ we have by induction on (3) that $\mathbf{A}_i\alpha_1 \in s$. Furthermore, since $(1, \mathbf{A}_i\alpha_2) <' (1, \mathbf{A}_i\alpha_1; \alpha_2)$, we have for the aforementioned t that $\mathbf{A}_i\alpha_2 \in t$. Hence we have some $t \in \mathrm{S}_\Lambda$ such that $t = \mathbf{r^0}(i, \alpha_1)(s)$ and $\mathbf{A}_i\alpha_2 \in t$ while also $\mathbf{A}_i\alpha_1 \in s$. Since t is maximal, $\neg\mathbf{A}_i\alpha_2 \notin t$. And, rephrasing (2b) to 'if $t = \mathbf{r^0}(i, \alpha)(s)$ and $\psi \notin t$ then $[\mathrm{do}_i(\alpha)]\psi \notin s$', we conclude by induction on (2b) that $[\mathrm{do}_i(\alpha_1)]\neg\mathbf{A}_i\alpha_2 \notin s$. Since s is maximal it follows that $\neg[\mathrm{do}_i(\alpha_1)]\neg\mathbf{A}_i\alpha_2 \in s$. Hence $\mathbf{A}_i\alpha_1 \in s$ and $\langle \mathrm{do}_i(\alpha_1)\rangle\mathbf{A}_i\alpha_2 \in s$, which, since s is a Λ-theory, implies that $\mathbf{A}_i\alpha_1; \alpha_2 \in s$, which was to be shown.

Having proved the truth-theorem both for **1**-logics and for **0**-logics, we can prove that deducibility for a logic Λ corresponds with validity in the canonical model M_Λ.

PROPOSITION 60. *For all* **b**-*logics* Λ *and all* $\varphi \in \mathrm{L}$ *we have:* $\vdash^\Lambda \varphi$ *iff* $\mathrm{M}_\Lambda \models^{\mathbf{b}} \varphi$.

Proof. Let $\varphi \in \mathrm{L}$ be arbitrary. Then we have:

$\vdash^\Lambda \varphi$ iff $\varphi \in s$, for all $s \in \mathrm{S}_\Lambda$
iff $\mathrm{M}_\Lambda, s \models^{\mathbf{b}} \varphi$ for all $s \in \mathrm{S}_\Lambda$
iff $\mathrm{M}_\Lambda \models^{\mathbf{b}} \varphi$ ■

Using the propositions and theorems shown above, we can now prove those given in Section 6. Note that Proposition 33 is already shown as the third item of Proposition 41.

30. THEOREM. *For* $\mathbf{b} \in$ bool *and all* $\varphi \in$ L *we have:*

- $\vdash^{\mathbf{b}} \varphi \Leftrightarrow \models^{\mathbf{b}} \varphi$

Proof. We prove the theorem by proving two implications.

'$\Leftarrow$' If $\models^{\mathbf{b}} \varphi$ then M $\models^{\mathbf{b}} \varphi$ for all M $\in$ **M**. Since $M_{LCap_\mathbf{b}} \in$ **M** it follows that $M_{LCap_\mathbf{b}} \models^{\mathbf{b}} \varphi$. By Proposition 60 it then follows that $\vdash^{\mathbf{b}} \varphi$.

'$\Rightarrow$' Suppose $\vdash^{\mathbf{b}} \varphi$ and let M $\in$ **M**. By Proposition 37 we have that $\{\psi \in L \mid M \models^{\mathbf{b}} \psi\}$ is a **b**-logic. Since $LCap_\mathbf{b}$ is the smallest **b**-logic, it follows that whenever $\varphi \in LCap_\mathbf{b}$ also $\varphi \in \{\psi \in L \mid M \models^{\mathbf{b}} \psi\}$, and hence M $\models^{\mathbf{b}} \varphi$. Since M is arbitrary, it follows that M $\models^{\mathbf{b}} \varphi$ for all M $\in$ **M** and thus $\models^{\mathbf{b}} \varphi$, which was to be shown. ■

34. PROPOSITION. *The proof systems* Σ_1 *and* Σ_0 *are strongly complete, i.e. every set* $\Phi \subseteq$ L *that is* $LCap_\mathbf{b}$*-consistent is* $\models^{\mathbf{b}}$*-satisfiable.*

Proof. The proposition follows, for arbitrary logics, directly from the proof of Proposition 49. For if Φ is Λ-consistent, then by the procedure given in the proof of Proposition 49 one constructs a maximal Λ-theory Γ that contains Φ. This Γ appears as a state in the canonical model for Λ, and by the truth-theorems, all formulas from Γ — and hence from Φ — are satisfied at this state. Hence every Λ-consistent set Φ is satisfied at some state of the canonical model for Λ, and since this canonical model is a well-defined one, the proposition follows. ■

ACKNOWLEDGEMENTS

The comments of an anonymous referee on a previous version of this paper were very useful. This work was partially supported by ESPRIT WG No. 23531 ('FIREworks').

This article was originally published as: B. van Linder, w. van der Hoek and J.-J. Ch. Meyer, Formalising abilites and opportunites of agents, *Fundamenta Informaticae*, **34** 53–101, 1998 and is reproduced here with permission from IOS Press.

Wiebe van der Hoek
Utrecht University, The Netherlands and Liverpool University, UK.

John-Jules Meyer
Utrecht University, The Netherlands.

Bernd van Linder
ABN-AMRO, Amsterdam, The Netherlands.

BIBLIOGRAPHY

[Benthem, 1984] J. van Benthem. Correspondence theory. In D.M. Gabbay and F. Guenthner, editors, *Handbook of Philosophical Logic*, volume 2, pages 167–247. Reidel, Dordrecht, 1984.

[Brown, 1988] M.A. Brown. On the logic of ability. *Journal of Philosophical Logic*, 17:1–26, 1988.

[Castelfranchi, 1995] C. Castelfranchi. Guarantees for autonomy in cognitive agent architecture. In M. Wooldridge and N.R. Jennings, editors, *Intelligent Agents – Agent Theories, Architectures, and Languages*, volume 890 of *Lecture Notes in Computer Science (subseries LNAI)*, pages 56–70. Springer-Verlag, 1995.

[Chellas, 1980] B.F. Chellas. *Modal Logic. An Introduction.* Cambridge University Press, Cambridge, 1980.

[Cohen and Levesque, 1990] P.R. Cohen and H.J. Levesque. Intention is choice with commitment. *Artificial Intelligence*, 42:213–261, 1990.

[CACM, 1994] *Communications of the ACM*, vol. 37, nr. 7. Special Issue on Intelligent Agents, 1994.

[Dershowitz and Jouannaud, 1990] N. Dershowitz and J.-P. Jouannaud. Rewrite systems. In J. van Leeuwen, editor, *Handbook of Theoretical Computer Science*, volume B, pages 243–320. Elsevier, 1990.

[Fiadeiro and Schobbens, 1996] J.L. Fiadeiro and P.-Y. Schobbens, editors. *Proceedings of the 2nd Workshop of the ModelAge Project*, 1996.

[Foner, 1993] L.N. Foner. What's an agent, anyway? A sociological case study. Technical report, MIT Media Laboratory, 1993.

[Goldblatt, 1982a] R. Goldblatt. *Axiomatising the Logic of Computer Programming*, volume 130 of *LNCS*. Springer-Verlag, 1982.

[Goldblatt, 1982b] R. Goldblatt. The semantics of Hoare's iteration rule. *Studia Logica*, 41:141–158, 1982.

[Goldblatt, 1992] R. Goldblatt. *Logics of Time and Computation*, volume 7 of *CSLI Lecture Notes*. CSLI, Stanford, 1992. Second edition.

[Halpern and Moses, 1992] J.Y. Halpern and Y. Moses. A guide to completeness and complexity for modal logics of knowledge and belief. *Artificial Intelligence*, 54:319–379, 1992.

[Halpern and Reif, 1983] J. Halpern and J. Reif. The propositional dynamic logic of deterministic, well-structured programs. *Theoretical Computer Science*, 27:127–165, 1983.

[Harel, 1984] D. Harel. Dynamic logic. In D.M. Gabbay and F. Guenthner, editors, *Handbook of Philosophical Logic*, volume 2, chapter 10, pages 497–604. D. Reidel, Dordrecht, 1984.

[Hilbert, 1931] D. Hilbert. Die Grundlegung der elementaren Zahlenlehre. *Mathematische Annalen*, 104:485–494, 1931.

[Hintikka, 1962] J. Hintikka. *Knowledge and Belief.* Cornell University Press, Ithaca, NY, 1962.

[Hoare, 1969] C.A.R. Hoare. An axiomatic basis for computer programming. *Communications of the ACM*, 12:576–580, 1969.

[van der Hoek *et al.*, 1998] W. van der Hoek, B. van Linder, and J.-J. Ch. Meyer. An integrated modal approach to rational agents. In M. Wooldridge and A. Rao, editors, *Foundations of Rational Agency*, volume 14 of *Applied Logic* series, pp. 133–168. Kluwer, Dordrecht, 1998.

[Hughes and Cresswell, 1968] G.E. Hughes and M.J. Cresswell. *An Introduction to Modal Logic.* Routledge, London, 1968.

[Hughes and Cresswell, 1984] G.E. Hughes and M.J. Cresswell. *A Companion to Modal Logic.* Methuen & Co. Ltd., London, 1984.

[Kenny, 1975] A. Kenny. *Will, Freedom and Power.* Basil Blackwell, Oxford, 1975.

[Klop, 1992] J.W. Klop. Term rewriting systems. In S. Abramsky, D.M. Gabbay, and T.S.E. Maibaum, editors, *Handbook of Logic in Computer Science*, volume 2, pages 1–116. Oxford University Press, New York, 1992.

[Kozen and Tiuryn, 1990] D. Kozen and J. Tiuryn. Logics of programs. In J. van Leeuwen, editor, *Handbook of Theoretical Computer Science*, volume B, pages 789–840. Elsevier, 1990.

[Kripke, 1963] S. Kripke. Semantic analysis of modal logic. *Zeitschrift für Mathematische Logik und Grundslagen der Mathematik*, 9:67–96, 1963.

[Kröger, 1980] F. Kröger. Infinite proof rules for loops. *Acta Informatica*, 14:371–389, 1980.

[Lespérance *et al.*, 1996] Y. Lespérance, H. Levesque, F. Lin, D. Marcu, R. Reiter, and R. Scherl. Foundations of a logical approach to agent programming. In M. Wooldridge, J.P. Müller, and M. Tambe, editors, *Intelligent Agents Volume II – Agent Theories, Architectures, and Languages*, volume 1037 of *Lecture Notes in Computer Science (subseries LNAI)*, pages 331–347. Springer-Verlag, 1996.

[Lesser, 1995] V. Lesser, editor. *Proceedings of the First International Conference on Multi-Agent Systems (ICMAS'95)*. MIT Press, 1995.

[van Linder *et al.*, 1994a] B. van Linder, W. van der Hoek, and J.-J. Ch. Meyer. Communicating rational agents. In B. Nebel and L. Dreschler-Fischer, editors, *KI-94: Advances in Artificial Intelligence*, volume 861 of *Lecture Notes in Computer Science (subseries LNAI)*, pages 202–213. Springer-Verlag, 1994.

[van Linder *et al.*, 1994b] B. van Linder, W. van der Hoek, and J.-J. Ch. Meyer. Tests as epistemic updates. In A.G. Cohn, editor, *Proceedings of the 11th European Conference on Artificial Intelligence (ECAI'94)*, pages 331–335. John Wiley & Sons, 1994.

[van Linder *et al.*, 1995] B. van Linder, W. van der Hoek, and J.-J. Ch. Meyer. Actions that make you change your mind. In A. Laux and H. Wansing, editors, *Knowledge and Belief in Philosophy and Artificial Intelligence*, pages 103–146. Akademie Verlag, 1995.

[van Linder *et al.*, 1997] B. van Linder, W. van der Hoek, and J.-J. Ch. Meyer. The dynamics of default reasoning. *Data and Knowledge Engineering*, 21(3):317–346, 1997.

[Maes, 1994] P. Maes. Agents that reduce work and information overload. *Communications of the ACM*, 37(7):30–40, July 1994.

[Maes, 1995] P. Maes. Intelligent software. *Scientific American*, 273(3):66–68, September 1995. Special Issue on Key Technologies for the 21st Century.

[Meyer and van der Hoek, 1995] J.-J. Ch. Meyer and W. van der Hoek. *Epistemic Logic for AI and Computer Science*. Cambridge University Press, 1995.

[Meyer *et al.*, 1999] J.-J. Ch. Meyer, W. van der Linder and B. van Hoek. A logical approach to the dynamics of commitments. *Artificial Intelligence*, 113:1–40, 1999.

[Moore, 1985] R.C. Moore. A formal theory of knowledge and action. In J.R. Hobbs and R.C. Moore, editors, *Formal Theories of the Commonsense World*, pages 319–358. Ablex, Norwood, NJ, 1985.

[Pörn, 1977] I. Pörn. *Action Theory and Social Science*. Reidel, Dordrecht, 1977.

[Rao and Georgeff, 1991a] A.S. Rao and M.P. Georgeff. Asymmetry thesis and side-effect problems in linear time and branching time intention logics. In J. Mylopoulos and R. Reiter, editors, *Proceedings of the Twelfth International Joint Conference on Artificial Intelligence (IJCAI'91)*, pages 498–504. Morgan Kaufmann, 1991.

[Rao and Georgeff, 1991b] A.S. Rao and M.P. Georgeff. Modeling rational agents within a BDI-architecture. In J. Allen, R. Fikes, and E. Sandewall, editors, *Proceedings of the Second International Conference on Principles of Knowledge Representation and Reasoning (KR'91)*, pages 473–484. Morgan Kaufmann, 1991.

[Rao and Georgeff, 1993] A.S. Rao and M.P. Georgeff. A model-theoretic approach to the verification of situated reasoning systems. In R. Bajcsy, editor, *Proceedings of the Thirteenth International Joint Conference on Artificial Intelligence (IJCAI'93)*, pages 318–324. Morgan Kaufmann, 1993.

[Riecken, 1994] D. Riecken. Intelligent agents. *Communications of the ACM*, 37(7):18–21, July 1994.

[Schütte, 1960] K. Schütte. *Beweistheorie*. Springer-Verlag, Berlin-Göttingen-Heidelberg, 1960.

[Segerberg, 1989] K. Segerberg. Bringing it about. *Journal of philosophical logic*, 18:327–347, 1989.

[Selker, 1994] T. Selker. Coach: A teaching agent that learns. *Communications of the ACM*, 37(7):92–99, July 1994.

[Shoham, 1993] Y. Shoham. Agent-oriented programming. *Artificial Intelligence*, 60:51–92, 1993.

[Spruit, 1993] P.A. Spruit. Henkin-style completeness proofs for Propositional Dynamic Logic. Manuscript.

[Wooldridge and Jennings, 1995a] M. Wooldridge and N. R. Jennings. Intelligent agents: Theory and practice. *The Knowledge Engineering Review*, 10(2):115–152, 1995.

[Wooldridge and Jennings, 1995b] M. Wooldridge and N.R. Jennings, editors. *Intelligent Agents – Agent Theories, Architectures, and Languages*, volume 890 of *Lecture Notes in Computer Science (subseries LNAI)*. Springer-Verlag, 1995.

[Wooldridge *et al.*, 1996] M. Wooldridge, J.P. Müller, and M. Tambe, editors. *Intelligent Agents Volume II – Agent Theories, Architectures, and Languages*, volume 1037 of *Lecture Notes in Computer Science (subseries LNAI)*. Springer-Verlag, 1996.

[Wright, 1963] G.H. von Wright. *Norm and Action*. Routledge & Kegan Paul, London, 1963.

[Wright, 1967] G.H. von Wright. The logic of action: A sketch. In N. Rescher, editor, *The Logic of Decision and Action*. University of Pittsburgh Press, 1967.

B. VAN LINDER, W. VAN DER HOEK
AND J.-J. CH. MEYER

SEEING IS BELIEVING AND SO ARE HEARING AND JUMPING

1 INTRODUCTION

The formalisation of rational agents is a topic of continuing interest in Artificial Intelligence. Research on this subject has held the limelight ever since the pioneering work of Moore [1980; 1984] in which knowledge and actions are considered. Over the years important contributions have been made on both *informational* aspects like knowledge and belief [Halpern and Moses, 1992; Meyer and van der Hoek, 1995], and *motivational*[1] aspects like commitments and obligations [Cohen, 1990]. Recent developments include the work on agent-oriented programming [Shoham, 1993; Thomas, 1993], the Belief-Desire-Intention architecture [Rao and Georgeff, 1991; Rao and Georgeff, 1991a; Rao and Georgeff, 1993], logics for the specification and verification of multi-agent systems [Wooldridge, 1994; Wooldridge and Fisher, 1992], logics for agents with bounded rationality [Huang, 1994; Huang *et al.*, 1992], and cognitive robotics [Lesperance *et al.*, 1994; Levesque, 1994].

In our research (van der Hoek *et al.* [1998; 2000]; van der Linder *et al.* [1994a; 1994; 2001; 1995; 1999]) we defined a *theorist* logic for rational agents, i.e., a logic that is used to *specify*, and to *reason about*, (various aspects of) the behaviour of rational agents. We consider both informational and motivational attitudes, and furthermore pay attention to various aspects of actions. In the basic architecture the *knowledge*, *belief* and *abilities* of agents, as well as the *opportunities* for and the *results* of their actions are formalised. In this framework it can, for instance, be modelled that an agent knows that some action is *correct* to achieve some state of affairs since it knows that performing the action will lead to this state, and that it knows that an action is *feasible* in the sense that the agent knows of its ability to perform the action. In subsequent research we extended our framework with nondeterministic actions [van der Hoek *et al.*, 2000], epistemic tests [van Linder *et al.*, 1994a], communicative actions [van Linder *et al.*, 1994], actions that model default reasoning [van Linder *et al.*, 2001], actions that model belief revision [van Linder *et al.*, 1995], and a formalisation of motivational attitudes [Meyer *et al.*, 1999]. Dunin-Keplicz & Radzikowska [1995] use our framework to reason about typical, as opposed to certain, effects of actions.

[1]The notions *informational* and *motivational* are both due to Shoham & Cousins [1994].

J.J.Ch. Meyer and J. Treur (eds.),
Handbook of Defeasible Reasoning and Uncertainty Management Systems, Vol. 7, 309–339.

The main contribution of this paper is a deeper investigation into *informative* actions, which correspond to the various ways in which agents can acquire information. We propose a formalisation of three different informative actions, viz. ***observations*** (these are basically the aforementioned epistemic tests), actions modelling *communication*, and actions that model the jumping to conclusions which is typical for *default reasoning.* Moreover, we show how these informative actions, constituting the basic informational attitudes of agents, can be modelled *into one coherent framework.* This is not just to complicate matters, but it is a crucial prerequisite if one wants to use a formalism like this to specify the behaviour of real, artificially constructed (multi-) intelligent agent systems. In the philosophical literature one is mostly occupied with the formalisation of one aspect of agency, but having a collection of incomparable or even incompatible models for various aspects of agency will not do when one wants to specify systems such as the above mentioned ones. Moreover, these aspects are not independent of each other: they interact, and this interaction can only be modelled adequately in an integrated framework. In this paper we are in particular interested in the various ways in which the different informative actions mentioned may interact, and how possible conflicts that may result from this interaction, can be solved. In order to resolve these possible information conflicts, we propose a classification of the information that an agent possesses according to credibility. Based on this classification, we formalise what it means for agents to have seen or heard something, or to believe something by default. Using the various informational attitudes that we thus introduced, we define the aforementioned informative actions. In the definitions of these actions an important part is played by a general belief revision action which satisfies the AGM postulates. The ability of agents to perform the three informative actions deals both with the limited capacities of agents to gather information and the preference of one (credible) source of information over another (less credible) one. The various definitions result in a framework in which agents can acquire information from various sources, solve conflicts between various information items, and attach degrees of credibility to their beliefs dependent on the way these beliefs were acquired. As such, the framework presented in this paper could be seen as an attempt to formalise the behaviour of ***intelligent information agents*** [Arens *et al.*, 1993; Levy *et al.*, 1994] which will help us sift our way through the information age.

The rest of the paper is organised as follows. In Section 2 we (re)introduce some of our ideas on knowledge, belief, abilities, opportunities, and results; furthermore the formal definitions of our framework are given. Those already familiar with our basic framework can safely skip this section. In Section 3 we present a classification by credibility of the information of an agent. In Section 4 we combine three informative actions into the basic framework. In 4.1 we formalise observations, in 4.2 communicative actions are formalised, and in 4.3 we formalise default jumps. In Section 5 we define the ability of agents with respect to the informative actions that we consider. In Section 6 we round off.

2 KNOWLEDGE, BELIEF, ABILITIES, OPPORTUNITIES, AND RESULTS

Here we restrict ourselves to informational attitudes and action aspects. At the informational level we consider both ***knowledge and belief.*** Formalising these notions has been a subject of continuing research both in analytical philosophy and in AI [Halpern and Moses, 1992; Hintikka, 1962]. In representing knowledge and belief we follow, both from a syntactical and a semantic point of view, the approach common in epistemic and doxastic logic: the formula $\mathbf{K}_i\varphi$ denotes the fact that agent i knows φ, and $\mathbf{B}_i\varphi$ that agent i believes φ. For the semantics we use Kripke-style possible worlds models.

At the action level we consider ***results***, ***abilities*** and ***opportunities.*** In defining the result of an action, we follow ideas of Von Wright [1963], in which the state of affairs brought about by execution of the action is defined to be its result. An important aspect of any investigation of action is the relation that exists between ability and opportunity. In order to successfully complete an action, both the opportunity and the ability to perform the action are necessary. Although these notions are interconnected, they are surely not identical [Kenny, 1975]: the ability of agents comprises mental and physical powers, moral capacities, and human and physical possibility, whereas the opportunity to perform actions is best described by the notion of circumstantial possibility. A nice example that illustrates the difference between ability and opportunity is that of a lion in a zoo [Elgesem, 1993]: although the lion will (ideally) never have the opportunity to eat a zebra, it certainly has the ability to do so. We propose that in order to make our formalisation of rational agents, like for instance robots, as accurate and realistic as possible, abilities and opportunities need also be distinguished in AI environments. The abilities of the agents are formalised via the $\mathbf{A}_i$ operator; the formula $\mathbf{A}_i\alpha$ denotes the fact that agent i has the ability to do α. When using the definitions of opportunities and results as given above, the framework of (propositional) dynamic logic provides an excellent means to formalise these notions. Using events $\mathrm{do}_i(\alpha)$ to refer to the performance of the action α by the agent i, we consider the formulae $\langle \mathrm{do}_i(\alpha)\rangle\varphi$ and $[\mathrm{do}_i(\alpha)]\varphi$. In our deterministic framework, $\langle \mathrm{do}_i(\alpha)\rangle\varphi$ is the stronger of these formulae; it represents the fact that the agent i has the opportunity to do α and that doing α results in φ being true. The formula $[\mathrm{do}_i(\alpha)]\varphi$ is noncommittal about the opportunity of the agent to do α but states that should the opportunity arise, only states of affairs satisfying φ would result. Besides the possibility to formalise both opportunities and results when using dynamic logic, another advantage lies in the compatibility of epistemic, doxastic and dynamic logic from a semantic point of view: a Kripke style semantics can be used to provide meaning to epistemic, doxastic and dynamic notions.

DEFINITION 1. Let a finite set $\mathcal{A} = \{1, \ldots, n\}$ of agents, and some denumerable sets Π of propositional symbols and At of atomic actions be given. The language $\mathcal{L}$ and the class of actions Ac are defined by mutual induction as follows.

- $\mathcal{L}$ is the smallest superset of Π such that
  - if $\varphi, \psi \in \mathcal{L}$ then $\neg\varphi, \varphi \vee \psi \in \mathcal{L}$
  - if $i \in \mathcal{A}$, $\alpha \in Ac$ and $\varphi \in \mathcal{L}$ then $\mathbf{K}_i\varphi, \mathbf{B}_i\varphi, \langle \mathrm{do}_i(\alpha)\rangle\varphi, \mathbf{A}_i\alpha \in \mathcal{L}$
- Ac is the smallest superset of At such that
  - if $\varphi \in \mathcal{L}$ then $\mathtt{confirm}\ \varphi \in Ac$
  - if $\alpha_1 \in Ac$ and $\alpha_2 \in Ac$ then $\alpha_1; \alpha_2 \in Ac$
  - if $\varphi \in \mathcal{L}$ and $\alpha_1, \alpha_2 \in Ac$ then $\mathtt{if}\ \varphi\ \mathtt{then}\ \alpha_1\ \mathtt{else}\ \alpha_2\ \mathtt{fi} \in Ac$
  - if $\varphi \in \mathcal{L}$ and $\alpha_1 \in Ac$ then $\mathtt{while}\ \varphi\ \mathtt{do}\ \alpha\ \mathtt{od} \in Ac$

The purely propositional fragment of $\mathcal{L}$ is denoted by $\mathcal{L}_0$.

The constructs $\wedge$, $\rightarrow, \leftrightarrow, \mathbf{tt}, \mathbf{ff}, \mathbf{M}_i$ and $[\mathrm{do}_i(\alpha)]\varphi$ are defined in the usual way. Other constructs are introduced by definitional abbreviation: `skip` is `confirm tt`, $\mathtt{if}\ \varphi\ \mathtt{then}\ \alpha_1\ \mathtt{fi}$ is $\mathtt{if}\ \varphi\ \mathtt{then}\ \alpha_1\ \mathtt{else\ skip\ fi}$, α^0 is `skip`, and α^{n+1} is $\alpha; \alpha^n$.

REMARK 2. The `confirm` action behaves essentially like the test actions in dynamic logic [Goldblatt, 1992; Harel, 1984]. As such this action differs substantially from tests as they are looked upon by humans: these genuine tests are usually assumed to contribute to the information of the agent that performs the test [van Linder *et al.*, 1994a], whereas by performing `confirm` φ it is just *confirmed* (verified, checked) that φ holds. The meaning of the other actions in Ac is respectively: the atomic action, sequential composition, conditional composition, and repetitive composition; `skip` denotes the void action.

In the following definitions it is assumed that some set $\mathbf{bool} = \{\mathbf{0}, \mathbf{1}\}$ of truth values is given.

DEFINITION 3. The class M of Kripke models consists of all tuples $\mathcal{M} = \langle \mathcal{S}, \pi, \mathrm{R}, \mathrm{B}, \mathtt{r}, \mathtt{c}\rangle$ such that

- $\mathcal{S}$ is a set of possible worlds, or states.

- $\pi : \Pi \times \mathcal{S} \to \mathbf{bool}$ is a total function that assigns a truth value to propositional symbols in possible worlds.
- $\mathrm{R} : \mathcal{A} \to \wp(\mathcal{S} \times \mathcal{S})$ is a function that yields the epistemic accessibility relations for a given agent. This function is such that $\mathrm{R}(i)$ is an equivalence relation for all i. For reasons of practical convenience we define $[s]_{\mathrm{R}(i)}$ to be $\{s' \in \mathcal{S} \mid (s, s') \in \mathrm{R}(i)\}$.
- $\mathrm{B} : \mathcal{A} \times \mathcal{S} \to \wp(\mathcal{S})$ is a function that yields the set of doxastic alternatives for a given agent in a given state. To model the kind of belief that we like to model it is demanded that for all agents i and for all possible worlds s and s' it holds that:
  - $\mathrm{B}(i, s) = \mathrm{B}(i, s')$ if $s' \in [s]_{\mathrm{R}(i)}$
  - $\mathrm{B}(i, s) \subseteq [s]_{\mathrm{R}(i)}$
  - $\mathrm{B}(i, s) \neq \emptyset$
- $\mathtt{r} : \mathcal{A} \times At \to \mathcal{S} \to \wp(\mathcal{S})$ is such that $\mathtt{r}(i, a)(s)$ yields the (possibly empty) state transition in s caused by the event $\mathrm{do}_i(a)$. This function is such that for all atomic actions a it holds that $|\mathtt{r}(i, a)(s)| \leq 1$ for all i and s, i.e., these events are ***deterministic***.
- $\mathtt{c} : \mathcal{A} \times At \to \mathcal{S} \to \mathbf{bool}$ is the capability function such that $\mathtt{c}(i, a)(s)$ indicates whether the agent i is capable of performing the action a in s.

DEFINITION 4. Let $\mathcal{M} = \langle \mathcal{S}, \pi, \mathrm{R}, \mathrm{B}, \mathtt{r}, \mathtt{c} \rangle$ be some Kripke model from M. For propositional symbols, negated formulae, and disjunctions, $\mathcal{M}, s \models \varphi$ is inductively defined as usual. For the other clauses $\mathcal{M}, s \models \varphi$ is defined as follows:

$$\begin{array}{ll}
(\mathcal{M}, s) \models \mathbf{K}_i\varphi & \Leftrightarrow \forall s' \in \mathcal{S}[(s, s') \in \mathrm{R}(i) \Rightarrow \mathcal{M}, s' \models \varphi] \\
(\mathcal{M}, s) \models \mathbf{B}_i\varphi & \Leftrightarrow \forall s' \in \mathcal{S}[s' \in \mathrm{B}(i, s) \Rightarrow \mathcal{M}, s' \models \varphi] \\
\mathcal{M}, s \models \langle \mathrm{do}_i(\alpha) \rangle \varphi & \Leftrightarrow \exists \mathcal{M}', s'[\mathcal{M}', s' \in \mathtt{r}(i, \alpha)(\mathcal{M}, s) \\
 & \quad \& \mathcal{M}', s' \models \varphi] \\
\mathcal{M}, s \models \mathbf{A}_i\alpha & \Leftrightarrow \mathtt{c}(i, \alpha)(\mathcal{M}, s) = \mathbf{1}
\end{array}$$

where $\mathtt{r}$ and $\mathtt{c}$ are extended as follows:

$$\begin{array}{ll}
\mathtt{r}(i, a)(\mathcal{M}, s) & = \{(\mathcal{M}, \mathtt{r}(i, a)(s))\} \\
\mathtt{r}(i, \mathtt{confirm}\ \varphi)(\mathcal{M}, s) & = \{(\mathcal{M}, s)\} \text{ if } (\mathcal{M}, s) \models \varphi \\
 & \quad \text{and } \emptyset \text{ otherwise} \\
\mathtt{r}(i, \alpha_1; \alpha_2)(\mathcal{M}, s) & = \mathtt{r}(i, \alpha_2)(\mathtt{r}(i, \alpha_1)(\mathcal{M}, s)) \\
\mathtt{r}(i, \mathtt{if}\ \varphi\ \mathtt{then}\ \alpha_1 & = \mathtt{r}(i, \alpha_1)(\mathcal{M}, s) \text{ if } (\mathcal{M}, s) \models \varphi \text{ and} \\
\quad \mathtt{else}\ \alpha_2\ \mathtt{fi})(\mathcal{M}, s) & \quad \mathtt{r}(i, \alpha_2)(\mathcal{M}, s) \text{ otherwise} \\
\mathtt{r}(i, \mathtt{while}\ \varphi & = \{(\mathcal{M}', s') \mid \exists k \in \mathbb{N} \exists \mathcal{M}_0, s_0 \ldots \exists \mathcal{M}_k, s_k \\
\quad \mathtt{do}\ \alpha\ \mathtt{od})((\mathcal{M}, s)) & \quad [\mathcal{M}_0, s_0 = (\mathcal{M}, s) \& \mathcal{M}_k, s_k = \mathcal{M}', s' \\
 & \quad \& \forall j < k[(\mathcal{M}_{j+1}, s_{j+1}) \in \\
 & \quad \mathtt{r}(i, \mathtt{confirm}\ \varphi; \alpha_1)(\mathcal{M}_j, s_j)] \\
 & \quad \& \mathcal{M}', s' \models \neg\varphi]\} \\
\text{where } \mathtt{r}(i, \alpha)(\emptyset) & = \emptyset
\end{array}$$

and

$$
\begin{array}{ll}
\mathtt{c}(i,a)((\mathcal{M},s)) & = \mathtt{c}(i,a)(s) \\
\mathtt{c}(i,\mathtt{confirm}\ \varphi)((\mathcal{M},s)) & = \mathbf{1} \text{ if } (\mathcal{M},s) \models \varphi \\
 & \quad \text{and } \mathbf{0} \text{ otherwise} \\
\mathtt{c}(i,\alpha_1;\alpha_2)((\mathcal{M},s)) & = \mathtt{c}(i,\alpha_1)((\mathcal{M},s)) = \mathbf{1} \\
 & \quad \&\ \mathtt{c}(i,\alpha_2)(\mathtt{r}(i,\alpha_1)((\mathcal{M},s))) = \mathbf{1} \\
\mathtt{c}(i,\mathtt{if}\ \varphi\ \mathtt{then}\ \alpha_1 & = \mathtt{c}(i,\mathtt{confirm}\ \varphi;\alpha_1)((\mathcal{M},s)) \text{ or} \\
\quad \mathtt{else}\ \alpha_2\ \mathtt{fi})(\mathcal{M},s) & \quad \mathtt{c}(i,\mathtt{confirm}\ \neg\varphi;\alpha_2)((\mathcal{M},s)) \\
\mathtt{c}(i,\mathtt{while}\ \varphi & = \mathbf{1} \text{ if } \exists k \in \mathbb{N}[\mathtt{c}(i,(\mathtt{confirm}\ \varphi;\alpha_1)^k; \\
\quad \mathtt{do}\ \alpha\ \mathtt{od}((\mathcal{M},s)) & \quad \mathtt{confirm}\ \neg\varphi)((\mathcal{M},s)) = \mathbf{1}] \\
 & \quad \text{and } \mathbf{0} \text{ otherwise} \\
\text{where } \mathtt{c}(i,\alpha)(\emptyset) & = \mathbf{1}
\end{array}
$$

Validity on and satisfiability in the class M of models are defined as usual.

REMARK 5. Definition 4 gives rise to the following remarks.

- When defining the R and B functions as in Definition 3, we end up with a notion of knowledge that satisfies an S5 axiomatisation, and a notion of belief that satisfies a KD45 axiomatisation (both according to the Chellas qualification [Chellas, 1980]). This means in particular that knowledge is veridical, that belief is non-absurd, and that agents have positive and negative introspection on both their knowledge and their belief. Knowledge and belief are related to each other as in the system of Kraus & Lehmann [1988].
- The notion of actions considered in Definition 4 generalises that of state-transformers as it is typical for dynamic logic [Harel, 1984], and allows for actions that transform pairs (Model, State). The reason for this generalisation lies in the fact that we account for non-standard actions like 'to observe' [van Linder *et al.*, 1994a], 'to commit' [Meyer *et al.*, 1999] and 'to inform' [van Linder *et al.*, 1994] in our framework, and these non-standard actions transform models rather than states.
- With regard to the abilities of agents, we justify Definition 4 as follows. The definition of $\mathtt{c}(i,\mathtt{confirm}\ \varphi)$ expresses that an agent is able to get confirmation for a formula φ if and only if φ holds. Note that the definitions of $\mathtt{r}(i,\mathtt{confirm}\ \varphi)$ and $\mathtt{c}(i,\mathtt{confirm}\ \varphi)$ imply that in circumstances such that φ holds, agents have both the opportunity and the ability to confirm φ. An agent is capable of performing a sequential composition $\alpha_1;\alpha_2$ if and only if it is capable of performing α_1 (now), and it is capable of executing α_2 after it has performed α_1. An agent is capable of performing a conditional composition, if either it is able to get confirmation for the condition and thereafter perform the then-part, or it is able to confirm the negation of the condition and perform the else-part afterwards. An agent is capable of performing a repetitive composition $\mathtt{while}\ \varphi\ \mathtt{do}\ \alpha\ \mathtt{od}$ if and only if it is able to perform the action $(\mathtt{confirm}\ \varphi;\alpha_1)^k;\mathtt{confirm}\ \neg\varphi$ for some natural number k.

3 A CLASSIFICATION OF INFORMATION

When modelling real-life situations it will hardly ever be the case that agents have complete information about the world and their place in it. In particular when planning it may be necessary for agents to try and acquire additional information from whatever source possible. Here we consider three possible sources of information for an agent, an *endogenous* one and two *exogenous* sources. The first exogenous source consists of the *observations* that an agent makes about the current world [van Linder *et al.*, 1994a]; the second exogenous source of information for a rational agent is *communication* with other agents [van Linder *et al.*, 1994]. The endogenous source of information that we consider is the possibility to *adopt assumptions by default* [van Linder *et al.*, 2001]. Other authors came up with more or less the same sources of information for rational agents. Castelfranchi [1994; 1994a] also considers three sources, viz. one endogenous source comprising reasoning and introspection (due to their perfect reasoning and introspection capacities this source is already implicitly present in the agents that we formalise), and perception and communication as exogenous sources. Dunin-Keplicz & Treur [1994] consider exactly the same sources of information as we do.

In any situation in which information is acquired from different sources, the possibility of conflicts exists. That is, it is possible that information acquired from one source contradicts information acquired from another source. To solve these information conflicts, it is important to define some strategies that prescribe what to do in the case of conflicts between new, incoming information and old, previously acquired information. Castelfranchi [1994] proposes that strategies which resolve information conflicts should take two main aspects into consideration. The first of these could be captured by the notion of *economical* — or *conservative* — aspects. That is, as long as there is no very good reason to favour new beliefs over old ones, an agent should stick to its old beliefs thereby avoiding the need for (costly) belief revisions. Having said so, the second point that should be taken into consideration is that more credible information should always be favoured over less credible information. So, in cases where the incoming information is strictly more credible than the already present information, a revision in favour of the new information should be performed, whereas in the case of peer conflicts economical aspects prevail, and therefore no revision takes place.

In order to compare various items of information, some degree of credibility should be attached to them. In particular, it is necessary to have the possibility to determine the credibility of each of the beliefs that an agent may have in a certain state. In order to be able to do so we propose to structure the information that an agent possesses into four sets, one inside the other. The innermost of these sets contains the *knowledge* of the agent. Knowledge can be seen as the information that an agent is born or built with. The set directly encompassing the knowledge set contains the *observational* beliefs of the agent. These are the beliefs that an agent has on the ground of its knowledge and the observations it has made. The third set contains the *communicational* beliefs of an agent. These are the combined beliefs that it either knows or acquired through observations and/or communication. The outermost set contains the *default beliefs*. These are the beliefs for which application of a default may have been a necessary condition. The credibility of a belief formula is determined by the smallest belief set that it is a member of, i.e., formulae that are known have the highest credibility whereas formulae that are believed by default have the lowest one. In order to syntactically account for this structuring of beliefs, we tag the belief modality as follows.

DEFINITION 6. The language $\mathcal{L}$ as given in Definition 1 is extended as follows:

- if $\varphi \in \mathcal{L}$ then $\mathbf{B}_i^o\varphi, \mathbf{B}_i^c\varphi, \mathbf{B}_i^d\varphi \in \mathcal{L}$

The intuitive interpretation of $\mathbf{B}_i^o\varphi$ is that φ belongs to the observational beliefs of agent i, $\mathbf{B}_i^c\varphi$ states that φ belongs to the communicational beliefs of i, and $\mathbf{B}_i^d$ refers to the default beliefs of agent i.

CONVENTION 7. Both for reasons of practical convenience and to keep our notation uniform, we sometimes use $\mathbf{B}_i^k$, for *known* beliefs, as a general way of writing $\mathbf{K}_i$.

For the semantics of the various tagged belief operators we propose the use of *belief clusters* [Meyer and van der Hoek, 1992], which are basically sets of worlds, that are situated within each other. That is, the knowledge cluster is the largest one, corresponding to the idea that knowledge is the most credible kind of belief. The observational belief cluster is situated within the knowledge cluster; this cluster contains the communicational belief cluster, which on its turn contains the default belief cluster. Hence the credibility of a belief is inversely proportional with the size of the cluster on which the belief is interpreted. Note furthermore that from the point of view of belief revision [Gärdenfors, 1988; Gärdenfors, 1992], *expansions* of the various belief sets correspond to *shrinkings* of the associated belief clusters and *contractions* of the belief sets correspond to *extensions* of the clusters (cf. [van Linder *et al.*, 1995]).

DEFINITION 8. The Kripke models given in Definition 3 are modified as follows: the function B is replaced by three functions $\mathrm{B}^o : \mathcal{A} \to \wp(\mathcal{S} \times \mathcal{S})$, $\mathrm{B}^c : \mathcal{A} \times \mathcal{S} \to \wp(\mathcal{S})$ and $\mathrm{B}^d : \mathcal{A} \times \mathcal{S} \to \wp(\mathcal{S})$. Defining the set $[s]_{\mathrm{B}^o(i)}$ analogously to $[s]_{\mathrm{R}(i)}$, these functions are such that for all $i \in \mathcal{A}$, $s, s' \in \mathcal{S}$:

- $\mathrm{B}^o(i)$ is an equivalence relation
- $\mathrm{B}^d(i,s) \neq \emptyset$
- $\mathrm{B}^d(i,s) \subseteq \mathrm{B}^c(i,s) \subseteq [s]_{\mathrm{B}^o(i)} \subseteq [s]_{\mathrm{R}(i)}$
- if $s' \in [s]_{\mathrm{R}(i)}$ then $[s']_{\mathrm{B}^o(i)} = [s]_{\mathrm{B}^o(i)}$, $\mathrm{B}^c(i,s') = \mathrm{B}^c(i,s)$ and

$\mathrm{B}^d(i, s') = \mathrm{B}^d(i, s)$

The tagged belief formulae are interpreted as expected:

DEFINITION 9. For a Kripke model $\mathcal{M}$ as given in Definition 8 and a state $s \in \mathcal{M}$, $\models$ is extended as follows:

$$(\mathcal{M}, s) \models \mathbf{B}_i^o\varphi \Leftrightarrow \forall s' \in \mathcal{S}[(s, s') \in \mathrm{B}^o(i) \Rightarrow \mathcal{M}, s' \models \varphi]$$
$$(\mathcal{M}, s) \models \mathbf{B}_i^c\varphi \Leftrightarrow \forall s' \in \mathrm{B}^c(i, s)[\mathcal{M}, s' \models \varphi]$$
$$(\mathcal{M}, s) \models \mathbf{B}_i^d\varphi \Leftrightarrow \forall s' \in \mathrm{B}^d(i, s)[\mathcal{M}, s' \models \varphi]$$

When defining knowledge and the various notions of belief as in Definition 9 we indeed obtain the credibility ordering we are aiming at as an ordinary material implication. According to this ordering, knowing some formula is the strongest informational attitude, and believing it by default is the weakest one.

PROPOSITION 10. *Define the ordering > on informational operators by* $\mathbf{B}_i^k > \mathbf{B}_i^o > \mathbf{B}_i^c > \mathbf{B}_i^d$, *and let* $\succeq$ *be the reflexive, transitive closure of* $>$. *Then for all* $\mathbf{X}, \mathbf{Y} \in \{\mathbf{B}_i^k, \mathbf{B}_i^o, \mathbf{B}_i^c, \mathbf{B}_i^d\}$, *and for all formulae* φ *:*

- $\mathbf{X} \succeq \mathbf{Y} \Rightarrow \models \mathbf{X}\varphi \rightarrow \mathbf{Y}\varphi$

The various notions of belief that we introduced validate different axiomatisations: knowledge and observational belief satisfy an S5-axiomatisation, whereas communicational and default belief validate an KD45-axiomatisation.

PROPOSITION 11. *Let* $\mathbf{X} \in \{\mathbf{B}_i^k, \mathbf{B}_i^o\}$ *and* $\mathbf{Y} \in \{\mathbf{B}_i^k, \mathbf{B}_i^o, \mathbf{B}_i^c, \mathbf{B}_i^d\}$. *For all formulae* φ *we have:*

1. $\models \mathbf{Y}(\varphi \rightarrow \psi) \rightarrow (\mathbf{Y}\varphi \rightarrow \mathbf{Y}\psi)$ K
2. $\models \neg(\mathbf{Y}\varphi \wedge \mathbf{Y}\neg\varphi)$ D
3. $\models \mathbf{X}\varphi \rightarrow \varphi$ T
4. $\models \mathbf{Y}\varphi \rightarrow \mathbf{Y}\mathbf{Y}\varphi$ 4
5. $\models \neg\mathbf{Y}\varphi \rightarrow \mathbf{Y}\neg\mathbf{Y}\varphi$ 5
6. $\models \varphi \Rightarrow \models \mathbf{Y}\varphi$ N

Although our system is a generalisation of the more common systems for knowledge and belief [van der Hoek, 1993; Kraus and Lehmann, 1988], it is also the case in our system that in certain situations multiple informational operators prefixing a formula collapse to one operator.

DEFINITION 12. A formula φ is *i-doxastic sequenced* if there is a formula ψ, and operators $\mathbf{X}_1, \ldots, \mathbf{X}_m \in \{\mathbf{B}_i^k, \mathbf{B}_i^o, \mathbf{B}_i^c, \mathbf{B}_i^d, \neg\mathbf{B}_i^k, \neg\mathbf{B}_i^o, \neg\mathbf{B}_i^c, \neg\mathbf{B}_i^d\}$ and $m > 0$ such that $\varphi = \mathbf{X}_1 \ldots \mathbf{X}_m\psi$.

PROPOSITION 13. *Let* φ *be an i-doxastic sequenced formula, and let* $\mathbf{Y} \in \{\mathbf{B}_i^k, \mathbf{B}_i^o, \mathbf{B}_i^c, \mathbf{B}_i^d\}$. *Then:*

- $\models \mathbf{Y}\varphi \leftrightarrow \varphi$

Given the four modal operators $\mathbf{B}_i^k, \mathbf{B}_i^o, \mathbf{B}_i^c, \mathbf{B}_i^d$ it is possible to model nine different informational attitudes of a given agent with regard to a given formula. If we define for $x \in \{k, o, c, d\}$, $\mathbf{Agn}_i^x\varphi =^{\text{def}} \neg\mathbf{B}_i^x\varphi \wedge \neg\mathbf{B}_i^x\neg\varphi$, representing the fact that agent i is *agnostic* with regard to φ on the level x, the attitudes with respect to a formula φ are the following:

1. $\mathbf{B}_i^k\varphi$ — *'i knows φ'*
2. $\mathbf{B}_i^k\neg\varphi$ — *'i knows $\neg\varphi$'*
3. $\mathbf{Agn}_i^k\varphi \wedge \mathbf{B}_i^o\varphi$ — *'i saw φ'*
4. $\mathbf{Agn}_i^k\varphi \wedge \mathbf{B}_i^o\neg\varphi$ — *'i saw $\neg\varphi$'*
5. $\mathbf{Agn}_i^o\varphi \wedge \mathbf{B}_i^c\varphi$ — *'i heard φ'*
6. $\mathbf{Agn}_i^o\varphi \wedge \mathbf{B}_i^c\neg\varphi$ — *'i heard $\neg\varphi$'*
7. $\mathbf{Agn}_i^c\varphi \wedge \mathbf{B}_i^d\varphi$ — *'i believes φ by default'*
8. $\mathbf{Agn}_i^c\varphi \wedge \mathbf{B}_i^d\neg\varphi$ — *'i believes $\neg\varphi$ by default'*
9. $\mathbf{Agn}_i^d\varphi$ — *'i is completely agnostic wrt φ'*

Given these attitudes and their intuitive interpretation, we introduce the following predicates by definitional abbreviation:

- $\mathbf{Saw}_i\varphi =^{\text{def}} \mathbf{Agn}_i^k\varphi \wedge \mathbf{B}_i^o\varphi (= \mathbf{B}_i^o\varphi \wedge \neg\mathbf{B}_i^k\varphi)$
- $\mathbf{Heard}_i\varphi =^{\text{def}} \mathbf{Agn}_i^o\varphi \wedge \mathbf{B}_i^c\varphi (= \mathbf{B}_i^c\varphi \wedge \neg\mathbf{B}_i^o\varphi)$
- $\mathbf{Jumped}_i\varphi =^{\text{def}} \mathbf{Agn}_i^c\varphi \wedge \mathbf{B}_i^d\varphi (= \mathbf{B}_i^d\varphi \wedge \neg\mathbf{B}_i^c\varphi)$

In fact, the $\mathbf{Heard}_i$ operator does not formalise hearing *per se*, but formalises *believing on the basis of hearing.* One cannot prevent an agent from being told inconsistencies. However, the agent will deal with these inconsistencies in such a way that at any given time its beliefs grounded in the things that it heard are consistent.

The following proposition formalises the behaviour of the derived belief operators with respect to the axioms of modal logic as given by Chellas [1980].

PROPOSITION 14. *Let φ and ψ be formulae. Let $\mathbf{X}$ be in the set $Bel = \{\mathbf{Saw}_i, \mathbf{Heard}_i, \mathbf{Jumped}_i \mid i \in \mathcal{A}\}$, and let $\mathbf{Y} \in Bel \cup \{\mathbf{K}_i\}$. Define the ordering $\succeq'$ to be the reflexive and transitive closure of $>'$ with $\mathbf{K}_i >' \mathbf{Saw}_i >' \mathbf{Heard}_i >' \mathbf{Jumped}_i$.*[2] *Then we have:*

1. $\models \mathbf{X}\varphi \wedge \mathbf{X}(\varphi \to \psi) \to \mathbf{X}\psi$ — K
2. $\models \neg(\mathbf{X}\varphi \wedge \mathbf{X}\neg\varphi)$ — D
3. $\models \mathbf{Saw}_i\varphi \to \varphi$ — T
4. $\not\models \mathbf{X}\varphi \to \mathbf{X}\mathbf{X}\varphi$ — 4
5. $\models \mathbf{X}\varphi \to \mathbf{K}_i\mathbf{X}\varphi$ — 4′
6. $\not\models \neg\mathbf{X}\varphi \to \mathbf{X}\neg\mathbf{X}\varphi$ — 5
7. $\models \neg\mathbf{X}\varphi \to \mathbf{K}_i\neg\mathbf{X}\varphi$ — 5′
8. $\models (\mathbf{X}\varphi \wedge \mathbf{X}\psi) \to \mathbf{X}(\varphi \wedge \psi)$ — C
9. $\not\models \mathbf{X}(\varphi \wedge \psi) \to (\mathbf{X}\varphi \wedge \mathbf{X}\psi)$ — M
10. $\models \mathbf{X}(\varphi \wedge \psi) \to (\bigvee_{\mathbf{Y} \succeq' \mathbf{X}} \mathbf{Y}\varphi \wedge \bigvee_{\mathbf{Y} \succeq' \mathbf{X}} \mathbf{Y}\psi) \wedge (\mathbf{X}\varphi \vee \mathbf{X}\psi)$ — M′
11. $\models \varphi \not\Rightarrow \models \mathbf{X}\varphi$ *and even* $\models \varphi \Rightarrow \models \neg\mathbf{X}\varphi$ — N

[2] Note that $\succeq'$ and $>'$ are the obvious modifications of $\succeq$ and $>$ to deal with the derived informational operators.

12. $\models \varphi \to \psi \not\Rightarrow \models \mathbf{X}\varphi \to \mathbf{X}\psi$ RM
13. $\models \varphi \to \psi \Rightarrow \models \mathbf{X}\varphi \to \bigvee_{\mathbf{Y} \succeq' \mathbf{X}} \mathbf{Y}\psi$ RM′
14. $\models \varphi \leftrightarrow \psi \Rightarrow \models \mathbf{X}\varphi \leftrightarrow \mathbf{X}\psi$ RE

REMARK 15. That the K-axiom holds for all three derived belief operators might look somewhat surprising at first sight. However, validity of this axiom has everything to do with the ***rationality*** of agents. Take the example of a rational agent grounding its beliefs of both φ and $\varphi \to \psi$ in its observations. Being the rational creature that it is, it is obvious that it in any case observationally believes ψ. Again given the rationality of the agent it cannot attach a higher credibility to ψ than to $\varphi \to \psi$: for the latter is implied by the former, and therefore the credibility attached to it is at least the credibility attached to ψ. Hence both ψ and $\varphi \to \psi$ are believed with the same strength. Beliefs are consistent (the D-axiom), and beliefs that are grounded in observations are furthermore veridical (the T-axiom). The first property is highly desirable for rational agents, the latter property is the essential characteristic of beliefs acquired through observations (cf. [Barwise, 1989], p. 12). None of the derived belief operators validates positive or negative introspection. However, all operators satisfy introspective properties with respect to knowledge: if an agent believes something on the ground of its observations (communicating, default jumps) then it ***knows*** that it does so. This is as one would intuitively expect: if an agent sees something, then it does not *see* that it sees it, but a rational agent ***knows*** that it sees it. All derived belief operators distribute over conjunctions (the R-axiom), and none of these satisfy necessitation (the N-rule). The reason for this latter fact is that valid propositions are already known, and are therefore never grounded in observations, communication or default jumps. The derived belief operators are in general not monotonic (the RM-rule), but satisfy some kind of 'upward monotonicity' (the RM′-rule), corresponding to the idea that whenever some proposition is believed to a certain degree, weaker propositions might already be believed to a higher degree (the degenerate case being one where the weaker propositions are validities that are known to the agent). Finally, all derived belief operators are closed under equivalence of (believed) propositions. Although Barwise argues against this property (cf. [Barwise, 1989], p. 18), it seems harmless for artificial agents with perfect reasoning capacities.

4 COMBINING INFORMATIVE ACTIONS

In this section we successively formalise three kinds of informative actions, which correspond to the various ways in which agents can acquire information. The first of these actions is the one that models the observations that agents may make. Through performing an **observe** φ action an agent observes whether φ holds in the state in which it is currently residing. The second action corresponds to information acquisition through communication. By performing an `inform` (φ, i) action, some agent j may inform agent i of the fact φ. The third and last action corresponds to the endogenous source of information acquisition, viz. the reasoning by default. By performing `try_jump` φ actions agents may try and jump to the default conclusion φ. In defining the semantics of these 'high-level' informative actions, we use a special 'low-level' action that causes the beliefs of an agent to be *revised* in the appropriate manner. Intuitively, a revision is the change of belief through which some formula is — whenever possible — consistently included in the belief set of an agent. If the formula that is to be included is consistent with the already present beliefs of the agent it is simply added, otherwise — in order to maintain consistency of the agent's belief set — some of the old beliefs of the agent are deleted before including the new formula. The revision implemented by the actions that we present validates the (well known and by now standard) AGM postulates for belief revision [Alchourrón *et al.*, 1985; Gärdenfors, 1988; Gärdenfors, 1992; van Linder *et al.*, 1995]. To deal adequately with the classification of the agents information, the low-level revision action is defined to stretch over the various information classes. To emphasise the distinction between the high and the low-level informative actions, we introduce in addition to the set of rational agents a constant e in our language. The intuitive reading of this constant could either be *external environment/observer*, or *supervisor*. The idea is that this external environment performs the low-level actions that are not available to the agents, but that appear in the implementation of actions that are available.

DEFINITION 16. The class of high-level informative actions Ac_i^h and the class of low-level informative actions Ac_i^l are defined by:

- $Ac_I^h = \{\texttt{observe}\ \varphi, \texttt{inform}\ (\varphi, i), \texttt{try\_jump}\ \varphi \mid \varphi \in \mathcal{L}_0, i \in \mathcal{A}\}$
- Ac_I^l is the smallest superset of $\{\texttt{revise}^x(\varphi, i) \mid x \in \{1, 2, 3, d, c, o\}, i \in \mathcal{A}, \varphi \in \mathcal{L}_0\}$ that is closed under sequential composition.

The class of actions Ac is extended with the class Ac_I^h.

Informally speaking, the $\texttt{revise}^x, x = d, c, o$ actions that we define below take care of a belief revision in the sense of the AGM postulates but *stretched over different levels.* For example, consider the case where a revision with some formula φ of the observational beliefs of an agent should take place. Given the classification of belief as presented in Section 3 this implies that not only the agent's observational beliefs, but also its communicational and default beliefs need to be updated, i.e., the revision should affect other kinds of the agent's beliefs. In the following definition, the $\texttt{revise}^k, k = 1, 2, 3$ actions take care of a revision per level, whereas the $\texttt{revise}^x, x = d, c, o$ actions force a belief revision on all appropriate levels.

DEFINITION 17. Let $\mathcal{M} = \langle \mathcal{S}, \pi, \mathrm{R}, \mathrm{B}^o, \mathrm{B}^c, \mathrm{B}^d, \mathtt{r}, \mathtt{c} \rangle$ be some Kripke model, let $s \in \mathcal{S}$, let $\mathcal{S}' \subseteq \mathcal{S}$, and let $\varphi \in \mathcal{L}_0$.

- $[s]^{\varphi+}_{\mathrm{B}^o(i)} =^{\mathrm{def}} [s]_{\mathrm{B}^o(i)} \cap \{s' \in \mathcal{S} \mid \mathcal{M}, s \models \varphi \Leftrightarrow \mathcal{M}, s' \models \varphi\}$
- $[s]^{\varphi-}_{\mathrm{B}^o(i)} =^{\mathrm{def}} [s]_{\mathrm{B}^o(i)} \setminus [s]^{\varphi+}_{\mathrm{B}^o(i)}$
- $\mathrm{Cl}_{\mathrm{eq}}(\mathcal{S}') =^{\mathrm{def}} \mathcal{S}' \times \mathcal{S}'$

DEFINITION 18. For $\mathcal{M} = \langle \mathcal{S}, \pi, \mathrm{R}, \mathrm{B}^o, \mathrm{B}^c, \mathrm{B}^d, \mathrm{r}, \mathrm{c}\rangle$ a Kripke model with $s \in \mathcal{S}$, i some agent, and φ some propositional formula we define:

- The function r is retyped to $(\mathcal{A} \times Ac) \cup (\{e\} \times Ac_I^i) \to (\mathsf{M} \times \mathcal{S}) \to \wp(\mathsf{M} \times \mathcal{S})$.
- $\mathtt{r}(e, \mathtt{revise}^1(i,\varphi))((\mathcal{M}, s)) = \mathcal{M}', s$ where
  $\mathcal{M}' = \langle \mathcal{S}, \pi, \mathrm{R}, \mathrm{B}^o, \mathrm{B}^c, \mathrm{B}^{d'}, \mathrm{r}, \mathrm{c}\rangle$ with
  $\mathrm{B}^{d'}(i', s') = \mathrm{B}^d(i', s')$ if $i \neq i'$ or $s' \notin [s]_{\mathrm{R}(i)}$
  $$\mathrm{B}^{d'}(i, s') = \begin{cases} \mathrm{B}^d(i,s) \cap [\![\varphi]\!] \text{ if } \mathrm{B}^d(i,s) \cap [\![\varphi]\!] \neq \emptyset, s' \in [s]_{\mathrm{R}(i)} \\ \mathrm{B}^c(i,s) \cap [\![\varphi]\!] \text{ if } \mathrm{B}^d(i,s) \cap [\![\varphi]\!] = \emptyset, s' \in [s]_{\mathrm{R}(i)} \end{cases}$$
- $\mathtt{r}(e, \mathtt{revise}^2(i,\varphi))((\mathcal{M}, s)) = \mathcal{M}', s$ where
  $\mathcal{M}' = \langle \mathcal{S}, \pi, \mathrm{R}, \mathrm{B}^o, \mathrm{B}^{c'}, \mathrm{B}^d, \mathrm{r}, \mathrm{c}\rangle$ with
  $\mathrm{B}^{c'}(i', s') = \mathrm{B}^c(i', s')$ if $i \neq i'$ or $s' \notin [s]_{\mathrm{R}(i)}$
  $$\mathrm{B}^{c'}(i, s') = \begin{cases} \mathrm{B}^c(i,s) \cap [\![\varphi]\!] \text{ if } \mathrm{B}^c(i,s) \cap [\![\varphi]\!] \neq \emptyset, s' \in [s]_{\mathrm{R}(i)} \\ [s]_{\mathrm{B}^o(i)} \cap [\![\varphi]\!] \text{ if } \mathrm{B}^c(i,s) \cap [\![\varphi]\!] = \emptyset, s' \in [s]_{\mathrm{R}(i)} \end{cases}$$
- $\mathtt{r}(e, \mathtt{revise}^3(i,\varphi))((\mathcal{M}, s)) = \mathcal{M}', s$ where
  $\mathcal{M}' = \langle \mathcal{S}, \pi, \mathrm{R}, \mathrm{B}^{o'}, \mathrm{B}^c, \mathrm{B}^d, \mathrm{r}, \mathrm{c}\rangle$ with
  $\mathrm{B}^{o'}(i') = \mathrm{B}^o(i')$ if $i' \neq i$
  $\mathrm{B}^{o'}(i) = (\mathrm{B}^o(i) \setminus \mathrm{Cl}_{\mathrm{eq}}([s]_{\mathrm{B}^o(i)})) \cup \mathrm{Cl}_{\mathrm{eq}}([s]^{\varphi+}_{\mathrm{B}^o(i)}) \cup \mathrm{Cl}_{\mathrm{eq}}([s]^{\varphi-}_{\mathrm{B}^o(i)})$
- $\mathtt{r}(e, \mathtt{revise}^d(i,\varphi))((\mathcal{M}, s)) = \mathtt{r}(e, \mathtt{revise}^1(i,\varphi))((\mathcal{M}, s))$
- $\mathtt{r}(e, \mathtt{revise}^c(i,\varphi))((\mathcal{M}, s)) = \mathtt{r}(e, \mathtt{revise}^2(i,\varphi);$
  $\mathtt{revise}^1(i,\varphi))((\mathcal{M}, s))$
- $\mathtt{r}(e, \mathtt{revise}^o(i,\varphi))((\mathcal{M}, s)) = \mathtt{r}(e, \mathtt{revise}^3(i,\varphi); \mathtt{revise}^2(i,\varphi);$
  $\mathtt{revise}^1(i,\varphi))((\mathcal{M}, s))$

REMARK 19. Definition 18 gives rise to the following remarks.

- The belief revision implemented in Definition 18 is the so called *All-is-Good* revision that we previously defined [van Linder *et al.*, 1995]. This kind of revision corresponds to a revision based on *full meet* contraction as it is defined by Gärdenfors [1988]. Full meet revision is the most ***rigorous*** way of revising beliefs under the AGM postulates. Although this rigour of revision is in general undesirable, for our goals it does not constitute much of a problem (see item 10 of Proposition 23).
- An important observation concerning Definition 18 is that the function r is in fact not well-defined, neither for the $\mathtt{revise}^m, m = 1,2,3$ actions nor for the $\mathtt{revise}^x, x = d,c,o$ actions , i.e., it is possible that the result from $\mathtt{r}(e, \mathtt{revise}^r(i,\varphi))((\mathcal{M}, s)), r \in \{1,2,3,d,c,o\}$ does not satisfy the demands given in Definition 8. For the $\mathtt{revise}^m, m = 1,2,3$ actions this is a consequence of the fact that

each of these actions works on its own particular level, and does not take 'lower' levels into account. Therefore it is for instance possible that performing the $\texttt{revise}^2$ action restricts the set of B^c worlds to those satisfying some formula φ whilst the set B^d still contains $\neg\varphi$ worlds, which conflicts the subset ordering imposed in Definition 8. In the case of the $\texttt{revise}^x, x = d, c, o$ actions the non well-definedness follows from the fact that this action could be executed in inappropriate circumstances. For instance, performing a $\texttt{revise}^c(i, \varphi)$ action in a situation in which $\mathbf{B}_i^o\neg\varphi$ holds results in the set $\mathrm{B}^c(i, s)$ being empty, thereby contradicting the demand for non-emptiness given in Definition 8. Both problems observed above are however not that serious: we will ensure that the revision actions behave correctly when applied in the implementation of a high-level informative action. Nevertheless, should one consider this deficiency to be insurmountable, one could always expand Definition 18 in such a way that the problem disappears (thereby ending up with a definition that is four or five times the size of Definition 18).

The following definition considers some properties that are more or less typical for informative actions.

DEFINITION 20. Let χ be an arbitrary formula. We distinguish the following properties of actions α:

- $\models \langle \mathrm{do}_i(\alpha)\rangle \mathbf{tt}$ *realisability*
- $\models \langle \mathrm{do}_i(\alpha)\rangle\chi \leftrightarrow [\mathrm{do}_i(\alpha)]\chi$ *determinism*
- $\models \langle \mathrm{do}_i(\alpha;\alpha)\rangle\chi \leftrightarrow \langle \mathrm{do}_i(\alpha)\rangle\chi$ *idempotence*

Realisability of an action implies that agents have the opportunity to perform the action regardless of the situation, determinism of an action means that performing the action results in a unique state of affairs, and idempotence of an action implies that sequentially performing the action an arbitrary number of times has the same effect as performing the action just once.

4.1 Formalising observations: seeing is believing

Through observations an agent ***learns whether*** some proposition is true of the state in which it is residing. For artificial agents it seems to be a reasonable assumption to demand that observations are ***truthful*** [van Linder *et al.*, 1994a]. That is, if some observation yields information that φ, then it should indeed be the case that φ.[3] Observations form the most trustworthy way of acquiring information: utterances like 'I've seen it with my own eyes' or 'Seeing is believing' support this claim. The formalisation that we propose

[3]Note that this property does not hold for human agents: magicians make a living out of this.

is therefore such that observations overrule any beliefs acquired by other means, i.e., observations will in general lead to belief revisions. The truth-value assigned to φ in the current world determines how the (observational) beliefs of the agent are revised: if φ holds, the agent's beliefs are revised with φ, otherwise a revision with $\neg\varphi$ takes place. Note that observations will never conflict with an agent's knowledge: since knowledge is veridical, and observations are truthful, it is not possible that an observation that some formula φ holds contradicts knowledge that $\neg\varphi$ holds.

DEFINITION 21. For all Kripke models $\mathcal{M}$ with state s, and for all agents i and propositional formulae φ we define:

$$\mathtt{r}(i, \mathtt{observe}\ \varphi)((\mathcal{M}, s)) = \begin{cases} (\mathcal{M}, s) & \text{if } (\mathcal{M}, s) \models \neg\mathbf{Agn}_i^k\varphi \\ \mathtt{r}(e, \mathtt{revise}^o(i, \varphi))((\mathcal{M}, s)) & \text{if } (\mathcal{M}, s) \models \mathbf{Agn}_i^k\varphi \wedge \varphi \\ \mathtt{r}(e, \mathtt{revise}^o(i, \neg\varphi))((\mathcal{M}, s)) & \text{if } (\mathcal{M}, s) \models \mathbf{Agn}_i^k\varphi \wedge \neg\varphi \end{cases}$$

When defining the function $\mathtt{r}$ for the $\mathtt{observe}$ action as done in Definition 21 it is indeed the case that the revision actions are applied correctly.

PROPOSITION 22. *For all Kripke models $\mathcal{M}$ with state s, and for all agents i and propositional formulae φ, $\mathtt{r}(i, \mathtt{observe}\ \varphi)((\mathcal{M}, s))$ is a well-defined Kripke model.*

PROPOSITION 23. *For all propositional formulae φ, ψ, for arbitrary formulae χ, and for all agents i, i' we have:*

1. $\mathtt{observe}\ \varphi$ *is deterministic, realizable and idempotent for all* $\varphi \in \mathcal{L}_0$
2. $\models \mathbf{K}_{i'}\psi \leftrightarrow \langle \mathrm{do}_i(\mathtt{observe}\ \varphi)\rangle \mathbf{K}_{i'}\psi$
3. $\models \langle \mathrm{do}_i(\mathtt{observe}\ \varphi)\rangle\chi \leftrightarrow \langle \mathrm{do}_i(\mathtt{observe}\ \neg\varphi)\rangle\chi$
4. $\models \langle \mathrm{do}_i(\mathtt{observe}\ \varphi)\rangle\neg\mathbf{Agn}_i^o\varphi$
5. $\models \varphi \rightarrow \langle \mathrm{do}_i(\mathtt{observe}\ \varphi)\rangle\mathbf{B}_i^o\varphi$
6. $\models \neg\varphi \rightarrow \langle \mathrm{do}_i(\mathtt{observe}\ \varphi)\rangle\mathbf{B}_i^o\neg\varphi$
7. $\models \varphi \wedge \mathbf{Agn}_i^k\varphi \rightarrow \langle \mathrm{do}_i(\mathtt{observe}\ \varphi)\rangle\mathbf{Saw}_i\varphi$
8. $\models \neg\varphi \wedge \mathbf{Agn}_i^k\varphi \rightarrow \langle \mathrm{do}_i(\mathtt{observe}\ \varphi)\rangle\mathbf{Saw}_i\neg\varphi$
9. $\models \varphi \wedge (\mathbf{Heard}_i\neg\varphi \vee \mathbf{Jumped}_i\neg\varphi) \rightarrow \langle \mathrm{do}_i(\mathtt{observe}\ \varphi)\rangle\mathbf{Saw}_i\varphi$
10. $\models \varphi \wedge \mathbf{B}_i^c\neg\varphi \rightarrow \langle \mathrm{do}_i(\mathtt{observe}\ \varphi)\rangle((\mathbf{B}_i^c\chi \leftrightarrow \mathbf{B}_i^o\chi) \wedge (\mathbf{B}_i^d\chi \leftrightarrow \mathbf{B}_i^c\chi))$

REMARK 24. Item 1 of Proposition 23 states that observations validate all of the properties introduced in Definition 20. Item 2 states that the knowledge fluents — the propositional formulae known to be true — of all agents persist under execution of an $\mathtt{observe}$ action by one of them. Item 3 states that the $\mathtt{observe}\ \varphi$ action formalises 'observing whether φ': observing whether φ is equivalent to observing whether $\neg\varphi$. Item 4 states that agents always have the opportunity to make observations, and that after observing whether φ, agents are no longer observationally agnostic with respect to φ. Items 5 and 6 follow by a combination of the property of truthfulness with the idea of agents learning whether some proposition is true: if φ is true, agents learn *that* φ, and analogously for $\neg\varphi$. Item 7 and 8 state that for knowledge-agnostic agents observations actually lead to *learning by seeing.* Item 9 — a special case of item 7 — is a very important one: it states that observations are the most credible source of information. Observations overrule other beliefs acquired through communication or adopted by default, i.e, incorrect communicational or default beliefs are *revised* in favour of observational beliefs. Item 10 sheds some more light on the way in which beliefs are revised: observing something that contradicts communicational beliefs leads to a *reset* of both the latter and the default beliefs of the agent, i.e., after such a revision all beliefs of the agent are at least grounded in its observations. Item 10 is a consequence of the use of All-is-Good revision, and indicates that our agents are rather rigorous in the maintenance of their beliefs.

4.2 Formalising communication: hearing is believing

Through communication agents can pass on some of the information that they possess to other agents which can use this incoming information to update their beliefs. The formalisation of communication that we present here is that of 'super-cooperative', didactic agents.[4] These agents always have the opportunity to pass on all of their belief to every other agent. Note that this action is in fact not a genuinely informative one: only the agent that receives the sent formula, and not the agent that performs the action, has the possibility to extend its information (with respect to the sent formula). It is however possible that the sending agent — the one who executes the `inform` action — does acquire information, for instance on the beliefs of the receiving agent.

In our opinion, when modelling communication in multi-agent systems, the concept of ***trust and dependence relations*** between agents deserves considerable attention. These relations model whether some agent is considered to be a credible source on a certain subject by another agent [Huang, 1990; van Linder *et al.*, 1994]. In communication this notion of credibility is very important: an agent i may or may not accept the information on φ that an agent j sends to it, depending on i's trust in j with respect to φ. We formalise the dependence relations between agents using the *dependent* operator $\mathbf{D}_{i,j}$, introduced by Huang [1990]. We like to think of $\mathbf{D}_{i,j}\varphi$ as expressing the fact that agent j is a *teacher* of agent i on the subject φ.

DEFINITION 25. We extend Definition 1 as follows:

- if $i, j \in \mathcal{A}$ and $\varphi \in \mathcal{L}$ then $\mathbf{D}_{i,j}\varphi \in \mathcal{L}$

[4]This kind of communication is among the ones we previously formalised [van Linder *et al.*, 1994]. Shoham's agents [Shoham, 1993] communicate in this way.

The Kripke models from Definition 3 and Definition 8 are extended with a function $\mathrm{D} : \mathcal{A} \times \mathcal{A} \times \mathcal{S} \rightarrow \wp(\mathcal{L})$. We define:

- $(\mathcal{M}, s) \models \mathbf{D}_{i,j}\varphi$ iff $\varphi \in \mathrm{D}(i, j, s)$

Besides credibility of the sending agent, formalised through the $\mathbf{D}_{i,j}$ operator, it is also very important *how* the sending agent acquired its information [Castelfranchi, 1994]. If the agent itself ***observed*** the truth of the formula that it is sending, then this formula should be considered more credible than when the agent ***heard*** the formula, or ***adopted it by default***. The formalisation that we propose is such that whenever an agent sends its observational beliefs then this will in general lead to an incorporation in the beliefs of the receiving agent. To ensure that the credibility ordering within the agent's beliefs remains meaningful, i.e., to ensure that communicational beliefs are indeed more credible than default beliefs, the source of these communicational beliefs needs to be credible. Now whenever some agent is sending out formulae that it heard, it is in general not possible to determine the credibility of these formulae, since one lost track of their source. Therefore, beliefs that the sending agent itself acquired through communication lead to a revision of the beliefs of the receiving agent only if this agent is completely agnostic with respect to these formulae. Formulae that the sending agent believes by default are considered to be too weak to be incorporated in the receiving agent's beliefs. Agents will therefore not react upon being told a default belief.

DEFINITION 26. For all Kripke models $\mathcal{M}$ with state s, and for all agents i, j and propositional formulae φ we define:

$$\mathtt{r}(j, \mathtt{inform}\ (\varphi, i))((\mathcal{M}, s)) = \begin{cases} \emptyset & \text{if } (\mathcal{M}, s) \models \neg \mathbf{B}_j^d \varphi \\ \mathtt{r}(e, \mathtt{revise}^c(i, \varphi))((\mathcal{M}, s)) & \text{if } (\mathcal{M}, s) \models \mathbf{D}_{i,j}\varphi \wedge ((\mathbf{B}_j^o\varphi \wedge \mathbf{Agn}_i^o\varphi) \vee \\ & \qquad (\mathbf{Heard}_j\varphi \wedge \mathbf{Agn}_i^d\varphi)) \\ (\mathcal{M}, s) & \text{otherwise} \end{cases}$$

The definition of $\mathtt{r}$ for the $\mathtt{inform}$ action also applies the revision actions correctly:

PROPOSITION 27. *For all Kripke models $\mathcal{M}$ with state s, and for all agents i, j and propositional formulae φ, if $\mathtt{r}(j, \mathtt{inform}\ (\varphi, i))((\mathcal{M}, s)) \neq \emptyset$ then it holds that $\mathtt{r}(j, \mathtt{inform}\ (\varphi, i))((\mathcal{M}, s))$ is a well-defined Kripke model.*

The following proposition shows that Definition 26 indeed takes care of a formalisation of the intuitive ideas that we previously exposed.

PROPOSITION 28. *For all propositional formulae φ, ψ, for arbitrary formulae χ, and for all agents i, i' and j we have:*

1. *$\mathtt{inform}\ (\varphi, i)$ is deterministic and idempotent for all $\varphi \in \mathcal{L}_0, i \in \mathcal{A}$*
2. $\models \mathbf{B}_{i'}^o\psi \rightarrow [\mathrm{do}_j(\mathtt{inform}\ (\varphi, i))]\mathbf{B}_{i'}^o\psi$

3. $\models \mathbf{B}^d_j\varphi \leftrightarrow \langle \mathrm{do}_j(\mathtt{inform}\ (\varphi, i))\rangle \mathbf{tt}$

4. $\models \mathbf{B}^d_j\varphi \wedge \neg\mathbf{D}_{i,j}\varphi \rightarrow (\langle \mathrm{do}_j(\mathtt{inform}\ (\varphi, i))\rangle\chi \leftrightarrow \chi)$

5. $\models \mathbf{D}_{i,j}\varphi \wedge \mathbf{B}^o_j\varphi \rightarrow \langle \mathrm{do}_j(\mathtt{inform}\ (\varphi, i))\rangle \mathbf{B}^c_i\varphi$

6. $\models \mathbf{D}_{i,j}\varphi \wedge \mathbf{B}^o_j\varphi \wedge \mathbf{Agn}^o_i\varphi \rightarrow \langle \mathrm{do}_j(\mathtt{inform}\ (\varphi, i))\rangle \mathbf{Heard}_i\varphi$

7. $\models \mathbf{D}_{i,j}\varphi \wedge \mathbf{Heard}_j\varphi \wedge \mathbf{Agn}^d_i\varphi \rightarrow \langle \mathrm{do}_j(\mathtt{inform}\ (\varphi, i)\rangle \mathbf{Heard}_i\varphi$

8. $\models \mathbf{D}_{i,j}\varphi \wedge \mathbf{Heard}_j\varphi \wedge \neg\mathbf{Agn}^d_i\varphi \rightarrow (\langle \mathrm{do}_j(\mathtt{inform}\ (\varphi, i))\rangle\chi \leftrightarrow \chi)$

9. $\models \mathbf{D}_{i,j}\varphi \wedge \mathbf{Jumped}_j\varphi \rightarrow (\langle \mathrm{do}_j(\mathtt{inform}\ (\varphi, i))\rangle\chi \leftrightarrow \chi)$

REMARK 29. Proposition 28 formalises the intuitive ideas concerning communication that we previously exposed. Item 1 deals with the properties introduced at the beginning of Section 4. Item 2 states that the observational belief fluents — and hence also the knowledge fluents — of the agents persist under execution of an `inform` action. Item 3 states that agents have the opportunity to inform other agents of all the formulae that they themselves believe. Attempts to send non-beliefs are doomed to fail. Our agents are therefore not allowed to *gossip*, i.e., it is not allowed for an agent to spread around rumours that it itself does not even believe. Note that item 3 also shows that the `inform` action is not realizable. Item 4 deals with the dependence relation between the agents: if the receiving agent does not trust the sending agent, it lets the information pass without updating its beliefs (and changing anything else for that matter). Item 5 states that if some trustworthy agent j tells another agent i some formula that it either knows or observed, this leads to a state of affairs in which the receiving agent believes φ at least with the credibility attached to communicational beliefs; whenever i is *a priori* agnostic with regard to this formula on the level of communicational beliefs, the receiving agent actually *learns* φ (item 6). Items 7 and 8 deal with the situation of *hearsay*: agent j tells i some formula that j itself has heard. In this case the beliefs of the receiving agent are updated only if the agent is agnostic on the default level with respect to the formula. The ratio behind this validity is that agent i does not necessarily consider the source that agent j used to acquire its belief in φ to be reliable. It might for instance be the case that j heard φ from some agent j' that is trusted by j, but not by i. Item 9 states that default beliefs of an agent are not transferable: this notion of belief is that weak that no revision takes place upon hearing one of these beliefs.

4.3 Formalising default jumps: jumping is believing

In a previous paper [van Linder *et al.*, 2001] we proposed a dynamic formalisation of default reasoning, in which agents attempt to jump to certain plausible formulae, called *defaults*. A remarkable aspect of this approach is our representation of defaults as *common possibilities*.[5] The idea underly-

[5]Note, however, that our formalisation of default reasoning does by no means depend on the representation of defaults as common possibilities. If one would wish to do so,

ing this representation is that defaults are founded in *common sense*, and in our multi-agent architecture, *common* sense is related to the knowledge and lack of knowledge of *all* agents. The modality of common possibility captures this idea of defaults as determined by the (lack of) knowledge of all agents. The intuitive interpretation of some formula φ being a common possibility is that it is considered epistemically possible by all agents, i.e., none of the agents knows φ to be false.

DEFINITION 30 ([van Linder *et al.*, 2001]). For all φ, the formula $\mathbf{N}\varphi$, for nobody knows not φ, is defined by:

$$\mathbf{N}\varphi =^{\mathrm{def}} \mathbf{M}_1\varphi \wedge \cdots \wedge \mathbf{M}_n\varphi$$

The formalisation of attempted jumps to defaults in the context of other sources of information is based on the idea that adopting formulae by default is the last resort: only if no other means of acquiring information are available a jump to a default is attempted. As such, default jumps are effective only for agents that are completely agnostic with respect to the default that is jumped to. If the agent is not completely agnostic, the attempted jump reduces to the empty action `skip`.

DEFINITION 31. For all Kripke models $\mathcal{M}$ with state s, and for all agents i and propositional formulae φ we define:

$$\mathtt{r}(i, \mathtt{try\_jump}\ \varphi)((\mathcal{M}, s)) = \begin{cases} \emptyset & \text{if } (\mathcal{M}, s) \not\models \mathbf{N}\varphi \\ \mathtt{r}(e, \mathtt{revise}^d(i, \varphi))((\mathcal{M}, s)) & \text{if } (\mathcal{M}, s) \models \mathbf{N}\varphi \wedge \mathbf{Agn}_i^d\varphi \\ (\mathcal{M}, s) & \text{otherwise} \end{cases}$$

Also for this third kind of informative action, the function $\mathtt{r}$ is defined correctly:

PROPOSITION 32. *For all Kripke models $\mathcal{M}$ with state s, and for all agents i and propositional formulae φ, if $\mathtt{r}(i, \mathtt{try\_jump}\ \varphi)((\mathcal{M}, s)) \neq \emptyset$ then it holds that $\mathtt{r}(i, \mathtt{try\_jump}\ \varphi)((\mathcal{M}, s))$ is a well-defined Kripke model.*

PROPOSITION 33. *For all propositional formulae φ, ψ, for arbitrary formulae χ, and for all agents i, i' we have:*

1. $\mathtt{try\_jump}\ \varphi$ *is deterministic and idempotent for all* $\varphi \in \mathcal{L}_0$
2. $\models \mathbf{B}_{i'}^c\psi \rightarrow [\mathrm{do}_i(\mathtt{try\_jump}\ \varphi)]\mathbf{B}_{i'}^c\psi$
3. $\models \mathbf{N}\varphi \leftrightarrow \langle \mathrm{do}_i(\mathtt{try\_jump}\ \varphi)\rangle \mathbf{tt}$
4. $\models \langle \mathrm{do}_i(\mathtt{try\_jump}\ \varphi)\rangle \mathbf{tt} \leftrightarrow \langle \mathrm{do}_i(\mathtt{try\_jump}\ \varphi)\rangle \neg\mathbf{Agn}_i^d\varphi$
5. $\models \mathbf{N}\varphi \wedge \mathbf{Agn}_i^d\varphi \rightarrow \langle \mathrm{do}_i(\mathtt{try\_jump}\ \varphi)\rangle \mathbf{Jumped}_i\varphi$
6. $\models \mathbf{N}\varphi \wedge \neg\mathbf{Agn}_i^d\varphi \rightarrow (\langle \mathrm{do}_i(\mathtt{try\_jump}\ \varphi)\rangle\chi \leftrightarrow \chi)$

---

one could easily predefine the agents' defaults in a syntactical way (see also [van Linder *et al.*, 2001]).

REMARK 34. The first item of Proposition 33 deals with the properties introduced in Definition 20. Again it is obvious that the property of realisability is not validated since an agent may attempt to jump to a non-default, and in item 3 it is formalised that such attempted jumps are doomed to fail. Item 2 states that the knowledge fluents, the observational belief fluents, and the communicational belief fluents of all agents persist under the attempted jump to a formula by one of them. Item 4 states that default jumps that are possible always result in the agent not being default-agnostic with regard to the formula that is jumped to. Item 5 formalises the idea that agents that are completely agnostic with respect to some formula φ jump to new default beliefs by applying the `try_jump` φ action. The last item states that attempted jumps to default conclusions yield information for completely agnostic agents only.

5 THE ABILITY TO GATHER INFORMATION

The formalisation of the abilities of intelligent information agents to execute informative actions is determined by two different notions. On the one hand, the abilities of an agent restrict its practical possibility to acquire information: only the actions that are within the agent's capacities can be used as means to extend its information. On the other hand, the agent's abilities steer the way in which it acquires its information: through its abilities an agent can be forced to acquire the most credible information possible. With respect to the latter point, we for instance demand it to be the case that agents resort to the use of default jumps only if they cannot acquire this information through observations.

Since by nature observations provide the most credible source of information, the ability to observe is determined strictly by the fact that the agents' information gathering is limited. For artificial agents these limits are simply a consequence of the way these agents were built. One can for instance think of a group of robots, each equipped with its personal set of sensors: one robot can observe whether its environment is radioactive, another whether the atmosphere contains oxygen, but no robot can sense both. Being more or less a construction decision, we assume that these observational capacities of robots are determined on beforehand, i.e., with respect to the agents' abilities observations are treated as atomic actions.

With regard to the capabilities of agents to inform other agents, we basically follow the approach we pursued previously [van Linder *et al.*, 1994]: an agent has the ability to inform all other agents of exactly the things it believes.

The ability to attempt to jump to a default captures two different aspects. Firstly, agents are able to jump to a default only if they *know* it to be a default. Secondly, the capability to jump depends on the observational capacities of the agent in the sense that an agent is (mentally) able to attempt a jump to a formula only if it knows that it is not able to observe whether the formula holds. In this way it is ensured that agents resort to default jumps only if the possibility of acquiring the information through observations is excluded.

DEFINITION 35. For Kripke models $\mathcal{M}$ with states s, for agents i, j and formula $\varphi \in \mathcal{L}_0$, the function c is modified such that

- it is predefined for all $\mathtt{observe}\ \varphi$ actions
- $\mathtt{c}(i, \mathtt{observe}\ \varphi)((\mathcal{M}, s)) = \mathtt{c}(i, \mathtt{observe}\ \neg\varphi)((\mathcal{M}, s))$
- $\mathtt{c}(j, \mathtt{inform}\ (\varphi, i))((\mathcal{M}, s)) = \mathtt{c}(j, \mathtt{confirm}\ \mathbf{B}^d_j\varphi)((\mathcal{M}, s))$
- $\mathtt{c}(i, \mathtt{try\_jump}\ \varphi)((\mathcal{M}, s)) =$
$$\mathtt{c}(i, \mathtt{confirm}\ \mathbf{K}_i(\neg\mathbf{A}_i\mathtt{observe}\ \varphi \wedge \mathbf{N}\varphi))((\mathcal{M}, s))$$

PROPOSITION 36. *For agents i, j and formulae $\varphi \in \mathcal{L}_0$ we have:*

1. $\models \mathbf{A}_j\mathtt{inform}\ (\varphi, i) \leftrightarrow \langle \mathrm{do}_j(\mathtt{inform}\ (\varphi, i))\rangle \mathtt{tt}$
2. $\models \mathbf{A}_i\mathtt{try\_jump}\ \varphi \rightarrow \langle \mathrm{do}_i(\mathtt{try\_jump}\ \varphi)\rangle \mathtt{tt}$
3. $\models \mathbf{A}_i\mathtt{observe}\ \varphi \rightarrow \neg\mathbf{A}_i\mathtt{try\_jump}\ \varphi$

REMARK 37. The first two items of Proposition 36 state that both the `inform` action and the `try_jump` action are *A-realizable* [van der Hoek *et al.*, 1998], i.e., being able to perform these action implies having the opportunity to do so. This property is typical for actions without circumstantial prerequisites, of which the `inform` and `try_jump` actions are typical examples. The last item formalises the idea that agents attempt jumps to a formula only if it is not within their capacities to observe whether the formula holds.

6 DISCUSSION

In this paper a formal framework is proposed in which various informative actions are combined, corresponding to the different ways in which rational agents can acquire information. In order to solve the various conflicts that could possibly occur when acquiring information from different sources, we propose a classification of the information that an agent possesses according to credibility. Based on this classification, we formalise what it means for agents to have seen or heard something, or to believe something by default. We present a formalisation of observations, communication actions, and the attempted jumps to conclusions that constitute default reasoning. To implement these informative actions we use a general belief revision action which satisfies the AGM postulates; dependent on the credibility of the incoming information this revision action acts on one or more parts of the classified belief set of the agents. Through the agents' abilities it is both formalised that the capacities of agents to acquire information are limited, and that agents prefer one kind of (credible) information acquisition to another (less credible) one. The framework presented in this paper might be seen as a first step towards a formalisation of intelligent information agents which will help us overcome the information explosion and will play an important part in life in the information age. More in particular, it is hoped that our framework may provide the basis for a specification tool for these agents which is a necessary thing to have when one wants to design them for practical purposes.

A SELECTED PROOFS

PROPOSITION 10. *Define the ordering > on informational operators by* $\mathbf{B}_i^k > \mathbf{B}_i^o > \mathbf{B}_i^c > \mathbf{B}_i^d$, *and let* $\succeq$ *be the reflexive, transitive closure of* >. *Then for all* $\mathbf{X}, \mathbf{Y} \in \{\mathbf{B}_i^k, \mathbf{B}_i^o, \mathbf{B}_i^c, \mathbf{B}_i^d\}$, *and for all formulae* φ :

- $\mathbf{X} \succeq \mathbf{Y} \Rightarrow \models \mathbf{X}\varphi \to \mathbf{Y}\varphi$

Proof. Let $i \in \mathcal{A}$ and $\mathbf{X}, \mathbf{Y} \in \{\mathbf{B}_i^k, \mathbf{B}_i^o, \mathbf{B}_i^c, \mathbf{B}_i^d\}$. Since $\mathrm{B}^d(i,s) \subseteq \mathrm{B}^c(i,s) \subseteq [s]_{\mathrm{B}^o(i)} \subseteq [s]_{\mathrm{R}(i)}$ it follows that whenever $\mathbf{X} > \mathbf{Y}$ the set of informational alternatives used to determine truth of $\mathbf{Y}$ is a subset of the set of informational alternatives used to determine truth of $\mathbf{X}$. Since $\subseteq$ is reflexive and transitive it follows that the same holds whenever $\mathbf{X} \succeq \mathbf{Y}$. Now let φ be an arbitrary formula, and assume that $\mathbf{X} \succeq \mathbf{Y}$. Then it holds for all models $\mathcal{M}$ with state s:

$(\mathcal{M}, s) \models \mathbf{X}\varphi$
$\Leftrightarrow$ φ holds at all $\mathbf{X}$-informational alternatives of s
$\Rightarrow$ φ holds at all $\mathbf{Y}$-informational alternatives of s
$\Leftrightarrow (\mathcal{M}, s) \models \mathbf{Y}\varphi$

which suffices to conclude the proof of the proposition. ■

PROPOSITION 13. *Let* φ *be an i-doxastic sequenced formula, and let* $\mathbf{Y} \in \{\mathbf{B}_i^k, \mathbf{B}_i^o, \mathbf{B}_i^c, \mathbf{B}_i^d\}$. *Then:*

- $\models \mathbf{Y}\varphi \leftrightarrow \varphi$

Proof. Let φ be an i-doxastic sequenced formula $\mathbf{X}\psi$, and let $\mathcal{M}$ be some Kripke model with state s. Slightly abusing notation we define $\mathrm{X}(i,s)$ to be $\mathrm{X}(i,s)$ for $\mathrm{X} = \mathrm{B}^c, \mathrm{B}^d$ and $[s]_{\mathrm{X}(i)}$ for $\mathrm{X} = \mathrm{R}, \mathrm{B}^o$. We distinguish two cases:

- $\mathbf{X} \in \{\mathbf{B}_i^k, \mathbf{B}_i^o, \mathbf{B}_i^c, \mathbf{B}_i^d\}$. In this case we have:

$$
\begin{aligned}
& (\mathcal{M}, s) \models \varphi \\
\Leftrightarrow\ & (\mathcal{M}, s) \models \mathbf{X}\psi \\
\Leftrightarrow\ & \forall s' \in \mathrm{X}(i,s)[\mathcal{M}, s' \models \psi] \\
\overset{*}{\Leftrightarrow}\ & \forall s'' \in \mathrm{Y}(i,s) \forall s' \in \mathrm{X}(i,s'')[\mathcal{M}, s' \models \psi] \\
\Leftrightarrow\ & \forall s' \in \mathrm{Y}(i,s)[\mathcal{M}, s' \models \mathbf{X}\psi] \\
\Leftrightarrow\ & (\mathcal{M}, s) \models \mathbf{YX}\psi \\
\Leftrightarrow\ & (\mathcal{M}, s) \models \mathbf{Y}\varphi
\end{aligned}
$$

- $\mathbf{X} \in \{\neg\mathbf{B}_i^k, \neg\mathbf{B}_i^o, \neg\mathbf{B}_i^c, \neg\mathbf{B}_i^d\}$. This case is analogous.

The equivalences tagged with $\star$ hold since for $\mathrm{X}, \mathrm{Y} \in \{\mathrm{R}, \mathrm{B}^o, \mathrm{B}^c, \mathrm{B}^d\}$, $\mathrm{Y}(i,s) \neq \emptyset$, $\mathrm{Y}(i,s) \subseteq [s]_{\mathrm{R}(i)}$ and $\mathrm{X}(i,s'') = \mathrm{X}(i,s)$ for all $s'' \in [s]_{\mathrm{R}(i)}$. The two items given above capture all possible i-doxastic sequenced formulae φ and thus we conclude that the proposition holds. ■

PROPOSITION 14. *Let φ and ψ be formulae. Let $\mathbf{X}$ be in the set $Bel = \{\mathbf{Saw}_i, \mathbf{Heard}_i, \mathbf{Jumped}_i \mid i \in \mathcal{A}\}$, and let $\mathbf{Y} \in Bel \cup \{\mathbf{K}_i\}$. Define the ordering $\succeq'$ to be the reflexive and transitive closure of $>'$ with $\mathbf{K}_i >' \mathbf{Saw}_i >' \mathbf{Heard}_i >' \mathbf{Jumped}_i$. Then we have:*

1. $\models \mathbf{X}\varphi \wedge \mathbf{X}(\varphi \to \psi) \to \mathbf{X}\psi$ K
2. $\models \neg(\mathbf{X}\varphi \wedge \mathbf{X}\neg\varphi)$ D
4. $\not\models \mathbf{X}\varphi \to \mathbf{X}\mathbf{X}\varphi$ 4
5. $\models \mathbf{X}\varphi \to \mathbf{K}_i\mathbf{X}\varphi$ 4′
7. $\models \neg\mathbf{X}\varphi \to \mathbf{K}_i\neg\mathbf{X}\varphi$ 5′
9. $\not\models \mathbf{X}(\varphi \wedge \psi) \to (\mathbf{X}\varphi \wedge \mathbf{X}\psi)$ M
10. $\models \mathbf{X}(\varphi \wedge \psi) \to (\bigvee_{\mathbf{Y}\succeq'\mathbf{X}} \mathbf{Y}\varphi \wedge \bigvee_{\mathbf{Y}\succeq'\mathbf{X}} \mathbf{Y}\psi) \wedge (\mathbf{X}\varphi \vee \mathbf{X}\psi)$ M′
11. $\models \varphi \not\Rightarrow \models \mathbf{X}\varphi$ *and even* $\models \varphi \Rightarrow \models \neg\mathbf{X}\varphi$ N
13. $\models \varphi \to \psi \Rightarrow \models \mathbf{X}\varphi \to \bigvee_{\mathbf{Y}\succeq'\mathbf{X}} \mathbf{Y}\psi$ RM′

Proof. Since the various case are highly analogous for the different operators, we restrict ourselves to proving the cases for the $\mathbf{Heard}_i$ operator. Let $\mathcal{M}$ be some Kripke model with state s, and let φ and ψ be formulae.

1. Suppose $(\mathcal{M},s) \models \mathbf{Heard}_i\varphi \wedge \mathbf{Heard}_i(\varphi \to \psi)$, i.e., $(\mathcal{M},s) \models \mathbf{B}_i^c\varphi \wedge \neg\mathbf{B}_i^o\varphi \wedge \mathbf{B}_i^c(\varphi \to \psi) \wedge \neg\mathbf{B}_i^o(\varphi \to \psi)$. Then since $\mathbf{B}_i^c$ validates the K axiom, it holds that $(\mathcal{M},s) \models \mathbf{B}_i^c\psi$, hence to show $(\mathcal{M},s) \not\models \mathbf{B}_i^o\psi$. Assume towards a contradiction that $(\mathcal{M},s) \models \mathbf{B}_i^o\psi$. Then also $(\mathcal{M},s) \models \mathbf{B}_i^o(\varphi \to \psi)$ since $\mathbf{B}_i^o$ validates the K-axiom. Since this contradicts the fact that $(\mathcal{M},s) \models \neg\mathbf{B}_i^o(\varphi \to \psi)$, we conclude that $(\mathcal{M},s) \not\models \mathbf{B}_i^o\psi$. Hence $(\mathcal{M},s) \models \mathbf{Heard}_i\psi$.
2. Since $\mathbf{B}_i^c$ satisfies the D-axiom, it holds that $(\mathcal{M},s) \models \neg(\mathbf{B}_i^c\varphi \wedge \mathbf{B}_i^c\neg\varphi)$. Hence also $(\mathcal{M},s) \models \neg(\mathbf{Heard}_i\varphi \wedge \mathbf{Heard}_i\neg\varphi)$.
4. This item is most easily proved as a direct consequence of the following one. For whenever $(\mathcal{M},s) \models \mathbf{Heard}_i\varphi$ it follows that $(\mathcal{M},s) \models \mathbf{K}_i\mathbf{Heard}_i\varphi$. Hence $(\mathcal{M},s) \models \mathbf{B}_i^o\mathbf{Heard}_i\varphi$ and thus we conclude that $(\mathcal{M},s) \not\models \mathbf{Heard}_i\mathbf{Heard}_i\varphi$.
5. Assume $(\mathcal{M},s) \models \mathbf{Heard}_i\varphi$, i.e., $(\mathcal{M},s) \models \mathbf{B}_i^c\varphi \wedge \neg\mathbf{B}_i^o\varphi$. By Proposition 13 it follows that $(\mathcal{M},s) \models \mathbf{K}_i\mathbf{B}_i^c\varphi \wedge \mathbf{K}_i\neg\mathbf{B}_i^o\varphi$ and since $\mathbf{K}_i$ satisfies the C-axiom we have $(\mathcal{M},s) \models \mathbf{K}_i(\mathbf{B}_i^c\varphi \wedge \neg\mathbf{B}_i^o\varphi)$ and thus $(\mathcal{M},s) \models \mathbf{K}_i\mathbf{Heard}_i\varphi$.
7. Assume $(\mathcal{M},s) \models \neg\mathbf{Heard}_i\varphi$. Then $(\mathcal{M},s) \models \neg\mathbf{B}_i^c\varphi \vee \mathbf{B}_i^o\varphi$. From Proposition 13 it follows that $(\mathcal{M},s) \models \mathbf{K}_i\neg\mathbf{B}_i^c\varphi \vee \mathbf{K}_i\mathbf{B}_i^o\varphi$. This implies $(\mathcal{M},s) \models \mathbf{K}_i(\neg\mathbf{B}_i^c\varphi \vee \mathbf{B}_i^o\varphi)$ and hence $(\mathcal{M},s) \models \mathbf{K}_i\neg\mathbf{Heard}_i\varphi$.

9. If ψ is a tautology, i.e., $\models \psi$, then also $\models \mathbf{K}_i\psi$. For ψ a tautology, it is easy to see that the formula $\mathbf{Heard}_i(\varphi \wedge \psi) \wedge \mathbf{Heard}_i\varphi \wedge \neg\mathbf{Heard}_i\psi$ is satisfiable for some contingency φ.
10. Assume $(\mathcal{M}, s) \models \mathbf{Heard}_i(\varphi \wedge \psi)$. Since $\models (\varphi \wedge \psi) \rightarrow \varphi$ and $\models (\varphi \wedge \psi) \rightarrow \psi$ both hold, the first of the conjuncts on the right-hand side follows more or less directly from the RM' rule. For the second conjunct note that $(\mathcal{M}, s) \models \neg\mathbf{Heard}_i\varphi \wedge \neg\mathbf{Heard}_i\psi$ together with $(\mathcal{M}, s) \models \mathbf{Heard}_i(\varphi \wedge \psi)$ would imply $(\mathcal{M}, s) \models \mathbf{B}^o_i(\varphi \wedge \psi)$ which contradicts $(\mathcal{M}, s) \models \mathbf{Heard}_i(\varphi \wedge \psi)$.
11. If $\models \varphi$ then also $\models \mathbf{K}_i\varphi$ and hence directly $\models \neg\mathbf{Saw}_i\varphi \wedge \neg\mathbf{Heard}_i\varphi \wedge \neg\mathbf{Jumped}_i\varphi$.
13. Assume that $\models (\varphi \rightarrow \psi)$ and $(\mathcal{M}, s) \models \mathbf{Heard}_i\varphi$, i.e., $(\mathcal{M}, s) \models \mathbf{B}^c_i\varphi \wedge \neg\mathbf{B}^o_i\varphi$. Then also $(\mathcal{M}, s) \models \mathbf{B}^c_i\psi$. Now if $(\mathcal{M}, s) \not\models \mathbf{B}^o_i\psi$, then $(\mathcal{M}, s) \models \mathbf{Heard}_i\psi$; otherwise if $(\mathcal{M}, s) \not\models \mathbf{K}_i\psi$ it holds that $(\mathcal{M}, s) \models \mathbf{Saw}_i\psi$, and else $(\mathcal{M}, s) \models \mathbf{K}_i\psi$. Hence $(\mathcal{M}, s) \models \bigvee_{\mathbf{Y} \succeq' \mathbf{Heard}_i} \mathbf{Y}\psi$. ■

PROPOSITION 22. *For all Kripke models $\mathcal{M}$ with state s, and for all agents i and propositional formulae φ, $\mathtt{r}(i, \mathtt{observe}\ \varphi)((\mathcal{M}, s))$ is a well-defined Kripke model.*

Proof. Assume that $\mathcal{M}$ is a (well-defined) Kripke model with state s, let i be an agent and φ some propositional formula. It is easily seen from Definitions 18 and 21 that $\mathtt{r}(i, \mathtt{observe}\ \varphi)((\mathcal{M}, s))$ yields a pair $\mathcal{M}', s$ where $(\mathcal{M}, s)$ is a tuple in shape similar to a Kripke model. We show that $\mathcal{M}'$ is indeed a well-defined Kripke model. We distinguish three cases, corresponding to the three cases in Definition 21.

- $(\mathcal{M}, s) \models \neg\mathbf{Agn}^k_i\varphi$. In this case $\mathcal{M}' = \mathcal{M}$ and hence $\mathcal{M}'$ is well-defined.
- $(\mathcal{M}, s) \models \mathbf{Agn}^k_i\varphi \wedge \varphi$. In this case $\mathcal{M}', s = \mathtt{r}(e, \mathtt{revise}^o(i, \varphi))((\mathcal{M}, s))$. From Definition 18 it follows that $\mathcal{M}'$ differs from $\mathcal{M}$ only in $\mathrm{B}^{o\prime}(i)$, $\mathrm{B}^{c\prime}(i, s')$ and $\mathrm{B}^{d\prime}(i, s')$, where $s' \in [s]_{\mathrm{R}(i)}$. To check well-definedness of $\mathcal{M}'$ it suffices to consider these three aspects, which we shall do successively.
  - By definition of $\mathtt{r}(e, \mathtt{revise}^3(\varphi, i))$ we have $\mathrm{B}^{o\prime}(i) = (\mathrm{B}^o(i) \setminus \mathrm{Cl}_{\mathrm{eq}}([s]_{\mathrm{B}^o(i)})) \cup \mathrm{Cl}_{\mathrm{eq}}([s]^{\varphi+}_{\mathrm{B}^o(i)}) \cup \mathrm{Cl}_{\mathrm{eq}}([s]^{\varphi-}_{\mathrm{B}^o(i)})$. Since $(\mathcal{M}, s) \models \varphi$ it follows that $[s]^{\varphi+}_{\mathrm{B}^o(i)} = [s]_{\mathrm{B}^o(i)} \cap [\![\varphi]\!]$ and $[s]^{\varphi-}_{\mathrm{B}^o(i)} = [s]_{\mathrm{B}^o(i)} \cap [\![\neg\varphi]\!]$. We show that $\mathrm{B}^{o\prime}(i)$ is an equivalence relation by successively showing that it is reflexive, symmetrical and transitive.
    - It is obvious that $(s', s') \in \mathrm{B}^{o\prime}(i)$ for all $s' \in \mathcal{S}$. For if $s' \notin [s]_{\mathrm{B}^o(i)}$, $(s', s') \in \mathrm{B}^o(i) \setminus \mathrm{Cl}_{\mathrm{eq}}([s]_{\mathrm{B}^o(i)})$, and otherwise s' is either in $[s]^{\varphi+}_{\mathrm{B}^o(i)}$ or in $[s]^{\varphi-}_{\mathrm{B}^o(i)}$, which implies that either

$(s', s') \in \mathrm{Cl_{eq}}([s]^{\varphi+}_{\mathrm{B}^o(i)})$ or $(s', s') \in \mathrm{Cl_{eq}}([s]^{\varphi-}_{\mathrm{B}^o(i)})$. In either case $(s', s') \in \mathrm{B}^{o\prime}(i)$ and hence $\mathrm{B}^{o\prime}(i)$ is reflexive.

- Let (s_1, s_2) in $\mathrm{B}^{o\prime}(i)$. If (s_1, s_2) is either in $\mathrm{Cl_{eq}}([s]^{\varphi+}_{\mathrm{B}^o(i)})$ or in $\mathrm{Cl_{eq}}([s]^{\varphi-}_{\mathrm{B}^o(i)})$ it follows by definition of $\mathrm{Cl_{eq}}$ that either $(s_2, s_1) \in \mathrm{Cl_{eq}}([s]^{\varphi+}_{\mathrm{B}^o(i)})$ or $(s_2, s_1) \in \mathrm{Cl_{eq}}([s]^{\varphi-}_{\mathrm{B}^o(i)})$, hence $(s_2,s_1) \in \mathrm{B}^{o\prime}(i)$. So assume that $(s_1,s_2) \in \mathrm{B}^o(i) \setminus \mathrm{Cl_{eq}}([s]_{\mathrm{B}^o(i)})$. Then $(s_1, s_2) \in \mathrm{B}^o(i)$ and (s_1, s_2) is not a member of $\mathrm{Cl_{eq}}([s]_{\mathrm{B}^o(i)})$. Since $\mathrm{B}^o(i)$ is symmetrical, it follows that $(s_2, s_1) \in \mathrm{B}^o(i)$ and by definition of $\mathrm{Cl_{eq}}$ we have that $(s_2, s_1) \notin \mathrm{Cl_{eq}}([s]_{\mathrm{B}^o(i)})$. Hence $(s_2, s_1) \in \mathrm{B}^o(i) \setminus \mathrm{Cl_{eq}}([s]_{\mathrm{B}^o(i)}) \subseteq \mathrm{B}^{o\prime}(i)$ which suffices to collude that $\mathrm{B}^{o\prime}(i)$ is symmetrical.
- Let $(s_1, s_2) \in \mathrm{B}^{o\prime}(i)$ and $(s_2, s_3) \in \mathrm{B}^{o\prime}(i)$. We distinguish three cases:
  1. $(s_1, s_2) \in \mathrm{Cl_{eq}}([s]^{\varphi+}_{\mathrm{B}^o(i)})$. Then $s_2 \in [s]^{\varphi+}_{\mathrm{B}^o(i)}$. Now $(s_2, s_3) \in \mathrm{B}^{o\prime}(i)$ implies $s_3 \in [s]^{\varphi+}_{\mathrm{B}^o(i)}$. Hence we have $(s_2, s_3) \in \mathrm{Cl_{eq}}([s]^{\varphi+}_{\mathrm{B}^o(i)})$. By definition of $\mathrm{Cl_{eq}}$ it follows that $(s_1, s_3) \in \mathrm{Cl_{eq}}([s]^{\varphi+}_{\mathrm{B}^o(i)}) \subseteq \mathrm{B}^{o\prime}(i)$.
  2. $(s_1, s_2) \in \mathrm{Cl_{eq}}([s]^{\varphi-}_{\mathrm{B}^o(i)})$. This case is completely analogous to the case where $(s_1, s_2) \in \mathrm{Cl_{eq}}([s]^{\varphi+}_{\mathrm{B}^o(i)})$.
  3. $(s_1, s_2) \in \mathrm{B}^o(i) \setminus \mathrm{Cl_{eq}}([s]_{\mathrm{B}^o(i)})$. In this case $(s_1, s_2) \in \mathrm{B}^o(i)$ and $(s_1, s_2) \notin \mathrm{Cl_{eq}}([s]_{\mathrm{B}^o(i)})$. Since $\mathrm{B}^o(i)$ is an equivalence relation this implies that $s_1 \notin [s]_{\mathrm{B}^o(i)}$ and $s_2 \notin [s]_{\mathrm{B}^o(i)}$: for if either one of them would be in $[s]_{\mathrm{B}^o(i)}$, they would both be due to the transitivity of $\mathrm{B}^o(i)$, and this contradicts $(s_1, s_2) \notin \mathrm{Cl_{eq}}([s]_{\mathrm{B}^o(i)})$. From $s_2 \notin [s]_{\mathrm{B}^o(i)}$ it follows that $(s_2, s_3) \notin \mathrm{Cl_{eq}}([s]_{\mathrm{B}^o(i)})$ and thus $(s_2, s_3) \in \mathrm{B}^o(i) \setminus \mathrm{Cl_{eq}}([s]_{\mathrm{B}^o(i)})$. Thus, $(s_1, s_3) \in \mathrm{B}^o(i)$ and since $s_1 \notin [s]_{\mathrm{B}^o(i)}$ it follows that $(s_1, s_3) \in \mathrm{B}^o(i) \setminus \mathrm{Cl_{eq}}([s]_{\mathrm{B}^o(i)}) \subseteq \mathrm{B}^{o\prime}(i)$.

  Hence $\mathrm{B}^{o\prime}(i)$ is transitive.

Thus $\mathrm{B}^{o\prime}(i)$ is reflexive, symmetrical and transitive, and hence an equivalence relation. Furthermore, $\mathrm{B}^{o\prime}(i) \subseteq \mathrm{B}^o(i) \subseteq [s]_{\mathrm{R}(i)} = [s]_{\mathrm{R}'(i)}$.

- Let $s' \in [s]_{\mathrm{R}(i)}$. We distinguish two cases:
  - If $\mathrm{B}^{c\prime}(i, s')$ is equal to $\mathrm{B}^c(i, s) \cap [\![\varphi]\!]$, it follows that $\mathrm{B}^{c\prime}(i, s') \neq \emptyset$. Since $\mathrm{B}^c(i, s') \subseteq [s]_{\mathrm{B}^o(i)}$, and $[s]_{\mathrm{B}^{o\prime}(i)} = [s]_{\mathrm{B}^o(i)} \cap [\![\varphi]\!]$, it follows that $\mathrm{B}^{c\prime}(i, s') \subseteq [s]_{\mathrm{B}^{o\prime}(i)}$. In this case $\mathrm{B}^{d\prime}(i, s')$ is either equal to $\mathrm{B}^d(i, s') \cap [\![\varphi]\!]$, which would imply by a similar argument as given above that $\mathrm{B}^{d\prime}(i, s') \neq \emptyset$ and $\mathrm{B}^{d\prime}(i, s') \subseteq \mathrm{B}^{c\prime}(i, s')$, or equal to $\mathrm{B}^{c\prime}(i, s')$, which would also imply that

$\mathrm{B}^{d'}(i,s') \neq \emptyset$ and $\mathrm{B}^{d'}(i,s') \subseteq \mathrm{B}^{c'}(i,s')$. In both cases the resulting tuple would indeed be a well-defined Kripke model.

– If $\mathrm{B}^{c'}(i,s')$ is equal to $[s]_{\mathrm{B}^{o'}(i)} \cap [\![\varphi]\!] = [s]_{\mathrm{B}^{o'}(i)}$, this implies that $\mathrm{B}^{c}(i,s) \cap [\![\varphi]\!] = \emptyset$, and hence also $\mathrm{B}^{d}(i,s) \cap [\![\varphi]\!] = \emptyset$. In this case $\mathrm{B}^{d'}(i,s') = \mathrm{B}^{c'}(i,s') = [s]_{\mathrm{B}^{o'}(i)}$, and hence $\mathcal{M}'$ would also be a well-defined Kripke model.

Since it holds in both cases, we conclude that $\mathcal{M}'$ is a well-defined Kripke model.

- $(\mathcal{M},s) \models \mathbf{Agn}_i^k\varphi \wedge \neg\varphi$. This case is completely analogous to the case where $(\mathcal{M},s) \models \mathbf{Agn}_i^k\varphi \wedge \varphi$.

Since in all three cases $\mathcal{M}'$ is well-defined, we conclude that the proposition holds. ■

PROPOSITION 23. *For all propositional formulae φ, ψ, for arbitrary formulae χ, and for all agents i, i' we have:*

4. $\models \langle \mathrm{do}_i(\mathtt{observe}\ \varphi)\rangle \neg \mathbf{Agn}_i^o\varphi$

10. $\models \varphi \wedge \mathbf{B}_i^c\neg\varphi \rightarrow \langle \mathrm{do}_i(\mathtt{observe}\ \varphi)\rangle((\mathbf{B}_i^c\chi \leftrightarrow \mathbf{B}_i^o\chi) \wedge (\mathbf{B}_i^d\chi \leftrightarrow \mathbf{B}_i^c\chi))$

Proof. Let $\mathcal{M}$ be some Kripke model with state s, let φ, ψ be propositional formulae, let χ be an arbitrary formula, and let i, i' be agents.

4. Let $\mathcal{M}', s = \mathtt{r}(i, \mathtt{observe}\ \varphi)((\mathcal{M}, s))$; $\mathcal{M}', s$ exists due to the realisability of $\mathtt{observe}\ \varphi$. If $(\mathcal{M},s) \models \neg\mathbf{Agn}_i^k\varphi$ then $\mathcal{M}', s = (\mathcal{M},s)$ and hence $\mathcal{M}', s \models \neg\mathbf{Agn}_i^k\varphi$ which implies that $\mathcal{M}', s \models \neg\mathbf{Agn}_i^o\varphi$. If $(\mathcal{M},s) \models \mathbf{Agn}_i^k\varphi \wedge \varphi$ then $\mathrm{B}^{o'}(i,s) = \mathrm{B}^{o}(i,s) \cap [\![\varphi]\!]$ by definition of $\mathtt{r}(e, \mathtt{revise}^o\varphi)$. Hence $\mathcal{M}', s' \models \varphi$ for all $s' \in \mathrm{B}^{o'}(i,s)$ and thus $\mathcal{M}', s \models \mathbf{B}_i^o\varphi$. The case where $(\mathcal{M},s) \models \mathbf{Agn}_i^k\varphi \wedge \neg\varphi$ is completely analogous to the case where $(\mathcal{M},s) \models \mathbf{Agn}_i^k\varphi \wedge \varphi$.
10. Suppose $(\mathcal{M},s) \models \varphi \wedge \mathbf{B}_i^c\neg\varphi$. Then obviously $(\mathcal{M},s) \models \varphi \wedge \mathbf{Agn}_i^k\varphi$. Let $\mathcal{M}', s = \mathtt{r}(i, \mathtt{observe}\ \varphi)((\mathcal{M},s))$. By definition of $\mathtt{r}(e, \mathtt{revise}^o (i,\varphi))$ it follows that
    - $\mathrm{B}^{o'}(i,s) = \mathrm{B}^{o}(i,s) \cap [\![\varphi]\!]$
    - $\mathrm{B}^{c'}(i,s) = \mathrm{B}^{o'}(i,s)$
    - $\mathrm{B}^{d'}(i,s) = \mathrm{B}^{o'}(i,s)$

    Hence $\mathcal{M}', s \models \mathbf{B}_i^o\chi$ iff $\mathcal{M}', s \models \mathbf{B}_i^c\chi$ and $\mathcal{M}', s \models \mathbf{B}_i^c\chi$ iff $\mathcal{M}', s \models \mathbf{B}_i^d\chi$ which suffices to conclude item 10. ■

PROPOSITION 28. *For all propositional formulae φ, ψ, for arbitrary formulae χ, and for all agents i, i' and j we have:*

5. $\models \mathbf{D}_{i,j}\varphi \wedge \mathbf{B}_j^o\varphi \rightarrow \langle \mathrm{do}_j(\mathtt{inform}\ (\varphi, i))\rangle \mathbf{B}_i^c\varphi$

7. $\models \mathbf{D}_{i,j}\varphi \wedge \mathbf{Heard}_j\varphi \wedge \mathbf{Agn}_i^d\varphi \rightarrow \langle \mathrm{do}_j(\mathtt{inform}\ (\varphi, i))\rangle \mathbf{Heard}_i\varphi$

Proof. Let $\mathcal{M}$ be some Kripke model with state s, let i, j be agents, let φ be some propositional and let χ be an arbitrary formula.

5. Assume that $(\mathcal{M}, s) \models \mathbf{D}_{i,j}\varphi \wedge \mathbf{B}^o_j\varphi$. From Definitions 18 and 26 it is obvious that a model $\mathcal{M}'$ exists with $\mathtt{r}(j, \mathtt{inform}\ (\varphi, i))((\mathcal{M}, s)) = \mathcal{M}', s$. Now let $\mathcal{M}'$ be such a model. We distinguish three cases:
   - $(\mathcal{M}, s) \models \neg\mathbf{B}^c_i\neg\varphi$. In this case $\mathcal{M}', s = \mathtt{r}(e, \mathtt{revise}^c(i, \varphi))((\mathcal{M}, s))$ is such that $\mathrm{B}^{c\prime}(i, s) = \mathrm{B}^c(i, s) \cap [\![\varphi]\!]$. Hence $\mathcal{M}', s' \models \varphi$ for all $s' \in \mathrm{B}^{c\prime}(i, s)$ and $\mathcal{M}', s \models \mathbf{B}^c_i\varphi$. From this, we may conclude that $(\mathcal{M}, s) \models \langle \mathrm{do}_j(\mathtt{inform}\ (i, \varphi))\rangle\mathbf{B}^c_i\varphi$.
   - $(\mathcal{M}, s) \models \mathbf{B}^c_i\neg\varphi \wedge \mathbf{Agn}^o_i\varphi$. Let $\mathtt{r}(e, \mathtt{revise}^c(i, \varphi))((\mathcal{M}, s)) =$ $\mathcal{M}', s$. Then $\mathcal{M}', s$ is such that $\mathrm{B}^{c\prime}(i, s) = \mathrm{B}^o(i, s) \cap [\![\varphi]\!]$. Hence $\mathcal{M}', s' \models \varphi$ for all $s' \in \mathrm{B}^{c\prime}(i, s)$ and $\mathcal{M}', s \models \mathbf{B}^c_i\varphi$. Thus, in $(\mathcal{M}, s)$ we may conclude $\langle \mathrm{do}_j(\mathtt{inform}\ (i, \varphi))\rangle\mathbf{B}^c_i\varphi$ to hold.
   - $(\mathcal{M}, s) \models \neg\mathbf{Agn}^o_i\varphi$. In this case $\mathcal{M}', s = (\mathcal{M}, s)$. Given the fact that $(\mathcal{M}, s) \models \mathbf{B}^o_j\varphi$ and that observational beliefs are veridical it follows that $(\mathcal{M}, s) \models \mathbf{B}^o_i\varphi$ and thus $(\mathcal{M}, s) \models \mathbf{B}^c_i\varphi$. Thus, in $(\mathcal{M}, s)$ we may conclude $\langle \mathrm{do}_j(\mathtt{inform}\ (\varphi, i))\rangle\mathbf{B}^c_i\varphi$ to hold.

   Since in all three cases $(\mathcal{M}, s) \models \langle \mathrm{do}_j(\mathtt{inform}\ (\varphi, i))\rangle\mathbf{B}^c_i\varphi$ we conclude that item 5 holds.
7. Assume that $(\mathcal{M}, s) \models \mathbf{D}_{i,j}\varphi \wedge \mathbf{Heard}_j\varphi \wedge \mathbf{Agn}^d_i\varphi$. Let $\mathcal{M}'$ be the model such that $\mathcal{M}', s = \mathtt{r}(j, \mathtt{inform}\ (\varphi, i))((\mathcal{M}, s))$. From clause 2 of Definition 26 it follows that $\mathcal{M}', s = \mathtt{r}(e, \mathtt{revise}^c(i, \varphi))((\mathcal{M}, s))$. Since $(\mathcal{M}, s) \models \mathbf{Agn}^d_i\varphi$ and hence $(\mathcal{M}, s) \models \neg\mathbf{B}^c_i\neg\varphi$, it follows that $\mathrm{B}^{c\prime}(i, s) = \mathrm{B}^c(i, s) \cap [\![\varphi]\!]$. Hence $\mathcal{M}', s \models \mathbf{B}^c_i\varphi$. Since $(\mathcal{M}, s) \models \mathbf{Agn}^d_i\varphi$ implies $(\mathcal{M}, s) \models \mathbf{Agn}^o_i\varphi$, it follows that $(\mathcal{M}, s) \models \mathbf{Agn}^o_i\varphi$, and since the $\mathtt{revise}^c$ action does not affect observational belief fluents it follows that $\mathcal{M}', s \models \mathbf{Agn}^o_i\varphi$. Hence $\mathcal{M}', s \models \mathbf{Heard}_i\varphi$ and therefore $(\mathcal{M}, s) \models \langle \mathrm{do}_j(\mathtt{inform}\ (\varphi, i))\rangle\mathbf{Heard}_i\varphi$. ■

PROPOSITION 33. *For all propositional formulae φ, ψ, for arbitrary formulae χ, and for all agents i, i' we have:*

4. $\models \langle \mathrm{do}_i(\mathtt{try\_jump}\ \varphi)\rangle\mathbf{tt} \leftrightarrow \langle \mathrm{do}_i(\mathtt{try\_jump}\ \varphi)\rangle\neg\mathbf{Agn}^d_i\varphi$

5. $\models \mathbf{N}\varphi \wedge \mathbf{Agn}^d_i\varphi \rightarrow \langle \mathrm{do}_i(\mathtt{try\_jump}\ \varphi)\rangle\mathbf{Jumped}_i\varphi$

Proof. Let $\mathcal{M}$ be some Kripke model with state s, let i be an agent, let φ be some propositional formula, and let χ be some arbitrary formula.

4. Suppose $(\mathcal{M}, s) \models \langle \mathrm{do}_i(\mathtt{try\_jump}\ \varphi)\rangle\mathbf{tt}$. Let $\mathcal{M}', s$ be such that $\mathcal{M}', s = \mathtt{r}(i, \mathtt{try\_jump}\ \varphi)((\mathcal{M}, s))$. We distinguish two cases:
   - If $(\mathcal{M}, s) \not\models \mathbf{Agn}^d_i\varphi$ then $\mathcal{M}' = \mathcal{M}$. Hence $\mathcal{M}', s \models \neg\mathbf{Agn}^d_i\varphi$ and thus $(\mathcal{M}, s) \models \langle \mathrm{do}_i(\mathtt{try\_jump}\ \varphi)\rangle\neg\mathbf{Agn}^d_i\varphi$.
   - If $(\mathcal{M}, s) \models \mathbf{Agn}^d_i\varphi$ then $\mathcal{M}', s = \mathtt{r}(e, \mathtt{revise}^d(i, \varphi))((\mathcal{M}, s))$. By definition of $\mathtt{r}(e, \mathtt{revise}^d(i, \varphi))$ it follows that $\mathrm{B}^{d\prime}(i, s) = \mathrm{B}^d(i, s) \cap [\![\varphi]\!]$. Then $\mathcal{M}', s' \models \varphi$ for all $s' \in \mathrm{B}^{d\prime}(i, s)$ and $\mathcal{M}', s \models \mathbf{B}^d_i\varphi$ implying $(\mathcal{M}, s) \models \langle \mathrm{do}_i(\mathtt{try\_jump}\ \varphi)\rangle\neg\mathbf{Agn}^d_i\varphi$.

Since in both cases $(\mathcal{M}, s) \models \langle \text{do}_i(\texttt{try\_jump}\ \varphi)\rangle \neg \mathbf{Agn}_i^d \varphi$ we conclude that the left-to-right implication of this item holds. The right-to-left implication is trivial.

5. Suppose $(\mathcal{M}, s) \models \mathbf{N}\varphi \wedge \mathbf{Agn}_i^d \varphi$. Let $\mathcal{M}', s = \mathtt{r}(i, \texttt{try\_jump}\ \varphi)((\mathcal{M}, s))$. From Definition 31 we infer $\mathcal{M}', s = \mathtt{r}(e, \texttt{revise}^d(i, \varphi))((\mathcal{M}, s))$. Then from the definition of $\mathtt{r}(e, \texttt{revise}^d(i, \varphi))$ we obtain $\mathcal{M}', s \models \mathbf{B}_i^d \varphi$, by a similar argument as given above. From $(\mathcal{M}, s) \models \mathbf{Agn}_i^d \varphi$ it follows that $(\mathcal{M}, s) \models \mathbf{Agn}_i^c \varphi$. Since execution of the $\texttt{revise}^d$ action does not affect the observational belief fluents it follows that $\mathcal{M}', s \models \mathbf{Agn}_i^c \varphi$ and thus $\mathcal{M}', s \models \mathbf{Jumped}_i \varphi$. Then also $(\mathcal{M}, s) \models \langle \text{do}_i(\texttt{try\_jump}\ \varphi)\rangle \mathbf{Jumped}_i \varphi$. ■

ACKNOWLEDGEMENTS

Thanks are due to Cristiano Castelfranchi for his very valuable suggestions on the psychological and philosophical aspects of the topics discussed in this paper; thanks are also due to Theo Huibers for his critical comments and remarks on a draft version of this paper. This research was partially supported by Esprit III BRWG project No.8319 'ModelAge', Esprit III BRA project No.6156 'Drums II', and the Vrije Universiteit Amsterdam; the third author was furthermore partially supported by the Katholieke Universiteit Nijmegen.

ACKNOWLEDGEMENT

This paper was originally published as "Seeing is believing (and so are hearing and jumping)", *Journal of Logic, Language and Information*, **6**, 33–61, and is reproduced here with kind permission from Kluwer Academic Publishers.

Wiebe van der Hoek
Utrecht University, The Netherlands and Liverpool University, UK.

John-Jules Meyer
Utrecht University, The Netherlands.

Bernd van Linder
ABN-AMRO, Amsterdam, The Netherlands.

BIBLIOGRAPHY

[Alchourrón *et al.*, 1985] C.E. Alchourrón, P. Gärdenfors, and D. Makinson. On the logic of theory change: partial meet contraction and revision functions. *Journal of Symbolic Logic*, 50:510–530, 1985.

[Arens *et al.*, 1993] Y. Arens, C.Y. Chee, C-N. Hsu, and C.A. Knoblock. Retrieving and integrating data from multiple information sources. *International Journal on Intelligent and Cooperative Information Systems*, 2(2):127–158, 1993.

[Barwise, 1989] J. Barwise. *The Situation in Logic*, volume 17 of *CSLI Lecture Notes*. CSLI, Stanford, 1989.

[Castelfranchi, 1994] C. Castelfranchi. Private communication. 1994.

[Castelfranchi, 1994a] C. Castelfranchi. Guaranties for autonomy in cognitive agent architecture. In M. Wooldridge and N.R. Jennings, editors, *Intelligent Agents. Proceedings of the 1994 ECAI Workshop on Agent Theories, Architectures, and Languages*, volume 890 of *Lecture Notes in Artificial Intelligence*, pages 56–70. Springer-Verlag, 1994.

[Chellas, 1980] B.F. Chellas. *Modal Logic. An Introduction.* Cambridge University Press, Cambridge, 1980.

[Cohen, 1990] P.R. Cohen and H.J. Levesque. Intention is choice with commitment. *Artificial Intelligence*, 42:213–261, 1990.

[Dunin-Keplicz and Radzikowska, 1995] B. Dunin-Keplicz and A. Radzikowska. Epistemic approach to actions with typical effects. In C. Froidevaux and J. Kohlas, editors, *Symbolic and Quantitative Approaches to Reasoning and Uncertainty (Proceedings of ECSQARU'95)*, volume 946 of *Lecture Notes in Computer Science (subseries LNAI)*, pages 180–188. Springer-Verlag, 1995.

[Dunin-Keplicz and Treur, 1994] B. Dunin-Keplicz and J. Treur. Compositional formal specification of multi-agent systems. In M. Wooldridge and N.R. Jennings, editors, *Intelligent Agents. Proceedings of the 1994 ECAI Workshop on Agent Theories, Architectures, and Languages*, volume 890 of *Lecture Notes in Artificial Intelligence*, pages 102–117. Springer-Verlag, 1994.

[Elgesem, 1993] D. Elgesem. *Action Theory and Modal Logic.* PhD thesis, Institute for Philosophy, University of Oslo, Oslo, Norway, 1993.

[Gärdenfors, 1988] P. Gärdenfors. *Knowledge in Flux: Modeling the Dynamics of Epistemic States.* The MIT Press, Cambridge, Massachusetts and London, England, 1988.

[Gärdenfors, 1992] P. Gärdenfors, editor. *Belief Revision.* Cambridge University Press, 1992.

[Goldblatt, 1992] R. Goldblatt. *Logics of Time and Computation*, volume 7 of *CSLI Lecture Notes.* CSLI, Stanford, 1992. Second edition.

[Halpern and Moses, 1992] J.Y. Halpern and Y. Moses. A guide to completeness and complexity for modal logics of knowledge and belief. *Artifical Intelligence*, 54:319–379, 1992.

[Harel, 1984] D. Harel. Dynamic logic. In D.M. Gabbay and F. Guenthner, editors, *Handbook of Philosophical Logic*, volume 2, chapter 10, pages 497–604. D. Reidel, Dordrecht, 1984.

[Hintikka, 1962] J. Hintikka. *Knowledge and Belief.* Cornell University Press, Ithaca, NY, 1962.

[van der Hoek, 1993] W. van der Hoek. Systems for knowledge and beliefs. *Journal of Logic and Computation*, 3(2):173–195, 1993.

[van der Hoek *et al.*, 1998] W. van der Hoek, B. van Linder, and J.-J. Ch. Meyer. An integrated modal approach to rational agents. In M. Wooldridge and A. Rao, editors, *Foundations of Rational Agency*, volume 14 of *Applied Logic* series, pp. 133–168. Kluwer, Dordrecht, 1998.

[van der Hoek *et al.*, 2000] W. van der Hoek, B. van Linder, and J.-J. Ch. Meyer. On agents that have the ability to choose. *Studia Logica*, 65:79–119, 2000.

[Huang, 1990] Z. Huang. Logics for belief dependence. In E. Börger, H. Kleine Büning, M.M. Richter, and W. Schönfeld, editors, *Computer Science Logic, 4th Workshop CSL'90*, volume 533 of *Lecture Notes in Computer Science*, pages 274–288. Springer-Verlag, 1991.

[Huang, 1994] Z. Huang. *Logics for Agents with Bounded Rationality.* PhD thesis, Universiteit van Amsterdam, 1994.

[Huang *et al.*, 1992] Z. Huang, M. Masuch, and L. Pólos. ALX, an action logic for agents with bounded rationality. Technical Report 92-70, Center for Computer Science in Organization and Management, University of Amsterdam, October 1992. To appear in Artificial Intelligence.

[Kenny, 1975] A. Kenny. *Will, Freedom and Power.* Basil Blackwell, Oxford, 1975.

[Kraus and Lehmann, 1988] S. Kraus and D. Lehmann. Knowledge, belief and time. *Theoretical Computer Science*, 58:155–174, 1988.

[Lesperance *et al.*, 1994] Y. Lesperance, H. Levesque, F. Lin, D. Marcu, R. Reiter, and R. Scherl. A logical approach to high-level robot programming – a progress report. To appear in Control of the Physical World by Intelligent Systems, Working Notes of the 1994 AAAI Fall Symposium, New Orleans, LA., 1994.

[Levesque, 1994] H. Levesque. Knowledge, action and ability in the situation calculus. Overheads from invited talk at TARK 1994.

[Levy *et al.*, 1994] A.Y. Levy, Y. Sagiv, and D. Srivastava. Towards efficient information gathering agents. In *Working Notes of the AAAI Spring Symposium on Software Agents*, pages 64–70, Stanford, California, 1994.

[van Linder *et al.*, 1994] B. van Linder, W. van der Hoek, and J.-J. Ch. Meyer. Communicating rational agents. In B. Nebel and L. Dreschler-Fischer, editors, *KI-94: Advances in Artificial Intelligence*, volume 861 of *Lecture Notes in Computer Science (subseries LNAI)*, pages 202–213. Springer-Verlag, 1994.

[van Linder *et al.*, 1994a] B. van Linder, W. van der Hoek, and J.-J. Ch. Meyer. Tests as epistemic updates. In A.G. Cohn, editor, *Proceedings of the 11th European Conference on Artificial Intelligence (ECAI'94)*, pages 331–335. John Wiley & Sons, 1994.

[van Linder *et al.*, 1995] B. van Linder, W. van der Hoek, and J.-J. Ch. Meyer. Actions that make you change your mind. In A. Laux and H. Wansing, editors, *Knowledge and Belief in Philosophy and Artificial Intelligence*, pages 103–146. Akademie Verlag, 1995.

[van Linder *et al.*, 2001] B. van Linder, W. van der Hoek, and J.-J. Ch. Meyer. The dynamics of default reasoning. In *DRUMS Handbook*, Volume 6, pp. 127–160. Kluwer, Dordrecht, 2001.

[Meyer and van der Hoek, 1992] J.-J. Ch. Meyer and W. van der Hoek. A modal logic for nonmonotonic reasoning. In W. van der Hoek, J.-J. Ch. Meyer, Y.H. Tan, and C. Witteveen, editors, *Non-Monotonic Reasoning and Partial Semantics*, pages 37–77. Ellis Horwood, Chichester, 1992.

[Meyer and van der Hoek, 1995] J.-J. Ch. Meyer and W. van der Hoek. *Epistemic Logic for AI and Computer Science.* Cambridge University Press, 1995.

[Meyer *et al.*, 1999] J.-J. Ch. Meyer, W. van der Hoek and B. van Linder. A logical approach to the dynamics of commitments. *Artificial Intelligence*, 113:1–40, 1999.

[Moore, 1980] R.C. Moore. Reasoning about knowledge and action. Technical Report 191, SRI International, 1980.

[Moore, 1984] R.C. Moore. A formal theory of knowledge and action. Technical Report 320, SRI International, 1984.

[Rao and Georgeff, 1991] A.S. Rao and M.P. Georgeff. Asymmetry thesis and side-effect problems in linear time and branching time intention logics. In *Proceedings of the Twelfth International Joint Conference on Artificial Intelligence (IJCAI91)*, pages 498–504, 1991.

[Rao and Georgeff, 1991a] A.S. Rao and M.P. Georgeff. Modeling rational agents within a BDI-architecture. In J. Allen, R. Fikes, and E. Sandewall, editors, *Proceedings of the Second International Conference on Principles of Knowledge Representation and Reasoning*, pages 473–484, San Mateo CA, 1991. Mogan Kaufmann.

[Rao and Georgeff, 1993] A.S. Rao and M.P. Georgeff. A model-theoretic approach to the verification of situated reasoning systems. In *Proceedings of the Thirteenth International Joint Conference on Artificial Intelligence (IJCAI'93)*, pages 318–324, 1993.

[Shoham, 1993] Y. Shoham. Agent-oriented programming. *Artificial Intelligence*, 60:51–92, 1993.

[Shoham and Cousins, 1994] Y. Shoham and S.B. Cousins. Logics of mental attitudes in AI. In G. Lakemeyer and B. Nebel, editors, *Foundations of Knowledge Representation and Reasoning*, volume 810 of *Lecture Notes in Computer Science (subseries LNAI)*, pages 296–309. Springer-Verlag, 1994.

[Thomas, 1993] S.R. Thomas. *PLACA, An Agent Oriented Programming Language*. PhD thesis, Department of Computer Science, Stanford University, Stanford CA, September 1993. Appeared as technical report STAN-CS-93-1487.

[Wooldridge, 1994] M. Wooldridge. *The Logical Modelling of Computational Multi-Agent Systems*. PhD thesis, Department of Computation, UMIST, Manchester, October 1994. Appeared as technical report MMU-DOC-94-01.

[Wooldridge and Fisher, 1992] M. Wooldridge and M. Fisher. A first-order branching time logic of multi-agent systems. In B. Neumann, editor, *Proceedings of the 10th European Conference on Artificial Intelligence (ECAI'92)*, pages 234–238. John Wiley & Sons, 1992.

[von Wright, 1963] G.H. von Wright. *Norm and Action*. Routledge & Kegan Paul, London, 1963.

J.-J. CH. MEYER, W. VAN DER HOEK
AND B. VAN LINDER

MOTIVATIONAL ATTITUDES IN THE KARO FRAMEWORK

1 INTRODUCTION

The formalisation of rational agents is a topic of continuing interest in Artificial Intelligence. Research on this subject has held the limelight ever since the pioneering work of Moore [1980] and Morgenstern [1986; 1987] in which knowledge and actions are considered. Over the years important contributions have been made on both *informational* attitudes like knowledge and belief [Halpern and Moses, 1992], and *motivational* attitudes like commitments and obligations [Cohen and Levesque, 1990]. Recent developments include the Belief-Desire-Intention architecture [Rao and Georgeff, 1991b], logics for the specification and verification of multi-agent systems [Wooldridge and Jennings, 1995], and cognitive robotics [Lespérance *et al.*, 1996].

In a series of papers [Linder *et al.*, 1998; Hoek *et al.*, 1994a; Linder *et al.*, 1994b; Hoek *et al.*, 1994b; Linder *et al.*, 1995b; Linder *et al.*, 1995a] we defined a *theorist* logic for rational agents, i.e. a formal system that may be used to *specify*, *analyse* and *reason about* the behaviour of rational agents. In the basic framework [Linder *et al.*, 1998; Hoek *et al.*, 1994a], the *knowledge*, *belief* and *abilities* of agents, as well as the *opportunities* for and the *results* of their actions are formalised. In this so-called KARO framework it can for instance be modelled that an agent knows that some action is *correct* to bring about some state of affairs since it knows that performing the action will lead to the state of affairs, and that it knows that an action is *feasible* in the sense that the agent knows of its ability to perform the action.

Having dealt with both informational attitudes and various aspects of action in previous work, this paper is aimed at providing a formalisation of the *motivational* attitudes of rational agents. In the last decade various formalisations of different kinds of motivational attitudes have been proposed [Cohen and Levesque, 1990; Rao and Georgeff, 1991a; Singh, 1994]. The approach presented in this paper makes three main contributions to the theory of formalising motivational attitudes. Firstly, we consider a fairly wide scope of motivational attitudes, situated at two different levels. At the *assertion* level, this is the level where operators deal with assertions, we consider *wishes* and *goals*. At the *practition* level, where operators range over actions, we define *commitments*. With respect to these commitments we introduce both an operator modelling the commitments that an agent

J.J.Ch. Meyer and J. Treur (eds.),
Handbook of Defeasible Reasoning and Uncertainty Management Systems, Vol. 7, 341–356.

has made, and an action which models the act of committing. The notions that we formalise avoid (most of) the well-known problems that plague formalisations of motivational attitudes. Secondly, our formalisation of the various notions is strictly ***bottom up***. That is, after defining the fundamental notion of wishes, goals are defined in terms of wishes, and commitments are introduced using the notion of goals. In this way, we provide a formalisation of motivational attitudes that does not have to resort to 'tricks' like (circularly) defining the intention to do an action in terms of the goal to have done it. Lastly, in our formalisation we will also try to connect to some relevant insights on motivational attitudes as they have been gained in the philosophical research on practical reasoning, in particular some ideas of the philosopher Von Wright [1963].

2 THE KARO FRAMEWORK

In this section we present a core language that is rich enough to reason about some of the agent's attitudes mentioned above–other operators will be defined on top of this– and indicate formal models for this language. Doing so, we will try not to lose ourselves in technical details (these can be found elsewhere, [Hoek *et al.*, 1994a; Linder *et al.*, 1994b; Linder *et al.*, 1994a; Linder *et al.*, 1995a]), but rather provide the reader with an intuitive grasp for the ideas underlying our formal definitions.

2.1 Language

The language $\mathcal{L}$ that we use to formalise these notions is based on a fixed set of propositional atoms, and the connectives $\wedge, \vee, \rightarrow, \neg$ to build formulas $\varphi, \psi, \ldots$ with their usual meaning; the canonical tautology $\top$ is defined to be $p \vee \neg p$ for p some arbitrary propositional atom, and the canonical contradiction $\perp$ is defined to be $\neg\top$. We denote the pure propositional language with $\mathcal{L}_0$. We extend this core language to deal with actions and epistemic and motivational attitudes.

Let i be a variable over a set of agents $\{1, \ldots, n\}$. Actions in the set Ac are either atomic actions (At $= \{a, b, \ldots\}$) or composed $(\alpha, \beta, \ldots)$ by means of confirmation of formulas ($\texttt{confirm}\ \varphi$), sequencing $(\alpha; \beta)$, conditioning ($\texttt{if}\ \varphi\ \texttt{then}\ \alpha\ \texttt{else}\ \beta$) and repetition ($\texttt{while}\ \varphi\ \texttt{do}\ \alpha$). (Actions that are either atomic or `confirm` actions will be called ***semi-atomic*** in this paper.) These actions α can then be used to build new formulas to express the possible ***result*** of the execution of α by agent i (the formula $\langle do_i(\alpha)\rangle\varphi$ denotes that φ is a result of i's execution of α), the ***opportunity*** for i to perform α ($\langle do_i(\alpha)\rangle\top$) and i's ***capability*** of performing the action α ($\mathbf{A}_i\alpha$). The formula $[do_i(\alpha)]\varphi$ is shorthand for $\neg\langle do_i(\alpha)\rangle\neg\varphi$, thus expressing that all possible results of performance of α by i imply φ, thereby being noncommittal about the agent's opportunity to perform α. We shall refer to the language built out of the constructs mentioned in this paragraph as the *core language*, denoted as $\mathcal{L}_c$. So $\mathcal{L}_c \subseteq \mathcal{L}$.

In order to successfully complete an action, both the opportunity and the ability to perform the action are necessary (we call this combination the ***practical possibility*** to do the action). Although these notions are interconnected, they are surely not identical: the abilities of agents comprise mental and physical powers, moral capacities, and physical possibility, whereas the opportunity to perform actions is best described by the notion of circumstantial possibility (cf. [Kenny, 1975]). To formalise the knowledge of agents on their practical (im)possibilities, we introduce the so-called Can-predicate and Cannot-predicate. These are binary predicates, pertaining to a pair consisting of an action and a proposition, and denoting that an agent knows that performing the action constitutes a practical (im)possibility to bring about the proposition. We consider practical possibility to consist of two parts, viz. correctness and feasibility: action α is *correct* with respect to φ iff $\langle \mathrm{do}_i(\alpha)\rangle\varphi$ holds and α is *feasible* iff $\mathbf{A}_i\alpha$ holds.

DEFINITION 1. The Can-predicate and the Cannot-predicate are, for all agents i, actions α and formulae φ, defined as follows in terms of practical possibilities.

- $\mathbf{PracPoss}_i(\alpha, \varphi) \triangleq \langle \mathrm{do}_i(\alpha)\rangle\varphi \wedge \mathbf{A}_i\alpha$
- $\mathbf{Can}_i(\alpha, \varphi) \triangleq \mathbf{K}_i\mathbf{PracPoss}_i(\alpha, \varphi)$
- $\mathbf{Cannot}_i(\alpha, \varphi) \triangleq \mathbf{K}_i\neg\mathbf{PracPoss}_i(\alpha, \varphi)$

Thus the Can-predicate and the Cannot-predicate express the agent's knowledge about its practical possibilities and impossibilities, respectively. Therefore these predicates are important for the agent's planning of actions.

In this paper we only consider ***deterministic*** actions, i.e. actions that have ***at most one*** successor state. In this case, the diamond-formula $\langle do_i(\alpha)\rangle\varphi$ is stronger than the box-formula$[do_i(\alpha)]\varphi$. To be more precise, we have the equivalence expressed by $\langle do_i(\alpha)\rangle\varphi \leftrightarrow ([do_i(\alpha)]\varphi \wedge \langle do_i(\alpha)\rangle\top)$. The diamond-formula expresses that agent i has the opportunity to perform α, and that φ is one of its effects, whereas the box-formula only asserts that agent i would end up in a situation in which φ holds if i would perform α, and as such it says nothing about the opportunity for i to do α.

Roughly speaking, the language $\mathcal{L}$ allows for ***objective*** formulas, i.e., formulas about an actual state of affairs, and we have ***operators*** to make assertions about the agents' (informational and motivational) attitudes, capacities and dynamics. Such operators can be ***practitional***, when their argument is an action (like in $\mathbf{A}_i\alpha$ saying that i is able of doing α, or $\mathbf{Com}_i\alpha$ which expresses that i is committed to α). But operators can also be ***assertional***, when their argument is a formula (like in $\mathbf{Goal}_i\varphi$—agent i has φ as a goal).

In such a case, one could also speak about just a *modal* operator. In this dichotomy, $\langle do_i(\alpha)\rangle\varphi$ (and, for that matter, α, $[do_i(\alpha)]\varphi$) lives in both worlds: it expresses the practitional fact that i has the opportunity to do α, and the assertional fact that φ is one of its possible effects. However, it is also important to note that the informational and motivational attitudes of the agents not only can be expressed at a *declarative* level (agent i believes φ, is committed to α), but are also considered at an *operational* level (agent i updates its beliefs with φ, or commits itself to α).

Whereas the compositional behaviours of results, abilities and opportunities are a main issue in our research, we feel that in order to dress up agents with cognitive features, one also has to come up with specific instances of atomic actions. Section 3 adds 'motivational' actions to the atomic kernel of actions. For any action α defined so far, we add the action $\texttt{commit\_to}\,\alpha$ to the set of actions. In some situations, agents cannot fulfil their promises. Therefore, we also consider $\texttt{uncommit}\,\alpha$ as a basic action. Our language is then completed with operators that deal with motivational attitudes. At the *assertion* level, we consider *wishes* and *goals*, represented as $\mathbf{W}_i$ and $\mathbf{Goal}_i$, respectively. At the *practition* level we define *commitments*, represented as $\mathbf{Com}_i$.

2.2 Semantics

Recalling that the full language described above is denoted with $\mathcal{L}$, we will now briefly indicate how to formally interpret formulas of $\mathcal{L}$. We use the following kind of Kripke models $M = \langle W, \pi, \mathtt{D}, \mathtt{M}\rangle$ where W is a non-empty set of worlds. In order to determine whether a formula $\varphi \in \mathcal{L}$ is true in w (if so, we write $(M, w) \models \varphi$, where (M, w) is called a *state*), π, $\mathtt{P}$, and $\mathtt{M}$ encode how to interpret the propositional atoms and the Dynamic and Motivational operators, respectively. For the propositional part, we stipulate $\mathcal{M}, s \models p$ iff $\pi(s)(p) = \mathit{true}$, and the logical connectives are interpreted in $\mathcal{M}, s$ as expected. We are going to extend the definition of $\mathcal{M}, s \models \varphi$ to arbitrary φ; if $\mathcal{M}$ is clear from context, we abbreviate $\{s \mid \mathcal{M}, s \models \varphi\}$ to $[\![\varphi]\!]$.

Let us agree to call a modal operator X to be a necessity operator for a relation R (or a function f) if the truth of $X\varphi$ at state M, w is defined as the truth of φ in all states s for which Rws (or $s \in f(w)$, respectively). Then, the $\mathtt{D}$-part of a model consists of two functions: a result function $\mathtt{r}$ such that, for each atomic action a and agent i, $\mathtt{r}(i, a)(\mathcal{M}, w)$ yields all resulting states (M', w') of i doing a in w. We will explain below why we generalize the function $\mathtt{r}$ from a world transformation (as it is usually defined in dynamic logic, cf. [Goldblatt, 1992]) to a transformation on states (M, w). $\mathtt{D}$ also contains a capability function $\mathtt{c}$ where $\mathtt{c}(i, a)(w)$ is true exactly for those atomic actions a that i is capable of in w. The function $\mathtt{r}$ is extended to arbitrary actions in a way standard for dynamic logic (cf. [Goldblatt, 1992]) and then $[do_i(\alpha)]$ is interpreted as the necessity operator for this extended

$\mathbf{r}^*$. In a similar fashion, c is extended for arbitrary actions. Then we have:
$\mathcal{M}, s \models [do_i(\alpha)]\varphi$ iff $\mathcal{M}, t \models \varphi$ for all $\mathcal{M}, t \in \mathbf{r}^*(\mathcal{M}, s)$
$\mathcal{M}, s \models \mathbf{A}_i\varphi$ iff $\mathbf{c}^*(\mathcal{M}, s) = \mathit{true}$

DEFINITION 2. We now give the extensions of r and c, respectively. Recall that we assume that all actions are *deterministic*. For convenience, we introduce the special state $\mathcal{E} = (\mathcal{M}, \epsilon)$, where ϵ is a world where all impossible actions lead to: from there, no other actions can be performed anymore.

$$
\begin{array}{ll}
\mathbf{r}^*(i,a)(\mathcal{M},s) & = \mathbf{r}(i,a)(\mathcal{M},s)\\
\mathbf{r}^*(i,\mathtt{confirm}\,\varphi)(\mathcal{M},s) & = (\mathcal{M},s) \text{ if } \mathcal{M},s\models\varphi \text{ and } \mathcal{E} \text{ otherwise}\\
\mathbf{r}^*(i,\alpha_1;\alpha_2)(\mathcal{M},s) & = \mathbf{r}^*(i,\alpha_2)(\mathbf{r}^*(i,\alpha_1)(\mathcal{M},s))\\
\mathbf{r}^*(i,\mathtt{if}\ \varphi\ \mathtt{then}\ \alpha_1 & = \mathbf{r}^*(i,\alpha_1)(\mathcal{M},s) \text{ if } \mathcal{M},s\models\varphi \text{ and}\\
\qquad \mathtt{else}\ \alpha_2\ \mathtt{fi})(\mathcal{M},s) & \quad \mathbf{r}^*(i,\alpha_2)(\mathcal{M},s) \text{ otherwise}\\
\mathbf{r}^*(i,\mathtt{while}\ \varphi & = (\mathcal{M}',s') \text{ iff } \exists k\in\mathbb{N}\exists\mathcal{M}_0,s_0\ldots\exists\mathcal{M}_k,s_k\\
\qquad \mathtt{do}\ \alpha_1\ \mathtt{od})(\mathcal{M},s) & \quad [\mathcal{M}_0,s_0=\mathcal{M},s\,\&\,\mathcal{M}_k,s_k=\mathcal{M}',s'\,\&\,\forall j<k\\
 & \quad [\mathcal{M}_{j+1},s_{j+1}=\mathbf{r}^*(i,\mathtt{confirm}\,\varphi;\alpha_1)(\mathcal{M}_j,s_j)]\\
 & \qquad \&\,\mathcal{M}',s'\models\neg\varphi]\\
\text{where } \mathbf{r}^*(i,\alpha)(\mathcal{E}) & = \mathcal{E}\\
\text{and} & \\
\mathbf{c}^*(i,a)(\mathcal{M},s) & = \mathbf{c}(i,a)(s)\\
\mathbf{c}^*(i,\mathtt{confirm}\,\varphi)(\mathcal{M},s) & = \mathit{true} \text{ if } \mathcal{M},s\models\varphi \text{ and } \mathit{false} \text{ otherwise}\\
\mathbf{c}^*(i,\alpha_1;\alpha_2)(\mathcal{M},s) & = \mathbf{c}^*(i,\alpha_1)(\mathcal{M},s)\,\&\,\mathbf{c}^*(i,\alpha_2)(\mathbf{r}^*(i,\alpha_1)(\mathcal{M},s))\\
\mathbf{c}^*(i,\mathtt{if}\ \varphi\ \mathtt{then}\ \alpha_1 & = \mathbf{c}^*(i,\mathtt{confirm}\,\varphi;\alpha_1)(\mathcal{M},s) \text{ or}\\
\qquad \mathtt{else}\ \alpha_2\ \mathtt{fi})(\mathcal{M},s) & \quad \mathbf{c}^*(i,\mathtt{confirm}\,\neg\varphi;\alpha_2)(\mathcal{M},s)\\
\mathbf{c}^*(i,\mathtt{while}\ \varphi & = \mathit{true} \text{ if } \exists k\in\mathbb{N}[\mathbf{c}^*(i,(\mathtt{confirm}\,\varphi;\alpha_1)^k;\\
\qquad \mathtt{do}\ \alpha_1\ \mathtt{od})(\mathcal{M},s) & \qquad \mathtt{confirm}\,\neg\varphi)(\mathcal{M},s)=\mathit{true}]\\
 & \quad \text{and } \mathit{false} \text{ otherwise}\\
\text{where } \mathbf{c}^*(i,\alpha)(\mathcal{E}) & = \mathit{true}
\end{array}
$$

Note that the clauses above only indicate how composed actions are interpreted. One main feature of our approach is the possibility to define actions that may effect the semantic units: actions may change the set of worlds that an agent considers epistemically possible, desirable, and also the sequence of actions it is planning to perform next. This will be exploited further in the following sections. With regard to the abilities of agents, the motivation for the choices made in Definition 2 is the following. The definition of $\mathbf{c}(i, \mathtt{confirm}\,\varphi)$ expresses that an agent is able to get confirmation for a formula φ iff φ holds. An agent is capable of performing a sequential composition $\alpha_1; \alpha_2$ iff it is capable of performing α_1 (now), and it is capable of executing α_2 after it has performed α_1. An agent is capable of performing a conditional composition, if either it is able to get confirmation for the condition and thereafter perform the then-part, or it is able to confirm the negation of the condition and perform the else-part afterwards. An agent is capable of performing a repetitive composition $\texttt{while}\,\varphi\,\texttt{do}\,\alpha\,\texttt{od}$ iff it is able to perform the action $(\texttt{confirm}\,\varphi; \alpha_1)^k; \texttt{confirm}\,\neg\varphi$ for some natural number k, i.e. it is able to perform the k-th unwinding of the while-loop. Also note that in the ';-clause' for $\mathtt{c}^*$ we have chosen to model *optimistic* agents: as a consequence, we have that any agent finds itself capable of performing $\alpha; \beta$, whenever it lacks the opportunity to perform α (for a discussion on this choice, cf. [Linder, 1996]).

Dealing with motivational attitudes of agents is the part where we depart significally from standard modal approaches. In the next section we will give the instantiation of the M-part of the model.

3 MOTIVATIONAL ATTITUDES

As explained in the introduction a rational agent does not only possess information processing capabilities; it should also be endowed with motivations and should be able to modify these. In this section we will discuss how these 'motivational attitudes' can be incorporated into our modal framework.

Informally speaking, we will equip an agent with an *agenda*, which contains the action(s) that the agent has committed itself to and is supposed to perform. Formally, this agenda will take the form of a function that yields for a given agent in a given possible world the set of actions that the agent is committed to in that world. The main technical difficulty that has to be solved is the proper representation in the model of the way the agent maintains its agenda while performing actions. Intuitively speaking, the agent will have to drop from its agenda the actions that it has performed already. Although this sounds rather commonsense and clear, the formal implementation of this idea turns out to be rather involved, as we shall see, mainly due to the need for considering the right mix of syntax and semantics of the actions in the agenda.

Formally, we go about as follows. We start with the formalisation of the concept of a *wish* or *desire*. We then continue with that of the notions of selecting wishes which may then be committed to (and next possibly also be uncommitted to). We will again interpret the acts of selecting, committing and uncommitting as model transformations rather than mere world transformations. The act of selecting changes a model by affecting the set of choices, and the act of (un)committing transforms the agent's agenda.

We extend the language as follows. We build the motivational part on top of the core language $\mathcal{L}_c$. (At the moment we have no need for more complicated formulas expressing, for example, that one wishes to believe something, or one is committed to a revision operator, but in principle this could be done if one likes.) The language $\mathcal{L}^{\mathrm{M}}$ is obtained by adding the following clauses:

- if $\varphi \in \mathcal{L}_c$ and $i \in \mathcal{A}$ then $\mathbf{W}_i\varphi \in \mathcal{L}^{\mathrm{M}}$
- if $\varphi \in \mathcal{L}_c$ and $i \in \mathcal{A}$ then $\Diamond_i\varphi \in \mathcal{L}^{\mathrm{M}}$
- if $\varphi \in \mathcal{L}_c$ and $i \in \mathcal{A}$ then $\mathbf{C}_i\varphi \in \mathcal{L}^{\mathrm{M}}$
- if $\alpha \in \mathrm{Ac}$ and $i \in \mathcal{A}$ then $\mathbf{Com}_i\alpha \in \mathcal{L}^{\mathrm{M}}$

We also consider an extended class Ac^{M} of actions that is the smallest superset of At closed under the clauses of the core language and such that

- if $\varphi \in \mathcal{L}_c$ then $\mathtt{select}\,\varphi \in \mathrm{Ac}^{\mathrm{M}}$
- if $\alpha \in \mathrm{Ac}$ then $\mathtt{commit\_to}\,\alpha \in \mathrm{Ac}^{\mathrm{M}}$
- if $\alpha \in \mathrm{Ac}$ then $\mathtt{uncommit}\,\alpha \in \mathrm{Ac}^{\mathrm{M}}$

3.1 Formalising wishes

In our approach we consider wishes to be the most primitive, fundamental motivational attitudes, that is to say, *in ultimo* agents are motivated to fulfil their wishes. Wishes are represented by means of a plain normal modal operator, i.e. wishes are straightforwardly interpreted as a necessity operator over an accessibility relation W.

Thus we instantiate the M-part of our models to cater for wishes as follows: The M-part of a model $\mathcal{M} = \langle W, \pi, \mathtt{D}, \mathtt{M}\rangle$ contains the functions $\mathrm{W} : \mathcal{A} \to \wp(W \times W)$, which determines the desirability relation of an agent in a state, and $\mathrm{C} : \mathcal{A} \times W \to \wp(\mathcal{L}_c)$ denoting the choices made by an agent in a state, and a function $\mathrm{Agenda} : \mathcal{A} \times W \to \wp(\mathrm{AcSeq})$, which records the commitments of agents, per state. Here AcSeq stands for the set of sequences of semi-atomic actions.

We can now interpret the $\mathbf{W}_i$ operator as usual:
$\mathcal{M}, s \models \mathbf{W}_i\varphi \Leftrightarrow \forall s' \in W((s,s') \in \mathrm{W}(i) \Rightarrow \mathcal{M}, s' \models \varphi)$

3.2 Selecting wishes

In order to transform wishes to goals, an agent has to first select candidate goals from its set of wishes on the basis of the criteria of unfulfilledness and implementability. In more ordinary language this means that an agent can choose a wish if it is (as yet) unfulfilled and implementable, i.e. the agent is able and has the opportunity to fulfil the wish by means of a finite sequence of atomic actions. Unfulfilledness of a formula φ is easily expressed in the language by means of the classical negation $\neg\varphi$. The notion of implementability is somewhat more involved. For this purpose we introduce an implementability operator $\Diamond_i$, which we interpret as follows:
$\mathcal{M}, s \models \Diamond_i\varphi \Leftrightarrow \exists k \in \mathbb{N} \exists a_1, \ldots, a_k \in \mathrm{At}(\mathcal{M}, s \models \mathbf{PracPoss}_i(a_1; \ldots; a_k, \varphi)),$

that is to say, φ is implementable by i if i has the practical possibility to perform a finite sequence of atomic actions yielding φ.

Having defined unfulfilledness and implementability, we can now formally introduce the `select` action.

DEFINITION 3. For model $\mathcal{M}$ with state s, $i \in \mathcal{A}$ and $\varphi \in \mathcal{L}_c$ we define:

$$\mathtt{r}(i, \mathtt{select}\,\varphi)(\mathcal{M}, s) = \begin{cases} \emptyset & \text{if } \mathcal{M}, s \models \neg \mathbf{W}_i\varphi \\ \mathtt{choose}(i, \varphi)(\mathcal{M}, s), s & \text{if } \mathcal{M}, s \models \mathbf{W}_i\varphi \end{cases}$$

where for $\mathcal{M} = \langle W, \pi, \mathtt{D}, \mathtt{M} \rangle$ with $\mathtt{M} = \langle \mathrm{W}, \mathrm{C}, \mathrm{Agenda} \rangle$ we define

$$\mathtt{choose}(i, \varphi)(\mathcal{M}, s) = \langle W, \pi, \mathtt{D}, \mathtt{M}' \rangle \text{ with } \mathtt{M}' = \langle \mathrm{W}, \mathrm{C}', \mathrm{Agenda} \rangle \text{ such that}$$
$$\mathrm{C}'(i', s') = \mathrm{C}(i', s') \text{ if } i \neq i' \text{ or } s \neq s'$$
$$\mathrm{C}'(i, s) \;\; = \mathrm{C}(i, s) \cup \{\varphi\}$$

$$\mathtt{c}(i, \mathtt{select}\,\varphi)(\mathcal{M}, s) = \mathbf{1} \Leftrightarrow \mathcal{M}, s \models \neg\varphi \wedge \Diamond_i\varphi$$

Finally we define the interpretation of the $\mathbf{C}_i$ operator:
$\mathcal{M}, s \models \mathbf{C}_i\varphi \Leftrightarrow \varphi \in \mathrm{C}(i, s)$

It can be shown [Linder *et al.*, 1997] the act of selecting a formula φ causes minimal change in the sense that the formula φ is marked to be chosen, and nothing else of the model is changed. This has as a corollary that, for example, wishes and implementability formulas remain true after selecting if they were so before.

3.3 Goals

Having defined wishes and selections, one might be tempted to straightforwardly define goals to be selected wishes, i.e. $\mathbf{Goal}_i\varphi \triangleq \mathbf{W}_i\varphi \wedge \mathbf{C}_i\varphi$. This definition is however not adequate to formalise the idea of goals being selected, ***unfulfilled, implementable*** wishes. In the selection operator the criteria of unfulfilledness and implementability have not been incorporated yet. So, an easy way to do this is just to add them. Therefore, goals are defined to be those wishes that are unfulfilled, implementable and selected.

DEFINITION 4. The $\mathbf{Goal}_i$ operator is for $i \in \mathcal{A}$ and $\varphi \in \mathcal{L}_c$ defined by:

- $\mathbf{Goal}_i\varphi \triangleq \mathbf{W}_i\varphi \wedge \neg\varphi \wedge \Diamond_i\varphi \wedge \mathbf{C}_i\varphi$

Below we state a few properties of wishes, selections and goals.

PROPOSITION 5. *For all $i \in \mathcal{A}$ and $\varphi \in \mathcal{L}_c$ we have:*

1. $\models \mathbf{W}_i\varphi \leftrightarrow \langle \mathrm{do}_i(\mathtt{select}\,\varphi)\rangle\top$
2. $\models \langle \mathrm{do}_i(\mathtt{select}\,\varphi)\rangle\top \leftrightarrow \langle \mathrm{do}_i(\mathtt{select}\,\varphi)\rangle\mathbf{C}_i\varphi$

3. $\models \neg \mathbf{A}_i \mathtt{select}\,\varphi \rightarrow [\mathrm{do}_i(\mathtt{select}\,\varphi)]\neg \mathbf{Goal}_i\varphi$

4. $\models \mathbf{PracPoss}_i(\mathtt{select}\,\varphi, \top) \leftrightarrow \langle \mathrm{do}_i(\mathtt{select}\,\varphi)\rangle \mathbf{Goal}_i\varphi$

5. $\models \varphi \Rightarrow \models \neg \mathbf{Goal}_i\varphi$

6. $(\varphi \rightarrow \psi) \rightarrow (\mathbf{Goal}_i\varphi \rightarrow \mathbf{Goal}_i\psi)$ *is not for all* $\varphi, \psi \in \mathcal{L}_c$ *valid*

7. $\mathbf{K}_i(\varphi \rightarrow \psi) \rightarrow (\mathbf{Goal}_i\varphi \rightarrow \mathbf{Goal}_i\psi)$ *is not for all* $\varphi, \psi \in \mathcal{L}_c$ *valid*

The first item of Proposition 5 states that agents have the opportunity to select all, and nothing but, their wishes. The second item formalises the idea that every choice for which an agent has the opportunity results in the selected wish being marked chosen. In the third item it is stated that whenever an agent is unable to select some formula, then selecting this formula will not result in it becoming one of its goals. The related item 4 states that all, and nothing but, practically possible selections result in the chosen formula being a goal. The fifth item states that no logically inevitable formula qualifies as a goal. Hence whenever a formula is valid this does not only not necessarily imply that it is a goal but it even necessarily implies that it is not. The last two items of Proposition 5 state that goals are neither closed under implications nor under known implications.

3.4 Commitments

After having defined goals in our framework, we go on with a formal description of how an agent can commit itself to a goal. To this end we introduce a $\mathtt{commit\_to}\,\alpha$ action that, when successful, will have as a result that the agent is committed to the action α. In order to accommodate our model to this action, we need some extra formal machinery.

In order to prepare grounds for the formal semantics of the $\mathtt{commit\_to}$ action we introduce the notion of a transition relation in the spirit of Structural Operational Semantics of Plotkin [1981]. This method is widely used in computer science and we can employ it here fruitfully to describe below what happens with the agent's agenda when it performs actions (from that agenda, so to speak).

In our set-up we consider transitions of the form $\langle \alpha, s\rangle \rightarrow_{i,a} \langle \alpha', s'\rangle$, where $\alpha, \alpha' \in \mathrm{Ac}, i \in \mathcal{A}, a$ semi-atomic and $s, s' \in W$. Such a transition expresses that if in state s, i has to perform the action α, after performing the (semi-) atomic action a, this leads to a state s' in which a remaining action α' has still to be performed. We use the symbol Λ for the empty action, with as property that $\Lambda; \alpha = \alpha; \Lambda = \alpha$. Furthermore, we use the projection function π_2, which is assumed to yield the second element of a pair.

Transitions are given by the following deductive system, often called a transition system:

DEFINITION 6. The transition system T is given by the following axioms:

- $\langle \alpha, s \rangle \rightharpoonup_{i,\alpha} \langle \Lambda, s' \rangle$ with $s' = \pi_2(\mathrm{r}(i,\alpha)(\mathcal{M}, s))$ if α is semi-atomic.
- $\langle \texttt{if}\ \varphi\ \texttt{then}\ \alpha_1\ \texttt{else}\ \alpha_2\ \texttt{fi}, s \rangle \rightharpoonup_{i,\texttt{confirm}\ \varphi} \langle \alpha_1, s \rangle$ if $s \models \varphi$
- $\langle \texttt{if}\ \varphi\ \texttt{then}\ \alpha_1\ \texttt{else}\ \alpha_2\ \texttt{fi}, s \rangle \rightharpoonup_{i,\texttt{confirm}\ \neg\varphi} \langle \alpha_2, s \rangle$ if $s \not\models \varphi$
- $\langle \texttt{while}\ \varphi\ \texttt{do}\ \alpha\ \texttt{od}, s \rangle \rightharpoonup_{i,\texttt{confirm}\ \varphi} \langle \alpha; \texttt{while}\ \varphi\ \texttt{do}\ \alpha\ \texttt{od}, s \rangle$ if $s \models \varphi$
- $\langle \texttt{while}\ \varphi\ \texttt{do}\ \alpha\ \texttt{od}, s \rangle \rightharpoonup_{i,\texttt{confirm}\ \neg\varphi} \langle \Lambda, s \rangle$ if $s \not\models \varphi$

and the following rule:

- $\frac{\langle \alpha_1, s \rangle \rightharpoonup_{i,a} \langle \alpha_1', s' \rangle}{\langle \alpha_1; \alpha_2, s \rangle \rightharpoonup_{i,a} \langle \alpha_1'; \alpha_2, s' \rangle}$

Next, we introduce for convenience's sake an Intend predicate, analogously to the Can predicate. This predicate expresses that the agent intends to do an action with some result if it can perform the action with this result and, moreover, he knows that this result is a goal.

DEFINITION 7. For $\alpha \in \mathrm{Ac}^{\mathrm{M}}, i \in \mathcal{A}$ and $\varphi \in \mathcal{L}_c$ we define:

- $\mathbf{Intend}_i(\alpha, \varphi) \triangleq \mathbf{Can}_i(\alpha, \varphi) \wedge \mathbf{K}_i\mathbf{Goal}_i\varphi$

Having established the formal prerequisites, we can now present the formal semantics of the `commit_to` action. Informally, a `commit_to` α is only successful if the agent intends to do α with some result φ. If it is successful, the agent's agenda is updated in the worlds it is in. Moreover, also in all other possible worlds that are related to this world, either by performing actions or by considering epistemic alternatives, the agenda is updated. The general idea is that if the agent performs (part of) an action that is in its agenda, the agenda in the resulting state will contain the remainder of the action that was in its agenda in the original state. Moreover, since we like to model that an agent knows what it is committed to, we also stipulate that epistemic alternatives of states contain the same agenda.

Finally, it is stipulated that an agent is able to commit to an action iff its agenda is empty, and is ready to receive another commitment, so to speak. This models a simple kind of agent which could be called a *single-minded* one. By varying this definition one may model other agents as well. However, for the sake of simplicity—it is very convenient to have to consider only at most one action sequence in the agenda—we have chosen this definition here.

DEFINITION 8. [1] For all models $\mathcal{M} = \langle W, \pi, \mathtt{D}, \mathtt{M} \rangle$ with state s, for all $i \in \mathcal{A}$ and $\alpha \in \mathrm{Ac}$ we define:

---

[1] In fact, in order for this definition to be well-defined regarding the agenda function some restrictions have to be put on the models regarding the function r_0 and the interpretation of the `confirm` actions. For ease of presentation these are omitted here. Details can be found in [Linder *et al.*, 1997].

$\mathtt{r}(i, \mathtt{commit\_to}\,\alpha)(\mathcal{M}, s) = \emptyset$ if $\mathcal{M}, s \models \neg\mathbf{Intend}_i(\alpha, \varphi)$ for all $\varphi \in \mathrm{C}(i, s)$
$\mathtt{r}(i, \mathtt{commit\_to}\,\alpha)(\mathcal{M}, s) = \mathcal{M}', s$
  with $\mathcal{M}' = \langle W, \pi, \mathtt{D}, \mathtt{M}'\rangle$ and $\mathtt{M}' = \langle \mathrm{W}, \mathrm{C}, \mathrm{Agenda}'\rangle$
  where Agenda$'$ is minimal such that it is closed under the following:
  for all $s' \in \mathrm{B}^k(i, s)$, $\mathrm{Agenda}'(i, s') = \mathrm{Agenda}(i, s') \cup \{\alpha\}$
  and for all $s', s'', s''' \in W, \alpha' \in \mathrm{Agenda}'(i, s')$
  such that, for some semi-atomic a,
  $\langle \alpha', s'\rangle \rightharpoonup_{i,a} \langle \alpha'', s''\rangle$ and $s''' \in \mathrm{B}^k(i, s'')$:
    $\mathrm{Agenda}'(i, s''') = \mathrm{Agenda}(i, s''') \cup \{\alpha''\}$
otherwise

$\mathtt{c}(i, \mathtt{commit\_to}\,\alpha)(\mathcal{M}, s) = \mathbf{1}$ iff $\mathrm{Agenda}(i, s) = \emptyset$

Again one can show [Linder *et al.*, 1997] that the commit_to action is minimal in the sense that only the agenda of agent i is updated and that only the agenda in the states that are affected in the sense described above are updated.

Next we define an operator $\mathbf{Com}_i$ that indicates that the agent i is committed to an action. We define the interpretation of this operator such that, in any state epistemically equivalent with the state it is in, the agent is committed to all actions that are (semantically equivalent to) 'initial parts' of the actions written in its agenda. In order to capture the notion of semantical equivalence of actions we use our transition systems again, and define the notion of a *computation run*.

DEFINITION 9. $\mathrm{CR}^{\mathrm{C}}_{\mathcal{M}}(i, \alpha, s) \ni a_1; a_2; \ldots ; a_n$ iff $\langle \alpha, s\rangle \rightharpoonup_{i,a_1} \langle \alpha_1, s_1\rangle \rightharpoonup_{i,a_2} \langle \alpha_2, s_2\rangle \rightharpoonup_{i,a_3} \ldots \rightharpoonup_{i,a_n} \langle \alpha_n, s_n\rangle$ for some $\alpha_1, \ldots, \alpha_n \in \mathrm{Ac}, s_1, \ldots, s_n \in W$, such that $\alpha_n = \Lambda$.

Note that due to the fact that our actions are deterministic, the set $\mathrm{CR}^{\mathrm{C}}_{\mathcal{M}}(i, \alpha, s)$ contains *at most one* element. This should be kept in mind while considering the definitions below. Actions that have the same computation run (with respect to a certain starting state) are considered to be semantically equivalent (in that state). So now we are able to give our interpretation of the $\mathbf{Com}_i$ operator.

$$\begin{aligned} &\mathcal{M}, s \models \mathbf{Com}_i\alpha \Leftrightarrow \\ &\quad \forall s' \in \mathrm{B}^k(i, s) \exists \alpha_1 \in \mathrm{CR}^{\mathrm{C}}_{\mathcal{M}}(i, \alpha, s') \exists \alpha_2 \in \mathrm{Agenda}(i, s') \\ &\quad \exists \alpha_2' \in \mathrm{CR}^{\mathrm{C}}_{\mathcal{M}}(i, \alpha_2, s')(\mathrm{Prefix}(\alpha_1, \alpha_2')) \end{aligned}$$

where Prefix stands for the prefix relation on computation runs (which are sequences of actions).

Finally, we note that to let an agent be really rational, it should also be capable in certain situations to abandon its commitments, for instance, when the goal is achieved or is not implementable any more. In the definition below this is put as follows: an $\mathbf{uncommit}\,\alpha$ action is only successful if the agent i was committed to the action α, and the agent is able to uncommit iff it does no longer intend to do α for any purpose φ.

DEFINITION 10. For all models $\mathcal{M} = \langle W, \pi, \mathtt{D}, \mathtt{M}\rangle$ with state s, for all $i \in \mathcal{A}$ and $\alpha \in \mathrm{Ac}$ we define:[2]

$$\mathtt{r}(i, \mathtt{uncommit}\,\alpha)(\mathcal{M}, s) = \emptyset \text{ if } \mathcal{M}, s \models \neg\mathbf{Com}_i\alpha$$
$$\mathtt{r}(i, \mathtt{uncommit}\,\alpha)(\mathcal{M}, s) = \mathcal{M}', s$$
with $\mathcal{M}' = \langle W, \pi, \mathtt{D}, \mathtt{M}'\rangle$ and $\mathtt{M}' = \langle \mathrm{W}, \mathrm{C}, \mathrm{Agenda}'\rangle$
where for all $s' \in \mathrm{B}^k(i, s)$,
$$\mathrm{Agenda}'(i, s') = \mathrm{Agenda}(i, s') \backslash \{\beta \mid \mathrm{Prefix}(\mathrm{CR}^{\mathrm{C}}_{\mathcal{M}}(i, \alpha, s'), \mathrm{CR}^{\mathrm{C}}_{\mathcal{M}}(i, \beta, s'))\}$$
and for all $s', s'', s''' \in W$ with $\alpha' \in \mathrm{Agenda}'(i, s')$ and such that, for some semi-atomic a, $\langle \alpha', s'\rangle \rightharpoonup_{i,a} \langle \alpha'', s''\rangle$ and $s''' \in \mathrm{B}^k(i, s'')$,
$$\mathrm{Agenda}'(i, s'') = \mathrm{Agenda}(i, s'') \backslash \{\beta \mid \mathrm{Prefix}(\mathrm{CR}^{\mathrm{C}}_{\mathcal{M}}(i, \alpha'', s''), \mathrm{CR}^{\mathrm{C}}_{\mathcal{M}}(i, \beta, s''))\}$$
otherwise

$$\mathtt{c}(i, \mathtt{uncommit}\,\alpha)(\mathcal{M}, s) = \mathbf{1} \text{ iff } \mathcal{M}, s \models \neg\mathbf{Intend}_i(\alpha, \varphi) \text{ for all } \varphi \in \mathrm{C}(i, s)$$

The complication in this definition, as compared to that of the $\mathtt{commit\_to}$ operator, is due to the fact that 'committedness' is closed under taking prefixes of (computation runs of) actions, so that in order to successfully uncommit to an action α also all actions that have α as a prefix (with respect to computation runs) should be removed from the agent's agenda.

Some properties of the operators treated in this section are given in the following proposition.

PROPOSITION 11. *For all $i \in \mathcal{A}$, $\alpha, \beta \in \mathrm{Ac}$ and $\varphi \in \mathcal{L}_c$ we have:*

1. $\models \mathbf{Intend}_i(\alpha, \varphi) \to \langle \mathrm{do}_i(\mathtt{commit\_to}\,\alpha)\rangle\top$
2. $\models \langle \mathrm{do}_i(\mathtt{commit\_to}\,\alpha)\rangle\top \leftrightarrow \langle \mathrm{do}_i(\mathtt{commit\_to}\,\alpha)\rangle\mathbf{Com}_i\alpha$
3. $\models \mathbf{Com}_i\alpha \to \neg\mathbf{A}_i\mathtt{commit\_to}\,\beta$
4. $\models [\mathrm{do}_i(\mathtt{commit\_to}\,\alpha)]\neg\mathbf{A}_i\mathtt{commit\_to}\,\beta$
5. $\models \mathbf{Com}_i\alpha \leftrightarrow \langle \mathrm{do}_i(\mathtt{uncommit}\,\alpha)\rangle\neg\mathbf{Com}_i\alpha$
6. $\models \mathbf{Intend}_i(\alpha, \varphi) \to \neg\mathbf{A}_i\mathtt{uncommit}\,\alpha$
7. $\models (\mathbf{C}_i\varphi \leftrightarrow \mathbf{K}_i\mathbf{C}_i\varphi) \to (\mathbf{A}_i\mathtt{uncommit}\,\alpha \leftrightarrow \mathbf{K}_i\mathbf{A}_i\mathtt{uncommit}\,\alpha)$

[2] Actually, in order to let this notion be well-defined, we need certain minimality conditions which we omit here for simplicity's sake (for a more rigorous treatment, see [Linder *et al.*, 1997])

8. $\models \mathbf{Com}_i\alpha \wedge \neg\mathbf{Can}_i(\alpha, \top) \rightarrow \mathbf{Can}_i(\mathtt{uncommit}\,\alpha, \neg\mathbf{Com}_i\alpha)$

In the third item it is stated that being committed prevents an agent from having the ability to (re)commit. The fourth item states that the act of committing is ability-destructive with respect to future commit actions, i.e. by performing a commitment an agent loses its ability to make any other commitments. Item 5 states that being committed is a necessary and sufficient condition for having the opportunity to uncommit; as mentioned above, agents have the opportunity to undo all of their commitments. In item 6 it is stated that agents are (morally) unable to undo commitments to actions that are still known to be correct and feasible to achieve some goal. In item 7 it is formalised that agents know of their abilities to uncommit to some action. The last item states that whenever an agent is committed to an action that is no longer known to be practically possible, it knows that it can undo this impossible commitment.

Finally, the following proposition states some intuitive properties of the $\mathbf{Com}_i$ operator with respect to complex actions.

PROPOSITION 12. *For all* $i \in \mathcal{A}$, $\alpha, \alpha_1, \alpha_2 \in \text{Ac}$ *and all* $\varphi \in \mathcal{L}_c$ *we have:*

1. $\models \mathbf{Com}_i\alpha \rightarrow \mathbf{K}_i\mathbf{Com}_i\alpha$

2. $\models \mathbf{Com}_i(\alpha_1;\alpha_2) \rightarrow \mathbf{Com}_i\alpha_1 \wedge \mathbf{K}_i[\text{do}_i(\alpha_1)]\mathbf{Com}_i\alpha_2$

3. $\models \mathbf{Com}_i\mathtt{if}\,\varphi\,\mathbf{then}\,\alpha_1\,\mathbf{else}\,\alpha_2\,\mathtt{fi} \wedge \mathbf{K}_i\varphi \rightarrow \mathbf{Com}_i(\mathtt{confirm}\,\varphi;\alpha_1)$

4. $\models \mathbf{Com}_i\mathtt{if}\,\varphi\,\mathbf{then}\,\alpha_1\,\mathbf{else}\,\alpha_2\,\mathtt{fi} \wedge \mathbf{K}_i\neg\varphi \rightarrow \mathbf{Com}_i(\mathtt{confirm}\,\neg\varphi;\alpha_2)$

5. $\models \mathbf{Com}_i\mathtt{while}\,\varphi\,\mathbf{do}\,\alpha\,\mathbf{od} \wedge \mathbf{K}_i\varphi \rightarrow$
$\mathbf{Com}_i((\mathtt{confirm}\,\varphi;\alpha);\mathtt{while}\,\varphi\,\mathbf{do}\,\alpha\,\mathbf{od})$

The first item of Proposition 12 states that commitments are known. The second item states that a commitment to a sequential composition $\alpha_1;\alpha_2$ of actions implies a commitment to the initial part α_1, and that the agent knows that after execution of this initial part α_1 it will be committed to the remainder α_2. The third and fourth item formalise the rationality of agents with regard to their commitments to conditionally composed actions. The last item concerns the unfolding of a while-loop: if an agent is committed to a while-loop while knowing the condition of the loop to be true, then the agent is also committed to the then-part of the while-loop.

4 CONCLUSION

In this paper we have presented a formalisation of motivational attitudes, the attitudes that explain why agents act the way they do. This formalisation concerns operators both on the assertion level, where operators range over propositions, and on the practition level, where operators range over actions. An important feature of our formalisation is the attention paid to the acts associated with selecting between wishes and with (un)committing to actions. Starting from the primitive notion of wishes, we defined goals to be selected, unfulfilled, implementable wishes. Commitments may be made to actions that are known to be correct and feasible with respect to some goal and may be undone whenever the action to which an agent has committed itself has either become impossible or useless. Both the act of making, and the act of undoing commitments are formalised as model-transforming actions in our framework. The actions that an agent is committed to are recorded in its agenda in such a way that commitments are closed under prefix-taking and under practical identity, i.e. having identical computation runs. On the whole our formalisation is a rather expressive one, which tries to be faithful to a certain extent to both commonsense intuition and philosophical insights.

There is an extensive literature on motivational attitudes. Probably the most influential account of motivational attitudes is due to Cohen & Levesque [1990]. Starting from the primitive notions of implicit goals and beliefs, Cohen & Levesque define so-called persistent goals, which are goals which agents give up only when they think they are either satisfied or will never be true, and intentions, both ranging over propositions and over actions. The idea underlying persistent goals is similar to that underlying our notion of goals. In the framework of Cohen & Levesque agents intend to bring about a proposition if they intend to do some action that brings about the proposition. An agent intends to do an action if it has the persistent goal to have done the action. In our approach we do not use such a reduction technique.

Another important formalisation of motivational attitudes is proposed by Rao & Georgeff [1991b] in their BDI-architecture. Treating desires and intentions as primitive, Rao & Georgeff focus on the process of intention revision rather than the 'commitment acquisition' which is essential in our formalisation. Both desires and intentions in their framework suffer from the problems associated with logical omniscience, which we have avoided in our formalisation of goals. (Not in our treatment of wishes, though, but these only play a subsidiary role to get to goals.)

ACKNOWLEDGEMENTS

This paper contains condensed and simplified material from [Meyer *et al.*, 1999].

J.-J. Ch. Meyer, W. van der Hoek and B. van Linder
Utrecht University, The Netherlands.

B. van Linder is currently at ABN AMRO Bank N.V.

BIBLIOGRAPHY

[Cohen and Levesque, 1990] P.R. Cohen and H.J. Levesque. Intention is choice with commitment. ***Artificial Intelligence***, 42:213–261, 1990.

[Goldblatt, 1992] R. Goldblatt. ***Logics of Time and Computation***, volume 7 of ***CSLI Lecture Notes***. CSLI, Stanford, 1992. Second edition.

[Halpern and Moses, 1992] J.Y. Halpern and Y. Moses. A guide to completeness and complexity for modal logics of knowledge and belief. ***Artificial Intelligence***, 54:319–379, 1992.

[Hoek *et al.*, 1994a] W. van der Hoek, B. van Linder, and J.-J. Ch. Meyer. A logic of capabilities. In A. Nerode and Yu. V. Matiyasevich, editors, ***Proceedings of the Third International Symposium on the Logical Foundations of Computer Science (LFCS'94)***, volume 813 of ***Lecture Notes in Computer Science***, pages 366–378. Springer-Verlag, 1994.

[Hoek *et al.*, 1994b] W. van der Hoek, B. van Linder, and J.-J. Ch. Meyer. Unravelling nondeterminism: On having the ability to choose (extended abstract). In P. Jorrand and V. Sgurev, editors, ***Proceedings of the Sixth International Conference on Artificial Intelligence: Methodology, Systems, Applications (AIMSA'94)***, pages 163–172. World Scientific, 1994.

[Kenny, 1975] A. Kenny. ***Will, Freedom and Power***. Basil Blackwell, Oxford, 1975.

[Lespérance *et al.*, 1996] Y. Lespérance, H. Levesque, F. Lin, D. Marcu, R. Reiter, and R. Scherl. Foundations of a logical approach to agent programming. In M. Wooldridge, J.P. Müller, and M. Tambe, editors, ***Intelligent Agents Volume II – Agent Theories, Architectures, and Languages***, volume 1037 of ***Lecture Notes in Computer Sciencc (subseries LNAI)***, pages 331–347. Springer-Verlag, 1996.

[Linder *et al.*, 1994a] B. van Linder, W. van der Hoek, and J.-J. Ch. Meyer. Communicating rational agents. In B. Nebel and L. Dreschler-Fischer, editors, ***KI-94: Advances in Artificial Intelligence***, volume 861 of ***Lecture Notes in Computer Science (subseries LNAI)***, pages 202–213. Springer-Verlag, 1994.

[Linder *et al.*, 1994b] B. van Linder, W. van der Hoek, and J.-J. Ch. Meyer. Tests as epistemic updates. In A.G. Cohn, editor, ***Proceedings of the 11th European Conference on Artificial Intelligence (ECAI'94)***, pages 331–335. John Wiley & Sons, 1994.

[Linder *et al.*, 1995a] B. van Linder, W. van der Hoek, and J.-J. Ch. Meyer. Actions that make you change your mind. In A. Laux and H. Wansing, editors, ***Knowledge and Belief in Philosophy and Artificial Intelligence***, pages 103–146. Akademie Verlag, 1995.

[Linder *et al.*, 1995b] B. van Linder, W. van der Hoek, and J.-J. Ch. Meyer. The dynamics of default reasoning (extended abstract). In C. Froidevaux and J. Kohlas, editors, ***Symbolic and Quantitative Approaches to Reasoning and Uncertainty***, volume 946 of ***Lecture Notes in Computer Science (subseries LNAI)***, pages 277–284. Springer-Verlag, 1995. Full version to appear in ***Data & Knowledge Engineering***.

[Linder *et al.*, 1998] B. van Linder, W. van der Hoek, and J.-J. Ch. Meyer. Formalizing abilities and opportunities of agents. ***Fundamenta Informaticae***, 34(1,2):53–101, 1998.

[Linder *et al.*, 1997] B. van Linder, J.-J. Ch. Meyer, and W. van der Hoek. Formalising motivational attitudes of agents using the karo framework. Technical Report UU-CS-1997-03, Utrecht University, 1997.

[Linder, 1996] B. van Linder. ***Modal Logics for Rational Agents***. PhD thesis, Utrecht University, 1996.

[Meyer *et al.*, 1999] J.-J. Ch. Meyer, W. van der Hoek and B. van Linder. A logical approach to the dynamics of commitments. ***Artificial Intelligence***, 113:1–40, 1999.

[Moore, 1980] R.C. Moore. Reasoning about knowledge and action. Technical Report 191, SRI International, 1980.

[Morgenstern, 1986] L. Morgenstern. A first order theory of planning, knowledge and action. In *Proceedings of the 1st Conference on Theoretical Aspects of Reasoning about Knowledge (TARK86)*,, pp. 99–114, 1986.

[Morgenstern, 1987] L. Morgenstern. Knowledge preconditions for actions and plans. In *Proceedings of the Tenth Internatinal Joint Conference on Artificial Intelligence (IJCAI-87)*, pp. 867–874, Milan, Italy, 1987.

[Plotkin, 1981] G. Plotkin. A structural approach to operational semantics. Technical Report DAIME FN-19, Aarhus University, 1981.

[Rao and Georgeff, 1991a] A.S. Rao and M.P. Georgeff. Asymmetry thesis and side-effect problems in linear time and branching time intention logics. In J. Mylopoulos and R. Reiter, editors, *Proceedings of the Twelfth International Joint Conference on Artificial Intelligence (IJCAI'91)*, pages 498–504. Morgan Kaufmann, 1991.

[Rao and Georgeff, 1991b] A.S. Rao and M.P. Georgeff. Modeling rational agents within a BDI-architecture. In J. Allen, R. Fikes, and E. Sandewall, editors, *Proceedings of the Second International Conference on Principles of Knowledge Representation and Reasoning (KR'91)*, pages 473–484. Morgan Kaufmann, 1991.

[Singh, 1994] M.P. Singh. *Multiagent Systems*, volume 799 of *LNAI*. Springer-Verlag, Berlin-Heidelberg, 1994.

[Wooldridge and Jennings, 1995] M. Wooldridge and N. R. Jennings. Intelligent agents: Theory and practice. *The Knowledge Engineering Review*, 10(2):115–152, 1995.

[Wright, 1963] G.H. von Wright. *Norm and Action.* Routledge & Kegan Paul, London, 1963.

FRANK DIGNUM AND BERND VAN LINDER

MODELLING SOCIAL AGENTS: TOWARDS DELIBERATE COMMUNICATION

1 INTRODUCTION

The formalization of rational agents is a topic of continuing interest in AI. Research on this subject has held the limelight ever since the pioneering work of Moore [1985] in which knowledge and actions are considered. Over the years contributions have been made on both *informational* attitudes like knowledge and belief [Meyer and van der Hoek, 1995] and *motivational* attitudes like intentions and commitments [Cohen and Levesque, 1990; Dignum *et al.*, 1996].

In our basic framework [van der Hoek *et al.*, 1994; van Linder, 1996] we modelled the informational attitudes of agents as well as various aspects of action by means of a theory about the *knowledge, belief* and *abilities* of agents, as well as the *opportunities* for, and the *results* of their actions. In this framework it can for instance be modelled that an agent knows that it is able to perform an action and that it knows that it is correct to perform that action to bring about some result.

In [Dignum *et al.*, 1996; van Linder *et al.*, 1996] we dealt with the motivational attitudes of agents. In these papers we defined the concepts of *wishes*, *goals*, *intentions*, and *commitments* or *obligations*. By combining this formalization with the basic framework it is for instance possible to model the fact that an agent prefers some situation to hold and it also knows that it is able to achieve that situation by performing a sequence of actions. Furthermore it can be modelled that after an agent commits itself to achieve a goal it is obliged to perform those actions that bring about its goal.

Finally, in [Dignum and Weigand, 1995; van Linder *et al.*, 1994] we formalized communication between agents. In this theory we can show both the communication itself as well as the consequences of communication. For instance, if some authorised agent gives orders to another agent to perform a certain action, the latter agent will be obliged to perform the action. Also if an authorised agent asserts a fact to another agent, the latter agent will believe this fact to be true. The examples show that the communication can change both the informational attitudes of agents as well as the behaviour of agents.

In this paper we intend to bring the different fragments of the framework together in one all-embracing formal system. That is, we will define a model for the following concepts: *belief, knowledge, action, wish, goal, decision,*

J.J.Ch. Meyer and J. Treur (eds.),
Handbook of Defeasible Reasoning and Uncertainty Management Systems, Vol. 7, 357–380.

intention, commitment, obligation and *communication.* Following [Dignum *et al.*, 1996; van Linder *et al.*, 1996] we base this model on dynamic logic [Harel, 1979], which is extended with epistemic, doxastic, temporal and deontic (motivational) operators. The semantics will be based on Kripke structures with a variety of relations imposed on the states. We characterise this integrated framework by pointing out some differences with other work.

In the area of Multi-Agent Systems much research is devoted to the coordination of the agents. Many papers have been written about protocols (like contract net) that allow agents to negotiate and cooperate (e.g. [Müller, 1996; Davis and Smith, 1983]). Most of the cooperation between agents is based on the assumption that they have some joint goal or intention. Such a joint goal enforces some type of cooperative behaviour on all agents (see e.g. [Cohen and Levesque, 1991; Jennings, 1993; Sandu, 1996]). The conventions according to which the agents coordinate their behaviour is hard-wired into the protocols that the agents use to react to the behaviour (cq. messages) of other agents.

This raises several issues. The first issue is that, although agents are said to be autonomous, they always react in a predictable way to each message. Namely their response will follow the protocol that was built-in. The question then arises how autonomous these agents actually are. How do these communication protocols fit into the goal directed behaviour of the agents? Of course the protocol will fit in case the agents have a joint goal to start with, but what happens when an agent gets an unexpected message? In order to react appropriately to such a message the agent has to be able to reason about the intended effect of it and it should be able to figure out a response that is in line with its own goals. This is what we would like to call *deliberate communication.*
Note that we do not argue that agents do not have to deliberate when they follow a protocol. The counterproposal to a proposal message can be made through very intricate reasoning about the possible strategy of the other party! However, no deliberation is needed about what type of message is sent (only an accept, reject or counterproposal is usually possible).

Besides autonomy, an important characteristic of agents is that they can react to a changing environment. However, if the protocols that they use to react to (at least some part of) the environment are fixed, they have no ways to respond to changes. For instance, if an agent notices that another agent is cheating it cannot switch to another protocol to protect itself. In general it is difficult (if not impossible) for agents to react to violations of the conventions by other agents.

As was also argued in [Norman *et al.*, 1996], autonomous agents need a richer communication protocol than contract net (or similar protocols) to be able to retain their autonomy. A greater autonomy of the agent places a higher burden on the communication. An autonomous agent might negotiate over every request it gets. In this paper we will describe a mechanism to avoid excessive communication. It is similar to the one employed in [Norman *et al.*, 1996], but defined more formally and still more generally applicable.

In this paper we will describe the different components and their relationships informally. Even though we will give a precise definition of our language and the models used to interpret this language, we only have room to sketch the actual semantics.

2 THE CONCEPTS

The concepts that we formalize can roughly be situated at four different levels: the informational level, the action level, the motivational level and the social level. We will introduce the concepts of each of these levels in the following subsections.

2.1 The Informational Level

At the informational level we consider both knowledge and belief. Many formalizations have been given of these concepts and we will follow the more common approach in epistemic and doxastic logic: the formula $K_i\phi$ denotes the fact that agent i knows ϕ and $B_i\phi$ that agent i believes ϕ. Both concepts are interpreted in a Kripke-style semantics, where each of the operators is interpreted by a relation between a possible world and a set of possible worlds determining the formulas that the agent knows respectively believes. We demand knowledge to obey an S5 axiomatisation, belief to validate a KD45 axiomatisation, and agents to believe all the things that they know. Of course we are aware that the above axiomatization of knowledge and beliefs can lead to some problems about logical omiscience on the one hand and to problems with conflicting beliefs on the other hand. However, we chose to use the most simple axiomatization here in order not to complicate the semantics too much. It is possible, without altering the other levels, to choose for more complex and/or realistic axiomatizations of knowledge and belief.

2.2 The Action Level

At the action level we consider both dynamic and temporal notions. The main dynamic notion that we consider is that of actions, which we interpret as functions that map some some state of affairs into another one. Following [van der Hoek *et al.*, 1994; Wieringa *et al.*, 1989] we use parameterised actions to describe the event consisting of a particular agent's execution of an action. We let $\alpha(i)$ indicate that agent i performs the action α. It is also possible to compose actions into an action expression using operators for the parallel execution of actions "&", the choice between actions "+", the sequential composition of actions ";" and the negation of an action "-". Action expressions do not play a major role in this paper where we want to lay out the basic concepts, but they can be used at any place where actions are used.

The results of actions are modelled using dynamic logic as described by Harel in [Harel, 1979]. We use $[\alpha(i)]\phi$ to indicate that *if* agent i performs the action indicated by α the result will be ϕ. Note that it does not state anything about whether the action will actually be performed. So, it might for instance be used to model a statement like: 'If I jump over 2.5m high I will be the world record holder'.

Besides these formulas that indicate the results of actions we also would like to express that an agent has the reliable opportunity to perform an action. This is done through the predicate OPP: $OPP(\alpha(i))$ indicates that agent i has the opportunity to do α, i.e. the event $\alpha(i)$ will possibly take place. In this paper we say that an agent i has the opportunity to perform α if there is some way for i to perform α. The agent not necessarily has the opportunity to perform α in any possible way! From this observation it follows that:

$$OPP(\alpha_1) \rightarrow OPP(\alpha_1 + \alpha_2)$$

but also that

$$OPP(\alpha_1) \not\rightarrow OPP(\alpha_1 \& \alpha_2)$$

In [van der Hoek *et al.*, 1994] we also used an *ability* predicate. This predicate was used to indicate that an agent has the inherent possibility to perform a certain action (at a certain place and time). So, whether an agent can perform an action depends on both its inherent ability and the external opportunity to perform the action. In the present paper the abilities of agents do not play a major role. Therefore we left them out of the formalization and assume (for the present) that the abilities of an agent are incorporated in the opportunities for an agent to perform an action.

Besides the OPP operator, which already has a temporal flavor to it, we introduce two genuinely temporal operators: $PREV$, denoting the events that actually just took place, and the "standard" temporal operator $NEXT$, which indicates, in our case, which event will actually take place next. Note that the standard dynamic logic operator "$\langle\rangle$" can only be used to indicate the *possible* next action. That is, we can use $\langle\alpha\rangle true$ to indicate from the present state a next state can be reached by performing α. However, to denote the *actual* next action a new operator is needed! See [Dignum, 1992] for a more in depth discussion of this issue. A similar argument holds for the $PREV$ operator for actions. We also define a more traditional temporal logic $NEXT$ operator on formulas in terms of the $NEXT$ operator on events.

$$NEXT(\phi) \text{ iff } NEXT(\alpha(i)) \wedge [\alpha(i)]\phi$$

This means that the formula ϕ is true in all next states iff when an action $\alpha(i)$ is performed next the formula ϕ is true after the performance of $\alpha(i)$.

For the $NEXT$ and $PREV$ predicates we assume the following relations to hold:

1. $NEXT(\alpha_1 + \alpha_2) = NEXT(\alpha_1) \vee NEXT(\alpha_2)$
2. $NEXT(\alpha_1 \& \alpha_2) = NEXT(\alpha_1) \wedge NEXT(\alpha_2)$
3. $NEXT(\overline{\alpha}) = \neg NEXT(\alpha)$
4. $NEXT(\alpha_1; \alpha_2) = NEXT(\alpha_1) \wedge [\alpha_1]NEXT(\alpha_2)$

1. $PREV(\alpha_1 + \alpha_2) = PREV(\alpha_1) \vee PREV(\alpha_2)$
2. $PREV(\alpha_1 \& \alpha_2) = PREV(\alpha_1) \wedge PREV(\alpha_2)$
3. $PREV(\overline{\alpha}) = \neg PREV(\alpha)$
4. $PREV(\alpha_1; \alpha_2) = PREV(\alpha_2) \wedge Before(\alpha_2)PREV(\alpha_1)$

We will not make the operator $Before(\alpha)$ explicit in this paper due to space and time constraints. However, it can be expressed quite naturally using standard temporal operators.

2.3 The Motivational Level

At the motivational level we consider a variety of concepts, ranging from wishes, goals and decisions to intentions and commitments. The most fundamental of these notions is that of wishes. Formally, wishes are defined as the combination of implicit and explicit wishes, which allows us to avoid all kinds of problems that plague other formalizations of motivational attitudes. A formula ϕ is wished for by an agent i, denoted by $W_i\phi$, iff ϕ is true in all the states that the agent considers desirable, and ϕ is an element of a predefined set of (explicitly wished) formulas.

Goals are not primitive in our framework, but instead defined in terms of wishes. Informally, a wish of agent i constitutes one of i's goals iff i knows the wish not to be brought about yet, but implementable, i.e. i has the opportunity to achieve the goal. In terms of [Cohen and Levesque, 1990] our agents only have achievement goals. To formalize this notion, we first introduce the operator *Achiev*; $Achiev_i\phi$ means that agent i has the opportunity to perform some action which leads to ϕ, i.e.

$$Achiev_i\phi \equiv \exists\beta : [\beta(i)]\phi \wedge OPP(\beta(i))$$

A goal is now formally defined as a wish that does not hold but is achievable:

$$Goal_i\phi \equiv W_i\phi \wedge \neg\phi \wedge Achiev_i\phi$$

Note that our definition implies that there are three ways for an agent to drop one of its goals: since it no longer considers achieving the goal to be desirable, since the wish now holds, or since it is no longer certain that it can achieve the goal. This implies in particular that our agents will not indefinitely pursue impossible goals.
Goals can either be known or unconscious goals of an agent. Most goals will be known, but we will later on see that goals can also arise from commitments and these goals might not be known explicitly.
Because the goals also depend on a set of explicit wishes we can avoid the problem of having all the consequences of a goal as goal as well.

Intentions are divided in two categories, viz. the intention to perform an action and the intention to bring about a proposition. We define the intention of an agent to perform a certain action as primitive. We relate intentions and goals in two ways. Firstly, the intention to reach a certain state is defined as the goal to reach that state. The second way is through ***decisions***. An intention to perform an action is based on the decision to try to bring about a certain proposition. We assume a (total) ordering between the explicit wishes of each agent in each world. On the basis of this ordering the agent can make a decision to try to achieve the goal that has the highest preference. Because the order of the wishes may differ in each world, this does not mean that once a goal has been fixed the agent will always keep on trying to reach that goal (at least not straight away). As the result of deciding to do α, denoted by $DEC(i,\alpha)$, the agent has the intention to do α, denoted by $INT_i\alpha$. I.e. $[DEC(i,\alpha)]INT_i\alpha$.
The precondition above is described formally by

$$OPP(DEC(i,\alpha)) \text{ iff } \exists\phi\exists\beta : Goal_i\phi \wedge [\alpha;\beta(i)]\phi \wedge \neg\exists\psi(Goal_i\psi \wedge \phi <_i \psi)$$

There is no direct relation between the intention to perform an action and the action that is actually performed next. We do, however, establish an indirect relation between the two through a binary ***implementation*** predicate, ranging over pairs of actions. The idea is that the formula $IMP_i(\alpha_1,\alpha_2)$ expresses that for agent i executing α_2 is a reasonable attempt at executing α_1. For example, if I intend to jump over 1.5m and I jump over 1.4m it can be said that I tried to fulfil my intention, i.e. the latter action is within the intention of performing the first action. However, if instead of jumping over 1.5m I killed a referee it cannot be said anymore that I performed that action with the intention of jumping over 1.5m.

Having defined the binary IMP predicate, we may now relate intended actions to the actions that are actually performed. We demand the action that is actually performed by an agent to be an attempt to perform one of its intentions. Formally, this amounts to the formula

$$NEXT(\alpha_2(i)) \rightarrow (\exists\alpha_2 : INT(\alpha_1(i)) \wedge IMP_i(\alpha_1,\alpha_2))$$

being valid.

The last concept that we consider at the motivational level is that of commitment. This concept is also part of the social level if the commitment is made towards another agent. As the result of i performing a $COMMIT(i, j, \alpha)$ action the formula $O_{ij}\alpha$ becomes true. (See [Dignum and Weigand, 1995] for more details.) I.e. by committing itself to an action, an agent i obliges itself towards j to perform the action α. The commitment can be a private one if j is the same as i. In that case the result is an obligation of the agent towards itself to perform the action. Although the obligation does not ensure the actual performance of the action by the agent, it does have two practical consequences. If an agent commits itself to an action and afterwards does not perform the action a *violation* condition is registered, i.e. the state is not ideal (anymore). The registration of the violation is done through the introduction of a deontic relation between the worlds. This relation connects each world with the set of ideal worlds with respect to that world. More details about the formal semantics of this deontic operator can be found in [Dignum *et al.*, 1996].

Secondly, an obligation to perform an action leads to the goal of having performed the action. Formally this is achieved with the following formula:

$$O_{ij}\alpha \rightarrow W_i(PREV(\alpha(i)))$$

Note that this is sufficient to create a goal, because $PREV(\alpha(i))$ does not hold currently (the action is not performed yet when the obligation arises) and it is achievable (by performing the action $\alpha(i)$. The above formalisation is actually also a shortcut on the more cognitive correct assumption that the agent actually wants to avoid the violation of the obligation. I.e. $W_i(\neg Violation(O_{ij}\alpha))$. Depending on how severe are the consequences of this violation for the agent this wish will get a higher priority.

We only consider sincere agents and therefore we assume that an agent can only commit itself to actions that it intends to do eventually, i.e. intention provides a precondition for commitment.

2.4 The social component

The $COMMIT$ described in the previous section is one of the four types of *speech acts* [Searle, 1969] that play a role in the social component. Speech acts are used to communicate between agents. The result of a speech act is a change in the doxastic or deontic state of an agent, or in some cases a change in the state of the world. A speech act always involves at least two agents; a speaker and a hearer. If an agent sends a message to another agent but that agent does not "listen" (does not receive the message) the speech act is not successful.

The most important feature in which our framework for speech acts differs from other frameworks for speech acts (based on the work of Searle) is that a speech act in our framework is not just the sending of the message by an agent but is the composition of sending and receiving of a message by two (or more) agents!

We distinguish the following speech act types: ***commitments, directions, declarations*** and ***assertions***. The idea underlying a direction is that of giving orders, i.e. an utterance like 'Pay the bill before next week'. A typical example of a declaration is the utterance 'Herewith you are granted permission to access the database', and a typical assertion is 'I tell you that the earth is flat'. Each type of speech act should be interpreted within the background of the relationship between the speaker and the hearer of the speech act. In particular for directions and declarations the agent uttering the statement should have some kind of basis of authority for the speech act to have any effect.

We distinguish three types of relations between agents: *peer* relation, *power* relation and *authorization* relation. The first two relations are similar to the ones used in the ADEPT system [Norman *et al.*, 1996; Jennings *et al.*, 1996]. The power relation is used to model hierarchical relations between agents. We assume that these relations are fixed during the lifecycle of the agents. Within such a relation less negotiation is possible about requests and demands. This reduces the amount of communication and therefore increases the efficiency of the agents.

The peer relation exists between all agents that have no prior contract or obligations towards each other (with respect to the present communication). This relation permits extensive negotiations to allow a maximum of autonomy for the agents.

The last relation between agents is the authorization relation which is a type of temporary power relation that can be built up by the agents themselves.

The power relation is formalized as a partial ordering between the agents, which is expressed as follows: $i \ll j$ means that j has a higher rank than i. The authority relation is formalized through a binary predicate *auth*; $auth(i, \alpha)$ means that agent i is authorised to perform α. It seems that this specifies a property of one agent; however, the other agent is usually part of the specification of α. Therefore the authorization to perform an action implicitly determines an authorization relation between the agents involved in that action as well.

One way to create the authorisation relations is by agent j giving an implicit authorisation to i to give him some directives. For example, when agent i orders a product from agent j it implicitly gives the authorisation to agent j for demanding payment from i for the product (after delivery). We will see later that most communicative actions have also implicit components and effects that are usually determined by the context and conventions within which the communication takes place.

Besides the implicit way to create authorizations, they can also be created explicitly by a separate speech act which is formally a declaration that the authorization is true.

The speech acts themselves are formalised as meta-actions (based on earlier work [Dignum and Weigand, 1995]):

- $DIR(x, i, j, \alpha)$ formalises that agent i directs agent j to perform α on the basis of x, where x can be either *peer*, *power* or *authority*.
- $DECL(i, j, f)$ models the declaration of i in the "presence" of j that f holds.
- $ASS(x, i, j, f)$ formalises the assertion of i to agent j that f holds. on the basis of x, where x can be either *peer*, *power* or *authority*.
- $COMMIT(i, j, \alpha)$ describes that i commits itself towards j to performα.

Note that the commit and the declarative do not take a relation parameter. This is basically because the effect of a commit is the same irrespective of the relation between the agents, while the declarative has to be authorized not by the other party but by some third party.

A directive from agent i to agent j to perform α results in an obligation of j towards i to perform that action *if* agent i was either in a power relation towards j or was authorized to give the order. In a similar way the assertion of proposition f by i to j results in the fact that j will believe f *if* i had authority over j. Creating the authorizations is an important part of the negotiation between agents when they are establishing some type of contract. On the basis of the authorizations that are created during the negotiation some protocol for the transactions between the agents can be followed quick and efficiently. (See [Weigand *et al.*, 1996] for more details on contracts between agents). Formally, the following formulas hold for the effects of commitments, orders and declaratives:

- $[COMMIT(i, j, \alpha)]K_j([DECL(j, i, P_{ij}(\alpha(i)))]O_{ij}\alpha)$
- $auth(i, DIR(authority, i, j, \alpha)) \rightarrow [DIR(authority, i, j, \alpha)]K_jO_{ji}\alpha$
- $j \ll i \rightarrow [DIR(power, i, j, \alpha)]K_jO_{ji}\alpha$
- $[DIR(x, i, j, \alpha)]K_jINT_i\alpha(j)$
- $auth(i, DECL(i, j, f)) \rightarrow [DECL(i, j, f)]K_jf$
- $[DECL(i, j, f)]K_jW_if$
- $[ASS(x, i, j, f)]K_jB_if$
- $auth(i, ASS(authority, i, j, f)) \rightarrow [ASS(authority, i, j, f)]B_jf$

- $j \ll i \rightarrow [ASS(power, i, j, f)]B_j f$

A commitment always results in a kind of conditional obligation. The obligation is conditional on the permission of the agent towards which the commitment is made. (This is very close to the ACCEPT action in other frameworks). It is important that j knows that giving this permission results in an obligation for i. It means that j can use this in the plan (and/or goal) formation! Because we assume that $K_i\phi \rightarrow \phi$ it is also true in the world. The giving of permission is formally described by $[DECL(j, i, P_{ij}(\alpha(i)))]$, where $P_{ij}(\alpha(i)) \equiv \neg O_{ij}(\alpha(i))$. I.e. the permission to perform α is equivalent to the fact that there is no obligation to perform the negation of α. The permission of j is necessary because j might play a (passive) role in the action α initiated by i. Of course j must be willing to play its part. It signifies this by giving the permission to i. In contrast to the other speech acts no precondition has to hold for a commitment to obtain its desired result.

A directive from agent i results in an obligation of agent j (towards i) if agent i was authorised to give the order or i has a power relation towards j. Also in this case it is important that j knows that it has this obligation! If i has no authority or power over j then the directive is actually a request. It results in the fact j knows that i wants him to perform α. If j does not mind to perform α it can commit himself to perform α and create an obligation.

Assertions can be used to transfer beliefs from one agent to another. Note that agent j does not automatically believe what agent i tells him. We do assume that agents are sincere and thus we have the following axiom:

$$OPP(ASS(x, i, j, f)) \rightarrow B_i f$$

That is, an agent can only assert facts that it believes itself.
The only way to directly transfer a belief is when agent i is authorised to make a statement. Usually this situation arises when agent j first requested some information from i. Such a request for information (modelled by a directive without authorisation) gives an implicit authorisation on the assertions that form the answer to the request.

A declaration can change the state of the world if the agent making the declaration is authorised to do so. (This is the only speech act that has a direct effect on the states other than a change of the mental attitudes of the agents! Although this might not be obvious it follows from the fact that j knows f afterwards, which can only be the case if f is actually true).
If agent i has no authority to declare the fact, then the only result of the speech act is that i establishes a preference for itself. It prefers the fact to be true. Because j is "listening" to this declaration it also knows of this preference.

Although we do not attempt to give a (complete) axiomatization, we want to mention the following axioms for the declaratives, because they are very fundamental for creating relationships between agents.

$$[DECL(i,j,auth(j,DIR(authority,j,i,\alpha(i))))]$$
$$auth(j,DIR(authority,j,i,\alpha(i)))$$

which states that an agent i can create authorisations for an agent j concerning actions that i has to perform.
The following axiom is important for the acceptance of offers:

$$[DECL(i,j,P_{ji}(\alpha(i)))]P_{ji}(\alpha(i))$$

which states that an agent can always give permission to another agent to perform some action.
Note that it may very well be that another agent forbids j to perform α! The permission is only with respect to i!

Before we give a sketch of a formalization of all the above concepts, in the next section we will first illustrate how the communicative acts as defined above can be used to model the basic messages in other frameworks.

3 FORMAL COMMUNICATION

In the previous section we gave a brief overview of the basic messages that agents can use in our framework. To show the power of our framework and to show the relation with other work on communication between agents we show how the basic illocutions that are used for the negotiating agents in the ADEPT system (and that also form the heart of many other negotiation systems) can be modelled within our framework. We only show this for the negotiation because it forms an important part of the communication between agents.

The negotiating agents in the ADEPT system use the four illocutions: PROPOSE, COUNTERPROPOSE, ACCEPT and REJECT. These four illocutions also form the basic elements of many other negotiation systems.

The PROPOSE is directly translated into a COMMIT. The obligation that follows from a proposal depends on the acceptance of the receiving party. However, the ACCEPT that is used as primitive in ADEPT and most other systems involves more than the giving of permission that we already indicated above.

The ACCEPT message has three components. That is, we consider the ACCEPT to be the simultaneous expression of three illocutions.

1. Giving permission to perform the action
2. Commitment to perform those actions that are necessary to make the proposal succeed

3. Giving (implicit) authority for subsequent actions (linked to the proposal by convention. e.g. accepting a delivery means that you authorize the other party to demand payment)

For example if agent i sends the following message to j:

> PROPOSE,i,j,
> I will deliver 20 computers (pentium, 32M, etc.) to you for \$1000,- per computer

then the ACCEPT message of j to i:

> ACCEPT,j,i,
> You will deliver 20 computers (pentium, 32M, etc.) to me for \$1000,- per computer

means:

1. You are permitted to deliver the computers: $DECL(j, P_{ij}(deliver))$
2. I will receive the computers (sign a receipt): $COMMIT(j, i, receive)$
3. I give you authority to ask for payment after delivery: $DECL(j, [deliver]auth(i, DIR(authority, i, j, pay)))$

It is important to notice that only the first component of the meaning of the ACCEPT message is fixed. The other two components depend on the action involved and the conventions (contracts) under which the transaction is negotiated.

The REJECT message is the denegation of the ACCEPT message. It means that the agent is either not giving permission for the action, not committing itself to its part of the action or not willing to give authority to subsequent actions. Formally this could be expressed as the disjunction of the negation of these three parts. However, usually it is seen as not giving permission for the action. If the permission is not given then the other two parts are of no consequence any more.

The COUNTERPROPOSE is a composition of a REJECT and PROPOSE message. Formally it can thus be expressed as the parallel execution of these two primitives.

Besides the formal representation of the illocution of the message we can also give some preconditions on the basic message types. Only the PROPOSE message type does not have preconditions. This is as expected because the PROPOSE is used to start the negotiation. The other types of messages are all used as answer to a PROPOSE (or COUNTERPROPOSE) message. We can formally describe the precondition that these message types can only be used after a PROPOSE or COUNTERPROPOSE as follows:

- $OPP(ACCEPT(j,i,\alpha)) \leftrightarrow$
  $(PREV(PROPOSE(i,j,\alpha)) \vee$
  $PREV(COUNTERPROPOSE(i,j,\alpha)))$
- $OPP(REJECT(j,i,\alpha)) \leftrightarrow$
  $(PREV(PROPOSE(i,j,\alpha)) \vee$
  $PREV(COUNTERPROPOSE(i,j,\alpha)))$
- $OPP(COUNTERPROPOSE(j,i,\beta)) \leftrightarrow$
  $\beta \neq \alpha \wedge (PREV(PROPOSE(i,j,\alpha)) \vee$
  $PREV(COUNTERPROPOSE(i,j,\alpha)))$

In the precondition of the COUNTERPROPOSE we included the fact that a counterproposal should differ from the proposal that it counters. (Although not mentioned in this paper, the semantics of actions does give an equivalence relation between actions). More elaborate conversation rules are needed to describe long term dependencies within protocols. E.g. one cannot repeat the same proposal later on if it already has been rejected. These rules should be incorporated within the protocols that the agents are using.

We do not want to give the formalisation of complete protocols at this place due to space limitations. However, we can indicate quite easily the results of the most common pairs of messages where agent i first proposes something to agent j after which agent j can accept it, reject it or counterpropose it. These moves are formally described as follows:

- $[PROPOSE(i,j,\alpha)(i)][ACCEPT(j,i,\alpha)(j)]O_{ij}(\alpha(i)) \wedge P_{ji}(\alpha(i))$ (accept)

  Furthermore, if the success of $\alpha(i)$ depends on the performance of $\beta(j)$ by j:
  $[PROPOSE(i,j,\alpha)(i)][ACCEPT(j,i,\alpha)(j)]O_{ji}(\beta(j))$
  And if conventions determine that i can perform $\beta(i)$ after acceptance of the proposal then:
  $[PROPOSE(i,j,\alpha)(i)][ACCEPT(j,i,\alpha)(j)][\alpha(i)]auth(i,\beta(i))$
- $[PROPOSE(i,j,\alpha)(i)][REJECT(j,i,\alpha)(j)]\neg O_{ij}(\alpha(i)) \wedge \neg P_{ji}(\alpha(i))$ (reject)
- $[PROPOSE(i,j,\alpha)(i)][COUNTERPROPOSE(j,i,\beta)(j)]$
  $\neg O_{ij}(\alpha(i)) \wedge \neg P_{ji}(\alpha(i)) \wedge OPP(ACCEPT(i,j,\beta)(i)+$
  $REJECT(i,j,\beta)(i))$ (counter)

Note that the counterproposal has no effect of itself yet. Only the reject component of the counterproposal has immediate effect. The proposal component of the counterproposal only takes effect after an appropriate answer of i.

For the reject we only indicated that the obligation does not arise and there is also no permission to perform the action. The rest of the effect depends on the context and is usually not of prime interest.

The formalisation of the basic messages in the ADEPT system shows two things.
First, that our framework is powerful enough to formally describe the negotiation in the ADEPT system including the effects of the communication.
Secondly, that seemingly simple message types, like ACCEPT, have complicated meanings that partly depend on the context in which they are used.

Although in this section we have shown how some messages as used in ADEPT and their effects can be formally described in our framework this is only the start of deliberative communication. The next interesting step that we intend to take is to show why a certain reaction is given to a message from another agent given the situation and the goals of the agent. In principle it is possible to check in the above example whether a proposal should be accepted, rejected or countered given the goals of the agent and the effects of these answers. We leave a detailed exploration of this very interesting area for future work.

4 A SKETCH OF A FORMALIZATION

In this section we precisely define the language that we use to formally represent the concepts described in the previous section, and the models that are used to interpret this language. We will not go into too much detail with regard to the actual semantics, but try to provide the reader with an intuitive grasp for the formal details without actually mentioning them.

The language that we use is a multi-modal, propositional language, based on three denumerable, pairwise disjoint sets: Π, representing the propositional symbols, Ag representing agents, and At containing atomic action expressions. The language $FORM$ is defined in four stages. Starting with a set of propositional formulas ($PFORM$), we define the action- and meta-action expressions, after which $FORM$ can be defined.

The set Act of regular action expressions is built up from the set At of atomic (parameterised) action expressions using the operators ; (sequential composition), + (nondeterministic composition), & (parallel composition), and $^-$ (action negation). The constant actions **any** and **fail** denote 'don't care what happens' and 'failure' respectively.

DEFINITION 1. The set Act of action expressions is defined to be the smallest set closed under:

1. $At \cup \{\mathbf{any}, \mathbf{fail}\} \subseteq Act$
2. $\alpha_1, \alpha_2 \in Act \Longrightarrow \alpha_1;\alpha_2, \alpha_1 + \alpha_2, \alpha_1 \& \alpha_2, \overline{\alpha_1} \in Act$

The set $MAct$ of general action expressions contains the regular actions and all of the special meta-actions informally described in the previous section. For these meta-actions it is not always clear whether they can be performed in parallel or what the result is of taking the negation of a meta-action. This area needs a more thorough study in the future. For simplicity, we restrict ourselves in this paper to closing the set $MAct$ under sequential composition.

DEFINITION 2. The set $MAct$ of general action expressions is defined to be the smallest set closed under:

1. $Act \subseteq MAct$
2. $\alpha \in Act, i, j \in Ag, x \in \{peer, authority, power\}$
$$\Longrightarrow DEC(i,\alpha), COMMIT(i,j,\alpha), DIR(x,i,j,\alpha) \in MAct$$
3. $\gamma\alpha_1, \gamma\alpha_2 \in MAct \Longrightarrow \gamma\alpha_1; \gamma\alpha_2 \in MAct$

The complete language $FORM$ is now defined to contain all the constructs informally described in the previous section. That is, there are operators representing informational attitudes, motivational attitudes, aspects of actions, and the social traffic between agents.

DEFINITION 3. The language $FORM$ of formulas is defined to be the smallest set closed under:

1. $PFORM \subseteq FORM$
2. $\phi, \phi_1, \phi_2 \in FORM \Longrightarrow \neg\phi, \phi_1 \wedge \phi_2 \in FORM$
3. $\phi \in FORM, i \in Ag \Longrightarrow K_i\phi, B_i\phi \in FORM$
4. $\gamma\alpha \in MAct, \phi \in FORM \Longrightarrow [\gamma\alpha]\phi \in FORM$
5. $\psi, \phi \in FORM, i, j \in Ag, x \in \{peer, authority, power\} \Longrightarrow$
$$[DECL(i,j,\psi)]\phi, [ASS(x,i,j,\psi)]\phi \in FORM$$
6. $[\gamma\alpha]\phi, [\gamma\beta]\psi, \theta \in FORM \Rightarrow [\gamma\alpha; \gamma\beta]\theta \in FORM$
7. $\alpha \in Act, \phi \in FORM \Longrightarrow PREV(\alpha), OPP(\alpha), NEXT(\phi) \in FORM$
8. $\phi, \psi \in FORM, i, j \in Ag, \alpha, \alpha_1, \alpha_2 \in Act \Longrightarrow W_i\phi, \psi <_i \phi, INT_i\alpha, i \ll j$
$$IMP_i(\alpha_1, \alpha_2), O_{ij}(\alpha), auth(i,\alpha) \in FORM$$

Because the assert and declare speech acts range over formulas they can only be introduced at this stage!

The models used to interpret $FORM$ are based on Kripke-style possible worlds models. That is, the backbone of these models is given by a set Σ of states, and a valuation π on propositional symbols relative to a state.

Various relations and functions on these states are used to interpret the various (modal) operators. These relations and functions can roughly be classified in four parts, dealing with the informational level, the action level, the motivational level and the social level, respectively. We assume tt and ff to denote the truth values 'true' and 'false', respectively.

DEFINITION 4. A model Mo for $FORM$ from the set CMo is a structure $(\Sigma, \pi, I, A, M, S)$ where

1. Σ is a non-empty set of states and $\pi : \Sigma \times \Pi \rightarrow \{tt, ff\}$.

2. $I = (Rk, Rb)$ with $Rk : Ag \rightarrow \wp(\Sigma \times \Sigma)$ denoting the epistemic alternatives of agents and $Rb : Ag \times \Sigma \rightarrow \wp(\Sigma)$ denoting the doxastic alternatives.

3. $A = (Sf, Mf, Ropp, Rprev, Rnext)$ with $Sf : Ag \times Act \times \Sigma \rightarrow \wp(\Sigma)$ yielding the interpretation of regular actions, $Mf : Ag \times MAct \times (CMo \times \Sigma) \rightarrow (CMo \times \Sigma)$ yielding the interpretation of meta-actions, $Ropp : Ag \times \Sigma \rightarrow \wp(Act)$ denoting opportunities, $Rprev : \Sigma \rightarrow Act$ yielding the action that has been performed last and $Rnext : \Sigma \rightarrow Act$ yielding the action that will be performed next.

4. $M = (Rp, Rep, <, Ri, Ria, Ro)$ with $Rp : Ag \times \Sigma \rightarrow \wp(\Sigma)$ denoting implicit wishes, $Rep : Ag \times \Sigma \rightarrow \wp(FORM)$ yielding explicit wishes, $< \subseteq Ag \times \Sigma \rightarrow FORM \times FORM$ which is a preference relation on wishes, $Ri : Ag \times \Sigma \rightarrow \wp(Act)$ denoting intended actions, $Ria : Ag \times \Sigma \rightarrow \wp(Act) \times \wp(Act)$ denoting implementation relations between actions and $Ro : Ag \times Ag \rightarrow \wp(\Sigma \times \Sigma)$ denoting obligations.

5. $S = (Auth, \prec)$ with $Auth : Ag \times \wp(MAct) \rightarrow \{tt, ff\}$ yielding authorisations and $\prec: Ag \times Ag \rightarrow \{tt, ff\}$ yielding hierarchical relations between agents.

such that the following constraints are validated:

1. $Rk(i)$ is an equivalence relation for all i, and $Rb(i, s) \neq \emptyset$, $Rb(i, s) \subseteq \{s' \mid (s, s') \in Rk(i)\}$ and $(s, s') \in Rk(i) \Longrightarrow Rb(i, s) = Rb(i, s')$, which ensures that knowledge validates an S5 axiomatisation and belief obeys a KD45 axiomatisation, while agents indeed believe all things they know.

2. Sf yields the state-transition interpretation for regular actions. This function satisfies the usual constraints ensuring an adequate interpretation of composite actions in terms of their constituents. The function Mf models the model-transforming interpretation of meta-action. Below we elaborate on the definition of Mf for the meta-actions introduced in the previous section.

3. $Rnext(s) \in Ropp(i,s) \subseteq \{\alpha \mid Sf(i,\alpha,s) \neq \emptyset\}$, which ensures that opportunities are a subset of the actions that are possible by virtue of the circumstances and that the next action performed is an opportunity. Furthermore, $Rprev(s) = \alpha$ iff $\alpha \in Ropp(i,s')$ for some s' with $s \in Sf(i,\alpha,s')$, which relates previously executed actions to past opportunities.

4. $Ri(i,s) \subseteq \{\alpha \mid Sf(i,\alpha,s) \neq \emptyset\}$ and for all $s \in \Sigma$ some $s' \in \Sigma$ exists with $(s,s') \in Ro$.

The complete semantics contains an algebraic semantics of action expresses, based on the action semantics of Meyer [Meyer, 1988]. In this paper we will abstract from the algebraic interpretation of actions and instead interpret actions as functions on states of affairs. For the meta-actions the state-transition interpretation is not adequate, because meta-actions do not change states but they change relations between states. For instance, in the case of an assertion, the effect is to change the doxastic state of the receiving agent, and nothing else. To formalize this behaviour, we interpret meta-actions as model-transforming functions. In the case of an assertion, the resulting model will differ from the starting model in the doxastic accessibility relation of the receiving agent.

DEFINITION 5. The binary relation $\models$ between an element of $FORM$ and a pair consisting of a model Mo in CMo and a state s in Mo is for propositional symbols, conjunctions and negations defined as usual. Epistemic formulas $K_i\phi$ and doxastic formulas $B_i\phi$ are interpreted as necessity operators over Rk and Rb respectively. For the other formulas $\models$ is defined as follows:

$$
\begin{array}{lcl}
Mo, s \models [\alpha(i)]\phi & \Longleftrightarrow & Mo, s' \models \phi \text{ for all } s' \in Sf(i,\alpha,s) \\
Mo, s \models [\gamma\alpha(i)]\phi & \Longleftrightarrow & Mo', s' \models \phi \text{ for all } Mo', s' \in Mf(i,\alpha,Mo,s) \\
Mo, s \models PREV(\alpha(i)) & \Longleftrightarrow & \alpha(i) \in Rprev(s) \\
Mo, s \models OPP(\alpha(i)) & \Longleftrightarrow & \alpha(i) \in Ropp(i,s) \\
Mo, s \models NEXT(\alpha(i)) & \Longleftrightarrow & \alpha(i) \in Rnext(s) \\
Mo, s \models W_i\phi & \Longleftrightarrow & Mo, s' \models \phi \text{ for all } s' \in Rp(i,s) \text{ and } \phi \in Rep(i,s) \\
Mo, s \models \psi <_i \phi & \Longleftrightarrow & (\psi,\phi) \in < (i,s) \\
Mo, s \models i \ll j & \Longleftrightarrow & i \prec j \\
Mo, s \models INT_i\alpha & \Longleftrightarrow & \alpha \in Ri(i,s) \\
Mo, s \models IMP_i(\alpha_1,\alpha_2) & \Longleftrightarrow & (\alpha_1,\alpha_2) \in Ria(i,s) \\
Mo, s \models O_{ij}(\phi) & \Longleftrightarrow & Mo, s' \models \phi \text{ for all } s' \text{ with } (s,s') \in Ro(i,j) \\
Mo, s \models O_{ij}(\alpha) & \Longleftrightarrow & Mo, s \models [\mathbf{any}(i)]O_{ij}(PREV(\alpha(i))) \\
Mos, \models auth(i,\alpha) & \Longleftrightarrow & Auth(i,\alpha,s) = tt
\end{array}
$$

The functions interpreting the special meta-actions are described below in terms of the preconditions and the postconditions for execution of the actions. The precondition describes on which models the model-transforming function has the desired effect and the postcondition describes the model yielded by the application of the meta-action.

DEC The precondition for execution of $DEC(i, \alpha)$ is that for some $\phi \in FORM$, $Goal_i\phi \wedge [\alpha(i); \beta(i)]\phi$ holds, for some $\alpha(i), \beta(i) \in Act$ and furthermore no ψ exists such that $P_i\psi$ and $\phi < \psi$ hold. Thus agents may only decide to intend to do those actions that fulfil some most preferred goal. As the result of execution of $DEC(i, \alpha)$ the model is changed in such a way that $INT_i\alpha$ holds in the resulting model.

$COMMIT$ There are no preconditions for execution of $COMMIT(i, j, \alpha)$ by agent i. The effect of the commitment is that the model is changed in such a way that $O_{ij}(\alpha)$ holds afterwards.

DIR The preconditions for execution of $DIR(authority, i, j, \alpha)$ by i are given by $auth(i, DIR(i, j, \alpha))$. This implies that agent i should have the authority over j before it can order it around. The effect of such an action is that j is obligated to i to perform α, which is implemented in a way similar to the implementation of the $COMMIT$ action. The same holds for a directive on the basis of power.

$DECL$ The action $DECL(i, j, f)$ has as precondition that i is authorised to declare f. (Only some civil servants can declare people to be married in the Netherlands). Execution of an action $DECL(i, j, f)$ in a certain state of a model will be a modification of the valuation π such that f is true in all the resulting states of the resulting models. Note that whenever f is inconsistent no model results.

ASS The precondition for $ASS(x, i, j, f)$ is that i is demanded to believe f, i.e. B_if should hold. This implies in particular that agents are not allowed to gossip, i.e. spread around rumors that they themselves do not even believe. As the result of executing $ASS(authority, i, j, f)$ by i in some state s, two cases arise. If $auth(i, ASS(authority, i, j, f))$ holds then the model under consideration is modified such that $Rb(j, s)$ contains only states in which f is true, which indeed implies that B_jf holds in s in the resulting model. Regardless whether i is authorised to make the assertion the model under consideration is modified such that $Rk(j, s)$ contains only states s' such that $Rb(i, s')$ contains only states in which f is true. This implies that K_jB_if holds in the resulting model.

5 RELATED APPROACHES

In this section we very briefly indicate the main differences between our approach and three other approaches to model rational agents, viz. the framework proposed by Cohen & Levesque [1990], the BDI-framework of Rao & Georgeff [1991] and the theoretical framework for multi-agent systems of M. Singh [1994]. After that we will shortly compare our approach to some other work on communicating agents.

The main difference between our approach and the one of Cohen & Levesque is that they define intentions in terms of goals and beliefs. We agree with this approach when it concerns intentions on propositions. However, we do not take a goal to be a primitive notion as is the case in the approach of Cohen & Levesque. Because they take a goal to be a primitive notion they have to define different types of goals in order to define the persistence of a goal, the achievability of a goal, etc. All these properties are direct consequences of our definition of a goal in terms of preferences and achievabilities. The distinction between goals and preferences allows for a bigger flexibility than possible in the approach of Cohen & Levesque. Furthermore, whereas we define the intention to perform an action as primitive, Cohen & Levesque define the intention to perform an action as the goal to reach a state where that action has been performed. Although both types of intentions of Cohen & Levesque are based on the notions of goals and beliefs it is not clear what is the relation between the intention to reach a certain state and the intention to perform an action; in fact these notions seem to be unrelated. However, it seems desirable that the intention to reach a certain goal induces the intention to perform an action which is needed to reach that goal. In our approach this relation is established through the notion of decisions. A goal can induce a decision. The decision then induces the intention to perform an action. Another relation that remains unclear in the theory of Cohen & Levesque is that between an intended action and the action that is actually performed. The only relation they give is that the intended action should be the same as the action whose goal it is to be performed. Which in itself does not mean anything for the actual course of events. In our approach we introduce the notion of an implementation relation between actions, which introduces a loose coupling between intended actions and the actions that are actually performed.

The last point also shows one of the main differences between our framework and that of Rao & Georgeff. In their framework it holds that if an agent intends to perform an action it will also actually perform the action. They weaken this assumption in [Kinny and Georgeff, 1991], where they investigate different strategies with respect to commitment to a goal. An agent can be single minded when it always performs its intended actions and open minded when it discards its goals on the basis of new information it receives. This work, however, is of a very practical nature and does not have a theoretical counterpart. Therefore it remains unclear how this would be reflected in the BDI framework.

In the BDI framework they avoided making the intention operator into a temporal operator by introducing the notion of a successful performance of an action and a failed performance of an action. However, the relation between a successful performed event and an event that failed to be performed is unclear. Can this be any other event? Can it include the event itself? At present the best one can say if an event has been performed (either successful or failed) is that *some* event has been performed.

A last, rather important, point of difference between our framework and the other two is the fact that we also include the social level, which we consider essential, but is only briefly mentioned by Cohen & Levesque, and not considered at all by Rao & Georgeff.

The social level is treated in the theory of Singh, in the form of communication between agents. The framework of Singh is based on CTL^*, a branching-time logic. The main difference with our framework is that he does not incorporate deontic relations between agents. It means that the success of a speech act is not defined in the same way as we do. A directive is successful if it results in the hearer intending to perform the action that it was ordered to perform. This is much stronger than in our framework. Of course, an obligation can lead to a goal, but only if the hearer places the directive above the plan it is currently performing.

How does our work compare with the other current work on communicating agents?

Bretier & Sadek [1996] formally define the effects of communicative actions. However, this theory is limited to effects on the beliefs and intentions of agents. We have shown that the concept of obligation also forms an important ingredient of the effects of some speech acts. That is, directives and commissives always result in obligations. It must be remarked that the application for which the theory in [Bretier and Sadek, 1996] is used does not need these illocutions.

Another difference between our work and that of [Bretier and Sadek, 1996] is that our theory is geared to communication between agents while their theory is geared towards the management of human-agent dialogue. The same holds for the theory of Traum [1996]. Although many aspects are compatible, there are also some differences. One of the main differences is that in the communication between humans and agents the detection of the illocution of messages is not trivial and important to steer the dialogue in a natural way. Agents will use a formal language in which the illocution of each message is clear.

Very interesting in the work of Traum is that he also recognizes the importance of obligations between the participants in a dialogue. Besides mutual knowledge and believe this is an important relation, because it indicates the expectations that the participants have of each other.

Finally the work of Noriega & Sierra [1996] comes very close in spirit to our work. They also try to give one formal framework for the different components of communicating agents. Their theory is more flexible than ours in that they allow different agents to use different languages and even the different components might use different inference rules. The relation between the components is given by so-called bridge rules. We assume a uniform language for all agents and components. This provides for an automatic integration of the components. The main point in which the theory in [Noriega and Sierra, 1996] is lacking compared to ours is the semantics of the speech acts themselves and their effects.

6 CONCLUSIONS

In this paper we presented an informal overview and a sketchy formalization of the concepts that we consider essential to model rational agents. In our very flexible and highly expressive framework we propose a variety of concepts, which are roughly situated at four different levels: the informational level, where knowledge and belief are considered, the action level, where we consider various aspects of action, the motivational level, where we dealt with preferences, goals, intentions, etc., and the social level, which is concerned with the social traffic between agents.

The resulting multi-modal logic is quite complex. However, we want to make two remarks about the logical formalism. First, it is not our aim to build an automated theorem prover that can prove theorems in this very rich logic. The use of a logical formalism gives the opportunity to automatically generate the logical effects of a sequence of steps in a protocol. These could be subsequently implemented in a more efficient formalism. The logical description, however, can be used as a very general and precise specification of that implementation.

Secondly, the use of logic forces a very precise formal description of the communication. It is very important that this is realized when the communication protocols are automatized. (As is the aim in communication between agents). If the communication is automatic it becomes very important to know the exact effects of the messages. What is the knowledge of each agent and what are its obligations (resp. expectations).

We admit that the logical formulas get very complicated and are not very readable. However, it is easy to define suitable abbreviations for standard formulas. At least, working this way, it is clear what these abbreviations mean exactly!

In subsequent work we want to show how communication protocols that are used in more practical work like [Müller, 1996; Norman *et al.*, 1996] can be given a formal semantics in our framework. For the basic illocutions used in the ADEPT system [Norman *et al.*, 1996] this has been done in this paper and has already led to the discovery that the seemingly simple ACCEPT message has unexpected results.
Also we want to define an agent architecture for communicating agents that adhere to our theory. Some groundwork in this respect has been done in

[Verhagen *et al.*, 1996]. What needs to be done in this respect is to show how communicative actions can form part of plans to reach a goal of the agent and also how goals of an agent can be adjusted under the influence of communication from other agents. We believe both parts can be done using the formalism described in this paper.

ACKNOWLEDGEMENTS

This article is a modified and extended version from [Dignum and van Linder, 1997]. We therefore like to thank the anonymous referees and the attendees of the ATAL workshop for their remarks on the basis of which this article has taken shape.

Frank Dignum
Utrecht University, The Netherlands.

Bernd van Linder
ABN AMRO, The Netherlands.

BIBLIOGRAPHY

[Bretier and Sadek, 1996] P. Bretier and M. D. Sadek. A rational agent as the kernel of a cooperative spoken dialogue system: Implementing a logical theory of interaction. In J. P. Müller, M. J. Wooldridge, and N. R. Jennings, editors, *Intelligent Agents III — Proceedings of the Third International Workshop on Agent Theories, Architectures, and Languages (ATAL-96)*, Lecture Notes in Artificial Intelligence 1193, pages 189-204, Springer-Verlag, Heidelberg, 1996.

[Cohen and Levesque, 1990] P. Cohen and H. Levesque. Intention is choice with commitment. *Artificial Intelligence*, vol.42, pages 213-261, 1990.

[Cohen and Levesque, 1991] P. Cohen and H. Levesque. Teamwork *Nous*, vol.35, pages 487-512, 1991.

[Davis and Smith, 1983] R. Davis and R. Smith. Negotiation as a metaphor for distributed problem solving. *Artificial Intelligence*, vol.20, pages 63-109, 1983.

[Dignum, 1992] F. Dignum. Using Transactions in Integrity Constraints: Looking forward or backwards, what is the difference? In *First International Workshop on Applied Logic: Logic at Work*, UVA, Amsterdam, 1992.

[Dignum and Weigand, 1995] F. Dignum and H. Weigand. Modelling communication between cooperative systems In J. Iivari et al., *Advanced information systems engineering*, pages 140-153, Springer-Verlag, Heidelberg, 1995.

[Dignum *et al.*, 1996] F. Dignum, J.-J.Ch. Meyer, R. Wieringa and R. Kuiper. A modal approach to intentions, commitments and obligations: intention plus commitment yields obligation. In M. Brown and J. Carmo (eds.),*Deontic logic, agency and normative systems*, Workshops in Computing, Springer-Verlag, pages 80-97, 1996.

[Dignum and van Linder, 1997] F. Dignum and B. van Linder. Modelling social agents: Communication as actions. In M. Wooldridge J. Muller and N. Jennings, (eds.), *Intelligent Agents III (LNAI-1193)*, Springer-Verlag, pages 205-218, 1997.

[Harel, 1979] D. Harel. First Order Dynamic Logic. LNCS 68 Springer-Verlag, Heidelberg, 1979.

[van der Hoek *et al.*, 1994] W. van der Hoek, B. van Linder and J.-J.Ch. Meyer. A logic of capabilities. In Nerode and Matiyasevich, eds, *Proceedings of LFCS'94*, LNCS 813, Springer-Verlag, Heidelberg, pages 366-378, 1994.

[Jennings, 1993] N. Jennings. Commitments and Conventions: The foundation of coordination in Multi-Agent systems. *Knowledge Engineering Review*, vol. 8(3), pages 223-250, 1993.

[Jennings *et al.*, 1996] N. Jennings, P. Faratin, M. Johnson, P. O'Brien and M. Wiegand. Using Intelligent Agents to Manage Business Processes. In *Proceedings The Practical Application of Intelligent Agents and Multi-A gent Technology*, pages 345-360, The practical application company, London, 1996.

[Kinny and Georgeff, 1991] D. Kinny and M. Georgeff. Commitment and Effectiveness of Situated Agents. In *Proceedings International Joint Conference on Artificial Intelligence*, Sydney, Australia, pages 82-88, Morgan Kaufman Publishers, 1991.

[van Linder *et al.*, 1994] B. van Linder, W. van der Hoek and J.-J.Ch. Meyer. Communicating rational agents. In Nebel and Dreschler-Fisher, eds, *Proceedings of KI'94*, LNCS 861, pages 202–213, Springer-Verlag, Heidelberg, 1994.

[van Linder *et al.*, 1996] B. van Linder, W. van der Hoek and J.-J.Ch. Meyer. How to motivate your agents. On making promises that you can keep. In Wooldridge, Müller and Tambe, eds, *Intelligent Agents II*, LNCS 1037, pages 17–32, Springer-Verlag, Heidelberg, 1996.

[van Linder, 1996] B. van Linder *Modal Logics for Rational Agents*, PhD Thesis, Utrecht University, 1996.

[Meyer, 1988] J.-J.Ch. Meyer. A different approach to deontic logic. In *Notre Dame Journal of Formal Logic*, vol.29, pages 109–136, 1988.

[Meyer and van der Hoek, 1995] J.-J.Ch. Meyer and W. van der Hoek. *Epistemic Logic for AI and computer science*, Cambridge University Press, Cambridge, 1995.

[Moore, 1985] R. Moore. A formal theory of knowledge and action. In J. Hobbs and R. Moore, eds, *Formal theories of the commonsense world*, pages 319-358, Ablex Publ. Comp., 1985.

[Müller, 1996] J. P. Müller. A cooperation model for autonomous agents. In J. P. Müller, M. J. Wooldridge, and N. R. Jennings, editors, *Intelligent Agents III — Proceedings of the Third International Workshop on Agent Theories, Architectures, and Languages (ATAL-96)*, Lecture Notes in Artificial Intelligence 1193, pages 245-260. Springer-Verlag, Heidelberg, 1996.

[Noriega and Sierra, 1996] P. Noriega and C. Sierra. Towards layered dialogical agents. In J. P. Müller, M. J. Wooldridge, and N. R. Jennings, editors, *Intelligent Agents III — Proceedings of the Third International Workshop on Agent Theories, Architectures, and Languages (ATAL-96)*, Lecture Notes in Artificial Intelligence 1193, pages 173-188. Springer-Verlag, Heidelberg, 1996.

[Norman *et al.*, 1996] T. J. Norman, N. R. Jennings, P. Faratin, and E. H. Mamdani. Designing and implementing a multi-agent architecture for business process management. In J. P. Müller, M. J. Wooldridge, and N. R. Jennings, editors, *Intelligent Agents III — Proceedings of the Third International Workshop on Agent Theories, Architectures, and Languages (ATAL-96)*, Lecture Notes in Artificial Intelligence 1193, pages 261-276. Springer-Verlag, Heidelberg, 1996.

[Rao and Georgeff, 1991] A.S. Rao and M.P. Georgeff. Modeling rational agents within a BDI-architecture. In J. Allen et al., eds, *Proceedings of KR'91*, pages 473-484, Morgan Kaufman Publishers, 1991.

[Sandu, 1996] G. Sandu. Reasoning about collective goals. In J. P. Müller, M. J. Wooldridge, and N. R. Jennings, editors, *Intelligent Agents III — Proceedings of the Third International Workshop on Agent Theories, Architectures, and Languages (ATAL-96)*, Lecture Notes in Artificial Intelligence 1193, pages 131-140. Springer-Verlag, Heidelberg, 1996.

[Searle, 1969] J.R. Searle. Speech Acts. Cambridge University Press, Cambridge, 1969.

[Singh, 1994] M. Singh. Multiagent Systems. LNAI 799, Springer-Verlag, Heidelberg,1994.

[Traum, 1996] D. R. Traum. A reactive-deliberative model of dialogue agency. In J. P. Müller, M. J. Wooldridge, and N. R. Jennings, editors, *Intelligent Agents III — Proceedings of the Third International Workshop on Agent Theories, Architectures, and Languages (ATAL-96)*, Lecture Notes in Artificial Intelligence 1193, pages 157-172. Springer-Verlag, Heidelberg, 1996.

[Verhagen *et al.*, 1996] E. Verhagen, F. Dignum and H. Weigand. A Language/Action Perspective on Cooperative Information Agents. In F. Dignum at al., eds, *Communication Modeling - The Language/Action Perspective (LAP-96)*, electronic Workshops in Computing, Springer-Verlag, London, 1996.

[Weigand *et al.*, 1996] H. Weigand, E. Verharen and F. Dignum. Interoperable Transactions in Business Models: A Structured Approach. In P. Constantopoulos, J. Mylopoulos and Y. Vassiliou, eds, *Advanced Infor mation Systems Engineering (LNCS 1080)*, pages 193-209, Springer-Verlag, 1996.

[Wieringa *et al.*, 1989] R. Wieringa, J.-J.Ch. Meyer and H. Weigand. Specifying dynamic and deontic integrity constraints. *Data & knowledge engineering*, vol.4, pages 157-189, 1989.

PART IIIC

FORMAL ANALYSIS:

REASONING ABOUT DYNAMICS

WITOLD ŁUKASZEWICZ AND EWA MADALIŃSKA-BUGAJ

REASONING ABOUT ACTION AND CHANGE USING DIJKSTRA'S SEMANTICS FOR PROGRAMMING LANGUAGES

1 INTRODUCTION

We apply Dijkstra's semantics for programming languages [Dijkstra, 1976; Dijkstra and Scholten, 1990] to formalization of reasoning about action and change. The basic idea is to specify effects of actions in terms of *formula transformers*, i.e. functions from formulae into formulae.[1] More specifically, with each action A we associate two formula transformers, called the *strongest postcondition for* A and the *weakest liberal precondition for* A. The former, when applied to a formula α, returns a formula representing the set of all states that can be achieved by starting execution of A in some state satisfying α. The latter, when applied to a formula α, returns a formula providing a description of all states such that whenever execution of A starts in any one of them and terminates, the output state satisfies α[2].

The idea of employing formula transformers to specify effects of actions is not new in the AI literature and goes back to STRIPS system [Fikes and Nilsson, 1971]. Waldinger [Waldinger, 1977] introduces a notion of a *regression operator* which corresponds closely to the weakest precondition transformer. Pednault [Pednault, 1986; Pednault, 1988; Pednault, 1989] employes regression operators in the context of plan synthesis. In [Pednault, 1986] a notion of progression operator, corresponding to Dijkstra's strongest postcondition, is introduced and analysed.

Formula transformers approach to reasoning about action and change has one major advantage and one major weakness when compared to purely logical formalisms such as Situation Calculus [McCarthy and Hayes, 1969; Lifschitz, 1988; Lifschitz and Rabinov, 1989; Gelfond *et al.*, 1991; Baker, 1991] or Features and Fluents [Sandewall, 1993; Sandewall, 1994]. On the positive side, describing effects of actions in terms of formula transformers decreases computational complexity. The price to pay for it is the loss of expressibility.

Our proposal combines computational effectiveness with expressibility. Although not so expressible as Situation Calculus or Feature and Fluents, the formalism specified here allows to deal with a broad class of actions, including those

[1] Formula transformers are often referred to as predicate transformers in the Computer Science literature.

[2] The reader familiar with Dijkstra's approach may wonder why we do not use the *weakest precondition* transformer (wp) which plays a prominent role in reasoning about programs. The reason is that, in general, the wp transformer is slightly too strong for our purposes. However, since all actions we consider in this paper terminate, the weakest precondition and the weakest liberal precondition coincide for them.

J.J.Ch. Meyer and J. Treur (eds.),
Handbook of Defeasible Reasoning and Uncertainty Management Systems, Vol. 7, 383–398.

with random and indirect effects. Also, both temporal prediction and postdiction reasoning tasks can be solved without restricting initial nor final states to completely specified.

The paper is organized as follows. Section 2 is a brief introduction to Dijkstra's semantics for a simple programming language. In section 3, we outline a general procedure to define action languages using Dijkstra's methodology, illustrate this procedure by specifying a simple "shooting" language, and introduce a notion of an *action scenario*. Section 4 defines the kind of reasoning we shall be interested in, and provides a simple method of realizing this type of inference. In section 5, we illustrate this method by considering a number of examples, well-known from the AI literature. Section 6 is devoted to actions with indirect effects. Finally, section 7 contains discussion and ideas for future work.

2 INTRODUCTION TO DIJKSTRA'S SEMANTICS

In [Dijkstra and Scholten, 1990] we are provided with a simple programming language whose semantics is specified in terms of formula trasformers. More specifically, with each command S there are associated three formula transformers, called the *weakest precondition*, the *weakest liberal precondition* and the *strongest postcondition*, denoted by wp, wlp and sp, respectively. Before providing the meaning of these transformers, we have to make some remarks and introduce some terminology.

First of all, we assume here that the programming language under consideration contains one type of variables only, namely Boolean variables. This assumption may seem overly restrictive, but as a matter of fact no other variables will be needed for our purpose.

Let V be a set of Boolean variables. A *state over* V is any function σ from the members of V into the truth-values $\{0, 1\}$.

An *assertion language* over a set V of Boolean variables, denoted by $\mathcal{L}(V)$, is the set of all formulae constructable in the usual way from members of V, sentential connectives $(\neg, \supset, \wedge, \vee, \equiv)$ and quantifiers $(\forall, \exists)$.[3] In what follows, the term 'formula' refers always to a formula of some fixed assertion language. A formula α is said to be a *Boolean expression* if it contains no quantifiers. If β is a formula, $\alpha_1, \ldots, \alpha_n$ are Boolean expressions and $x_1, \ldots, x_n$ are variables, then we write $\beta[x_1 \leftarrow \alpha_1, \ldots, x_n \leftarrow \alpha_n]$ to denote the formula which obtains from β by simultaneously replacing all free occurrences of $x_1, \ldots, x_n$ by $\alpha_1, \ldots, \alpha_n$, respectively.

The *value of a formula* $\alpha \in \mathcal{L}(V)$ *in a state* σ, written $\| \alpha \|_\sigma$, is an element from $\{0, 1\}$ specified by the usual rules for classical propositional logic, together with:

(i) $\| \forall x.\alpha \|_\sigma = 1$ iff $\| \alpha \|_{\sigma'} = 1$, for each σ' identical to σ except perhaps on the variable x.

[3]Note that quantifiers can be applied to Boolean variables only.

(ii) $\| \exists x.\alpha \|_\sigma = 1$ iff $\| \alpha \|_{\sigma'} = 1$, for some σ' identical to σ except perhaps on the variable x.

It follows from the clauses (i)–(ii) above that quantified formulae can be reduced into equivalent Boolean expressions. (Note that formulae of the form $\forall x.\alpha$ can be always replaced by $\alpha[x \leftarrow T] \wedge \alpha[x \leftarrow F]$, whereas formulae of the form $\exists x.\alpha$ by $\alpha[x \leftarrow T] \vee \alpha[x \leftarrow F]$.)[4]

A state σ is said to *satisfy* a formula α iff $\| \alpha \|_\sigma = 1$. If σ satisfies α, then σ is called a *model* of α. When we say that σ satisfies α, we always implicitly assume that σ provides an interpretation for all Boolean variables occurring in α.

The formula transformers mentioned above are to be understood as follows. For each command S and each formula α:

- $wp(S, \alpha)$ is the formula whose models are precisely all states such that execution of S begun in any one of them is guaranteed to terminate in a state satisfying α.

- $wlp(S, \alpha)$ is the formula whose models are precisely all states such that whenever execution of S starts in any one of them and terminates, the output state satisfies α.

- $sp(S, \alpha)$ is the formula whose models are precisely all states such that each of them can be reached by starting execution of S in some state satisfying α.

For example, let S be the assignment command $x := T$, and consider the formula $x \vee y$, where x and y are Boolean variables of the programming language under consideration.[5] Clearly, $wp(S, x \vee y) = T$, for if σ is any state, then execution of S begun in σ terminates in a state satisfying $x \vee y$. Since the command $x := T$ always terminates, $wlp(S, x \vee y) = wp(S, x \vee y)$. On the other hand, $sp(S, x \vee y) = x$ because the states satisfying x are precisely those that can be reached by starting execution of S in some state satisfying $x \vee y$.

2.1 *List of commands*

The considered language consists of *skip* command, *assignment* to simple variables, *alternative* command and *sequential composition* of commands[6]. Semantics of these commands is specified as follows.[7]

[4] It follows, therefore, that assertion languages we consider here are in fact propositional.

[5] In this paper the letters T and F denote the propositional constants $True$ and $False$, respectively.

[6] The original Dijkstra's language contains *abort* command and *iterative* commands as well, but they are not needed for our purpose.

[7] In what follows, we do not specify the wp formula tranformers for the considered language, because they will not be needed in the sequel.

1. **The *skip* command.** This is the "empty" command in that its execution does not change the computation state. The semantics of *skip* is thus given by

$$wlp(skip, \alpha) = sp(skip, \alpha) = \alpha.$$

2. **The *assignment* command.** This command is of the form $x := e$, where x is a (Boolean) variable and e is a (Boolean) expression. The effect of the command is to replace the value of x by the value of e. The weakest liberal precondition of this command is given by

$$wlp(x := e, \ \ \alpha) = \alpha[x \leftarrow e].$$

The strongest postcondition for assignment command is a bit more complex. In general, it is given by

(1) $sp(x := e, \alpha) = \exists y.((x \equiv e[x \leftarrow y]) \wedge \alpha[x \leftarrow y]).$

If the variable x does not occur in the expression e, the equation (1) can be simplified. In this case

$$sp(x := e, \alpha) = (x \equiv e) \wedge \exists x.\alpha$$

or equivalently

(2) $sp(x := e, \alpha) = (x \equiv e) \wedge (\alpha[x \leftarrow T] \vee \alpha[x \leftarrow F]).$

In the sequel we shall often deal with assignment commands, $x := e$, where e is T or F. In this case the equation (2) can be replaced by

(3) $sp(x := e, \alpha) = \begin{cases} x \wedge (\alpha[x \leftarrow T] \vee \alpha[x \leftarrow F]) & \text{if e is } T \\ \neg x \wedge (\alpha[x \leftarrow T] \vee \alpha[x \leftarrow F]) & \text{if e is } F \end{cases}$

Reconsider the example given earlier, where α is $x \vee y$ and the assignment command is $x := T$. Applying (3), we get $sp(x := T, x \vee y) = x \wedge (T \vee y \vee F \vee y)$ which reduces to x.

3. **The *sequential composition* command.** This command is of the form $S_1; S_2$, where S_1 and S_2 are any commands. It is executed by first executing S_1 and then executing S_2. Its semantics is given by

$$wlp(S_1; S_2, \alpha) = wlp(S_1, wlp(S_2, \alpha));$$

$$sp(S_1; S_2, \alpha) = sp(S_2, sp(S_1, \alpha)).$$

4. **The *alternative* command.** This command is of the form

(4) **if** $B_1 \rightarrow S_1 \ \ \| \ \ \cdots \ \ \| \ \ B_n \rightarrow S_n$ **fi**

where $B_1, \ldots, B_n$ are Boolean expressions and $S_1, \ldots, S_n$ are any commands. $B_1, \ldots, B_n$ are called *guards* and expressions of the form $B_i \rightarrow S_i$ are called *guarded commands*. In the sequel, we refer to the general command (4) as IF. The command is executed as follows. If none of the guards is true, then the execution aborts. Otherwise, one guarded command $B_i \rightarrow S_i$ with true B_i is *randomly* selected and S_i is executed.[8] The semantics of IF is given by

$$wlp(\mathrm{IF}, \alpha) = \bigwedge_{i=1}^{n} (B_i \supset wlp(S_i, \alpha))$$

$$sp(\mathrm{IF}, \alpha) = \bigvee_{i=1}^{n} (sp(S_i, B_i \wedge \alpha).$$

2.2 *Main results*

Dijkstra and Scholten [1990] consider various classes of computations. The class we are primarily interested in here is called by these authors *initially α and finally β under control of S*, where α and β are formulae and S is a command. This class, which will be denoted by $[S]^{\alpha}_{\beta}$, represents the set of all computations under control of S that start in a state satisfying α and terminate in a state satisfying β.

Suppose that $c \in [S]^{\alpha}_{\beta}$. Obviously, since S terminates in a state satisfying β and $wlp(S, \neg\beta)$ represents the set of all states such that S begun in any of them either terminates in a state satisfying $\neg\beta$ or loops forever, it must be the case that the initial state of c satisfies $\alpha \wedge \neg wlp(S, \neg\beta)$. Similarly, since S starts in a state satisfying α and $sp(S, \alpha)$ represents the set of all states such that any of them can be reached by starting execution of S in some state satisfying α, we conclude that the final state of c satisfies $\beta \wedge sp(S, \alpha)$. An interesting question is whether the formulae $\alpha \wedge \neg wlp(S, \neg\beta)$ and $\beta \wedge sp(S, \alpha)$ provide a complete description of the initial and final states of the computations from the class $[S]^{\alpha}_{\beta}$. That the answer is positive follows from the following result which can be found in [Dijkstra and Scholten, 1990].

THEOREM 1. The formula $\alpha \wedge \neg wlp(S, \neg\beta)$ (resp. $\beta \wedge sp(S, \alpha)$) holds in a state σ iff there exists a computation c from $[S]^{\alpha}_{\beta}$ such that σ is the initial (resp. final) state of c.

Consider now the class of computations $\mathcal{C} = [S]^{\alpha}_{\beta}$, where S is the sequence $S_1; S_2; \ldots; S_n$. Let $ST_{\mathcal{C}}(i)$, $0 \leq i \leq n$, be the set of all states satisfying the following condition: for each $\sigma \in ST_{\mathcal{C}}(i)$, there exists a computation $c \in \mathcal{C}$ such that σ is reached by c after executing $S_1; \ldots; S_i$. In what follows, the members of $ST_{\mathcal{C}}(i)$ will be referred to as *i-states* of $\mathcal{C}$. Clearly, 0-states are initial states and n-states are final states. The following theorem provides a complete characterization of i-states.

[8]Note that when more than one guard is true, the selection of a guarded command is nondeterministic.

THEOREM 2. The formula $sp(S_1;\ldots;S_i,\alpha)\wedge\neg wlp(S_{i+1};\ldots;S_n,\neg\beta)$ holds in a state σ iff there exists a computation from the class $C=[S_1;S_2;\ldots;S_n]^{\alpha}_{\beta}$ such that $\sigma\in ST_C(i)$, $0\leq i\leq n$.[9]

Proof. If $i=0$ or $i=n$, the proof immediately follows from Theorem 1. Assume therefore that $0<i<n$.

($\Rightarrow$) Assume that the formula $sp(S_1;\ldots;S_i,\alpha)\wedge\neg wlp(S_{i+1};\ldots;S_n,\neg\beta)$ holds in a state σ. We have to show that there is a computation c from the class $[S_1;S_2;\ldots;S_n]^{\alpha}_{\beta}$ such that $\sigma\in ST_C(i)$. Since σ satisfies $sp(S_1;\ldots;S_i,\alpha)$, σ is the final state of some computation c', under control of $S_1;\ldots;S_i$, which starts in some state satisfying α. Similarly, since σ satisfies $\neg wlp(S_{i+1};\ldots;\neg\beta)$, σ is the initial state of some computation c'', under control of $S_{i+1};\ldots;S_n$, which terminates in some state satisfying β. Taking c to be the composition of c' and c'', it is immediately seen that c is the member of the class $C=[S_1;S_2;\ldots;S_n]^{\alpha}_{\beta}$ and $\sigma\in ST_C(i)$.

($\Leftarrow$) Assume now that $\sigma\in ST_C(i)$, for some computation c from the class $[S_1;S_2;\ldots;S_n]^{\alpha}_{\beta}$. We have to show that σ satisfies the formula $sp(S_1;\ldots;S_i,\alpha)\wedge\neg wlp(S_{i+1};\ldots;S_n,\neg\beta)$. Since the initial state of c satisfies α and $sp(S_1;\ldots;S_i,\alpha)$ is true in any state that can be reached from some state satisfying α (by performing $S_1;\ldots;S_i$), it must be the case that $sp(S_1;\ldots;S_i,\alpha)$ is true in σ. Similarly, since the final state of c satisfies β and $wlp(S_{i+1};\ldots;S_n,\neg\beta)$ holds in all states such that execution of $S_{i+1};\ldots;S_n$ begun in any one of them either terminates in a state satisfying $\neg\beta$ or loops forever, it must be the case that $\neg wlp(S_{i+1};\ldots;S_n,\neg\beta)$ is true in σ. In conclusion, σ satisfies $sp(S_1;\ldots;S_i,\alpha)\wedge\neg wlp(S_{i+1};\ldots;S_n,\neg\beta)$. ■

There is another class of computations that we shall be interested in. This class, denoted by $[S_1;\ldots;S_n]^{\alpha}_{\beta}(\gamma)$, represents the set of all computations under control of $S_1;\ldots;S_n$ that start in a state satisfying α, terminate in a state satisfying β and, in addition, any state of any computation from this class that can be reached after executing $S_1;\ldots;S_i$ $(0\leq i\leq n)$ satisfies γ.[10]

The next theorem provides a complete characterization of i-states of the introduced class of computations.

THEOREM 3. The formula $sp^{\gamma}(S_1;\ldots;S_i,\alpha\wedge\gamma)\wedge\neg wlp^{\gamma}(S_{i+1};\ldots;S_n,\neg(\beta\wedge\gamma))$ holds in a state σ iff there exists a computation from the class $C=[S_1;\ldots;S_n]^{\alpha}_{\beta}(\gamma)$ such that $\sigma\in ST_C(i)$. Here sp^{γ} and wlp^{γ} are specified by the following recursive definitions (δ is any formula).

$$sp^{\gamma}(S_1,\delta) = \gamma\wedge sp(S_1,\delta) \tag{5}$$
$$sp^{\gamma}(S_1;\ldots;S_i,\delta) = \gamma\wedge sp(S_i,sp^{\gamma}(S_1;\ldots;S_{i-1},\delta)) \tag{6}$$
$$wlp^{\gamma}(S_n,\delta) = \neg\gamma\vee wlp(S_n,\delta) \tag{7}$$
$$wlp^{\gamma}(S_{i+1};\ldots;S_n,\delta) = \neg\gamma\vee wlp(S_{i+1},wlp^{\gamma}(S_{i+2};\ldots;S_n,\delta)). \tag{8}$$

[9] We assume here that $sp(S_1;\ldots;S_i,\alpha)$ is α if $i=0$ and $\neg wlp(S_{i+1};\ldots;S_n,\neg\beta)$ is β if $i=n$.

[10] Note that if γ is T, then the class $[S_1;\ldots;S_n]^{\alpha}_{\beta}(\gamma)$ reduces to the class $[S_1;\ldots;S_n]^{\alpha}_{\beta}$.

Proof. We start with a few observations whose proofs can be easily obtained by induction on i ($0 \leq i \leq n$).

(1) If σ satisfies $sp^{\gamma}(S_1; \ldots; S_i, \alpha \wedge \gamma)$, then σ is the final state of some computation from the class $[S_1; \ldots; S_i]^{\alpha}_{T}(\gamma)$.

(2) If σ satisfies $\neg wlp^{\gamma}(S_{i+1}; \ldots; S_n, \neg(\beta \wedge \gamma))$, then σ is the initial state of some computation from the class $[S_{i+1}; \ldots; S_n]^{T}_{\beta}(\gamma)$.

(3) If $\sigma \in ST_{\mathcal{C}}(i)$ for some computation c from the class $\mathcal{C} = [S_1; \ldots; S_n]^{\alpha}_{\beta}(\gamma)$, then σ satisfies $sp^{\gamma}(S_1; \ldots; S_i, \alpha \wedge \gamma)$.

(4) If $\sigma \in ST_{\mathcal{C}}(i)$ for some computation c from the class $\mathcal{C} = [S_1; \ldots; S_n]^{\alpha}_{\beta}(\gamma)$, then σ satisfies $\neg wlp^{\gamma}(S_{i+1}; \ldots; S_n, \neg(\beta \wedge \gamma))$.

Given the observations (1)–(4), the proof of the theorem proceeds exactly like that of Theorem 2. ■

3 ACTION LANGUAGES AND ACTION SCENARIOS

In this section, we show how to specify action languages using Dijkstra's semantics, illustrate the procedure by defining a simple "shooting" language and introduce a notion of an action scenario.

To define an action language one proceeds in three steps.

(1) First, we choose an assertion language to represent the effect of actions. For the shooting language we take the language with two Boolean variables: a and l, standing for *alive* and *loaded*, respectively. To be in accord with the AI terminology, these variables will be referred to as ***fluents***.

(2) The next step is to provide action symbols representing the actions under consideration. In the shooting language we have four such symbols: *load* (a gun), *wait*, *spin* (a chamber) and *shoot* (a turkey). The intention is that *load* makes the gun loaded, *wait* does not cause any changes in the world, the effect of *spin* is that randomly the gun is loaded or not after the action, regardless of whether it was loaded before or not, and *shoot* makes the gun unloaded and the turkey dead, provided that the gun was loaded before.

(3) The final step is to define Dijkstra-style semantics for the choosen actions. To perform this step for the shooting language, it suffices to note that the considered actions can be easily translated into the programming language specified in the previous section. More specifically, *load* corresponds to the assignment command $l := T$, *wait* is just the *skip* command, whereas *spin*

and *shoot* are translated into alternative commands

spin		*shoot*
if		**if**
$T \to l := T$ ‖	and	$l \to a := F; l := F$ ‖
$T \to l := F$		$\neg l \to skip$
fi		**fi**

respectively. Given the above translations, the chosen actions can be provided with Dijkstra-style semantics. Performing routine calculations one easily obtains:

- $wlp(load, \alpha) = \alpha[l \leftarrow T]$;
- $sp(load, \alpha) = l \wedge (\alpha[l \leftarrow T] \vee \alpha[l \leftarrow F])$;
- $wlp(wait, \alpha) = sp(wait, \alpha) = \alpha$;
- $wlp(spin, \alpha) = \alpha[l \leftarrow T] \wedge \alpha[l \leftarrow F]$;
- $sp(spin, \alpha) = \alpha[l \leftarrow T] \vee \alpha[l \leftarrow F]$;
- $wlp(shoot, \alpha) = (l \supset \alpha[a \leftarrow F, l \leftarrow F]) \wedge (\neg l \supset \alpha)$;
- $sp(shoot, \alpha) = (\neg a \wedge \neg l \wedge \alpha[a \leftarrow T, l \leftarrow T]) \vee (\neg a \wedge \neg l \wedge \alpha[a \leftarrow F, l \leftarrow T]) \vee (\neg l \wedge \alpha)$.

The objects we shall be primarily interested in are *action scenarios.* These are expressions of the form

(9) $[\alpha]\ A_1; \ldots; A_n\ [\beta]$

where α and β are formulae and $A_1, \ldots, A_n$ are actions. The scenario has the following intuitive interpretation: α was observed to hold in the initial state, then the actions $A_1, \ldots, A_n$ were sequentially performed, and then β was observed to hold in the final state. Formally, the scenario (9) is to be viewed as representing the class of computations initially α and finally β under control of $A_1; \ldots; A_n$.

EXAMPLE 4 (Yale Shooting Scenario). Below is the famous Yale Shooting Scenario [Hanks and McDermott, 1987].

$$[a \wedge \neg l] load; wait; shoot[T]$$

4 REASONING ABOUT SCENARIOS

In this section, we provide a method to reason about scenarios.

We shall be interested in the following reasoning task: "Given a scenario $[\alpha]\ A_1; \ldots; A_n\ [\beta]$, a formula γ and an integer k such that $0 \leq k \leq n$, determine whether

γ is assured to hold after performing the actions $A_1, \ldots, A_k$".[11] Viewing a scenario as representing a class of computations, the above reasoning task can be stated more formally: "Given a scenario $[\alpha]\ A_1; \ldots; A_n\ [\beta]$, a formula γ and an integer k such that $0 \leq k \leq n$, determine whether γ is true in all k-states of the class of computations $[A_1; \ldots; A_n]^{\alpha}_{\beta}$.

In what follows, the symbol $\vdash$ stands for the provability relation of classical propositional logic.

Let $SC = [\alpha]\ A_1; \ldots; A_n\ [\beta]$ be a scenario and let $0 \leq k \leq n$. The *description of* the kth-*state of* SC, written $DS_k(SC)$, is the formula given by

$$DS_k(SC) = \begin{cases} \alpha \wedge \neg wlp(A_1; \ldots; A_n, \neg\beta) & \text{if } k = 0 \\ sp(A_1; \ldots; A_k, \alpha) \wedge \neg wlp(A_{k+1}; \ldots; A_n, \neg\beta) & \text{if } 0 < k < n \\ \beta \wedge sp(A_1; \ldots; A_n, \alpha) & \text{if } k = n \end{cases}$$

The next result follows immediately from Theorem 2.

THEOREM 5. Let $SC = [\alpha]\ A_1; \ldots; A_n\ [\beta]$ be a scenario. A formula γ is assured to hold after performing the actions $A_1; \ldots; A_k$ $(0 \leq k \leq n)$ iff $DS_k(SC) \vdash \gamma$.

REMARK 6. All actions considered in this paper are guaranteed to terminate. If $A_1; \ldots; A_k$ is any sequence of such actions, then $wlp(A_1; \ldots; A_k, F) = F$. Accordingly, if SC is a scenario of the form $[\alpha]\ A_1; \ldots; A_n\ [T]$, then $DS_0(SC) = \alpha$ and, for $0 < k < n$, $DS_k(SC) = sp(A_1; \ldots; A_k, \alpha)$. We shall often make use of this fact in the sequel.

5 EXAMPLES

In this section we consider a number of well-known examples to illustrate the method introduced in the previous section.

EXAMPLE 4 (continued) The Yale Shooting Scenario (YSS, for short) is an example of the temporal prediction. We have a description of the initial state and we are interested what is guaranteed to hold in the final state. The intended conclusion is that after performing the actions the turkey is dead and the gun is unloaded.

We calculate $DS_3(YSS)$.[12]

[11] For $k = 0$, i.e. when we want to determine whether γ is true in the initial state, the above reasoning task is known as the *temporal postdiction problem*. For $k = n$, i.e. when we want to determine whether γ holds in the final state, the task is usually referred to as the *temporal prediction problem*.

[12] Note the use of the symbols "=" and "≡" during the calculation. We write $X = Y$ if Y is obtained from X by employing the semantics of wlp or sp, whereas $X \equiv Y$ indicates that X and Y are logically equivalent.

$$\begin{array}{rcl} DS_3(YSS) & = & sp(load; wait; shoot, a \wedge \neg l) \\ & = & sp(shoot, sp(wait, sp(load, a \wedge \neg l))) \\ & = & sp(shoot, sp(wait, l \wedge ((a \wedge T) \vee (a \wedge F)))) \\ & \equiv & sp(shoot, sp(wait, l \wedge a)) \\ & = & sp(shoot, l \wedge a) \\ & = & (\neg a \wedge \neg l \wedge T \wedge T) \vee (\neg a \wedge \neg l \wedge T \wedge F) \vee (\neg l \wedge l \wedge a) \\ & \equiv & \neg a \wedge \neg l. \end{array}$$

Since $DS_3(YSS) \vdash \neg a \wedge \neg l$, we conclude that in the final state the turkey is dead and the gun is unloaded.

EXAMPLE 7 (Russian Turkey Scenario). The Russian Turkey Scenario (RTS, for short) is an example of the temporal prediction, where actions with random effects are allowed.[13] The world of the scenario is the same as for the Yale Shooting Scenario, but *wait* is replaced by *spin*. Recall that the effect of this latter action is that randomly the gun may or may not be loaded after its execution, regardless of whether it was loaded before or not. The scenario is given by

$$[a \wedge \neg l]\ load; spin; shoot\ [T].$$

The intended conclusion is that nothing can be said whether the turkey is alive or not in the final state.

We calculate $DS_3(RTS)$.

$$\begin{array}{rcl} DS_3(RTS) & = & sp(load; spin; shoot, a \wedge \neg l) \\ & = & sp(shoot, sp(spin, sp(load, a \wedge \neg l))) \\ & = & sp(shoot, sp(spin, l \wedge ((a \wedge T) \vee (a \wedge F)))) \\ & \equiv & sp(shoot, sp(spin, l \wedge a)) \\ & = & sp(shoot, (l \wedge a)[l \leftarrow T] \vee (l \wedge a)[l \leftarrow F]) \\ & \equiv & sp(shoot, a) \\ & = & (\neg a \wedge \neg l \wedge T) \vee (\neg a \wedge \neg l \wedge F) \vee (\neg l \wedge a) \\ & \equiv & \neg l. \end{array}$$

Since $DS_3(RTS) \not\vdash a$ and $DS_3(RTS) \not\vdash \neg a$, no conclusion can be derived with respect to whether a or $\neg a$ holds in the final state.

EXAMPLE 8 (Stanford Murder Mystery). Consider the following problem, known in the AI literature as the Stanford Murder Mystery (SMM).[14] The turkey is alive in the initial state, and after the actions *shoot* and *wait* are successively performed, it is dead. The story can be represented by the following scenario:

$$[a]\ shoot; wait\ [\neg a].$$

[13] This example is from [Sandewall, 1994].

[14] This example is from [Baker, 1991].

The question we are interested in is when the turkey died and whether the gun was originally loaded (the temporal postdiction). The intended conclusion is that the gun was loaded in the initial state and the turkey died during the shooting. We calculate $DS_0(SMM)$.

$$\begin{aligned} DS_0(SMM) &= a \wedge \neg wlp(shoot; wait, a) \\ &= a \wedge \neg wlp(shoot, wlp(wait, a)) \\ &= a \wedge \neg wlp(shoot, a) \\ &= a \wedge \neg((l \supset F) \wedge (\neg l \supset a)) \\ &\equiv a \wedge l. \end{aligned}$$

Since $DS_0(SMM) \vdash l$, we immediately conclude that the gun was loaded in the initial state.

Now, we calculate $DS_1(SMM)$.

$$\begin{aligned} DS_1(SMM) &= sp(shoot, a) \wedge \neg wlp(wait, a) \\ &= ((\neg a \wedge \neg l \wedge T) \vee (\neg a \wedge \neg l \wedge F) \vee (\neg l \wedge a)) \wedge \neg a \\ &\equiv \neg a \wedge \neg l. \end{aligned}$$

Since $DS_1(SMM) \vdash \neg a$, we infer that t'he turkey was dead after performing the action *shoot*.

6 RAMIFICATION PROBLEM

The *ramification problem* concerns efficient representation of the indirect effects of actions. In this paper we limit ourselves to the simplest class of ramifications, namely those introduced by *domain constraint axioms*.

The domain constraint axioms describe general facts that are assumed to hold in any state of the dynamically changing world under consideration. If α is such an axiom, then some fluents occurring in α may change their values even if these fluents are not included in the action's description. Consider, for instance, the Yale Shooting Scenario, augmented with the domain constraint axiom $a \equiv \neg d$, where d stands for *dead*. Given this axiom, the action *shoot* makes the turkey not only not alive, but also dead (provided, of course, that the gun is loaded).

To deal with the domain constraint axioms in our approach, one can try to take them as a part of the description of the kth-state of a given action scenario. While this sometimes works (for instance, for the Yale Shooting Scenario, supplied with $a \equiv \neg d$), this method is generally inappropriate. To see why, consider an illustrating example.

EXAMPLE 9. Consider the shooting language supplied with a new fluent w, standing for $walking$. Let DC be a domain constraint axiom given by $w \supset a$. Let WSS be the scenario

$$[w \wedge l] \; shoot \; [T].$$

The intended conclusion is that the turkey is not alive and not walking in the final state.

Calculating $DS_1(WSS)$, we get

$$\begin{aligned} DS_1(WSS) &= sp^{DC}(shoot, w \wedge l \wedge DC) \\ &= sp(shoot, w \wedge l \wedge (w \supset a)) \wedge (w \supset a) \\ &= ((\neg a \wedge \neg l \wedge w \wedge T \wedge (w \supset T) \vee (\neg a \wedge \neg l \wedge w \wedge T \\ &\quad \wedge (w \supset F) \vee (\neg l \wedge w \wedge l \wedge (w \supset a)) \wedge (w \supset a) \\ &\equiv ((\neg a \wedge \neg l \wedge w) \vee (\neg a \wedge \neg l \wedge w \wedge \neg w)) \wedge (w \supset a) \\ &\equiv (\neg a \wedge \neg l \wedge w) \wedge (w \supset a) \\ &\equiv F \end{aligned}$$

As we see, the description of the final state we obtained is inconsistent.

The inconsistency we have arrived at is due to the fact that the fluent w obeys the law of inertia. In our approach, all fluents that are not explicitly affected by an action retain their values when the action is performed. While this is a nice property from the standpoint of the frame problem, it leads to some difficulties when actions with indirect effects are involved.

To deal with domain constraint axioms, we need a mechanism which, for a given action A, releases chosen fluents from obeying the law of inertia when the action A is executed.[15] Fortunately, the release mechanism can be easily implemented in our formalism. Suppose that $f_1, \ldots, f_n$ are fluents which are to be released during executing an action A. To achieve the desired effect, A should be replaced by

$$A; release(f_1); \ldots; release(f_n)$$

where

(10) $release(f_i) \;=\; \mathbf{if}\ T \rightarrow f_i := T \ [\!]\ T \rightarrow f_i := F\ \mathbf{fi}.$

The composition of $release(f_1); \ldots; release(f_n)$ may be written for short as $release(f_1, \ldots, f_n)$.

There is another point that should be emphasized when domain constraint axioms are involved. These axioms should be taken as a part of the description of both the initial and final state.

EXAMPLE 9 (new solution). To properly deal with WSS scenario, the fluent w should be released when the action *shoot* is performed. Accordingly, we replace *shoot* by *shoot** given by $shoot; release(w)$. The description of the initial state is $w \wedge l \wedge DC$. We calculate $DS_1(WSS) \wedge DC$:

$$\begin{aligned} DS_1(WSS) \wedge DC &= sp(shoot^*, (w \wedge l) \wedge DC) \wedge DC \\ &= sp(shoot; release(w), (w \wedge l) \wedge DC) \wedge DC \\ &\equiv sp(release(w), \neg a \wedge \neg l \wedge w) \wedge (w \supset a) \\ &\equiv \neg a \wedge \neg l \wedge (w \supset a). \end{aligned}$$

[15] Compare [Kartha and Lifschitz, 1994].

Since $DS_1(YSS) \wedge DC \vdash \neg a \wedge \neg w$, we conclude that the turkey is not alive and not walking in the final state.

We now systematize our observations by providing a general method to reason about action scenarios with domain constraint axioms. We assume here that the number of these axioms is finite, so that they can be always regarded as a single formula.

DEFINITION 10. We say that an action A is *atomic* iff A is represented by a sequence $S_1; \ldots; S_n$, $n \geq 1$, of assignment commands.

DEFINITION 11. Let A be an action and the fluents $f_1, \ldots, f_m$ are to be released from the law of inertia when the action A is executed. By A^* we denote the following action:

1. if A is represented by *skip* then $A^* = A$.
2. if A is atomic then $A^* = A; \mathit{release}(f_1, \ldots, f_m)$.
3. if A is represented by alternative command of the form

$$\textbf{if} \quad B_1 \rightarrow S_1 \quad \| \quad \cdots \quad \| \quad B_n \rightarrow S_n \quad \textbf{fi}$$

where S_i, $i = 1, \ldots, n$, is a command representing an auxiliary action A^i, then A^* is

$$\textbf{if} \quad B_1 \rightarrow (A^1)^* \quad \| \quad \cdots \quad \| \quad B_n \rightarrow (A^n)^* \quad \textbf{fi}.$$

Let $SC = [\alpha]\ A_1; \ldots; A_n\ [\beta]$ be an action scenario and suppose that DC is the conjunction of all domain constraint axioms assiociated with SC. Assume further that the fluents $f_i^1, \ldots, f_i^{m_i}$ are to be released from the law of inertia when the action A_i is executed.

The *description of* the kth*-state of* SC with respect to DC, written $DS_k^{DC}(SC)$, is the formula given by

$$\begin{cases} \alpha \wedge \neg wlp^{DC}(A_1^*; \ldots; A_n^*, \neg(\beta \wedge DC)) & \text{if } k = 0 \\ sp^{DC}(A_1^*; \ldots; A_k^*, \alpha) \wedge \neg wlp^{DC}(A_{k+1}^*; \ldots; A_n^*, \neg(\beta \wedge DC)) & \text{if } 0 < k < n \\ \beta \wedge sp^{DC}(A_1^*; \ldots; A_n^*, \alpha) & \text{if } k = n \end{cases}$$

where sp^{DC} and wlp^{DC} are defined by the equations (5) – (8), with γ replaced by DC and S_i $(0 < i \leq n)$ replaced by A_i^*.

Theorem 3 immediately implies:

THEOREM 12. Let $SC = [\alpha]\ A_1; \ldots; A_n\ [\beta]$ be an action scenario, DC be the conjunction of domain constrain axioms and suppose that A_i^* $(0 \leq i \leq n)$ is specified as before. A formula γ is assured to hold after performing the actions $A_1; \ldots; A_k$ if $DS_k^{DC}(SC) \vdash \gamma$.

The different but more detailed approach to ramification problem we present in [Madalińska-Bugaj, 1997].

EXAMPLE 13. Consider a variant of the Yale Shooting Scenario. There are four fluents: a, l, b and u, standing for *alive*, *loaded*, *broken* (gun) and *usable* (gun), respectively. The actions are: *load*, and *shoot*. Finally, we have a domain constraint axiom DC given by $l \wedge \neg b \equiv u$. The meaning of the actions is as before with one proviso: to successfully perform the action *shoot*, the gun must be usable. The translation of the new version of *shoot* is

$$\textbf{if } u \rightarrow a, l := F, F \;[\!]\; \neg u \rightarrow skip \textbf{ fi}$$

and its semantics is given by

- $wlp(shoot, \alpha) = (u \supset \alpha[a \rightarrow F, l \rightarrow F]) \wedge (\neg u \supset \alpha)$
- $sp(shoot, \alpha) = sp(shoot, \alpha) = \neg a \wedge \neg l \wedge \exists a, l.(u \wedge \alpha) \vee \neg u \wedge \alpha.$

Because the fluent u is indirectly affected by *load* and *shoot*, it must be released when any of these actions is performed. Consider the scenario $VYSS$ given by

$$[\neg b] \; load; shoot \; [T].$$

The intended conclusion is that the turkey is not alive and the gun is unloaded, unusable and unbroken in the final state.

Since the fluent u is released during performing both the actions, we put $load^* = load; release(u)$ and $shoot^* = shoot; release(u)$. Performing straightforward calculations, one gets

(11) $DS_2^{DC}(VYSS) \equiv \neg a \wedge \neg l \wedge \neg u \wedge \neg b.$

In view of (11), we immediately conclude that the turkey is not alive and the gun is unloaded, unusable and unbroken in the final state.

7 CONCLUSIONS

We have applied Dijkstra's semantics for programming languages to formalization of reasoning about action and change. We believe that the results reported here are interesting and worth of further investigation. The presented approach can be employed to represent a broad class of action scenarios, including those where actions with random and indirect effects are permitted. In addition, both temporal prediction and postdiction tasks can be properly dealt with, without requiring initial or final situations to be completely specified. The major advantage of our proposal is that it is very simple and more effective than many other approaches directed at formalizing reasoning about action and change.

The actions considered in this paper are very simple in that they are represented by identifiers. In practice, it is often convenient to have action schemata rather, each representing a class of similar actions applicable to various tuples of individuals. For example, we may have the action schema $Put(x, y)$ with the intended meaning that a block x is to be put on the top of a block y. There is no technical difficulty to admit action schemata in our system, but this topic will be disscussed in a future paper.

As we remarked earlier, recent work of Sandewall [Sandewall, 1993; Sandewall, 1994] provides a very general framework to study logics of action and change. Obviously, the question of how our proposal fits in this framework should be investigated and will be pursued in the future. It is also interesting to compare our approch with $\mathcal{AR}$ language introduced recently by Kartha and Lifschitz [Kartha and Lifschitz, 1994].

The task of implementation is another point of interest. Calculating $DS_k(SC)$, for a given scenario SC, amounts to simple syntactic manipulations on formulae and can be performed very efficiently. The only computational problem is to determine whether a given formula can be derived from the description of the initial situation. This task can be realized by a theorem prover appropriate for the logic in which the effects of actions are described. For the shooting language all we need is a theorem prover for classical propositional logic. In more complex applications, a theorem prover for first-order logic will be required. In any case, we work in the framework of classical monotonic logic which makes our approach simpler and computationally more efficient than the approaches employing non-monotonic forms of reasoning.

ACKNOWLEDGEMENTS

We would like to thank Władysław M. Turski and Włodek Drabent for their comments on the previous draft of this paper.

This is a revised and extended version of the paper [Łukaszewicz and Madalińska-Bugaj, 1995]. The authors were supported by the ESPRIT Basic Research Action No. 6156 - DRUMS II.

Witold Łukaszewicz and Ewa Madalińska-Bugaj
Institute of Informatics, Warsaw University, Poland.

BIBLIOGRAPHY

[Baker, 1991] A. B. Baker. Nonmonotonic Reasoning in the Framework of Situation Calculus. *Artificial Intelligence*, **49**, 5–23, 1991.

[Dijkstra, 1976] E. W. Dijkstra. *A Discipline of Programming*. Prentice Hall, 1976.

[Dijkstra and Scholten, 1990] E. W. Dijkstra and C. S. Scholten. *Predicate Calculus and Program Semantics*. Springer-Verlag, 1990.

[Fikes and Nilsson, 1971] R. E. Fikes and N. J. Nilsson. STRIPS: A New Approach to the Application of Theorem Proving to Problem Solving. *Artifficial Intelligence*, **2**, 189–208, 1971.

[Gelfond *et al.*, 1991] M. Gelfond, V. Lifschitz and A. Rabinov. What Are the Limitations of Situation Calculus? *AAAI Symposium of Logical Formalization of Commonsense Reasoning*, pp. 55–69, Stanford, 1991.

[Hanks and McDermott, 1987] S. Hanks and D. McDermott. Nonmonotonic Logic and Temporal Projection. *Artificial Intelligence*, **33**, 379–412, 1987.

[Kartha and Lifschitz, 1994] G. N. Kartha and V. Lifschitz. Actions with Indirect Effects (Preliminary Report). In *Proc. KR-94*, Bonn, Germany, p. 341–350. Morgan Kaufmann Publishers, San Francisco, 1994.

[Lifschitz, 1988] V. Lifschitz. Formal Theories of Action. In *Readings in Nonmonotonic Reasoning*, M. Ginsberg, ed. pp. 35–57. Morgan Kaufmann Publishers, Palo Alto, 1988.

[Lifschitz and Rabinov, 1989] V. Lifschitz and A. Rabinov. Miracles in Formal Theories of Action. *Artificial Intelligence*, **38**, 225–237, 1989.

[Łukaszewicz and Madalińska-Bugaj, 1995] W. Łukaszewicz and E. Madalińska-Bugaj. Reasoning about Action and Change Using Dijkstra's Semantics for Programming Languages: Preliminary Report. *Proceedings of 14th International Joint Conference on Artificial Intelligence*, pp. 1950–1955, 1995.

[Madalińska-Bugaj, 1997] E. Madalińska-Bugaj. How to Solve Qualification and Ramification Using Dijkstra's Semantics for Programming Languages. *AI*IA-97: Advances in Artificial Intelligence, Proceedings of 5th Congress of the Italian Association for Artificial Intelligence*, pp. 381–392. LNAI 1321, Springer-Verlag, 1997.

[McCarthy and Hayes, 1969] J. McCarthy and P. J. Hayes. Some Philosophical Problems from the Standpoint of Artificial Intelligence. In *Machine Intelligence* **4**, B. Meltzer and D. Michie, eds. pp. 463–502, 1969.

[Pednault, 1986] E. P. D. Pednault.*Toward a Mathematical Theory of Plan Synthesis*. Ph. D. Thesis, Dept. of Electrical Engineering , Stanford University, Stanford, 1986.

[Pednault, 1988] E. P. D. Pednault. Synthesizing Plans that Contain Actions with Contex-Dependent Effects. *Computational Intelligence*, **4**, 356–372, 1988.

[Pednault, 1989] E. P. D. Pednault. ADL: Exploring the Middle ground between STRIPS and the Situation Calculus. In *Proc. KR-89*, R. Brachman, H. Levesque and R. Reiter, eds. pp. 324–333.1989, 324-333.

[Sandewall, 1993] E. Sandewall. The Range of Applicability of Nonmonotonic Logics for the Inertia Problem. In: *Proc. IJCAI-93*, pp. 738–743, 1993.

[Sandewall, 1994] E. Sandewall. *Features and Fluents. The Representation of Knowledge about Dynamical Systems*. Oxford Science Publications, 1994.

[Waldinger, 1977] R. Waldinger. Achieving Several Goals Simultaneusly. In: *Machine Intelligence* **8**. E. Ellock and D. Michie, eds. pp. 94–136. Ellis Horwood, Edinburgh, 1977.

WITOLD ŁUKASZEWICZ AND EWA MADALIŃSKA-BUGAJ

REASONING ABOUT ACTION AND CHANGE: ACTIONS WITH ABNORMAL EFFECTS

1 INTRODUCTION

Most of the research devoted to reasoning about action and change has been based on the assumption that each action behaves in a fixed way. More specifically, to each action A there is assigned a unique specification S describing the effects of A in terms of a state in which A is performed.[1] For instance, the well-known action *shoot* is usually defined as making a gun unloaded and a turkey dead, provided that a gun was loaded. Accordingly, each time the action is executed in a state in which the gun is loaded, it is taken for granted that the turkey is made dead.

In this paper, we generalize the above assumption by admitting actions that may exhibit abnormal behaviour. More precisely, with each action A we associate a pair of specifications, S_1 and S_2, corresponding respectively to a normal and an abnormal behaviour of A. The intention, of course, is that A behaves according to S_1 unless the contrary follows from observations. Reconsider the action *shoot*. We may define its normal behaviour as making a gun unloaded and a turkey dead, provided that a gun is loaded, and the abnormal one as making only the gun unloaded. Now, if all we know is that the action was performed in a state in which the gun was loaded, we infer that the turkey was made dead. However, if we additionally observe that the turkey is alive after executing the action, we are forced to assume that the action behaved abnormally.

To formalize effects of actions, we use Dijkstra's approach, originally developed for reasoning about programs [Dijkstra, 1976; Dijkstra and Scholten, 1990]. The strength of Dijkstra's proposal is its effectiveness. It has been showed in our earlier papers [Łukaszewicz and Madalińska-Bugaj, 1994; Łukaszewicz and Madalińska-Bugaj, 1995], where Dijkstra's methodology was employed to formalize conventional forms of inference about action and change. To distinguish between normal and abnormal behaviour of actions, Dijkstra's formalism is combined here with Reiter's default logic [Reiter, 1980].[2]

The paper is organized as follows. Section 2 is a brief introduction to Dijkstra's semantics for a simple programming language. In Section 3, we show how to define action languages using Dijkstra's methodology, illustrate this by specifying a simple action language, and introduce a notion of action scenario. Section 4

[1]Depending on whether A is deterministic or not, S can be formally viewed as an unary function or a binary relation, specified on a space of all possible states.

[2]The problem of formalizing actions with abnormal effects has been recently addressed by Radzikowska [1995], who works in the framework of Sandewall's Features and Fluents [Sandewall, 1994].

J.J.Ch. Meyer and J. Treur (eds.),
Handbook of Defeasible Reasoning and Uncertainty Management Systems, Vol. 7, 399–409.

defines the kind of inference we shall be interested in, whereas section 5 provides a method realizing this type of reasoning. Section 6 contains a number of examples. Finally, in section 7 we provide conclude remarks and discuss the future work.

We assume that the reader is familiar with Reiter's default logic.

2 INTRODUCTION TO DIJKSTRA'S SEMANTICS

In [Dijkstra and Scholten, 1990] we are provided with a simple programming language whose semantics is specified in terms of formula trasformers. More specifically, with each command there are associated two formula transformers, called the the *weakest liberal precondition* and the *strongest postcondition*, denoted by wlp and sp, respectively.[3] A programming language we use here is that considered in [Łukaszewicz and Madalińska-Bugaj, 1995]. The above paper provides also specifications of the formula transformers for the commands under consideration.

Dijkstra and Scholten [1990] consider various classes of computations. One of them is called by these authors *initially* α *and finally* β *under control of* S, where α and β are formulae and S is a command. This class, written $[S](\alpha, \beta)$, represents the set of all computations under control of S that start in a state satisfying α and terminate in a state satisfying β.

Suppose that $c \in [S](\alpha, \beta)$. Obviously, since S terminates in a state satisfying β and $wlp(S, \neg\beta)$ represents the set of all states such that S begun in any of them either terminates in a state satisfying $\neg\beta$ or loops forever, it must be the case that the initial state of c satisfies $\alpha \wedge \neg wlp(S, \neg\beta)$. Similarly, since S starts in a state satisfying α and $sp(S, \alpha)$ represents the set of all states such that any of them can be reached by starting execution of S in some state satisfying α, we conclude that the final state of c satisfies $\beta \wedge sp(S, \alpha)$. An interesting question is whether the formulae $\alpha \wedge \neg wlp(S, \neg\beta)$ and $\beta \wedge sp(S, \alpha)$ provide a complete description of the initial and final states of the computations from the class $[S](\alpha, \beta)$. That the answer is positive follows from the following result which can be found in [Dijkstra and Scholten, 1990].

THEOREM 1. *The formula* $\alpha \wedge \neg wlp(S, \neg\beta)$ *(resp.* $\beta \wedge sp(S, \alpha)$*) holds in a state* σ *iff there exists a computation* c *from* $[S](\alpha, \beta)$ *such that* σ *is the initial (resp. final) state of* c.

In this paper, we shall be interested in a more general class of computations. This class, written $[S_1; \ldots; S_n](\alpha_0, \ldots, \alpha_n)$, where $S_1, \ldots, S_n$ are commands and $\alpha_0, \ldots, \alpha_n$ are formulae, represents the class of all computations under control of $S_1; \ldots; S_n$ that start in a state satisfying α_0, terminate in a state satisfying α_n and, in addition, any state of any computation from this class that can be reached after executing $S_1; \ldots; S_i$ $(0 < i < n)$ satisfies α_i.

---

[3]Dijkstra and Scholten consider also a third formula transformer, called the *weakest precondition*, but this will not be used in our paper.

Suppose that $\mathcal{C}= [S_1; S_2; \ldots; S_n](\alpha_0, \ldots, \alpha_n)$. We define two sequences of formulae, $SP_1^{\mathcal{C}}, \ldots, SP_n^{\mathcal{C}}$ and $WLP_n^{\mathcal{C}}, \ldots, WLP_0^{\mathcal{C}}$, by the following induction:

$$SP_0^{\mathcal{C}} = \alpha_0; \quad WLP_n^{\mathcal{C}} = \neg\alpha_n$$

and for $0 < i \leq n$

$$SP_i^{\mathcal{C}} = \alpha_i \wedge sp(S_i, SP_{i-1}^{\mathcal{C}}); \quad WLP_{n-i}^{\mathcal{C}} = \neg\alpha_{n-i} \vee wlp(S_{n-i+1}, WLP_{n-i+1}^{\mathcal{C}}).$$

Let $\mathcal{C}=[S_1; \ldots; S_n](\alpha_0, \ldots, \alpha_n)$. For each $0 \leq i \leq n$, we write $ST_{\mathcal{C}}(i)$ to denote the set of all states satisfying the following condition: for each $\sigma \in ST_{\mathcal{C}}(i)$, there exists a computation $c \in \mathcal{C}$ such that σ is reached by c after executing $S_1; \ldots; S_i$. In what follows, the members of $ST_{\mathcal{C}}(i)$ will be referred to as *i-states* of $\mathcal{C}$. Clearly, 0-states are initial states and n-states are final states.

The following result, which is an obvious generalization of Theorem 2 from [Łukaszewicz and Madalińska-Bugaj, 1995], provides a complete characterization of i-states.

THEOREM 2. *The formula $SP_i^{\mathcal{C}} \wedge \neg WLP_i^{\mathcal{C}}$ holds in a state σ iff there exists a computation from $\mathcal{C}$ such that $\sigma \in ST_{\mathcal{C}}(i)$, $0 \leq i \leq n$.*

Let $\mathcal{C}=[S_1; \ldots; S_n](\alpha_0, \ldots, \alpha_n)$. We write $D_i(\mathcal{C})$, $0 \leq i \leq n$, to denote the formula $SP_i^{\mathcal{C}} \wedge \neg WLP_i^{\mathcal{C}}$. In view of Theorem 2, $D_i(\mathcal{C})$ provides the complete characterization of i-states. Accordingly, it will be referred to as the *description of the i-states of* $\mathcal{C}$.

By a *subcommand* S of a command S' we mean a substring S of S' which is also a command. In particular, for any command S, S is a subcommand of S.

We say that a variable x is *inaffected* by a command S iff S does not contain a subcommand of the form $x := e$.

The following result will be useful.

PROPOSITION 3. *Let σ and σ' be initial and final states, respectively, of some computation under control of S. If x is inaffected by S, then $\sigma(x) = \sigma'(x)$.*

Proof. Follows from the definitions of *sp* and *wlp*. ■

COROLLARY 4. *Let $\mathcal{C}=[S_1; \ldots; S_n](\alpha_0, \ldots, \alpha_n)$. Suppose that $\sigma \in ST_{\mathcal{C}}(i)$ and $\sigma' \in ST_{\mathcal{C}}(j)$ $(0 \leq i, j \leq n)$. Assume further that σ and σ' are members of the same computation $c \in \mathcal{C}$. If x is inaffected by any of $S_1, \ldots, S_n$, then $\sigma(x) = \sigma'(x)$.*

3 ACTION LANGUAGES

In this section, we show how to define action languages using Dijkstra's methodology and illustrate this by specifying a simple action language $\mathcal{L}$ that will be used in examples.

To define an action language, one starts with an alphabet consisting of primitive symbols from the following pairwise disjoint classes: (1) A countable set of Boolean variables, called *fluents* (these serve to describe the application domain under consideration); (2) A denumerable set of auxiliary Boolean variables, called *abnormality variables*: ab_1,ab_2,...; (3) A countable set of *action symbols*; (4) Two truth-constants: T and F; (5) Usual sentential connectives.

The classes (2), (4)–(5) are fixed, whereas the classes (1) and (3) varies from an alphabet into an alphabet. Accordingly, each alphabet is uniquely determined by its fluents and action symbols.

The alphabet of our action language $\mathcal{L}$ contains three fluents, a, l and s, standing for $alive$, $loaded$ and $stolen$, respectively, and three action symbols *load* (a gun), $shoot$ (a turkey) and $leave$ (a car overnight in a garage).

A *formula* is a Boolean combination of fluents, abnormality variables and truth-constants. A formula containing no abnormality variables is said to be an *observation*.

The objects we shall be primarily interested in are (*action*) *scenarios*. These are expressions of the form

(1) $SC = \langle\alpha_0\rangle\ A_1; \langle\alpha_1\rangle\ A_2; \ldots; \langle\alpha_{n-1}\rangle\ A_n\ \langle\alpha_n\rangle$

where $\alpha_0, \ldots, \alpha_n$ are observations and $A_1, \ldots, A_n$ are action symbols. The scenario has the following intuitive interpretation: α_0 was observed to hold in the initial state, then the actions $A_1, \ldots, A_n$ were sequentially performed, and α_i, for $0 < i \leq n$, was observed to hold after performing the action A_i. In particular, α_n is the formula observed in the final state. In intermediate states, if no observation is made, i.e. $\alpha_i \equiv T$, we omit this element in the scenario.

EXAMPLE 5. Below is a variant of the famous Yale Shooting Scenario [Hanks and McDermott, 1987].

$$\langle a \wedge \neg l\rangle\ load; shoot\ \langle a\rangle$$

As we remarked earlier, our approach is based on the assumption that to each action symbol A, there are assigned two specifications, describing normal and abnormal performance of A. For the action symbols from the language $\mathcal{L}$, these specifications are the following. The action *load*, when performed normally, makes the gun loaded; when performed abnormally, does not cause any changes in the world. The action $shoot$, when performed normally, makes the gun unloaded and the turkey dead, provided that the gun was loaded before and, when performed abnormally, makes the gun unloaded. Finally, $leave$, when performed normally does not cause any changes in the world; when performed abnormally, it makes the car stolen.

The crucial point in using Dijkstra's methodology is to interpret action symbols occurring in a scenario as commands of the programming language defined in the previous paper of this volume (pp. 385–400). The construction of such a command is a two step process. Firstly, with each action symbol A we associate two commands, denoted $S_1(A)$ and $S_2(A)$, describing respectively normal and abnormal performance of A.[4] For action symbols occurring in the language $\mathcal{L}$, $S_1(load)$ is $l := T$, $S_2(load)$ is $skip$, $S_1(shoot)$ is **if** $l \rightarrow a := F; l := F \; [\!] \; \neg l \rightarrow skip$ **fi**, $S_2(shoot)$ is $l := F$, $S_1(leave)$ is $skip$ and $S_2(leave)$ is $s := T$. Given $S_1(A)$ and $S_2(A)$, we define a *command schema* corresponding to A, written $S(A)$, by

$$\textbf{if } \neg ab \rightarrow S_1(A) \; [\!] \; ab \rightarrow S_2(A) \textbf{ fi}$$

where ab is a parameter which can be replaced by any abnormality variable from $\{ab_1, ab_2, \ldots\}$.[5] Accordingly,

$S(load)$ is **if** $\neg ab \rightarrow l := T \; [\!] \; ab \rightarrow skip$ **fi**,

$S(shoot)$ is **if** $\neg ab \rightarrow$ **if** $l \rightarrow a := F; l := F \; [\!] \; \neg l \rightarrow skip$ **fi** $[\!] \; ab \rightarrow l := F$ **fi**

$S(leave)$ is **if** $\neg ab \rightarrow skip \; [\!] \; ab \rightarrow s := T$ **fi**.

Given the above command schemata, the chosen action symbols can be provided with Dijkstra-style semantics. Performing routine calculations one easily obtains:

- $wlp(S(load), \alpha) = (\neg ab \supset \alpha[l \leftarrow T]) \wedge (ab \supset \alpha)$;
- $sp(S(load), \alpha) = \neg ab \wedge l \wedge (\alpha[l \leftarrow T] \vee \alpha[l \leftarrow F]) \vee ab \wedge \alpha$;
- $wlp(S(shoot), \alpha) = (\neg ab \supset (l \supset \alpha[a \leftarrow F, l \leftarrow F]) \wedge (\neg l \supset \alpha)) \wedge (ab \supset \alpha[l \leftarrow F])$;
- $sp(S(shoot), \alpha) = \neg ab \wedge ((\neg a \wedge \neg l \wedge \alpha[a \leftarrow T, l \leftarrow T]) \vee (\neg a \wedge \neg l \wedge \alpha[a \leftarrow F, l \leftarrow T]) \vee (\neg l \wedge \alpha)) \vee ab \wedge \neg l \wedge (\alpha[l \leftarrow T] \vee \alpha[l \leftarrow F])$;
- $wlp(S(leave), \alpha) = (\neg ab \supset \alpha) \wedge (ab \supset \alpha[s \leftarrow T])$
- $sp(S(leave), \alpha) = \neg ab \wedge \alpha \vee ab \wedge (\alpha[s \leftarrow T] \vee \alpha[s \leftarrow F])$.

4 ACTION SCENARIOS AS CLASSES OF COMPUTATIONS

Let SC be an action scenario given by (1). We write $\mathcal{C}(SC)$ to denote the class of computations

(2) $[S^1(A_1); \ldots; S^n(A_n)](\alpha_0, \ldots, \alpha_n)$

[4] The ability to represent the effects (normal and abnormal) of chosen action symbols as commands of the programming language is the necessary condition to use Dijkstra's approach to reasoning about action and change. We do not claim here that this is always possible. However, most of the actions that have been considered in the AI literature enjoy this property.

[5] It should be stressed that normality/abnormality does not concern an action, but rather its performance. Accordingly, we should use different abnormality variables not only for different action symbols, but also for different occurrences of the same action symbol.

where $S^i(A_i)$ is $S(A_i)$ with ab replaced by ab_i.

We would like to identify the scenario SC with the class of computations $\mathcal{C}(SC)$. Unfortunately, $\mathcal{C}(SC)$ is too large to properly represent the scenario SC. The reason is that replacing an action symbol A, occurring in the i-th position of SC, by the command **if** $\neg ab_i \rightarrow S_1(A) \;[\!]\; ab_i \rightarrow S_2(A)$ **fi** we make no distinction between normal and abnormal performance of the action represented by A.[6] On the other hand, our approach is based on the implicite assumption that normal performances of actions are to be preferred over abnormal ones. In the rest of this section, we specify a subclass of $\mathcal{C}(SC)$ that captures this intuition.

We start with some preliminary terminology.

Let c be a computation from the class $\mathcal{C}(SC)$. We write $c(i)$, $0 \le i \le n$, to denote the state that is reached by c after performing the commands $S^1(A_1); \ldots; S^i(A_i)$. We say that an action A_i is *realized normally in a computation* c iff the variable ab_i is assigned the value F in the state $c(i-1)$.

PROPOSITION 6. *An action A_i from a scenario SC is realized normally in a computation $c \in \mathcal{C}(SC)$ iff the variable ab_i is assigned the value F in all states of c.*

Proof. Follows from Corollary 4 and the fact that the variables $ab_1, \ldots, ab_n$ are inaffected by any of $S^1(A_1), \ldots, S^n(A_n)$. ■

Let c be a computation from the class $\mathcal{C}(SC)$, where SC is given by (1). We write $AB(c)$ to denote the set of these members of $\{ab_1, \ldots, ab_n\}$ which are assigned the value T in all states from c.[7]

Let c and c' be two computations from the class $\mathcal{C}(SC)$. We write $c \le_{AB} c'$ iff $AB(c) \subseteq AB(c')$. We say that c is *AB-minimal* in the class $\mathcal{C}(SC)$ iff for every $c' \in \mathcal{C}(SC)$, if $c' \le_{AB} c$ then $AB(c) = AB(c')$.

Clearly, the intended computations are those in which as many actions as possible are realized normally:

DEFINITION 7. Let SC be an action scenario given by (1). The *intended class of computations corresponding to* SC, written $\mathcal{IC}(SC)$, is the class of all AB-minimal elements from $\mathcal{C}(SC)$.

5 REASONING ABOUT SCENARIOS

In this section, we provide a method to reason about action scenarios.

We shall be interested in the following reasoning task: "Given an action scenario $SC = \langle\alpha_0\rangle\ A_1; \langle\alpha_1\rangle\ A_2; \ldots; \langle\alpha_{n-1}\rangle A_n\ \langle\alpha_n\rangle$, a formula γ and an integer k such that $0 \le k \le n$, determine whether γ holds after performing the actions

[6]In the sequel we shall not distinguish between an action symbol and the action it represents.

[7]By Proposition 6, $AB(c)$ characterizes these actions from $\{A_1, \ldots, A_n\}$ which are realized abnormally in a computation c.

$A_1, \ldots, A_k$".[8] Identifying a scenario SC with the class of computations $\mathcal{IC}(SC)$, the above reasoning task can be stated more formally: "Given a scenario $SC = \langle\alpha_0\rangle\ A_1; \langle\alpha_1\rangle\ A_2; \ldots; \langle\alpha_{n-1}\rangle A_n\ \langle\alpha_n\rangle$, a formula γ and an integer k such that $0 \leq k \leq n$, determine whether γ holds in all k-states of $\mathcal{IC}(SC)$.[9]

Consider a scenario $SC = \langle\alpha_0\rangle\ A_1; \langle\alpha_1\rangle\ A_2; \ldots; \langle\alpha_{n-1}\rangle A_n\ \langle\alpha_n\rangle$ and the class of computations $\mathcal{C}(SC)$ given by (2). In view of Theorem 2, we know that the set of all k-states of $\mathcal{C}(SC)$ is characterized by the formula

$$(3)\quad D_k(\mathcal{C}(SC)) = SP_k^{\mathcal{C}(SC)} \wedge \neg WLP_k^{\mathcal{C}(SC)}$$

where the formulae $SP_k^{\mathcal{C}(SC)}$ and $WLP_k^{\mathcal{C}(SC)}$ have been defined in section 2. To characterize the set of k-states of the class $\mathcal{IC}(SC)$, we use Reiter's default logic [Reiter, 1980]. More precisely, we define a default theory $\mathcal{T}_k(SC) = \langle A, \Delta\rangle$, where A (the set of axioms) consists of the formula $D_k(\mathcal{C}(SC))$ and Δ (the set of defaults) is specified by $\{: \neg ab_i / \neg ab_i | 1 \leq i \leq n\}$. The next theorem shows that extensions of the theory $\mathcal{T}_k(SC)$ provide a complete description of k-states of the class $\mathcal{IC}(SC)$.

THEOREM 8. *Let $SC = \langle\alpha_0\rangle\ A_1; \langle\alpha_1\rangle\ A_2; \ldots; \langle\alpha_{n-1}\rangle A_n\ \langle\alpha_n\rangle$ be an action scenario. A formula γ holds in all $k -$ states of the class $\mathcal{IC}(SC)$, $0 \leq k \leq n$, iff γ is the member of all extensions of the default theory $\mathcal{T}_k(SC)$.*

Proof. Let $\mathcal{IC}(SC) = \bigcup_{i=1}^{m} \mathcal{IC}_i$, where $\mathcal{IC}_i \cap \mathcal{IC}_j = \emptyset$, for $i \neq j$, and for all $i = 1, \ldots, m$, if computations $c_1, c_2 \in \mathcal{IC}_i$ then $AB(c_1) = AB(c_2)$. It is enough to show that the computation $c \in \mathcal{C}(SC)$ is $AB - minimal$, i.e. $c \in \mathcal{IC}_i$, for some i, $(1 \leq i \leq m)$ iff for all $1 \leq k \leq n$, there exists an extension E of a theory $\mathcal{T}_k(SC)$ such that $Th(D_k(\mathcal{IC}_i)) = E$.

$\Rightarrow$ Let $AB = \{ab_1, \ldots, ab_n\}$ be a set of all variables ab_i appearing in the scenario. If c is a $AB - minimal$ computation in a class $\mathcal{C}(SC)$ then $c \in \mathcal{IC}_i$ for some $1 \leq i \leq m$. Let $AB(c) = \{ab_1, \ldots, ab_l\}$. Denote by $AB'(c)$ the set $AB - AB(c)$. $AB'(c) = \{ab_{l+1}, \ldots, ab_n\}$. By Definition 7 a description of k-states for a subclass of computations $\mathcal{IC}_i$ is a formula $D_k(\mathcal{IC}_i) = D_k(\mathcal{C}(SC)) \wedge \neg ab_{l+1} \wedge \ldots \wedge \neg ab_n \not\equiv F$.

Let $\mathcal{T}_k(SC) = \langle D_k(\mathcal{C}(SC)), \{: \neg ab_i / \neg ab_i \mid 1 \leq i \leq n\}\rangle$. We show that there exists an extension E of a theory $\mathcal{T}_k(SC)$ such that $E = Th(D_k(\mathcal{C}(SC)) \cup \{\neg ab_{l+1}, \ldots, \neg ab_n\}) = Th(D_k(\mathcal{IC}_i))$.

1. $Th(D_k(\mathcal{IC}_i)) \subseteq E$. Since $D_k(\mathcal{IC}_i) \in E$ and $D_k(\mathcal{IC}_i) \not\equiv F$, we conclude $\{\neg ab_{l+1}, \ldots, \neg ab_n\} \subseteq CONSEQ(GD(E))$.

[8] For $k = 0$, i.e. when we want to determine whether γ is true in the initial state, the above reasoning task is known as the *temporal postdiction problem*. For $k = n$, i.e. when we want to determine whether γ holds in the final state, the task is usually referred to as the *temporal prediction problem*.

[9] The set of k-states of $\mathcal{IC}(SC)$ is specified as that of $\mathcal{C}(SC)$ (see section 2) restricted to the states occurring in the computations from $\mathcal{IC}(SC)$.

2. $E \subseteq Th(D_k(\mathcal{IC}_i))$. Assume to the contrary that $E \not\subset Th(D_k(\mathcal{IC}_i))$. Then there exists a default $d = (: \neg ab_j / \neg ab_j) \in GD(E)$, where $j < l + 1$. Hence $D_k(\mathcal{IC}_i) \wedge \neg ab_j \not\equiv F$. Thus there exists a computation c'' such that ab_j is assinged a value F and $AB(c'') \subseteq \{ab_1, \ldots, ab_{j-1}, ab_{j+1}, \ldots, ab_l\}$, hence $c'' \leq_{AB} c$ and $AB(c) \neq AB(c'')$, so c is not $AB - minimal$. A contradiction.

$\Leftarrow$ Let E be an extension of a theory $\mathcal{T}_k(SC)$. Let $GD(E) = \{d_1, \ldots, d_l\}$, where $d_i = (: \neg ab_i / \neg ab_i)$. Thus $D_k(\mathcal{C}(SC)) \wedge \neg ab_1 \wedge \ldots \wedge \neg ab_l \not\equiv F$. Then there exists a computation c such that $AB(c) = \{ab_{l+1}, \ldots, ab_n\}$. We show that c is $AB - minimal$.
Assume to the contrary that c is not $AB - minimal$. Hence there exists c' such that $AB(c') \subseteq \{ab_{l+1}, \ldots, ab_{j-1}, ab_{j+1}, \ldots, ab_n\}$. Let $\mathcal{IC}' = \{c'' \mid c'' \in \mathcal{C}(SC)$ and $AB(c') = AB(c'')\}$. Then $D_k(\mathcal{IC}') = D_k(\mathcal{C}(SC)) \wedge \neg ab_1 \wedge \ldots \wedge \neg ab_l \wedge \neg ab_j \not\equiv F$, so there exists an extension E_1 of a theory $\mathcal{T}_k(SC)$ such that $E \subset E_1$, but this is impossible. ■

COROLLARY 9. *Let SC be specified as before. A formula γ holds after performing the actions $A_1; \ldots; A_k$, $0 \leq k \leq n$, iff γ is the member of all extensions of the default theory $\mathcal{T}_k(SC)$.*

REMARK 10. A scenario $SC = \langle \alpha_0 \rangle\ A_1; \langle \alpha_1 \rangle\ A_2; \ldots; \langle \alpha_{n-1} \rangle A_n\ \langle \alpha_n \rangle$ gives rise to $n + 1$ different default theories: $\mathcal{T}_0(SC)$, $\mathcal{T}_1(SC)$, $\ldots \mathcal{T}_n(SC)$. It can be shown, however, that the extensions of all these theories are based on the same sets of generating defaults.[10] Accordingly, when sets of generating defaults for $\mathcal{T}_i(SC)$, $0 \leq i \leq n$, are computed, they can be used for all default theories corresponding to the scenario SC.

6 EXAMPLES

In this section we consider a few examples to illustrate the method introduced in the previous section. Recall that a command $S^i(action)$ is a command schema $S(action)$ in which the parameter ab is replaced by ab_i. The clauses specifying $sp(S(action), \alpha)$ and $wlp(S(action), \alpha)$ have been provided in section 3.

EXAMPLE 5 (continued). This scenario (SC, for short) is an example of the temporal prediction. We have a complete description of the initial state, a partial description of the final state and the question we are interested in is whether the gun was loaded in the final state. The scenario is given by

$$\langle a \wedge \neg l \rangle\ load; shoot\ \langle a \rangle.$$

First, we calculate $D_2(\mathcal{C}(SC))$, where $\mathcal{C}(SC) = [S^1(load); S^2(shoot)](a \wedge \neg l, T, a)$.[11]

[10] See [Reiter, 1980], for the definition of this notion.

[11] Note the use of the symbols "=" and "≡" during the calculation. We write $X = Y$ if Y is

$$
\begin{aligned}
D_2(\mathcal{C}(SC)) &= SP_2^{\mathcal{C}(SC)} \wedge \neg WLP_2^{\mathcal{C}(SC)}\\
&= a \wedge sp(S^2(shoot), sp(S^1(load), a \wedge \neg l))\\
&= a \wedge sp(S^2(shoot), \neg ab_1 \wedge l \wedge a \vee ab_1 \wedge a \wedge \neg l)\\
&= a \wedge (\neg ab_2 \wedge ((\neg a \wedge \neg l \wedge \neg ab_1) \vee (\neg l \wedge ab_1 \wedge a)) \vee ab_2 \wedge \neg l \wedge a)\\
&\equiv \neg ab_2 \wedge ab_1 \wedge a \wedge \neg l \vee ab_2 \wedge \neg l \wedge a.
\end{aligned}
$$

The default theory $\mathcal{T}_2(SC)$ consists of $D_2(\mathcal{C}(SC))$ as its unique axiom and the pair of defaults $\{: \neg ab_1/\neg ab_1, : \neg ab_2/\neg ab_2\}$. $\mathcal{T}_2(SC)$ has two extensions, E_1 and E_2, given by $Th(\{D_2(\mathcal{C}(SC)), \neg ab_1\})$ and $Th(\{D_2(\mathcal{C}(SC)), \neg ab_2\})$, respectively.[12] Since $\neg l$ is the member of both E_1 and E_2, we conclude that the gun was unloaded in the final state.

It is worth noting that the extensions E_1 and E_2 correspond to two possible courses of events: either *load* was performed normally and *shoot* abnormally (E_1) or *load* was performed abnormally and *shoot* normally (E_2).

Consider now the scenario SC' given by $\langle a \wedge \neg l\rangle$ *load*; $\langle l\rangle$ *shoot* $\langle a\rangle$. We leave it to the reader to check that the default theory $\mathcal{T}_2(SC')$ has one extension containing $\neg ab_1$, ab_2 and $\neg l$. Accordingly, we conclude that *load* was performed normally, *shoot* was performed abnormally and the gun was unloaded in the final state.

EXAMPLE 11. Consider now the scenario SC given by

$$\langle a \wedge \neg l\rangle\ load; shoot; load\ \langle a\rangle.$$

The inteded conclusion is that in the final state the gun is loaded. We start by calculating $D_3(\mathcal{C}(SC))$, where $\mathcal{C}(SC) = [S^1(load); S^2(shoot); S^3(load)](a \wedge \neg l, T, T, a)$.

$$
\begin{aligned}
D_3(\mathcal{C}(SC)) &= SP_3^{\mathcal{C}(SC)} \wedge \neg WLP_3^{\mathcal{C}(SC)}\\
&= a \wedge sp(S^3(load), sp(S^2(shoot), sp(S^1(load), a \wedge \neg l)))\\
&= a \wedge sp(S^3(load), sp(S^2(shoot), \neg ab_1 \wedge l \wedge a \vee ab_1 \wedge a \wedge \neg l))\\
&= a \wedge sp(S^3(load), \neg ab_2 \wedge (\neg a \wedge \neg l \wedge \neg ab_1 \vee \neg l \wedge ab_1 \wedge a)\\
&\quad \vee ab_2 \wedge (\neg l \wedge \neg ab_1 \wedge a \vee \neg l \wedge ab_1 \wedge a))\\
&\equiv a \wedge sp(S^3(load), \neg ab_2 \wedge (\neg a \wedge \neg l \wedge \neg ab_1 \vee \neg l \wedge ab_1 \wedge a)\\
&\quad \vee ab_2 \wedge \neg l \wedge a)\\
&= a \wedge (\neg ab_3 \wedge l \wedge (\neg ab_2 \wedge (\neg a \wedge \neg ab_1 \vee a \wedge ab_1) \vee (ab_2 \wedge a)) \vee\\
&\quad ab_3 \wedge (\neg ab_2 \wedge (\neg a \wedge \neg l \wedge \neg ab_1 \vee \neg l \wedge ab_1 \wedge a) \vee ab_2 \wedge \neg l \wedge a))\\
&\equiv \neg ab_3 \wedge l \wedge \neg ab_2 \wedge ab_1 \wedge a \vee \neg ab_3 \wedge l \wedge ab_2 \wedge a\\
&\quad \vee ab_3 \wedge \neg ab_2 \wedge ab_1 \wedge a \wedge \neg l \vee ab_3 \wedge ab_2 \wedge \neg l \wedge a.
\end{aligned}
$$

The default theory $\mathcal{T}_3(SC) = \langle\{D_3(\mathcal{C}(SC))\}, \{: \neg ab_1/\neg ab_1,\ : \neg ab_2/\neg ab_2,\ : \neg ab_3/\neg ab_3\}\rangle$ has two extensions, $E_1 = Th(\{D_3(\mathcal{C}(SC)), \neg ab_1, \neg ab_3\})$ and

obtained from X by employing the semantics of *wlp* or *sp*, whereas $X \equiv Y$ indicates that X and Y are logically equivalent.

[12] Note that E_1 contains ab_2 and E_2 contains ab_1.

$E_2 = Th(\{D_3(\mathcal{C}(SC)), \neg ab_2, \neg ab_3\})$. Since l belongs to both E_1 and E_2, we conclude that the gun was loaded in the final state.

Observe that the scenario SC gives rise to two possible courses of events: either both occurrences of $load$ were performed normally and $shoot$ was performed abnormally (E_1) or the first occurrence of $load$ was performed abnormally and $shoot$ and the second occurrence of $load$ were performed normally (E_2).

EXAMPLE 12 (Stolen Car Scenario). This example was first described without solution by Kautz [Kautz, 1986] and then solved by Baker [Baker, 1991]. There is one action symbol, $leave$, for leaving a car overnight in a garage. The car is left for two successive nights and after the second night it turns out that the car is stolen. The scenario is given by

$$SC = \langle \neg s \rangle\ leave; leave\ \langle s \rangle.$$

No conclusion is intended with respect to if the car was stolen after the first night.

We first calculate $D_1(\mathcal{C}(SC))$, where $\mathcal{C}(SC) = [S^1(leave); S^2(leave)] (\neg s, T, s)$.

$$\begin{aligned}
D_1(\mathcal{C}(SC)) &= SP_1^{\mathcal{C}(SC)} \wedge \neg WLP_1^{\mathcal{C}(SC)} \\
&= sp(S^1(leave), \neg s) \wedge \neg wlp(S^2(leave), \neg s) \\
&= (\neg ab_1 \wedge \neg s \vee ab_1) \wedge \neg((\neg ab_2 \supset \neg s) \wedge (ab_2 \supset F)) \\
&\equiv (\neg ab_1 \wedge \neg s \vee ab_1) \wedge \neg((ab_2 \vee \neg s) \wedge (\neg ab_2)) \\
&\equiv (\neg ab_1 \wedge \neg s \vee ab_1) \wedge (\neg ab_2 \wedge s \vee ab_2) \\
&\equiv (\neg ab_1 \wedge \neg s \wedge ab_2) \vee (\neg ab_2 \wedge s \wedge ab_1) \vee (ab_1 \wedge ab_2).
\end{aligned}$$

The default theory $\mathcal{T}_1 = \langle \{D_1(\mathcal{C}(SC))\}, \{: \neg ab_1 / \neg ab_1,\ : \neg ab_2 / \neg ab_2\} \rangle$ has two extensions $E_1 = Th(\{D_1(\mathcal{C}(SC)), \neg ab_1\})$ and $E_2 = Th(\{D_1(\mathcal{C}(SC)), \neg ab_2\})$. Since $\neg s \in E_1$ and $s \in E_2$, no conclusion can be derived with respect to whether the car was stolen after the first night.

7 CONCLUSIONS

We have combined Dijkstra's semantics for programming languages with Reiter's default logic to formalization of reasoning about action and change, where actions with abnormal effects are permitted. The presented approach can be used to represent a broad class of action scenarios, including those where actions with random effects are allowed. In addition, both temporal prediction and postdiction tasks can be properly dealt with, without requiring initial or final states to be completely specified.

In [Łukaszewicz and Madalińska-Bugaj, 1995] we showed how the ramification problem can be solved using Dijkstra's formalism. There is no difficulty to deal with this problem in the presented framework, but the topic will be discussed in the future paper.

The task of implementation is another point of interest. For a given scenario SC, calculating $D_k(\mathcal{C}(SC))$ amounts to simple syntactic manipulations on formulae and can be performed very efficiently. The only computational problem is to compute extensions of a default theory $\mathcal{T}_k(SC)$. However, for simple action scenarios, as those considered in this paper, we deal with propositional default logic which is decidable.

ACKNOWLEDGEMENTS

This is an extended and revised version of the paper [Łukaszewicz and Madalińska-Bugaj, 1995a]. The authors were partially supported by the ESPRIT Basic Research Action No. 6156 - DRUMS II.

Witold Łukaszewicz and Ewa Madalińska-Bugaj
Institute of Informatics, Warsaw University, Poland.

BIBLIOGRAPHY

[Baker, 1991] A. B. Baker. Nonmonotonic Reasoning in the Framework of Situation Calculus. *Artificial Intelligence*, **49**, 5–23, 1991.

[Dijkstra, 1976] E. W. Dijkstra. *A Discipline of Programming*. Prentice Hall, 1976.

[Dijkstra and Scholten, 1990] E. W. Dijkstra and C. S. Scholten. *Predicate Calculus and Program Semantics*. Springer-Verlag, 1990.

[Hanks and McDermott, 1987] S. Hanks and D. McDermott. Nonmonotonic Logic and Temporal Projection. *Artificial Intelligence*, **33**, 379–412, 1987.

[Kautz, 1986] H. A. Kautz. The Logic of Persistence. In: *Proc. AAAI-86*, pp. 401–405, 1986.

[Łukaszewicz and Madalińska-Bugaj, 1994] W. Łukaszewicz and E. Madalińska-Bugaj. Program Verification Techniques as a Tool for Reasoning about Action and Change. In *KI-94: Advances in Artificial Intelligence, Proceedings of 18th German Conference on Artificial Intelligence*, pp. 226–236. LNAI 861, Springer-Verlag, 1994.

[Łukaszewicz and Madalińska-Bugaj, 1995] W. Łukaszewicz and E. Madalińska-Bugaj. Reasoning about Action and Change Using Dijkstra's Semantics for Programming Languages. In this volume.

[Łukaszewicz and Madalińska-Bugaj, 1995a] W. Łukaszewicz and E. Madalińska-Bugaj. Reasoning about Action and Change: Actions with Abnormal Effects. In: *KI-95: Advances in Artificial Intelligence, Proceedings of 19th German Conference on Artificial Intelligence*, pp. 209–220. LNAI 981, Springer-Verlag, 1995.

[Radzikowska, 1995] A. Radzikowska. Reasoning about Action with Typical and Atypical Effects. In *KI-95: Advances in Artificial Intelligence, Proceedings of 19th German Conference on Artificial Intelligence*, pp. 197–208. LNAI 981, Springer-Verlag, 1995.

[Reiter, 1980] R. Reiter. A Logic for Default Reasoning. *Artificial Intelligence Journal*, **13**, 81–132, 1980.

[Sandewall, 1994] E. Sandewall. *Features and Fluents: The Representation of Knowledge about Dynamical Systems*. Oxford Logic Guides, 30, Oxford Science Publications, 1994.

JOHN-JULES CH. MEYER AND PATRICK DOHERTY

PREFERENTIAL ACTION SEMANTICS

1 INTRODUCTION

Reasoning about action and change has long been of special interest to AI and issues of knowledge representation (see [Sandewall and Shoham, 1994]). In particular, the issue of representing changes caused by actions in an efficient and economic way without the burden of explicitly specifying what is *not* affected by the actions involved and is left unchanged has been a major issue in this area, since typically this specification is huge and in some cases *a priori* not completely known. In a similar vein, one would also like to avoid explicitly stating all qualifications to actions and all secondary effects of actions. Most of the proposed solutions impose a so-called *law of inertia* on changes caused by actions which states that properties in the world tend to remain the same when actions occur unless this is known to be otherwise. Formally, the inertia assumption in AI has been treated as some kind of default reasoning which in turn has triggered a host of theories about this specific application and defeasible and nonmonotonic theories in general.

The problem that tends to arise with many of the proposed solutions is that application of the inertia assumption is generally too global, or coarse, resulting in unwanted or unintended side effects. One would like to invoke a more local or fine-grained application of inertia to the scenarios at hand and recent proposals tend to support this claim. One explanation for this *coarseness* is that typically one represents an action theory as a set of axioms and then considers a subclass of the models, the preferred models, as the theories intended meaning. This means that the effects of actions are represented or obtained in a slightly roundabout way: the action theory contains axioms from which the behavior of the actions can be deduced using the preferred models of these axioms which somehow have to capture or respect the law of inertia concerning these actions. In simple situations, this approach works fine, but it is well known that in more complex situations finding the right kinds of preferences on one's models is not only very difficult, but even claimed not to be possible.

Our claim is that this is due to the fact that the instrument of considering preferred models of theories that describe complete action scenarios is too coarse because of the fact that these models employ preference relations that stem from 'global' and not action-specific frame assumptions. The specification of preferred outcomes of actions is a delicate matter depending on the actions (and the environment) at hand, and should be handled at the action semantics level rather than the global logical theory describing the whole system. So, what we will do in this paper is to put preferences at the place they should be put, viz. the semantics of

J.J.Ch. Meyer and J. Treur (eds.),
Handbook of Defeasible Reasoning and Uncertainty Management Systems, Vol. 7, 411–426.

actions. On this level we can more succinctly *fine-tune* these preferences incorporating the mode of inertia that is needed for a particular action given a particular context (environment). For each action occurring in a scenario one can thus state the way the variables are known/expected to be affected: are they distinctly 'set' by the action to certain values, are they expected to be not affected, or do we know nothing about this at all, so that anything could happen with them? From this information one can deduce both the possible and the expected behaviour of actions in a scenario, which can be reasoned about in an action logic like dynamic logic [Harel, 1984].[1]

We call this way of assigning meaning to actions *preferential action semantics*, which may be contrasted with traditional preferential semantics, which in contrast can be referred to as preferential *theory* (or *assertion*) semantics. Our claim is that preferential action semantics provides us with a flexible framework in which the subtleties of the (expected) behaviour of actions can be expressed and handled in a straightforward and adequate manner. In this paper we will support this claim with some interesting examples which require such subtlety in representation. Interestingly, but very naturally, this view will lead us very close to what is studied in the area of so-called concurrency semantics, i.e. that area of computer science where models of concurrent or parallel computations are investigated. We see for instance that in this framework proposals from the AI literature dealing with action and change which use constructs such as occlusion/release [Doherty, 1994; Sandewall, 1994; Kartha and Lifschitz, 1994] get a natural interpretation with respect to the aspect of concurrency.

Finally, in this introduction, we want to discuss the following possible objection to our approach of coping with the frame problem. One might think that our solution is not a solution to the frame problem at all, since the above might give the impression that one has to specify exactly what happens for each action. However, this is not exactly true. The only thing that has to be specified for each action is to which class the variables involved belong: definitely set, framed (i.e. expected to remain the same) or completely free. The semantics decides then the rest. In fact, this also holds for preferential assertion semantics, where variables must also be classified with respect to their "mode of affectedness". It is well-known by now, that this is really needed; one cannot expect to devise some kind of 'magical' preference relation to work in all cases without this kind of information about the variables involved. Hard things cannot be expected to be obtained for free! The only difference is that in preferential action semantics this needs to(or rather, put more positively, may) be done on the level of an individual action. Our point is that specifying these things at a global level might be too much to ask from a (global, assertion-based) preferential entailment relation, which is then supposed to supply the 'right' outcomes in complicated situations, in one blow, so to speak.

[1]To be fair, of course, it might be the case that this action-specific treatment can be encoded into one global preference relation in traditional preferential (theory) semantics, but this will inevitably lead to cumbersome and very intricate models.

2 PREFERENTIAL SEMANTICS OF ACTIONS

In this section, we define a very simple language of actions[2] with which we illustrate our ideas on preferential semantics of actions. Of course, for representing *real* systems this simple language should be extended, but the current simplification will give the general idea.

We start with the set $\mathcal{FVAR}$ of feature variables and $\mathcal{FVAL}$ of feature values. Elements of $\mathcal{FVAL}$ are typically denoted by the letter *d*, possibly marked or subscripted.[3] Next, we define a system state σ as a function of feature variables to features values: $\sigma : \mathcal{FVAR} \to \mathcal{FVAL}$. So, for $x \in \mathcal{FVAR}$, $\sigma(x)$ yields it value. The set of states is denoted by Σ. To denote changes of states we require the concept of a variant of a state. The state $\sigma\{d/x\}$ is defined as the state such that $\sigma\{d/x\}(x) = d$ and $\sigma\{d/x\}(y) = \sigma(y)$ for $y \neq x$.

Let a set $\mathcal{A}$ of atomic actions be fixed. An atomic action $a \in \mathcal{A}$ comes with a signature indicating what variables are *framed*, which of these may nevertheless vary (are *released* from inertia) and which are definitely *set*: $a = a(\text{set}_a, \text{frame}_a, \text{release}_a)$, where $\text{set}_a, \text{frame}_a, \text{release}_a \subseteq \mathcal{FVAR}$, such that $\text{release}_a \subseteq \text{frame}_a$ and $\text{set}_a \cap \text{frame}_a = \emptyset$. We also define $\text{inert}_a = \text{frame}_a \setminus \text{release}_a$ and $\text{var}_a = \mathcal{FVAR} \setminus (\text{set}_a \cup \text{frame}_a)$.[4] The inert variables are those subject to inertia, so that it is preferred that they retain the same value; the var variables are those not subjected to inertia and are really variables in the true sense of the word. The distinction between var and released variables is a subtle one: typically when describing an action scenario some of the framed variables (which are normally subject to inertia) are temporarily released, while some variables are considered truly variable over the whole scenario. Sandewall [1995] describes the three classes of frame-released, frame-unreleased (inert), and var variables as *occluded*, *remanent*, and *dependent*. Kartha and Lifschitz [1994] were probably the first to recognize this three-tiered distinction, while Sandewall [1991] was the first to use the frame/occluded distinction to deal properly with nondeterministic actions and actions with duration.

Given the set of atomic actions, complex actions can be formed as follows:

$$\alpha = a \mid \text{if } b \text{ then } \alpha_1 \text{ else } \alpha_2 \text{ fi} \mid \alpha_1 \oplus \alpha_2 \mid \alpha_1 + \alpha_2 \mid \alpha_1 \parallel \alpha_2 \mid \text{fail}.$$

Here, $a \in \mathcal{A}$; if b then α_1 else α_2 fi , where b is a boolean test on feature variables, represents a conditional action with the obvious meaning; $\alpha_1 \oplus \alpha_2$ stands for *restricted choice* between actions α_1 and α_2, where the release mechanism is applied to the actions α_1 and α_2 separately; $\alpha_1 + \alpha_2$ stands for an *open* or *liberal choice* between α_1 and α_2, where the release mechanism induced by the two actions α_1

[2] Actually, these are action expressions/descriptions rather than actions, but we will use the term rather loosely here.

[3] For convenience, we will assume that all feature variables range over the same set of feature values, mostly the booleans, but of course this restriction can be lifted.

[4] When it is convenient, we may also specify the inert and var variables in an action, such as e.g. $a = a(\text{set}_a, \text{inert}_a, \text{var}_a)$.

and α_2 is employed for α_1 and α_2 in a joint fashion (to be explained later on); $\alpha_1 \| \alpha_2$ stands for the *parallel (simultaneous) performance* of both α_1 and α_2; fail denotes the *failing* action, possessing no successor states. The class of all actions is denoted by $\mathcal{A}ct$. We now introduce the class of *preferred actions* (or rather the class of preferred behaviors of actions) denoted by $\mathcal{P}ref\mathcal{A}ct = \{\alpha_\sharp \mid \alpha \in \mathcal{A}ct\}$, where $\alpha_\sharp$ expresses the *preferred* behavior of α.[5]

The formal semantics of actions is given by functions which essentially describe the way actions change states. We define a semantical function $[\cdot] : \mathcal{A}ct \to \Sigma \to (2^\Sigma \times 2^\Sigma)$ for $\alpha \in \mathcal{A}ct, \sigma \in \Sigma$. $[\alpha](\sigma)$ denotes the set of states that computation of action α may result in, together with information about which of these states are preferred (or expected). So, $[\alpha](\sigma) = (S, S')$, where $S' \subseteq S \subseteq \Sigma$, and S' are the preferred (expected) outcome states of α. If $[\alpha](\sigma) = (S, S')$, we refer to S and S' by means of $([\alpha](\sigma))_\flat$ (or $[\alpha]_\flat(\sigma)$) and $([\alpha](\sigma))_\sharp$ (or $[\alpha]_\sharp(\sigma)$), respectively. If $S' = S$, this means that there is no preferred strict subset. In this case, we will just write $[\alpha](\sigma) = S$.

We allow placing constraints Φ on the set of states, so that effectively, the function $[\cdot]$ is constrained: $[\cdot] : \mathcal{A}ct \to \Sigma_\Phi \to (2^{\Sigma_\Phi} \times 2^{\Sigma_\Phi})$, where $\Sigma_\Phi = \{\sigma \in \Sigma \mid \sigma \models \Phi\}$.[6]

We are now ready to define the semantics for atomic and complex actions in terms of the functions described above.

Atomic Actions

For atomic action $a = a(\mathrm{set}_a, \mathrm{frame}_a, \mathrm{release}_a)$, we define its semantics as follows. First, we determine the effect of a on the variables in set_a. We assume that this is deterministic; let us denote the (unique) state yielded by this effect by σ_a. We may e.g. write $\mathrm{set}_a = \{+x, -y\}$ when we want to express that x is set to true and y is set to false. For instance, if σ is a state containing boolean information about the feature l ("the gun is loaded or not"), and a is the action $\mathsf{load}(\mathrm{set}_{\mathsf{load}} = \{+l\})$, then $\sigma_{\mathsf{load}} = \sigma\{T/l\}$, representing that the load action sets the variable l to true.

$$[a(\mathrm{set}_a, \mathrm{frame}_a, \mathrm{release}_a)](\sigma) = (S, S')$$

where (supposing $\mathrm{frame}_a = \{x_1, x_2, \ldots, x_m\}$, $\mathrm{release}_a = \{x_1, x_2, \ldots, x_n\} \subseteq \mathrm{frame}_a$, so $n \leq m$, and $\mathrm{var}_a = \{y_1, y_2, \ldots, y_k\}$):

$$\begin{aligned} S = \{&\sigma_a\{d_1/x_1, d_2/x_2, \ldots, d_m/x_m, d_1'/y_1, d_2'/y_2, \ldots, d_k'/y_k\} \in \\ &\Sigma_\Phi \mid d_1, d_2, \ldots, d_m, d_1', d_2', \ldots, d_k' \in \mathcal{FVAL}\} \\ (&= \{\sigma' \in \Sigma_\Phi \mid \sigma'(z) = \sigma_a(z) \text{ for all } z \in \mathrm{set}_a\}) \end{aligned}$$

[5]Note that it is senseless to talk about $(\alpha_\sharp)_\sharp$. This is not allowed by the syntax. We leave the question to future research whether nestings of preference regarding action behavior can be useful in some way.

[6]Constraints will be used to treat the ramification problem in a later section.

and

$$S' = \{\sigma_a\{d_1/x_1, d_2/x_2, \ldots, d_n/x_n, d'_1/y_1, d'_2/y_2, \ldots, d'_k/y_k\} \in \Sigma_\Phi \mid d_1, d_2, \ldots, d_n, d'_1, d'_2, \ldots, d'_k \in \mathcal{FVAL}\}$$
$$(= \{\sigma' \in \Sigma_\Phi \mid \sigma'(z) = \sigma_a(z) \text{ for all } z \in \text{set}_a \cup \text{inert}_a\}).$$

Note that indeed $S' \subseteq S\ (\subseteq \Sigma_\Phi)$.

Although the definition looks fairly complicated, it simply states formally that the usual semantics of an action $a(\text{set}_a, \text{frame}_a, \text{release}_a)$ consists of those states that apart from the definite effect of the action on the variables in set_a, both var and frame variables may be set to any possible value, whereas the preferred semantics (capturing inertia) keeps the inert variables the same, although the var and release variables are still allowed to vary.

Let's, by way of an example, consider the action load again, now also in a context where the variable a, denoting being alive, plays a role. (You see, we are heading towards the inevitable Yale Shooting.) Suppose that load $=$ load$(\text{set}_{\text{load}} = \{+l\}, \text{frame}_{\text{load}} = \{a\}, \text{release}_{\text{load}} = \emptyset)$. Let's consider a state σ in which a is true (I'm alive) and l is false (unloaded gun). Now the formal semantics of the load action in this state gives us: $[\![\text{load}]\!](\sigma) = (S, S')$ with $S = \{\sigma\{T/l, T/a\}, \sigma\{T/l, F/a\}\}$ and $S' = \{\sigma\{T/l\}\} = \{\sigma\{T/l, T/a\}\}$, which means that apart from setting l to true (the gun becomes loaded), it is possible that both one stays alive and one dies, but that the former is preferred (expected). If one now, for some reason, would release the variable a from the frame (assumption), the expectation that a remains true is dropped.

Complex Actions

In the sequel, it will sometimes be convenient to use the notation $\alpha(\text{set}_\alpha = X, \text{frame}_\alpha = Y, \text{release}_\alpha = Z)$, or simply $\alpha(\text{set} = X, \text{frame} = Y, \text{release} = Z)$, or even $\alpha(\text{set}\ X, \text{frame}\ Y, \text{release}\ Z)$, for the action $\alpha(\text{set}_\alpha, \text{frame}_\alpha, \text{release}_\alpha)$, with $\text{set}_\alpha = X$, $\text{frame}_\alpha = Y$, and $\text{release}_\alpha = Z$. In addition, the set-theoretical operators are, when needed, extended to pairs in the obvious way: $(S_1, S'_1) \bullet (S_2, S'_2) = (S_1 \bullet S_2, S'_1 \bullet S'_2)$.

The conditional and fail actions are given the following meanings:

$$[\![\text{if } b \text{ then } \alpha_1 \text{ else } \alpha_2 \text{ fi}]\!](\sigma) = [\![\alpha_1]\!](\sigma) \text{ if } b(\sigma) = T; \text{ and } [\![\alpha_2]\!](\sigma) \text{ otherwise.}$$

$$[\![\text{fail}]\!](\sigma) = (\emptyset, \emptyset).$$

Let's now consider the choice operators. The difference between restricted and liberal choice is illustrated by the following example. Suppose we have the constraint that shower on (o) is equivalent to either a hot shower (h) or a cold shower (c), i.e. $o \leftrightarrow h \vee c$. Let ho stand for the action of putting the hot shower on ($h := T$), and co for the action of putting the cold shower on ($c := T$). In the case where the restricted choice action ho $\oplus$ co is performed in a state where $\neg o$

$(= \neg h \wedge \neg c)$ holds, we either choose to do ho in this state resulting in a state where $h \wedge o \wedge \neg c$ holds (so inertia is applied to $\neg c$), or co is chosen resulting in a state where $c \wedge o \wedge \neg h$ holds (so inertia is applied to $\neg h$). In contrast, if the liberal choice action ho + co is performed in a state where $\neg o$, we just look at the possibilities of doing ho, co, and *possibly both*, resulting in one of the states $\{h \wedge o \wedge \neg c, \neg h \wedge o \wedge c, h \wedge o \wedge c\}$. So one may view this as if every atom o, h, or c is allowed to change value and is not subject to any inertia.

The semantics of the restricted choice operator can be stated as follows. Let the function Constrain$_\Phi$ be such that it removes all states that do not satisfy the constraints Φ: $\text{Constrain}_\Phi(S) = \{\sigma \in S \mid \sigma \models \Phi\}$. When no confusion arises, we may omit the subscript Φ.

$$
\begin{aligned}
&[\alpha(\text{set}_\alpha, \text{frame}_\alpha, \text{release}_\alpha) \oplus \beta(\text{set}_\beta, \text{frame}_\beta, \text{release}_\beta)](\sigma) = \\
&\quad \text{Constrain}_\Phi([\alpha(\text{set}_\alpha, \text{frame}_\alpha, \text{release}_\alpha)](\sigma) \cup \\
&\qquad [\beta(\text{set}_\beta, \text{frame}_\beta, \text{release}_\beta)](\sigma)).
\end{aligned}
$$

The definition states that the restricted choice between α and β regards the actions α and β more or less separately. In particular, the release mechanism works separately for both actions α and β.

The semantics of the liberal choice operator can be stated as follows.

$$
\begin{aligned}
&[\alpha(\text{set}_\alpha, \text{frame}_\alpha, \text{release}_\alpha) + \beta(\text{set}_\beta, \text{frame}_\beta, \text{release}_\beta)](\sigma) = \\
&\quad \text{Constrain}_\Phi([\alpha(\text{set}_\alpha, \text{frame} = (\text{frame}_\alpha \cup \text{frame}_\beta \cup \text{set}_\beta) \setminus \text{set}_\alpha, \\
&\qquad \text{release} = (\text{release}_\alpha \cup \text{release}_\beta \cup \text{set}_\beta) \setminus \text{set}_\alpha)](\sigma) \cup \\
&\qquad [\beta(\text{set}_\beta, \text{frame} = (\text{frame}_\alpha \cup \text{frame}_\beta \cup \text{set}_\alpha) \setminus \text{set}_\beta, \\
&\qquad \text{release} = (\text{release}_\alpha \cup \text{release}_\beta \cup \text{set}_\alpha) \setminus \text{set}_\beta)](\sigma)).
\end{aligned}
$$

In this case, the situation for the liberal choice operator is considered much more uniformly in the sense that not only the set of frame variables is taken together, but also the release mechanism works in a much more uniform manner. For both actions the sets of release and set variables is added, so that inertia is less potent and more possibility of variability (also with respect to preferred outcomes) is introduced by considering joint effects of the two actions α and β.

The semantics of the parallel operator can be stated as follows.

$$
\begin{aligned}
&[\alpha(\text{set}_\alpha, \text{frame}_\alpha, \text{release}_\alpha) \parallel \beta(\text{set}_\beta, \text{frame}_\beta, \text{release}_\beta)](\sigma) = \\
&\quad \text{Constrain}_\Phi([\alpha(\text{set}_\alpha, \text{frame} = (\text{frame}_\alpha \cup \text{frame}_\beta \cup \text{set}_\beta) \setminus \text{set}_\alpha, \\
&\qquad \text{release} = (\text{release}_\alpha \cup \text{release}_\beta \cup \text{set}_\beta) \setminus \text{set}_\alpha)](\sigma) \cap \\
&\qquad [\beta(\text{set}_\beta, \text{frame} = (\text{frame}_\alpha \cup \text{frame}_\beta \cup \text{set}_\alpha) \setminus \text{set}_\beta, \\
&\qquad \text{release} = (\text{release}_\alpha \cup \text{release}_\beta \cup \text{set}_\alpha) \setminus \text{set}_\beta)](\sigma)).
\end{aligned}
$$

Note the similarity with the liberal choice operator. In fact, the only thing that has changed with respect to the latter is that now *only the joint* effects of both actions are taken into consideration, where the release mechanism for both actions is again taken as liberal as possible allowing for as much interaction as possible.

Finally, we consider the preferred behavior operator $\sharp$:

$$[\![\alpha_\sharp]\!](\sigma) = ([\![\alpha]\!](\sigma))_\sharp.$$

Example

Let us consider the shower example again. The actions ho and co can be described more precisely as ho(set$\{+h\}$, frame$\{o, c\}$, release$\{o\}$) and co(set$\{+c\}$, frame$\{o, h\}$, release$\{o\}$). Recall that we have $o \leftrightarrow h \vee c$ as a domain constraint (Φ). Let σ be such that $\sigma = \{F/h, F/c, F/o\}$. Now, $[\![(\text{ho} \oplus \text{co})_\sharp]\!](\sigma)$ becomes

$$\begin{aligned}
&(\text{Constrain}_\Phi([\![\text{ho}(\text{set}\{+h\}, \text{frame}\{o,c\}, \text{release}\{o\})]\!](\sigma) \cup \\
&\qquad [\![\text{co}(\text{set}\{+c\}, \text{frame}\{o,h\}, \text{release}\{o\})]\!](\sigma)))_\sharp = \\
&\{\sigma\{T/h, F/c, T/o\}, \sigma\{F/h, T/c, T/o\}\}, \text{while } [\![(\text{ho} + \text{co})_\sharp]\!] = \\
&(\text{Constrain}_\Phi([\![\text{ho}(\text{set}\{+h\}, \text{frame} = \text{release} = \{o,c\})]\!](\sigma) \cup \\
&\qquad [\![\text{co}(\text{set}\{+c\}, \text{frame} = \text{release} = \{o,h\})]\!](\sigma)))_\sharp = \\
&\{\sigma\{T/h, F/c, T/o\}, \sigma\{F/h, T/c, T/o\}, \sigma\{T/h, T/c, T/o\}\},
\end{aligned}$$

as expected.

In addition, consider the action $h \parallel c$ in the same setting. Intuitively, one would expect that this action should have the effect of putting the shower on with both cold and hot water. $[\![(\text{ho} \parallel \text{co})_\sharp]\!] = (\text{Constrain}_\Phi([\![\text{ho}(\text{set}\{+h\}, \text{frame} = \text{release} = \{o,c\})]\!](\sigma) \cap [\![\text{co}(\text{set}\{+c\}, \text{frame} = \text{release} = \{o,h\})]\!](\sigma)))_\sharp$ which is equivalent to $\{\sigma\{T/h, T/c, T/o\}\}$, as desired.

Remark on Semantical Entities

The observing reader may have noticed that in the above definitions we have abused our language slightly by mixing syntax and semantics. This is due to the fact that, although the signature of an action consisting of a specification of the set, framed and released variables has a very syntactic ring to it, it nevertheless conveys semantical information. When one is more rigorous, one should consider semantical entities of the following type: sets of tuples of the form $(S, S', (set, frame, release))$, where the S and S' with $S' \subseteq S$ are sets of states (denoting the possible resulting states and the preferred subset of these, respectively), and set, $frame$ and $release$ are sets of variables expressing the status of the variables with respect to the sets S and S'. Of course, this information is implicit in the sets S and S', but for the sake of defining the interpretation of the operators it is very convenient to have this information explicitly available in the denotations of results. Now we may define our operators on these enhanced semantical elements: on tuples they read as follows:

$$(S_1, S_1', (set_1, frame_1, release_1)) \oplus (S_2, S_2', (set_2, frame_2, release_2)) = \\ \{(S_1, S_1', (set_1, frame_1, release_1)), (S_2, S_2', (set_2, frame_2, release_2))\}$$

$$(S_1, S_1', (set_1, frame_1, release_1)) + \\ (S_2, S_2', (set_2, frame_2, release_2)) = \\ \{(S_1, S_1', (set_1, (frame_1 \cup frame_2 \cup set_2) \setminus \\ set_1, release_1 \cup release_2 \cup set_2) \setminus set_1), \\ (S_2, S_2', (set_2, (frame_1 \cup frame_2 \cup set_1) \setminus \\ set_2, release_1 \cup release_2 \cup set_1) \setminus set_2)\}$$

$$(S_1, S_1', (set_1, frame_1, release_1)) \parallel (S_2, S_2', (set_2, frame_2, release_2)) = \\ \{(S_1 \cap S_2, S_1' \cap S_2', (set_1 \cup set_2, (frame_1 \cup frame_2) \setminus (set_1 \cup set_2), \\ (release_1 \cup release_2) \setminus (set_1 \cup set_2))\}$$

Finally, we extend the definition to sets of tuples T_1 and T_2 in the obvious way: $T_1 \Delta T_2 = \bigcup_{t_1 \in T_1, t_2 \in T_2} t_1 \Delta t_2$ for $\Delta = \oplus, +, \parallel$. This shows how one can do the previous definitions more formally. However, we have chosen not to do this in the remainder of the paper in order to keep things more intelligible, and to focus on the main ideas.

3 PREFERENTIAL ACTION DYNAMIC LOGIC (PADL)

In order to define a logic for reasoning about actions which includes their preferred interpretations, we simply take the (ordinary) dynamic logic formalism which is well known from the theory of imperative programming [Harel, 1984]. Formulas in the class $\mathcal{F}orm$ are of the form $[\alpha]\phi$, where $\alpha \in \mathcal{A}ct \cup \mathcal{P}ref\mathcal{A}ct$, $\phi \in \mathcal{F}orm$, closed under the usual classical connectives.

The semantics of formulas is given by the usual Kripke-style semantics. A Kripke model is a structure $M = (\Sigma, \{R_\alpha \mid \alpha \in \mathcal{A}ct \cup \mathcal{P}ref\mathcal{A}ct\})$, where the accessibility relations R_α are given by $R_\alpha(\sigma, \sigma') \Leftrightarrow_{\text{def}} \sigma' \in [\alpha]_\flat(\sigma)$.

Formulas of the form $[\alpha]\phi$ are now interpreted as usual: $M, \sigma \models [\alpha]\phi \Leftrightarrow$ for all $\sigma' : R_\alpha(\sigma, \sigma') \Rightarrow M, \sigma' \models \phi$. The other connectives are dealt with as usual. Note the special case involving formulas with preferred actions where $[\alpha_\sharp]\phi$ is interpreted as: $M, \sigma \models [\alpha_\sharp]\phi \Leftrightarrow$ (for all $\sigma' : R_{\alpha_\sharp}(\sigma, \sigma') \Rightarrow M, \sigma' \models \phi$) $\Leftrightarrow$ (for all $\sigma' : \sigma' \in [\alpha_\sharp](\sigma) \Rightarrow M, \sigma' \models \phi$) $\Leftrightarrow$ (for all $\sigma' : \sigma' \in ([\alpha](\sigma))_\sharp \Rightarrow M, \sigma' \models \phi$). Validity in a model, $M \models \phi$, is defined as $M, \sigma \models \phi$ for all σ. Validity of a formula, $\models \phi$, is defined as $M \models \phi$ for all models M.

Some useful validities (here we assume the set Φ of constraints to be finite and abuse our language slightly and let Φ stand for the conjunction of its elements as well):
$\models [\alpha](\phi \rightarrow \psi) \rightarrow ([\alpha]\phi \rightarrow [\alpha]\psi)$
$\models$ [if b then α_1 else α_2 fi]$\phi \leftrightarrow ((b \wedge [\alpha_1]\phi) \vee (\neg b \wedge [\alpha_2]\phi))$
$\models [\alpha]\phi \rightarrow [\alpha_\sharp]\phi$
$\models [\alpha]\phi \rightarrow [\alpha \parallel \beta]\phi$
$\models ([\alpha_\sharp]\Phi \wedge [\beta_\sharp]\Phi) \rightarrow ([(\alpha \oplus \beta)_\sharp]\phi \leftrightarrow [\alpha_\sharp]\phi \wedge [\beta_\sharp]\phi)$
$\models [(\alpha + \beta)_\sharp]\phi \rightarrow [(\alpha \oplus \beta)_\sharp]\phi$

Note, by the way, that regarding non-preferred behaviour we have that $\models [\alpha + \beta]\phi \leftrightarrow [\alpha \oplus \beta]\phi\ (\leftrightarrow [\alpha]\phi \wedge [\beta]\phi)$. Furthermore, as usual in dynamic logic we have that: $\models \phi \Rightarrow \models [\alpha]\phi$.

However, some notable non-validities are:
$\not\models [\alpha_\sharp]\phi \rightarrow [(\alpha \parallel \beta)_\sharp]\phi$
$\not\models [(\alpha \oplus \beta)_\sharp]\phi \rightarrow [(\alpha + \beta)_\sharp]\phi$

4 SKIP VS. WAIT: CONCURRENCY

Let us now briefly examine the difference between a wait action in the AI context and a skip action in imperative programming. A strong monotonic inertia assumption is implicitly built into the state transitions of imperative programming where the meaning of the skip action for example is just the identity function; $[\![\text{skip}]\!] = \lambda\sigma.\sigma$. For the wait action, it also holds that $[\![\text{wait}_\sharp]\!] = \lambda\sigma.\sigma$, but in this case, the inertia assumption is weaker in the sense that the action may itself show any behavior, due to additional effects in the environment. Our approach offers the possibility of specifying this weaker notion which will even work properly in the context of unspecified concurrent actions. For example, if wait $=$ wait(set $=$ frame $=$ release $= \emptyset$), load $=$ load(set$\{+l\}$), and we consider the action wait $\parallel$ load, we obtain $[\![\text{wait} \parallel \text{load}]\!](\sigma) = [\![\text{wait}(\text{set} = \text{frame} = \text{release} = \emptyset) \parallel \text{load}(\text{set}\{+l\})]\!](\sigma) = [\![\text{wait}(\text{frame}\{l\}, \text{release}\{l\})]\!](\sigma) \cap [\![\text{load}(\text{set}\{+l\})]\!](\sigma)$ $D\{\sigma\{T/l\}, \sigma\{F/l\}\} \cap \{\sigma\{T/l\}\} = \{\sigma\{T/l\}\} = [\![\text{load}]\!](\sigma)$.

More interestingly, if we also consider the propositional fluent a, we see how the release and the law of inertia work together. Suppose wait $=$ wait(frame$\{a, l\}$), load $=$ load(set$\{+l\}$). $[\![\text{wait} \parallel \text{load}]\!](\sigma) = [\![\text{wait}(\text{frame}\{a, l\}) \parallel \text{load}(\text{set}\{+l\})]\!]$ $(\sigma) = [\![\text{wait}(\text{frame}\{a, l\}, \text{release}\{l\})]\!](\sigma) \cap \text{load}(\text{set}\{+l\}\text{frame}\{a\})]\!](\sigma)$. It follows that $\models (\neg l \wedge a) \rightarrow [\text{wait} \parallel \text{load}]l$, while $\models (\neg l \wedge a) \rightarrow [(\text{wait} \parallel \text{load})_\sharp]l \wedge a$, as would be expected.

The upshot of all this is that although preferably the wait action has the same effect as the skip action, nevertheless due to the (non-specified) *concurrent* actions that are done in parallel with the wait, and of which we do not have any control, additional effects might occur.

5 OTHER EXAMPLES

We will start with a number of standard examples and move towards larger and more complex examples which combine the frame and ramification problems with concurrent actions.

Yale Shooting Scenario:

Initially Fred is alive, then the gun is loaded, we wait for a moment and then shoot. Of course (under reasonable conditions), it is expected that Fred is dead after shooting. In our approach, this example is represented as follows: we have the features loaded (l), alive (a), and the actions load $=$ load(set$\{+l\}$, frame$\{a\}$), wait $=$ wait(frame$\{a, l\}$), and shoot $=$ if l then kill(set$\{-l, -a\}$) else wait(frame$\{a, l\}$)fi. Now we have that $(\neg l \wedge a) \rightarrow [\mathsf{load}_{\sharp}](l \wedge a)$; $(l \wedge a) \rightarrow [\mathsf{wait}_{\sharp}](l \wedge a)$; and finally $(l \wedge a) \rightarrow [\mathsf{kill}_{\sharp}]\neg a$, and hence also $(l \wedge a) \rightarrow [\mathsf{shoot}_{\sharp}]\neg a$, so that $\models (\neg l \wedge a) \rightarrow [\mathsf{load}_{\sharp}][\mathsf{wait}_{\sharp}][\mathsf{shoot}_{\sharp}]\neg a$.

Russian Turkey Shoot:

The scenario is more or less as before, but now the wait action is replaced by a spin action: spin $=$ spin(frame$\{a\}$), leaving the variable l out of the frame, which may then vary arbitrarily. Clearly, $\not\models (\neg l \wedge a) \rightarrow [\mathsf{load}_{\sharp}][\mathsf{spin}_{\sharp}][\mathsf{shoot}_{\sharp}]\neg a$, since $\not\models (l \wedge a) \rightarrow [\mathsf{spin}_{\sharp}]l$, although it is the case that $\models (l \wedge a) \rightarrow [\mathsf{spin}_{\sharp}]a$,

The Walking Turkey Shoot (Ramification):

Similar to the Yale Shooting Scenario, but now we also consider the feature walking (w) and the constraint that walking implies alive: $\Phi = \{w \rightarrow a\}$. So now we consider the action
shoot $=$ if l then kill(set$\{-l, -a\}$, release$\{w\}$) else wait(frame$\{a, l\}$) fi, and obtain $\models (l \wedge a) \rightarrow [\mathsf{shoot}_{\sharp}](\neg a \wedge \neg w)$. In this case, inertia on w is not applied.

We now proceed to some more complicated scenarios.

Jumping into the Lake Example [Crawford, 1994; Giunchiglia and Lifschitz, 1995]:

Consider the situation in which one jumps into a lake, wearing a hat. Being in the lake (l) implies being wet (w). So we have as a constraint $\Phi = \{l \rightarrow w\}$. If one is initially not in the lake, not wet and wearing a hat, the preferred result using inertia would be that after jumping into the lake, one is in the lake and wet, but no conclusions concerning wearing a hat after the jump can be derived. We do not want to apply inertia to the feature of wearing a hat , since it is conceivable that while jumping, one could lose one's hat. So technically, this means that the feature variable hat-on (h) is left out of the frame. (Another way of representing this, which one might prefer and which will give the same result, is viewing the frame constant over the whole scenario, including h, and then releasing h in the present situation.)

If one is in the lake and wet, we would expect that after getting out of the lake, one is not in the lake, but *still wet* in the resulting state. So, inertia would be applied to the feature wet. Furthermore, we may assume that getting out of the lake is much less violent than jumping into it, so that we may also put h in the frame. Finally, if one is out of the lake and wet, then putting on a hat would typically result in a state where one has a hat on, while remaining out of the lake and wet.

Formally, we can treat this relatively complicated scenario by means of our semantics as follows. Consider the feature variables l (being in the lake), w (being wet), h (wearing a hat), and the constraint $\Phi = \{l \to w\}$. In addition, we would need three actions.

- jump-into-lake = jil(set$\{+l\}$, frame$\{w\}$, release$\{w\}$), where w must be released in view of the constraint $l \to w$.
- get-outof-lake = gol(set$\{-l\}$, frame$\{w, h\}$); although l is set, w is not released, since l is set to false and this does not enforce anything in view of the constraint $l \to w$.
- put-on-hat = poh(set$\{+h\}$, frame$\{l, w\}$,).

Now, applying the logic gives the desired results: $(\neg l \wedge \neg w \wedge h) \to [\text{jil}_\sharp](l \wedge w)$, and $(\neg l \wedge \neg w \wedge h) \to [\text{jil}](l \wedge w)$; $(l \wedge w) \to [\text{gol}_\sharp](\neg l \wedge w)$, (even $(l \wedge w \wedge h) \to [\text{gol}_\sharp](\neg l \wedge w \wedge h)$), and $(l \wedge w) \to [\text{gol}]\neg l$; $(\neg l \wedge w) \to [\text{poh}_\sharp](\neg l \wedge w \wedge h)$, and $(\neg l \wedge w) \to [\text{poh}]h$.

What this example shows is that one still has to choose the signature of actions: what is put in the frame and what is not. This is not done automatically by the framework. We claim this to be an advantage because it provides enormous flexibility in its use, while at the same time it calls for exactness, so that the specifying of agents forces one to specify per action how things should be handled. The law of inertia (applied on non-released frame variables) takes care of the rest, so to speak.

It is important to emphasize that some of the newer approaches for dealing with directed ramification which introduce explicit causal axioms [Lin, 1995; Thielscher, 1995] essentially encode the same types of behavior, but at the same time rule out similar flexibility in specification of actions. Thielscher [1995] for example, claims that the frame/released approaches are limited and provides the extended circuit example as a counterexample. One should rather view frame/released approaches as the result of a compilation process which compiles causal dependencies of one form or another [Gustafsson and Doherty, 1996]. The distinction to keep in mind is whether one's formalism is capable of specifying frame/released constraints differently from state to state. This deserves further analysis in the context of this approach.

Lifting a Bucket of Water

One can also use preferential action semantics in cases where one has certain default behavior of actions *on other grounds than the law of inertia.* Consider the lifting of a bucket filled with water with a left and right handle by means of a robot with two arms. Let lift-left (ll) be the action of the robot's lifting the left handle of the bucket with its left arm and lift-right (lr) be the analogous action of the robot's right arm. Obviously, when only one of the two actions are performed separately, water will be spilled. On the other hand, when the two actions are done concurrently, things go alright and no water is spilled. We place a constraint on the scenario that $\neg s \leftrightarrow (l \leftrightarrow r)$.

Now, we can say that normally when lift-right is performed, water gets spilled. However, in the extraordinary case when lift-right is performed in a context where (coincidentally) lift-left is also performed, water is not spilled. This example can be represented clearly and succinctly with our semantics. We assume that initially, in state σ, neither arm is lifted, and no water is spilled (yet), i.e. the variables l, r and s are all false. One can associate with lift-right the semantics:

$$[\![\mathsf{lr}(\text{set}\{r\}, \text{frame}\{l\})]\!](\sigma) = (\{\sigma\{T/r\}\{T/s\}, \sigma\{T/r\}\{F/s\}\}, \{\sigma\{T/r\}\{T/s\}\}),$$

expressing that performance of lift-right leads to a state where the right arm is raised (r) and either water gets spilled or not, but that the former is preferred (on other grounds than inertia: note that s is not framed). Analogously, we can define this for lift-left, where instead of the variable r, a variable l is set to indicate the left arm is raised. So, in our dynamic logic, the result is $\models [\mathsf{lr}]r$ and $\models [\mathsf{ll}]l$, but $\not\models [\mathsf{lr}]s$ and $\not\models [\mathsf{ll}]s$. On the other hand, we do have $\models [\mathsf{lr}_\sharp]s$ and $\models [\mathsf{ll}_\sharp]s$. Furthermore, since $[\![\mathsf{ll} \parallel \mathsf{lr}]\!](\sigma) =$

$$\begin{aligned} &[\![\mathsf{ll}(\text{set}\{+l\}, \text{frame} = \text{release} = \{r\})]\!](\sigma) \cap [\![\mathsf{lr}(\text{set}\{+r\}, \text{frame} = \text{release} = \\ &\qquad \{l\})]\!](\sigma) = \\ &\qquad \{\sigma\{T/l\}\{T/r\}\{F/s\}, \sigma\{T/l\}\{F/r\}\{T/s\}\} \cap \\ &\{\sigma\{T/r\}\{T/l\}\{F/s\}, \sigma\{T/r\}\{F/l\}\{T/s\}\} = \{\sigma\{T/r\}\{T/l\}\{F/s\}), \end{aligned}$$

we also obtain that $\models [\mathsf{ll} \parallel \mathsf{lr}](r \wedge l \wedge \neg s)$, as desired.

6 DIRECTIONS FOR FUTURE WORK

We would like to investigate the possibility of introducing sequences of actions by considering the class *ActSeq* given by $\beta = \alpha \mid \beta_1; \beta_2$. This would allow one to write down the outcome of a scenario such as the Yale Shooting problem as: $(\neg l \wedge a) \rightarrow [\mathsf{load}_\sharp; \mathsf{wait}_\sharp; \mathsf{shoot}_\sharp]\neg a$, instead of having to resort to the (equivalent) slightly roundabout representation $(\neg l \wedge a) \rightarrow [\mathsf{load}_\sharp][\mathsf{wait}_\sharp][\mathsf{shoot}_\sharp]\neg a$, as we did earlier. Note that by this way of defining action sequences, we (purposely) prohibit considering preferred sequences. Thus, something like $(\beta_1; \beta_2)_\sharp$ would now be ill-formed in our syntax, while $\alpha_{1\sharp}; \alpha_{2\sharp}$ is allowed. It remains subject to further research whether something like $(\beta_1; \beta_2)_\sharp$ could be given a clear-cut semantics and whether it would be a useful construct to have.

Surprises [Sandewall, 1991; Sandewall, 1994] can also be expressed in preferential action semantics. A surprise is some outcome of an action which was not expected, so formally we can express this as follows: ϕ is a surprise with respect to action α (denoted surprise(α, ϕ)) iff it holds that $[\alpha_\sharp]\neg\phi \wedge \langle\alpha\rangle\phi$. This states that although it is expected that $\neg\phi$ will hold after performing α, ϕ is nevertheless (an implausible but possible) outcome of α. For instance, in a state where Fred is alive (a), it would come as a surprise that after a wait action, he would be *not* alive: $a \rightarrow ([\mathsf{wait}(\mathsf{frame}\{a\})_\sharp]a \wedge \langle\mathsf{wait}(\mathsf{frame}\{a\})\rangle\neg a)$ is indeed true with respect to our semantics.

An interesting question, raised by one of the anonymous referees, is whether for some applications it would be useful or even required to extend the 'two-level' semantics (viz. possible and expected behaviour) into a more fine-grained one with multiple levels. We do not see the need for this at the moment. It might be possible that our approach is already sufficiently fine-grained due to the fact that we consider these two levels for any action in the scenario, which in total yields an enormous flexibility.

Other interesting issues to be studied are delayed effects of actions and prediction. It will be interesting to see whether modeling delay by using a wait action with a specific duration *in parallel* with other actions would give adequate results, while prediction seems to be very much related to considering expected results of (longer) chains of actions as compared to chains of preferred actions (as briefly indicated above). Perhaps a notion of *graded typicality* of behavior might be useful in this context. We surmise that by the very nature of the $[\alpha]$ modality (related to weakest preconditions) the framework so far seems to fit for prediction but is not very suitable for postdiction or explanation of scenarios [Sandewall, 1994]. Perhaps extending it with the notion of strongest postconditions [Dijkstra and Scholten, 1990; Łukaszewicz and Madalińska-Bugaj, 1995a; Łukaszewicz and Madalińska-Bugaj, 1995] would be helpful here.

Finally, although we made a plea for using preferential *action* semantics rather than preferential *assertion* semantics to describe action scenarios, it would, of course, be interesting to investigate the relation between the two, hopefully substantiating our claim that the former is more flexible or easier to use than the latter. We expect that systematic studies of relations between underlying (ontological and epistemological) assumptions of action/agent systems and (assertion) preferential models such as [Sandewall, 1994] will be useful guidelines in this investigation.

7 RELATED WORK

We were much inspired by work by [Łukaszewicz and Madalińska-Bugaj, 1995a; Łukaszewicz and Madalińska-Bugaj, 1995]. In this work the authors also attempted to employ proven verification and correctness methods and logics from imperative programming for reasoning about action and change in AI. In particular Dijkstra's wp-formalism is used. This formalism is based on the notion of weakest preconditions (and strongest postconditions) of actions and is in fact very close to the dynamic logic framework: formulas of the form $[\alpha]\phi$ are actually the same as the wlp (weakest liberal precondition) of action α with respect to postcondition ϕ. In [Łukaszewicz and Madalińska-Bugaj, 1995a; Łukaszewicz and Madalińska-Bugaj, 1995] a central role is played by the following theorem from Dijkstra and Scholten [Dijkstra and Scholten, 1990] which says that a state $\sigma \models \alpha \wedge \neg wlp(S, \neg\beta)$ iff there is a computation c under control of S starting in a state satisfying α and terminating in a state satisfying β such that σ is the initial state of c.

What all this amounts to is that when in [Łukaszewicz and Madalińska-Bugaj, 1995a], weakest (liberal) preconditions and the above theorem are used, something is stated of the form that after execution of an action α ϕ may possibly be true, which in dynamic logic is expressed as $\langle\alpha\rangle\phi(= \neg[\alpha]\neg\phi)$. Typically, this leads to too weak statements: one does not want to say that there is *some* execution of α that leads to ϕ, but that the set of *all expected* (but of course not *all*) output states satisfy some property. This is exactly what we intend to capture by means of our preferential action semantics. Another aspect that we disagree with, as the reader might suspect from the above, is that [Łukaszewicz and Madalińska-Bugaj, 1995a] uses the skip statement to express the wait action. In our view this is equating *a priori* the action of waiting with its preferred behavior (in view of the law of inertia).

Finally, we mention that the work reported in [Dunin-Keplicz and Radzikowska, 1995] is similar in spirit to ours. Here also, a distinction between typical (preferred) and possible behavior of actions is made within a dynamic logic setting. Our approach is more concrete in the sense that we directly incorporate aspects of inertia into the semantics, and, moreover, have an explicit preference operator (applied to actions) in the language. This implies that we can also speak about preferred versus possible behavior in the object language. On the other hand, we have not (yet) considered preferred paths of executions of actions as in [Dunin-Keplicz and Radzikowska, 1995]

ACKNOWLEDGEMENTS

John-Jules Meyer is partially supported by ESPRIT BRWG project No. 8319 (MODELAGE). This research was initiated during the author's leave to Linköping University (IDA), the hospitality of which is gratefully acknowledged. Moreover, this author wishes to dedicate this paper to the memory of his father B. John Meyer(1917-1996).

Patrick Doherty is supported by the Swedish Research Council for Engineering Sciences (TFR) and the Knut and Alice Wallenberg Foundation.

The authors are grateful for the very useful suggestions of the anonymous referees to improve the paper. Also the comments of the attendants of Modelage'97 on the presentation of this paper are greatly appreciated.

This paper originally appeared as "Preferential action semantics" in J.-J. Ch. Meyer and P.-Y. Schobbens, eds, *Formal Models of Agents*, LNAI 1760, pp. 187–201, Springer, Berlin, 1999, and is reproduced here with kind permission from Springer.

John-Jules Meyer
Dept. of Computer Science, Utrecht University, The Netherlands.

Patrick Doherty
Dept. of Computer and Information Science, University of Linköping, Sweden.

BIBLIOGRAPHY

[Crawford, 1994] J. Crawford. Three issues in action. Unpublished note for the 5th Int. Workshop on Nonmonotonic Reasoning, 1994.

[Dijkstra and Scholten, 1990] E. W. Dijkstra and C. S. Scholten. *Predicate Calculus and Program Semantics*. Springer-Verlag, 1990.

[Doherty, 1994] P. Doherty. Reasoning about action and change using occlusion. In *Proc. of the 11th European Conference on Artificial Intelligence, Amsterdam*, pages 401–405, 1994.

[Dunin-Keplicz and Radzikowska, 1995] B. Dunin-Keplicz and A. Radzikowska. Epistemic approach to actions with typical effects. In Chr. Froideveaux and J. Kohlas, editors, *Symbolic and Quantitative Approaches to Reasoning and Uncertainty, Proc. ECSQARU'95*, Lecture Notes in Artificial Intelligence, pages 180–188. Springer-Verlag, 1995.

[Giunchiglia and Lifschitz, 1995] E. Giunchiglia and V. Lifschitz. Dependent fluents. In *Proc. IJCAI-95, Montreal*, pages 1964–1969, 1995.

[Gustafsson and Doherty, 1996] J. Gustafsson and P. Doherty. Embracing occlusion in specifying the indirect effects of actions. In *Proc. of the 5th Int'l Conf. on Principles of Knowledge Representation and Reasoning, (KR-96)*, 1996.

[Harel, 1984] D. Harel. Dynamic logic. In D. M. Gabbay and F. Guenthner, editors, *Handbook of Philosophical Logic*, volume 2, pages 496–604. Reidel, Dordrecht, 1984.

[Kartha and Lifschitz, 1994] G. N. Kartha and V. Lifschitz. Actions with indirect effects (preliminary report). In *Proc. of the 4th Int'l Conf. on Principles of Knowledge Representation and Reasoning, (KR-94)*, pages 341–350, 1994.

[Lin, 1995] F. Lin. Embracing causality in specifying the indirect effects of actions. In *Proc. IJCAI-95, Montreal*, 1995.

[Łukaszewicz and Madalińska-Bugaj, 1995] W. Łukaszewicz and E. Madalińska-Bugaj. Reasoning about action and change : Actions with abnormal effects. In I. Wachsmuth, C.-R Rollinger, and W. Brauer, editors, *Proc. KI-95: Advances in Artificial Intelligence*, volume 981 of *Lecture Notes in Artificial Intelligence*, pages 209–220. Springer-Verlag, Berlin, 1995.

[Łukaszewicz and Madalińska-Bugaj, 1995a] W. Łukaszewicz and E. Madalińska-Bugaj. Reasoning about action and change using Dijkstra's semantics for programming languages: Preliminary report. In *Proc. IJCAI-95, Montreal*, pages 1950–1955, 1995.

[Sandewall, 1991] E. Sandewall. Features and fluents. Technical Report LITH-IDA-R-91-29, Department of Computer and Information Science, Linköping University, 1991.

[Sandewall, 1994] E. Sandewall. *Features and Fluents: A Systematic Approach to the Representation of Knowledge about Dynamical Systems*. Oxford University Press, 1994.

[Sandewall, 1995] E. Sandewall. Systematic comparison of approaches to ramification using restricted minimization of change. Technical Report LiTH-IDA-R-95-15, Dept. of Computer and Information Science, Linköping University, May 1995.

[Sandewall and Shoham, 1994] E. Sandewall and Y. Shoham. Nonmonotonic temporal reasoning. In D. M. Gabbay, C. J. Hogger, and J. A. Robinson, editors, *Epistemic and Temporal Reasoning*, volume 4 of *Handbook of Artificial Intelligence and Logic Programming*. Oxford University Press, 1994.

[Thielscher, 1995] M. Thielscher. Computing ramifications by postprocessing. In *Proc. IJCAI-95, Montreal*, pages 1994–2000, 1995.

CATHOLIJN M. JONKER, JAN TREUR, WIEKE DE VRIES

REUSE AND ABSTRACTION IN VERIFICATION: AGENTS ACTING IN DYNAMIC ENVIRONMENTS

1 INTRODUCTION

Verification of agent systems is generally not an easy task. As agents may operate in a world that is constantly changing, and agent systems can consist of a number of interacting but independent agents, expressing behavioural requirements may lead to complex formulae. Nevertheless, verification is important, because it is the only way to guarantee that demands made on aspects of the system behaviour are satisfied. The high degree of complexity of agent system behaviour is as much the reason as the problem here: by simply checking the code of the agent system or by testing, proper behaviour can never be sufficiently established. Proper functioning is often crucial, because agent systems are increasingly employed in circumstances where mistakes have important consequences, for example in electronic commerce. But in practice, verification of agent systems is hardly ever done, because it is intricate.

So, means are needed to make verification of agent systems manageable. Developers of agent systems should be enabled to verify the system they are building, assisted by tools, even if they are not specialists in formal theory. Properties and proofs have to be intuitively clear to the verifier and even, at least to some degree, to the stakeholder(s) of the system, as verification results are part of the design rationale of the system. Also, time complexity of the verification process has to be controlled. This chapter discusses some principles that contribute to the support of verification of agent systems. These principles can be used for all agent systems, but here, they are applied in the context of a single agent that performs actions in a dynamic environment.

In [Jonker and Treur, 1998] a compositional verification method was introduced; see also [Engelfriet *et al.*, 1999] for its relation to temporal multi-epistemic logic. This method is extended in this chapter, and is briefly described here. Desired system behaviour is formalised in system requirements. In compositional verification, proving that a detailed design of an agent system properly respects the behavioural requirements (i.e., required properties of the system's behaviour) demands that these properties are refined across different process abstraction levels. For each of the agents as well as for the environment of the system, a specific set of the refined properties is imposed to ensure that the combined system satisfies the overall requirements. Also further refinement of the properties imposed on an agent leads to the identification of required properties of components within the agent.

The use of compositionality in verification has the following advantages:

J.J.Ch. Meyer and J. Treur (eds.),
Handbook of Defeasible Reasoning and Uncertainty Management Systems, Vol. 7, 427–454.

- reuse of verification results is supported (refining an existing verified compositional model by further decomposition, leads to verification of the refined system in which the verification proof of the original system can be reused).

- process hiding limits the complexity of the verification per process abstraction level.

In [Brazier *et al.*, 1998] it was shown how this method can be applied to prove properties of a system of negotiating agents.

However, this does not solve all problems. Even if compositionality is exploited, for nontrivial examples verification is a tedious process. Also, properties and proofs can still be very complex to read and to explain. To manage the complexity of the proofs, and to make their structure more transparent, additional structuring means and reuse facilities are necessary. This chapter contributes two manners to support proof structuring and reuse.

On the one hand a notion of abstraction is introduced that provides means to describe the global structure of proofs and properties in a formal manner. To this end, to the language to describe properties of systems, agents and agent components a formalism for abstraction is added. Parts of formulas can be given an intuitively enlightening name. This leads to a more informal look and feel for properties and proofs, without losing any formal rigour. The abstracted notions form a higher-level language to describe system behaviour. The terminology of this language abstracts away from details of the system design, and is closer to the way human verifiers conceptualise system behaviour. There are a number of benefits:

- Properties and proofs are more readable and easier to understand.

- Coming up with properties and proofs becomes easier, as the words chosen for the abstracted formulas guide and focus the cognitive verification process of the verification engineer, providing clean-cut central concepts.

- Verification becomes explainable, as part of the design rationale documentation of a system.

On the other hand, support of reuse requires that a library of predefined templates of properties and proofs is available for often occurring application characteristics and classes. By identifying generic elements in the structure of proofs and properties, reusable systems of properties and proofs can be constructed. To illustrate this, this chapter proposes a system of co-ordination property templates for applications of agents acting in dynamic environments. The properties and proofs of this system are an example of the contents of the verification library. The reusable templates for this application exploit the shorthanding mechanism. Some advantages of reuse are:

- Verification becomes faster. Often, the verification engineer only has to look up suitable properties and proofs from the verification library and customise these by instantiation.

- Verification becomes easier. The contents of the library are usually phrased using abstraction, so properties and proofs are more intuitively clear, making them more easy to use.

In the following section, the generic system consisting of an agent acting in a dynamic environment is sketched. For this application, a system of co-ordination properties is given in Section 6. This section also presents some auxiliary properties. But first, in Section 3 temporal models of agent systems descriptions are presented, as well some principles necessary to conduct verification when time is dense. This section also presents the two languages to describe system behaviour, the detailed language and the abstract language, and the connection between them. Section 4 sketches the requirements proved in the system of co-ordination properties, and the approach taken in proving these requirements. In Section 5, the abstraction mechanism is applied; abstract predicates are introduced for parts of properties, yielding an abstract language. In Section 7, proofs are briefly discussed. Finally, Section 8 contains some conclusions, comparisons with other work and directions for future research.

2 THE DOMAIN OF AGENTS ACTING IN A DYNAMIC ENVIRONMENT

In this section the characteristics of the application class of an agent in interaction with a dynamic environment are briefly discussed. A reusable system of properties for this class will be presented later on, describing correct co-ordination of the agent with its environment.

Agents that can perceive a dynamic environment and act in this environment are quite common. Realistic characteristics of such domains are:

- perceptions take time

- the generation of actions takes time

- execution of actions in the world takes time

- unexpected events can occur in the environment

To be more specific, often a proof is required that actions executed in the system are successful, that is, yield all their effects. This can be quite intricate because an action can only succeed when its execution is not disturbed too much. Examples of disturbances are:

- In the world a situation arises that is observed by the agent, and based on this observation, the agent decides to do an action A. It is very well possible that at the time of execution the world has changed in such a way during this period of observing, deliberation and starting to execute, that the action can no longer be successfully executed.

- During an action execution in the world, events happen that interfere with the action, causing the action to fail.

- Action executions overlap: a next action is being executed while at the same time, the previous action is still being executed, and as a result the expected consequences are not reached: failure.

In the literature, varying attitudes towards these disturbances can be found. In one part of the literature (e.g., standard situation calculus, as described in [McCarthy and Hayes, 1969; Reiter, 1991]), these disturbances are excluded in a global manner, e.g., action generation and execution have no duration at all and no events occur at all. The problem with these global assumptions is that they violate the characteristics of most of the application domains. Other literature attaches default effects to actions, thus incorporating non-monotonic logic in action execution (e.g., [Łukasiewicz and Madalinska-Bugaj, 2000]). Some literature takes into account duration of action execution (e.g., [Sandewall, 1994]). Literature that also takes into account the reasoning and decision processes in action generation is very rare. Another lack in the literature is that most authors don't try to verify implemented systems; they only state theories regarding actions, without relating them to practical system design or software specification.

Preliminary work on the analysis of this domain was reported in [Jonker *et al.*, 1998]. In continuation of this work we were led to the claims that:

- for realistic applications properties are required that do not (completely) exclude one of the characteristics above, but still impose enough structure on the process to prove that the agent is effective in its behaviour.

- these properties are explicitly related to practical system designs in a formal manner

- such properties exist: a system of co-ordination properties is introduced that fulfils the above requirements.

These claims will be supported by the next sections.

The specification of a generic architecture for a single agent in a dynamic world depicted in Figure 1 consists of two components in interaction: an agent (Ag) and the external world (EW). This specification is very generic; many concrete single agent systems adhere to it, while differing in other respects. Only a few aspects of the functioning of the system are specified by the architecture. The agent generates actions that are transferred from its output interface to the external world, and the external world generates observation results that are transferred to the input interface of the agent. Based on the observation results the agent is to

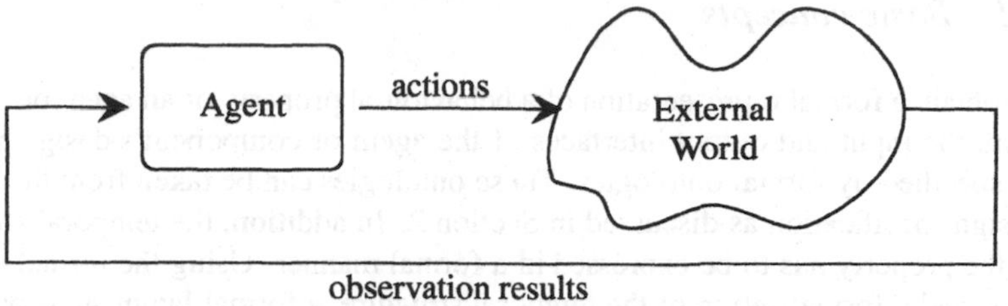

Figure 1. Agent and external world in interaction

decide which actions are to be performed. The agent only receives information on observation results; it has no direct access to the world state.

Formal ontologies are necessary to represent observations and decisions to do actions. The *input interface* of the agent is defined by the formal ontology observation results based on (in order-sorted predicate logic) the two sorts WORLD_INFO_ELEMENT and SIGN, and a binary relation observation_result on these sorts. Formulae that can be expressed using the information type observation results are, for example, observation_result(at_position(self, p0), pos), or observation_result(at_position(self, p1), neg). The content of the sort WORLD_INFO_ELEMENT is defined by terms denoting the ground atoms expressed using the application-specific ontology world info.

The *output interface* of the agent is defined by the formal ontology actions to be performed based on (in order-sorted predicate logic) the sort ACTION and the unary relation to_be_performed. For example the statement to_be_performed(goto(p)) can be expressed in this ontology. For the external world the input and output interfaces are the opposite of the agent's interfaces.

A number of applications of this architecture have been made using the compositional development method for multi-agent systems DESIRE; see [Brazier *et al.*, 2000] for the underlying principles.

3 TEMPORAL MODELS AND TEMPORAL LANGUAGES

For phrasing properties, a language is needed. Behaviour is described by properties of the execution traces of the system. In this section, the language used for this is introduced. Also, this section describes two principles which are very useful in proving, the pinpoint principle and the induction principle. These principles are applicable if all system traces obey the finite variability assumption. Finally, this section introduces the language abstraction formalism.

3.1 Basic concepts

To obtain a formal representation of a behavioural property of an agent or component, the input and output interfaces of the agent or component's design have to be specified by formal ontologies. These ontologies can be taken from the formal design specification as discussed in Section 2. In addition, the temporal structure of the property has to be expressed in a formal manner. Using the formal ontologies, and a formalisation of the temporal structure, a formal language is obtained to formulate behavioural properties. The semantics are based on compositional information states which evolve over time.

The state language $\mathsf{SL}(\mathsf{D})$ of a system component D is the (order-sorted) predicate logic language based on the interface ontologies of D. The formulae of this language are called *state formulae.* An *information state* M of a component D is an assignment of truth values $\{\mathsf{true}, \mathsf{false}, \mathsf{unknown}\}$ to the set of ground atoms in $\mathsf{SL}(\mathsf{D})$. The compositional structure of D is reflected in the structure of the information state. The set of all possible information states of D is denoted by $\mathsf{IS}(\mathsf{D})$.

The time frames are assumed *linear with initial time point 0.* A time frame may be discrete or dense. The approach introduced here works for dense linear time models as well as for discrete linear time models. A *trace* $\mathcal{M}$ of a component D over a time frame T is a sequence of information states $(\mathsf{M}^{\mathsf{t}})_{\mathsf{t}\in\mathsf{T}}$ in $\mathsf{IS}(\mathsf{D})$. Given a trace $\mathcal{M}$ of component D, the information state of the input interface of component C at time point t is denoted by $\mathsf{state}(\mathcal{M}, \mathsf{t}, \mathsf{input}(\mathsf{C}))$ where C is either D or a component with D. Analogously, $\mathsf{state}(\mathcal{M}, \mathsf{t}, \mathsf{output}(\mathsf{C}))$ denotes the information state of the output interface of component C at time point t.

These formalised information states can be related to formulae via the formally defined satisfaction relation $\models$. If φ is a state formula expressed in the input ontology for component C, then

$$\mathsf{state}(\mathcal{M}, \mathsf{t}, \mathsf{input}(\mathsf{C})) \models \varphi$$

denotes that φ is true in this state at time point $\mathsf{t} \in \mathsf{T}$.

These statements can be compared to *holds*-statements in situation calculus, described in [McCarthy and Hayes, 1969]. (A difference, however, apart from notational differences, is that we refer to a trace and time point, and that we explicitly focus on part of the system.) Based on these statements, which only use predicate symbols $\models$ and $<$ and $=$ (to compare moments in time), behavioural properties can be formulated in a formal manner in a sorted predicate logic with sorts T for time points, $\mathsf{Traces}(\mathsf{C})$ for traces of component C and F for state formulae. The usual logical connectives such as $\neg, \wedge, \Rightarrow, \forall, \exists$ are employed to construct formulae. Note that arbitrary combinations of quantification over time are allowed in this language. Additional sorts to define terms within the sort F are allowed. The language defined in this manner is denoted by $\mathsf{TL}(\mathsf{D})$ (Temporal Language of D).

An example of a formula of $\mathsf{TL(S)}$, where S refers to the whole system, is:

$$\begin{array}{l} \forall \mathcal{M} \in \mathsf{Traces(S)} : \\ \quad \forall \mathsf{t1} : \mathsf{state}(\mathcal{M}, \mathsf{t1}, \mathsf{output(Ag)}) \models \mathsf{to\_be\_performed(A)} \quad \Rightarrow \\ \quad \exists \mathsf{t2} > \mathsf{t1} : \mathsf{state}(\mathcal{M}, \mathsf{t2}, \mathsf{output(Ag)}) \models \mathsf{to\_be\_performed(B)} \end{array}$$

This expresses that for each trace of the system S if for some point in time the agent has generated the information $\mathsf{to\_be\_performed(A)}$ at its output interface, i.e., it has decided to do action A, there will be a later point in time that it generated $\mathsf{to\_be\_performed(B)}$ at its output interface, i.e., it has decided to do action B.

The languages $\mathsf{TL(D)}$ are built around constructs that enable the verifier to express properties in a detailed manner, staying in direct relation to the semantics of the design specification of the system. For example, the state formulae are directly related to information states of system components. But the detailed nature of the language also has disadvantages; properties tend to get long and complex. The formalism of abstraction, described in Section 3.3, alleviates this considerably.

3.2 Change pinpoint principle and induction principle

Within the logic defined in Section 3.1, some additional principles are needed in order to successfully prove properties. These principles restrict the class of temporal models. The proof methods used require the following principles:

Change pinpoint principle

This principle expresses that any change of a property of the state can be related to a unique time point: for each change of a statement that occurs in the trace it should be possible to point at a time point *where this change has just occurred.* In other words, for each state in the trace occurring at some point in time t3, such that a state property φ is true (resp. false or unknown), a first moment $\mathsf{t2} \leq \mathsf{t3}$ exists such that in the states for all time points t with $\mathsf{t2} \leq \mathsf{t} \leq \mathsf{t3}$ the statement φ is true (resp. false or unknown), and if t2 is not the initial time point, there exists a time point $\mathsf{t1} < \mathsf{t2}$ such that for all $\mathsf{t'}$ with $\mathsf{t1} \leq \mathsf{t'} < \mathsf{t2}$ the statement φ is not true (resp. not false or not unknown) at $\mathsf{t'}$:

$$\begin{array}{l} \forall \varphi\, \forall \mathsf{t3} : \mathcal{M}, \mathsf{t3} \models \varphi \quad \Rightarrow \\ \exists \mathsf{t2} \leq \mathsf{t3}\, \forall \mathsf{t} \in [\mathsf{t2}, \mathsf{t3}] : \mathcal{M}, \mathsf{t} \models \varphi \wedge \\ (\mathsf{t2} \neq 0 \Rightarrow \exists \mathsf{t1} < \mathsf{t2}\, \forall \mathsf{t'} \in [\mathsf{t1}, \mathsf{t2}\rangle : \mathcal{M}, \mathsf{t'} \not\models \varphi) \end{array}$$

For the cases that φ is false or unknown at t3, similar formulations can be given. This principle allows introducing the following abbreviations:

$$\begin{array}{l} \oplus\mathsf{state}(\mathcal{M}, \mathsf{t1}, \mathit{interface}) \models \varphi \;\equiv\; \mathsf{state}(\mathcal{M}, \mathsf{t1}, \mathit{interface}) \models \varphi \quad \wedge \\ \quad \exists \mathsf{t2} < \mathsf{t1}\, \forall \mathsf{t} : (\mathsf{t2} \leq \mathsf{t} < \mathsf{t1} \Rightarrow \mathsf{state}(\mathcal{M}, \mathsf{t}, \mathit{interface}) \not\models \varphi) \end{array}$$

$$\begin{array}{l} \oplus\mathsf{state}(\mathcal{M}, \mathsf{t1}, \mathit{interface}) \not\models \varphi \;\equiv\; \mathsf{state}(\mathcal{M}, \mathsf{t1}, \mathit{interface}) \not\models \varphi \quad \wedge \\ \quad \exists \mathsf{t2} < \mathsf{t1}\, \forall \mathsf{t} : (\mathsf{t2} \leq \mathsf{t} < \mathsf{t1} \Rightarrow \mathsf{state}(\mathcal{M}, \mathsf{t}, \mathit{interface}) \models \varphi) \end{array}$$

The ⊕-notation, pronounced as *just*, is used to denote a change to a certain information state. Closely related is the ⊗t1, t2⊕–notation, defined as follows:

$$\otimes \mathrm{t1}, \mathrm{t2} \oplus \mathrm{state}(\mathcal{M}, \mathrm{t2}, \mathit{interface}) \models \varphi \equiv \oplus \mathrm{state}(\mathcal{M}, \mathrm{t2}, \mathit{interface}) \models \varphi \wedge \forall \mathrm{t} : (\mathrm{t1} < \mathrm{t} < \mathrm{t2} \Rightarrow \neg \oplus \mathrm{state}(\mathcal{M}, \mathrm{t}, \mathit{interface}) \models \varphi)$$

$$\otimes \mathrm{t1}, \mathrm{t2} \oplus \mathrm{state}(\mathcal{M}, \mathrm{t2}, \mathit{interface}) \not\models \varphi \equiv \oplus \mathrm{state}(\mathcal{M}, \mathrm{t2}, \mathit{interface}) \not\models \varphi \wedge \forall \mathrm{t} : (\mathrm{t1} < \mathrm{t} < \mathrm{t2} \Rightarrow \neg \oplus \mathrm{state}(\mathcal{M}, \mathrm{t}, \mathit{interface}) \not\models \varphi)$$

This notation can be used to say that the information state has just changed in some way at t2, for the first time since t1. Sometimes it can be useful to also have the dual operator; this states that the information state has just changed at t2 and that this is the most recent change like that prior to t1. For this, we use the notation −⊗t1, t2⊕, defined as:

$$-\otimes \mathrm{t1}, \mathrm{t2} \oplus \mathrm{state}(\mathcal{M}, \mathrm{t2}, \mathit{interface}) \models \varphi \equiv \oplus \mathrm{state}(\mathcal{M}, \mathrm{t2}, \mathit{interface}) \models \varphi \wedge \forall \mathrm{t} : (\mathrm{t2} < \mathrm{t} < \mathrm{t1} \Rightarrow \neg \oplus \mathrm{state}(\mathcal{M}, \mathrm{t}, \mathit{interface}) \models \varphi)$$

$$-\otimes \mathrm{t1}, \mathrm{t2} \oplus \mathrm{state}(\mathcal{M}, \mathrm{t2}, \mathit{interface}) \not\models \varphi \equiv \oplus \mathrm{state}(\mathcal{M}, \mathrm{t2}, \mathit{interface}) \not\models \varphi \wedge \forall \mathrm{t} : (\mathrm{t2} < \mathrm{t} < \mathrm{t1} \Rightarrow \neg \oplus \mathrm{state}(\mathcal{M}, \mathrm{t}, \mathit{interface}) \not\models \varphi)$$

Induction principle

In proofs it turned out effective to use an induction principle of the following form.

If

1. a property P is true for all t within a certain initial interval from 0 to t0,

and

2. the truth of this property P for all t up to a time point t1 implies that there exists a time point t2 > t1 such that the property holds for all t′ up to t2,

then

3. the property P holds for all time points.

The formalisation of this principle for the logic defined in Section 3.1 is as follows:

$$\begin{aligned}
&\forall P(\mathcal{M}, \mathrm{t}) : \\
&\quad [\exists \mathrm{t0}\, \forall \mathrm{t} \leq \mathrm{t0} : P(\mathcal{M}, \mathrm{t}) \quad \wedge \\
&\quad \forall \mathrm{t1} : [\forall \mathrm{t} \leq \mathrm{t1} : P(\mathcal{M}, \mathrm{t}) \Rightarrow \exists \mathrm{t2} : [\mathrm{t1} < \mathrm{t2} \wedge \forall \mathrm{t} \leq \mathrm{t2} : P(\mathcal{M}, \mathrm{t})]] \Rightarrow \\
&\quad \forall \mathrm{t} : P(\mathcal{M}, \mathrm{t})
\end{aligned}$$

Induction principles for discrete linear time temporal logics are well-known (e.g., [Benthem, 1983]). An induction principle for situation calculus and its use is described in [Reiter, 1993]. The two principles are nontrivial constraints on the traces: linear time temporal models may exist for which this induction principle is not valid. For example, the intervals over which states are constant can become smaller and smaller, so that a certain time point will never be reached during the induction. For traces like this, the induction principle doesn't hold.

In order to guarantee the two principles expressed above, adherence to the following *finite variability assumption* is enough (for a slightly different variant see [Barringer *et al.*, 1986]):

Finite variability assumption

The finite variability assumption states that between any two time points, for each statement only a finite number of changes can occur:

$$\begin{aligned}&\forall s,t\,\forall\varphi : s < t \;\Rightarrow\; \exists n\in\mathbb{N}\,\exists t_1,\ldots t_n :\\ &\quad s = t_1 < \ldots < t_n = t \;\;\wedge\\ &\quad \forall i\in\{1,2,\ldots,n-1\}\,\forall t' : [t_i \leq t' < t_{i+1} \Rightarrow \;\; [\mathcal{M},t_i \models \varphi \Leftrightarrow \mathcal{M},t' \models \varphi] \wedge\\ &\qquad\qquad\qquad\qquad\qquad\qquad\qquad\qquad [\mathcal{M},t_i \models \neg\varphi \Leftrightarrow \mathcal{M},t' \models \neg\varphi]]\end{aligned}$$

The finite variability assumption expressed above defines a specific subset of all linear time traces. All traces in this subset satisfy both the change pinpoint principle and the induction principle.

3.3 *The language abstraction formalism*

Experience in nontrivial verification examples has taught us that the temporal expressions needed in proofs can become quite complex and unreadable. It turned out to be very useful to add new language elements as abbreviations of complex temporal formulae. These new language elements are defined within a language AL(D) (meaning Abstract Language of component D) with generic sorts T for time points, Traces(D) for traces and F for state formulae; additional sorts are allowed. As a simple example, for the property that there is an action execution starting in the world at t a new predicate `ActionExStarts` can be introduced. Then the property can be expressed in the abstracted language:

$$\texttt{ActionExStarts}(A, t, EW, \mathcal{M})$$

which is interpreted as:

$$\oplus\text{state}(\mathcal{M}, t, \text{input}(EW)) \models \text{to\_be\_performed}(A)$$

Semantics of these new language elements is defined as the semantics of the detailed formulae they abstract from. In logic the notion of *interpretation mapping* has been introduced to describe the interpretation of one logical language in another logical language, for example geometry in algebra (cf. [Hodges, 1993,

Ch. 5]). The languages AL(D) and TL(D) can be related to each other by a fixed interpretation mapping from the formulae in AL(D) onto formulae in TL(D).

The language AL(D) abstracts from details of the system design and enables the verifier to concentrate on higher level concepts. At the same time, each abstract formula has the same semantics as the related detailed formula, such that the relation to design specification details isn't lost. Proofs can be expressed either at the detailed or at the abstract level, and the results can be translated to the other level. Because formulae in the abstract level logic can be kept much simpler than the detail level logic, the proof relations expressed on that level are much more transparent.

Both languages AL(D) and TL(D) can be taken together in the language TL*(D). Within this language the new formulae of AL(D) are definable in terms of formulae of TL(D) [Chang and Keisler, 1973]. A definition of formula F in AL(D) then is of the form

$$\mathsf{F} \equiv D(\mathsf{F})$$

where D(F) is a fixed indicated formula of TL(D) (the definition of F).

4 APPROACHING THE PROBLEM OF CO-ORDINATION OF ACTIONS

For the application class described in Section 2, the aim is to prove that under the specified assumptions all actions executed in the agent system are successful, that is, yield all of their effects. To arrive at a reusable and intuitively pleasing proof, it was necessary and illuminating to separate the different aspects into a number of properties. These will constitute the *system of co-ordination properties*. In this section, some important aspects are described informally.

An action succeeds when the appropriate effects are obtained before or at the end of the execution of that action. To be able to talk of the end of an action execution, the ended-atoms are introduced. Without these atoms, the only way to detect the end of an execution is by looking at the (visible) effects of the action. But when effects are the only clue, it is impossible to say:

> "When an execution is not disturbed from its beginning to its end, then the appropriate effect will happen during this execution."

Without ended-atoms, indicating when an execution is over, this property is circular, because effects are the only indicators that an action has ended.

But just having ended-atoms is not enough. The idea is to identify an action execution of A by detecting a to_be_performed(A)-atom (henceforth sometimes abbreviated as tbp(A)-atom) at the beginning and an ended(A)-atom at the end. But this is impossible when action executions are allowed to overlap in an arbitrary way. Every action can be executed several times, and when these executions overlap it becomes impossible to identify pairs of tbp(A)- and ended(A)-atoms that belong to each other.

One of the properties in the system of co-ordination properties solves this problem. In section 6, these properties are formally defined. Here, an informal exposition suffices. One of the conditions for action success identified in the system of co-ordination properties, is non-overlapping of action executions. Property COORD0 states this. The property can be illustrated by this picture:

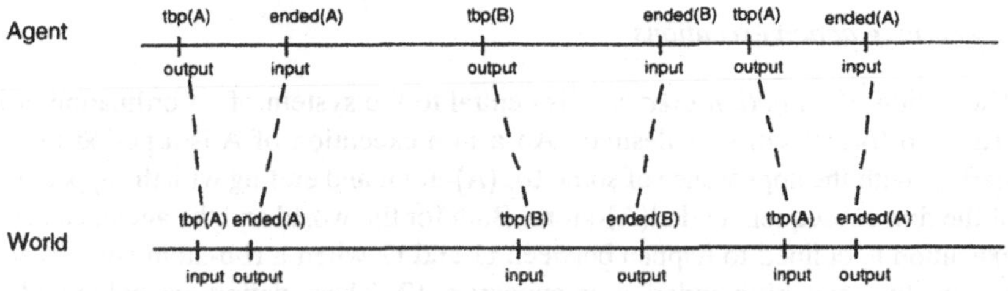

Figure 2.

When tbp-atoms and ended-atoms alternate in the depicted way, it is very easy to form pairs of tbp's and ended's that demarcate an action instance execution: you simply take the first ended(A) following the tbp(A). In the figure, there are action executions in the world as well as in the agent; these terms will be used frequently. COORD0 says that when one action is executing in the world, it is always the only one. It is also forbidden that two action executions entirely coincide. And because effects of an action are defined as expected outcomes that happen during execution of the action, no action execution can fail because of effects of other actions.

But action executions can also fail due to events. Events (in its vaguest meaning) are things the agent does not expect. Of course, the first association with the word "events" is things changing in the external world because of the dynamics of the world itself, independent of an action. But also changes due to failed actions are defined to be events. So, all changes happening in the world that the agent doesn't cause by successfully doing an action, are events.

A third reason for action failure is that the world can change prior to an action execution starting in the world. Between the moment the world situation arises that gives the agent reason to decide to do an action and the start of the execution of this action, events or other action effects could occur, disrupting the applicability of the action.

All aspects that influence the successfulness of actions will be formalised in the system of co-ordination properties.

5 ABSTRACT FORMULATIONS

In this section, a number of predicates of the abstract language are defined. The abstract language enables the verifier to express temporal properties of system behaviour using a vocabulary of clean-cut concepts. To be able to distinguish elements of the abstract language, a `different font` is used to denote them.

Concerning action executions

The notion of an *action execution* is central to the system of co-ordination properties, so formalisation is desired. An action execution of A is a period in time starting with the appearance of some tbp(A)-atom and ending with the appearance of the first subsequent ended(A)-atom. Both for the world and the agent, an action execution is defined to happen between t1 and t2 when a tbp-atom appears at t1 and the first matching ended-atom appears at t2. These definitions only yield the right intuitions when property COORD0 holds, as explained in the previous section. If not, then it is not reasonable to take on the first matching ended-atom as belonging to the tbp-atom, as it could be the end of an earlier executed instance of the same action.

Action executions are defined for the world as well as for the agent. Also, notions for executions starting or ending are defined. First, new predicates are introduced and explained in informal terms. Next, formal interpretations in terms of the detailed language are given.

Let A ∈ ACTION and t1, t2 > t1 be moments in time. Then, the abstract formula

ActionEx(A, t1, t2, EW, $\mathcal{M}$) denotes that there is an execution of A in the world starting at t1 and ending at t2, and

ActionEx(A, t1, t2, Ag, $\mathcal{M}$) denotes that there is an execution of A in the agent starting at t1 and ending at t2.

Interpretation in terms of the detailed language:

ActionEx(A, t1, t2, EW, $\mathcal{M}$) ≡ ⊕state($\mathcal{M}$, t1, input(EW)) ⊨ to_be_performed(A) ∧ ⊗t1,t2⊕state($\mathcal{M}$, t2, output(EW)) ⊨ ended(A)

ActionEx(A, t1, t2, Ag, $\mathcal{M}$) ≡ ⊕state($\mathcal{M}$, t1, output(Ag)) ⊨ to_be_performed(A) ∧ ⊗t1,t2⊕state($\mathcal{M}$, t2, input(Ag)) ⊨ ended(A)

Let A ∈ ACTION and t1 be a moment in time. Then, the abstract formula

ActionExStarts(A, t1, EW, $\mathcal{M}$)	denotes that there is an execution of A in the world starting at t1, and
ActionExStarts(A, t1, Ag, $\mathcal{M}$)	denotes that an execution of A in the agent starts at t1.

Interpretation in terms of the detailed language:

ActionExStarts(A, t1, EW, $\mathcal{M}$) $\equiv$ $\oplus$state($\mathcal{M}$, t1, input(EW)) $\models$ to_be_performed(A)

ActionExStarts(A, t1, Ag, $\mathcal{M}$) $\equiv$ $\oplus$state($\mathcal{M}$, t1, output(Ag)) $\models$ to_be_performed(A)

Let A $\in$ ACTION and t2 be a moment in time. Then, the abstract formula

ActionExEnds(A, t2, EW, $\mathcal{M}$)	denotes that an execution of A in the world ends at t2, and
ActionExEnds(A, t2, Ag, $\mathcal{M}$)	denotes that an execution of A in the agent ends at t2.

Interpretation in terms of the detailed language:

ActionExEnds(A, t2, EW, $\mathcal{M}$) $\equiv$ $\oplus$state($\mathcal{M}$, t2, output(EW)) $\models$ ended(A)

ActionExEnds(A, t2, Ag, $\mathcal{M}$) $\equiv$ $\oplus$state($\mathcal{M}$, t2, input(Ag)) $\models$ ended(A)

Concerning applicability

Actions can only be successfully executed in certain world states. There must be nothing obstructing the execution of the action. For each action A, the existence of a formula appl(A) is assumed, describing exactly the world situations in which the action can be fruitfully executed. It is not excluded that the effects of the action are already present in these world situations. Now, applicability can be defined straightforwardly:

Let A $\in$ ACTION and t1 be a moment in time. Then, the abstract formula

Appl(A, t1, $\mathcal{M}$) denotes that action A is applicable in the world at t1.

Interpretation in terms of the detailed language:

Appl(A, t1, $\mathcal{M}$) $\equiv$ state($\mathcal{M}$, t1, output(EW)) $\models$ appl(A)

Concerning situation specific effects

When an action is executed in a certain world state, there (possibly) will be some effects. By an effect is meant a single literal that becomes true as a result of an action. An action can have multiple effects, each happening at some moment during the execution of the action. The exact nature of the effects depends on the world situation; in this way, the effects are situation specific. A world situation is formalised as a partial assignment of definite truth values to atoms of world info. Each world situation can be refined to a number of complete assignments. A world situation can be represented as a conjunction of literals of world info. A world situation W′ is a refinement of a world situation W, written as $W \leq W'$, if their conjunction representations obey: $W' \Rightarrow W$.

The effects of actions are domain- and system dependent notions. For each system S, a relation SitSpecEffects is defined to represent the situation specific effects of actions. Given three arguments (an action A, a world situation W and Eff, a groundliteral of type world info), SitSpecEffects(A, W, Eff) means that Eff is an effect of the action A in the world situation W. This system-dependent relation must always satisfy the following demand (expressed in the abstract language):

> For some A ∈ ACTION, groundliteral Eff of type world info and some world situations W and W′:
> $W \leq W' \wedge \text{SitSpecEffect}(A, W, Eff) \Rightarrow \text{SitSpecEffect}(A, W', Eff)$.

This property is monotonicity of effects with respect to situations: when more facts are available concerning the state of the world, the set of predicted effects will be larger or the same.

Concerning expected effects

The predicate SitSpecEffect described above is static; it doesn't have a trace ($\mathcal{M}$) or moment in time as an argument. Using SitSpecEffect, a notion is defined for effects that are expected as a consequence of performing an action in the world at a certain moment in time. When an execution of an action A starts at t1 in the world, the effects expected depend on the factual world situation at t1. So, the following definition takes into account the output information state of EW, constrained to the factual information type world info.

Let A ∈ ACTION, l ∈ groundliterals(world info) and t be a moment in time. Then, the abstract formula

ExpEffect(l, A, t1, EW, $\mathcal{M}$) denotes that l is expected to happen as a result of executing A in the world at t1.

This formula is defined within the abstract language by:

$$\text{ExpEffect}(l, A, t1, EW, \mathcal{M}) \equiv \text{SitSpecEffect}(A, \text{state}(\mathcal{M}, t1, \text{output}(EW))|_{\text{world info}}, l)$$

Concerning effects of actions and events

A literal is defined to be an *effect of an action* when the literal is an expected outcome that becomes true during execution of the action. Note that this doesn't mean that the literal becomes true as a result of the action, though this will be usually the case. But when during an action execution an event happens, which causes changes that are also expected effects of the action being executed, these changes will be seen as effects of the action. This choice is made because there is no means by which an external observer can distinguish changes caused by actions from changes caused by events. A literal is defined to be an *effect of an event* when it is not an effect of any action.

Let A ∈ ACTION, I ∈ groundliterals(world info) and t be a moment in time. Then, the abstract formula

`ActionEff`(A, I, t, EW, $\mathcal{M}$) denotes that at t, I becomes true as a result of executing A.

Interpretation in terms of the detailed language:

`ActionEff`(A, I, t, EW, $\mathcal{M}$) ≡ ∃t1 < t ∃t2 ≥ t:
⊕state($\mathcal{M}$, t, output(EW)) ⊨ I ∧
`ActionEx`(A, t1, t2, EW, $\mathcal{M}$) ∧
`ExpEffect`(I, A, t1, EW, $\mathcal{M}$)

Let I ∈ groundliterals(world info) and t be a moment in time. Then, the abstract formula

`EventEff`(I, t, EW, $\mathcal{M}$) denotes that at t, I becomes true as a result of some event.

Interpretation:

`EventEff`(I, t, EW, $\mathcal{M}$) ≡ ⊕state($\mathcal{M}$, t, output(EW)) ⊨ I ∧
¬∃A ∈ ACTION:
`ActionEff`(A, I, t, EW, $\mathcal{M}$)

The next abstract formula is used to state that during an interval in time there are no effects of events.

Let *int* be an interval in time. Then, the abstract formula

`NoEventsDuring`(*int*, EW, $\mathcal{M}$) denotes that there are no events taking place in the world during *int*.

Definition within the abstract language:

`NoEventsDuring`(*int*, EW, $\mathcal{M}$) ≡ ∀I ∈ groundliterals(world info)
∀t ∈ *int* :
¬`EventEff`(I, t, EW, $\mathcal{M}$)

Concerning successful actions

The following formula of the abstract language states that an execution of A is successful, meaning that all expected effects are achieved during the execution:

Let A ∈ ACTION and t1, t2 > t1 be moments in time. Then, the abstract formula

`ExpEffectsHappen`(A, t1, t2, EW, $\mathcal{M}$) denotes that all expected effects of doing A between t1 and t2 are achieved.

Definition within the abstract language:

$\texttt{ExpEffectsHappen}(\mathsf{A}, \mathtt{t1}, \mathtt{t2}, \mathsf{EW}, \mathcal{M}) \equiv \texttt{ActionEx}(\mathsf{A}, \mathtt{t1}, \mathtt{t2}, \mathsf{EW}, \mathcal{M}) \wedge$
$\forall \mathsf{I} \in \text{world info}$
$\exists \mathtt{t3} \in \langle \mathtt{t1}, \mathtt{t2}]:$
$\texttt{ExpEffect}(\mathsf{I}, \mathsf{A}, \mathtt{t1}, \mathsf{EW}, \mathcal{M}) \Rightarrow$
$\texttt{ActionEff}(\mathsf{A}, \mathsf{I}, \mathtt{t3}, \mathsf{EW}, \mathcal{M})$

Concerning overlapping executions

The first two abstract formulae state that a certain execution, in the world or the agent respectively, doesn't overlap with any other execution.

Let A ∈ ACTION and t1, t2 > t1 be moments in time. Then, the abstract formula
$\texttt{NoOverlapDuringExecuting}(\mathsf{A}, \mathtt{t1}, \mathtt{t2}, \mathsf{EW}, \mathcal{M})$
denotes that in the world, between the beginning of an execution of A at t1 and the end of an execution of A at t2, no other executions begin or end.

Definition within the abstract language:
$\texttt{NoOverlapDuringExecuting}(\mathsf{A}, \mathtt{t1}, \mathtt{t2}, \mathsf{EW}, \mathcal{M}) \equiv$
$\texttt{ActionExStarts}(\mathsf{A}, \mathtt{t1}, \mathsf{EW}, \mathcal{M}) \wedge$
$\texttt{ActionExEnds}(\mathsf{A}, \mathtt{t2}, \mathsf{EW}, \mathcal{M}) \Rightarrow$
$(\forall \mathsf{B} \neq \mathsf{A} \in \mathsf{ACTION}\ \forall \mathtt{t} \in [\mathtt{t1}, \mathtt{t2}]:$
$\neg\, \texttt{ActionExStarts}(\mathsf{B}, \mathtt{t}, \mathsf{EW}, \mathcal{M}) \wedge$
$\neg\, \texttt{ActionExEnds}(\mathsf{B}, \mathtt{t}, \mathsf{EW}, \mathcal{M})) \wedge$
$(\forall \mathtt{t} \in \langle \mathtt{t1}, \mathtt{t2}] : \neg \texttt{ActionExStarts}(\mathsf{A}, \mathtt{t}, \mathsf{EW}, \mathcal{M})) \wedge$
$(\forall \mathtt{t} \in [\mathtt{t1}, \mathtt{t2}\rangle : \neg \texttt{ActionExEnds}(\mathsf{A}, \mathtt{t}, \mathsf{EW}, \mathcal{M}))$

Let A ∈ ACTION and t1, t2 > t1 be moments in time. Then, the abstract formula
$\texttt{NoOverlapDuringExecuting}(\mathsf{A}, \mathtt{t1}, \mathtt{t2}, \mathsf{Ag}, \mathcal{M})$
denotes that in the agent, between the decision to begin an execution of A at t1 and the end of an execution of A at t2, no other executions begin or end.

Definition within the abstract language:
$\texttt{NoOverlapDuringExecuting}(\mathsf{A}, \mathtt{t1}, \mathtt{t2}, \mathsf{Ag}, \mathcal{M}) \equiv$
$\texttt{ActionExStarts}(\mathsf{A}, \mathtt{t1}, \mathsf{Ag}, \mathcal{M}) \wedge$
$\texttt{ActionExEnds}(\mathsf{A}, \mathtt{t2}, \mathsf{Ag}, \mathcal{M}) \Rightarrow$
$(\forall \mathsf{B} \neq \mathsf{A} \in \mathsf{ACTION}\ \forall \mathtt{t} \in [\mathtt{t1}, \mathtt{t2}]:$
$\neg\, \texttt{ActionExStarts}(\mathsf{B}, \mathtt{t}, \mathsf{Ag}, \mathcal{M}) \wedge$
$\neg\, \texttt{ActionExEnds}(\mathsf{B}, \mathtt{t}, \mathsf{Ag}, \mathcal{M})) \wedge$
$(\forall \mathtt{t} \in \langle \mathtt{t1}, \mathtt{t2}] : \neg \texttt{ActionExStarts}(\mathsf{A}, \mathtt{t}, \mathsf{Ag}, \mathcal{M})) \wedge$
$(\forall \mathtt{t} \in [\mathtt{t1}, \mathtt{t2}\rangle : \neg \texttt{ActionExEnds}(\mathsf{A}, \mathtt{t}, \mathsf{Ag}, \mathcal{M}))$

Next, four abstract formulae are introduced that apply to a number of executions, namely the executions during a certain interval in time. There are two world notions and two agent notions. The abbreviations state that all executions started during the interval will also end, without being overlapped by other executions and that all executions ended during the interval have started during the interval, without being overlapped.

Let *int* be an interval in time. Then, the abstract formula
ExecutionsStartedEndWithoutOverlapping(EW, *int*, $\mathcal{M}$)
denotes that all actions executions started in the world during *int* end in the interval, and there is no overlapping of executions during this interval.

Definition within the abstract language:
ExecutionsStartedEndWithoutOverlapping(EW, *int*, $\mathcal{M}$) ≡
∀A ∈ ACTION ∀t1 ∈ *int* :
ActionExStarts(A, t1, EW, $\mathcal{M}$) ⇒
∃t2 > t1 : t2 ∈ *int* ∧
ActionExEnds(A, t2, EW, $\mathcal{M}$) ∧
NoOverlapDuringExecuting(A, t1, t2, EW, $\mathcal{M}$)

Let *int* be an interval in time. Then, the abstract formula
ExecutionsEndedStartWithoutOverlapping(EW, *int*, $\mathcal{M}$)
denotes that all actions executions ended in the world during *int* have started in the interval, and there is no overlapping of executions during this interval.

Definition within the abstract language:
ExecutionsEndedStartWithoutOverlapping(EW, *int*, M) ≡
∀A ∈ ACTION ∀t2 ∈ *int* :
ActionExEnds(A, t2, EW, $\mathcal{M}$) ⇒
∃t1 < t2 : t1 ∈ *int* ∧
ActionExStarts(A, t1, EW, $\mathcal{M}$) ∧
NoOverlapDuringExecuting(A, t1, t2, EW, $\mathcal{M}$)

Let *int* be an interval in time. Then, the abstract formula
ExecutionsStartedEndWithoutOverlapping(Ag, *int*, $\mathcal{M}$)
denotes that all actions executions started by the agent during *int* end in the interval, and there is no overlapping of executions during this interval.

Definition within the abstract language:
ExecutionsStartedEndWithoutOverlapping(Ag, *int*, $\mathcal{M}$) ≡
∀A ∈ ACTION ∀t1 ∈ *int*:
ActionExStarts(A, t1, Ag, $\mathcal{M}$) ⇒
∃t2 > t1 : t2 ∈ *int* ∧
ActionExEnds(A, t2, Ag, $\mathcal{M}$) ∧
NoOverlapDuringExecuting(A, t1, t2, Ag, $\mathcal{M}$)

Let *int* be an interval in time. Then, the abstract formula

ExecutionsEndedStartWithoutOverlapping(Ag, *int*, $\mathcal{M}$)
denotes that all actions executions ended in the agent during *int* have started in the interval, and there is no overlapping of executions during this interval.

Definition within the abstract language:

$$\begin{aligned}&\texttt{ExecutionsEndedStartWithoutOverlapping}(\text{Ag}, \textit{int}, \mathcal{M}) \equiv\\ &\quad \forall \text{A} \in \text{ACTION}\ \forall \text{t2} \in \textit{int}:\\ &\quad \texttt{ActionExEnds}(\text{A}, \text{t2}, \text{Ag}, \mathcal{M}) \Rightarrow\\ &\quad \exists \text{t1} < \text{t2} : \text{t1} \in \textit{int} \wedge\\ &\quad \texttt{ActionExStarts}(\text{A}, \text{t1}, \text{Ag}, \mathcal{M}) \wedge\\ &\quad \texttt{NoOverlapDuringExecuting}(\text{A}, \text{t1}, \text{t2}, \text{Ag}, \mathcal{M})\end{aligned}$$

When, during some interval, every execution started during the interval ends without being overlapped and every execution ended has started without being overlapped, it is guaranteed that all executions in the interval don't overlap in any way. So:

Let *int* be an interval in time. Then, the abstract formula
NoOverlappingInWorld(*int*, $\mathcal{M}$)
denotes that there is no overlapping of action executions in the world during *int*.

Definition within the abstract language:

$$\begin{aligned}&\texttt{NoOverlappingInWorld}(\textit{int}, \mathcal{M}) \equiv\\ &\quad \texttt{ExecutionsStartedEndWithoutOverlapping}(\text{EW}, \textit{int}, \mathcal{M}) \wedge\\ &\quad \texttt{ExecutionsEndedStartWithoutOverlapping}(\text{EW}, \textit{int}, \mathcal{M})\end{aligned}$$

Let *int* be an interval in time. Then, the abstract formula
NoOverlappingInAgent(*int*, $\mathcal{M}$)
denotes that there is no overlapping of action executions in the agent during *int*.

Definition within the abstract language:

$$\begin{aligned}&\texttt{NoOverlappingInAgent}(\textit{int}, \mathcal{M}) \equiv\\ &\quad \texttt{ExecutionsStartedEndWithoutOverlapping}(\text{Ag}, \textit{int}, \mathcal{M}) \wedge\\ &\quad \texttt{ExecutionsEndedStartWithoutOverlapping}(\text{Ag}, \textit{int}, \mathcal{M})\end{aligned}$$

The following two abstract formulae are instantiations of the previous two formulae. By $[0, \rightarrow\rangle$, the whole time line is denoted.

The abstract formula
NoOverlappingInWorld($\mathcal{M}$) denotes that there is no overlapping of action executions in the world.

Definition within the abstract language:

$$\texttt{NoOverlappingInWorld}(\mathcal{M}) \equiv \texttt{NoOverlappingInWorld}([0, \rightarrow\rangle, \mathcal{M})$$

The abstract formula

`NoOverlappingInAgent(`$\mathcal{M}$`)` denotes that there is no overlapping of action executions in the agent.

Definition within the abstract language:

$$\texttt{NoOverlappingInAgent}(\mathcal{M}) \equiv \texttt{NoOverlappingInAgent}([0,\rightarrow\rangle,\mathcal{M})$$

6 PROPERTIES

In Section 6.1, some basic properties are introduced, that are used in proving the system of co-ordination properties. The system of co-ordination properties is described in Section 6.2. And in Section 6.3, some additional properties that are necessary to do the proof by induction of COORD0 (which is a property from the system) are sketched.

6.1 Basic properties

A number of properties are needed which describe the basic behaviour of the agent and the world and the information transferral between them. In this section, some of these properties are introduced, while others are just sketched. Properties like these describe proper functioning of small parts of the system. In the proofs made for this chapter, they formed the premises of the proofs. The properties can often be proved directly from the semantics of the design specification. Four classes of properties exist.

Properties concerning starting and ending actions

There are two basic demands made on the world regarding the action executions starting and ending. These are that every start of an execution of action A has to be followed by an end of such an execution, and that every end of an execution of A may only appear when there has been an execution of A starting earlier.

$$\begin{array}{l}\text{END1}:\\ \forall \mathcal{M} \in \text{Traces(EW)}\ \forall \text{A} \in \text{ACTION}\ \forall \texttt{t1}:\\ \qquad \texttt{ActionExStarts}(\text{A}, \texttt{t1}, \text{EW}, \mathcal{M}) \quad \Rightarrow\\ \quad \exists \texttt{t2} > \texttt{t1} : \texttt{ActionExEnds}(\text{A}, \texttt{t2}, \text{EW}, \mathcal{M})\end{array}$$

$$\begin{array}{l}\text{END2}:\\ \forall \mathcal{M} \in \text{Traces(EW)}\ \forall \text{A} \in \text{ACTION}\ \forall \texttt{t2}:\\ \qquad \texttt{ActionExEnds}(\text{A}, \texttt{t2}, \text{EW}, \mathcal{M}) \quad \Rightarrow\\ \quad \exists \texttt{t1} < \texttt{t2} : \texttt{ActionExStarts}(\text{A}, \texttt{t1}, \text{EW}, \mathcal{M})\end{array}$$

Note that these properties don't state that for every start of an action there is a matching end. For example, these properties allow ten consecutive start of executions of A, followed by a single end.

Rationality properties

When the agent in the system satisfies the rationality properties presented in this section, this means that it makes the right decisions to perform actions. More specifically, there are three properties. Their formal definitions are left out of this chapter.

RAT2 states that the agent will always decide to do an action when it observes an occasion for the action and it is not still busy with another action execution. So, the agent doesn't simply decide to do an action whenever it observes an occasion for it; only when earlier actions have been fully executed, a new action can be started. This property is needed to prove COORD0, which states that action executions don't overlap. This is not the only way COORD0 can be proved; when the agent has a different acting-policy, different properties need to be phrased. For example, in a system with a purely reactive agent, that decides to do an action whenever it sees an occasion for it, COORD0 can also be proved when more strict demands are made on the dynamics of the world.

RAT3 states that when an agent decides to do an action it must have observed an occasion. This excludes behaviour in which the agent randomly decides to do things, which would certainly invalidate COORD0.

RAT1 says that occasionality implies applicability. This last property is necessary, because the agent could have wrong knowledge on when actions are executable. For example, if the agent is a pedestrian walking in busy city traffic, it may have the knowledge: 'When the traffic light goes red, I decide to perform the action of crossing the street'. So, in this system, an occasion for the action of crossing the street is a red traffic light. But this is not at all an applicable situation: in the situation that the traffic light is red, the action of crossing the street will probably not be executed successfully. The agent is not rational in its knowledge concerning occasions.

Properties concerning facts and observations

The factual world information is assumed to be two-valued; this is laid down in 2VAL. All factual world information will be observed, and all observed information corresponds to world facts. To state this, there are properties OBS1 and OBS2. Formal definitions are left out again.

Properties of interaction

As can be seen in Figure 1, there is information transfer from the world to the agent and from the agent to the world, respectively. This transferral of information needs to function properly. This comes down to three demands:

- Information at the source must be present at the destination some time later.
- Information at the destination must have been present at at least one source some time earlier.

- Pieces of information do not overtake each other while being transferred.

Formalisations again have been left out.

6.2 *The system of co-ordination properties*

One of the main objectives of this chapter is to establish a set of properties that enables the verifier to prove that all actions executed will succeed. Successfulness of action execution is influenced by a number of aspects. The objective is to give a clear separation of all aspects involved and to formalise demands on world- and the agent behaviour that guarantee overall action success. To obtain these goals, a system of co-ordination properties is devised. The properties of this system are discussed in this section.

The system is structured in the following way: COORD1 is the topmost property, proved from all other properties, COORD0 is the foundational property, parts of which are frequently used as condition in other co-ordination properties, COORD2, -3, -4 and -5 are used to prove COORD1. Actually, this is only part of the system of co-ordination properties; the other part is left out. These other properties serve to prove COORD5.

COORD0 is the foundation of the system of co-ordination properties. It enables the verifier to identify action executions, by formalising Figure 2 (in Section 4). The property states that action executions don't overlap, not in the world and neither in the agent. The agent- and the world-part will be part of the conditions of many properties to come, to enable identification of action executions. This is the abstract formula:

COORD0 :
$$\forall \mathcal{M} \in \text{Traces}(S)\quad \texttt{NoOverlappingInWorld}(\mathcal{M}) \;\wedge\; \texttt{NoOverlappingInAgent}(\mathcal{M})$$

COORD1 is the topmost property of the system of co-ordination properties. It simply states that all action executions of the system are applicable and yield all expected effects. This is the abstract formalisation:

COORD1 :
$$\forall \mathcal{M} \in \text{Traces}(S)\ \forall A \in \text{ACTION}\ \forall \texttt{t1}\ \forall \texttt{t2} > \texttt{t1} : \texttt{ActionEx}(A, \texttt{t1}, \texttt{t2}, \text{EW}, \mathcal{M}) \Rightarrow \texttt{Appl}(A, \texttt{t1}, \mathcal{M})) \;\wedge\; \texttt{ExpEffectsHappen}(A, \texttt{t1}, \texttt{t2}, \text{EW}, \mathcal{M})$$

It is essential that all action executions are applicable at the moment they start in the world. When actions are applicable at the moment the execution starts, the expected effects of the action are the desired effects. When an action is not applicable, there might be no effects at all, or unwanted ones.

COORD2 states that all action executions started at times that the action is applicable will be successful. Informally:

> 'When an action execution in the world begins at t1 and ends at t2,
> and
> when the action is applicable at t1 in the world
> then
> all expected effects of the action in that world situation will be realised during the execution.'

And this is its abstract formalisation:

$$\begin{array}{l}\text{COORD2}:\\ \forall \mathcal{M} \in \text{Traces(S)}\ \forall \text{A} \in \text{ACTION}\ \forall \texttt{t1}\ \forall \texttt{t2} > \texttt{t1}:\\ \qquad \texttt{ActionEx}(\text{A}, \texttt{t1}, \texttt{t2}, \text{EW}, \mathcal{M}) \quad \wedge\\ \qquad \texttt{Appl}(\text{A}, \texttt{t1}, \mathcal{M}) \qquad\qquad\qquad \Rightarrow\\ \qquad \texttt{ExpEffectsHappen}(\text{A}, \texttt{t1}, \texttt{t2}, \text{EW}, \mathcal{M})\end{array}$$

COORD3 is a world property that states that there are no events happening during action executions. Informally:

> 'If there is an action execution in the world
> and
> action executions do not overlap
> then
> no events happen during the execution.'

And this is the formalisation in the abstract language:

$$\begin{array}{l}\text{COORD3}:\\ \forall \mathcal{M} \in \text{Traces(EW)}\ \forall \text{A} \in \text{ACTION}\ \forall \texttt{t1}\ \forall \texttt{t2} > \texttt{t1}:\\ \qquad \texttt{ActionEx}(\text{A}, \texttt{t1}, \texttt{t2}, \text{EW}, \mathcal{M}) \quad \wedge\\ \qquad \texttt{NoOverlappingInWorld}(\mathcal{M}) \qquad\qquad \Rightarrow\\ \qquad \texttt{NoEventsDuring}([\texttt{t1}, \texttt{t2}], \text{EW}, \mathcal{M})\end{array}$$

COORD4 is a world property that says that an action execution in the world will be successful when the action is applicable and there are no disturbances caused by overlapping executions or events. These are all conditions for action success. Informally:

'If an action execution in the world begins at t1 and ends at t2
and
action executions do not overlap
and
no events happen during the execution
and
the action is applicable at t1
then
all effects of the action will be realised during the execution.'

This is the formalisation:

COORD4 :
$\forall \mathcal{M} \in \text{Traces(EW)}\ \forall A \in \text{ACTION}\ \forall t1\ \forall t2 > t1 :$
$\text{ActionEx}(A, t1, t2, EW, \mathcal{M}) \quad \wedge$
$\text{NoOverlappingInWorld}(\mathcal{M}) \quad \wedge$
$\text{NoEventsDuring}([t1, t2], EW, \mathcal{M}) \quad \wedge$
$\text{Appl}(A, t1, \mathcal{M}) \quad \Rightarrow$
$\text{ExpEffectsHappen}(A, t1, t2, EW, \mathcal{M})$

COORD5 simply states that an action is applicable at the moment its execution starts. This is a necessary condition for success of this action. This is its formalisation:

COORD5 :
$\forall \mathcal{M} \in \text{Traces(S)}\ \forall A \in \text{ACTION}\ \forall t1\ \forall t2 > t1 :$
$\text{ActionEx}(A, t1, t2, EW, \mathcal{M}) \quad \Rightarrow$
$\text{Appl}(A, t1, \mathcal{M})$

Because the abstraction formalism is exploited, these properties are relatively easy to read and understand. Technical details are hidden beneath intuitively clear notions. This system of properties is applicable for many systems with a single agent that performs actions in a changing world. By simple instantiation of the system specific details, such as the set of actions, the conditions of applicability and the effects of these actions, the system can be customised.

6.3 *Properties for proving* COORD0

An induction proof is used to establish COORD0. The reason for using induction is that induction allows you to focus on a slice of the timeline the property applies to and to make strong assumptions regarding the validity of the property during the rest of the time. In this case, induction over the number of action executions is used. Focussing on a certain action execution, all executions taking place earlier are assumed to be well behaved (that is, COORD0 holds for them). This is the induction hypothesis. In the induction step, only the results for the nth execution have to be proved, assuming all earlier executions are proper.

In order to make the induction step, a number of properties of the agent and of the world, respectively, are needed. As induction requires, the conditions of these properties focus at some occurrence during the execution of the system, and suppose that everything that has happened before that occurrence is in line with COORD0. Here, these properties are informally sketched, to show the assumptions on the behaviour of the agent and the world on which the system of co-ordination properties is built.

First, there is a property describing the demand on the agent behaviour in order to guarantee non-overlapping. Informally, this property states:

> 'When the agent is executing an action
> and
> all earlier executions don't overlap
> then
> the agent doesn't decide to do another action during the execution.'

Next, two properties describing the demands on the world behaviour are used. The first of the two is about the world not producing disturbing ended-atoms during executions:

> 'When there is an action execution in the world
> and
> all earlier executions don't overlap
> and
> no other execution starts during some initial interval of the execution
> then
> the world produces no disturbing ended-atoms during the initial interval.'

The second world property forbids a sequence of two ended-atoms without a tbp-atom in between:

> 'When there are two consecutive endings of actions A and B
> and
> all executions up and until the moment of the first ending don't overlap
> then
> an execution of B has started sometime between the two endings.'

Finally, one additional system property is necessary, stating that there never are two simultaneous decisions to do different actions.

7 PROOFS

In this section, a complete proof tree of COORD1 is given. In order to prove COORD1, it is possible to stay entirely within the abstract language; no abstractions need to be expanded into the detailed language. This makes the proof very easy.

To prove COORD1, all that is needed is performing simple modus ponens on a subset of the system of co-ordination properties. This is the proof tree:

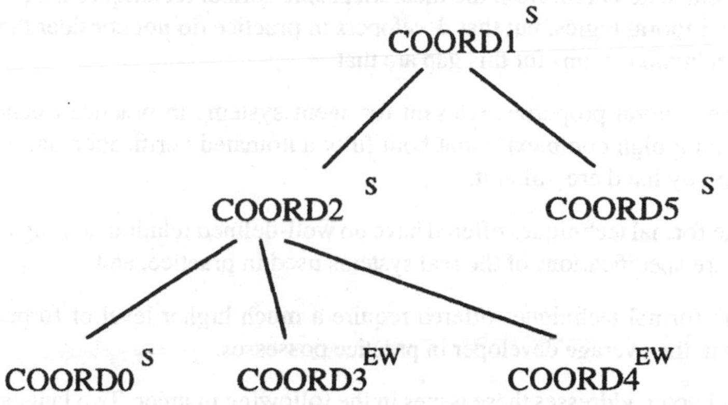

Figure 3.

The proofs of COORD0 and COORD5 are left out. This is a survey of the most important assumptions that are used in proving COORD1:

- The verifier must be able to completely specify
  - the situations in which actions are applicable
  - the effects of actions depending on conditions of the world situation
- The agent must be rational: it should decide to do an action when it sees an occasion for this action and it is not busy executing another action, it should never decide to do an action when it didn't see an occasion for this action, and whenever it thinks there is an occasion for an action, this action must be applicable in the world.
- The agent has perfect observations of the world state.
- The world behaves properly:
  - it correctly signals that action executions have ended
  - no events occur during action executions

– when an action execution is not disturbed (by events or effects of other action executions), it will succeed

8 DISCUSSION

One of the challenges to improve development methods for agent systems is to provide appropriate support for verification of agent systems being built in practice. The current state is that from the theoretical side formal techniques are proposed, such as temporal logics, but that developers in practice do not consider them useful. Three main reasons for this gap are that

- behavioural properties relevant for agent systems in practice usually have such a high complexity that both fully automated verification and verification by hand are difficult,
- the formal techniques offered have no well-defined relation to design or software specifications of the real systems used in practice, and
- the formal techniques offered require a much higher level of formal skills than the average developer in practice possesses.

This chapter addresses these issues in the following manner. Two languages are proposed: a detailed language, with a direct relation to the system design specification, and an abstract language in which properties can be formulated in a more conceptual manner. Both languages have been defined formally; moreover, well-defined relationships exist between the two languages, and between the detailed language and the system design specification. Proof strucures can be made visible within the abstract language; by this abstraction, complexity of the property and proof structures are reduced considerably. More detailed parts of proofs and properties can be hidden in the detailed language, and show up in the abstract language only in the form of, more abstractly formulated, reusable lemmas.

Two roles are distinguished within the verification engineering process: the verification support developer, who defines libraries of reusable properties in the abstract language, and their related properties in the detailed language, and the verification engineer, who uses these libraries to actually perform verification for a system being developed in practice. This means that the verification support developer needs to possess a high level of formal skills, whereas the requirements for the verification engineer are less demanding.

The approach has been illustrated by addressing the case of co-ordination of actions. Under realistic assumptions, such as action generation and action execution with duration, it is a complex problem to guarantee the successfulness of action executions. This problem was addressed and clarified using the approach described above. A set of reusable co-ordination properties has been defined both in the detailed language and in the abstract language. Indeed, the abstract formulations are much more accessible and explainable than their detailed counterparts. It has been shown that the abstractly formulated relationships between these properties can be expressed within the abstract languages. The co-ordination properties found have become part of a library of reusable properties that is being developed.

The languages used in this paper are similar to the approach in situation calculus [McCarthy and Hayes, 1969]. A difference is that explicit references are made to temporal traces and time points. In [Reiter, 1993], proving properties in situation calculus is addressed. A difference with our approach is that we incorporate arbitrary durations in the decision process of the agent, and in the interaction with the world. Another difference is that in our case the induction principle works for a larger class of temporal models than the discrete models in his case.

Catholijn M. Jonker, Jan Treur and Wieke de Vries
Vrije Universiteit Amsterdam, The Netherlands.

BIBLIOGRAPHY

[Barringer *et al.*, 1986] H. Barringer, R. Kuiper and A. Pnueli. A really abstract concurrent model and its temporal logic. In *Conference Record of the 15th ACM Symposium on Principles of Programming Languages, POPL'86*, pp. 173–183, 1986.

[Benthem, 1983] J. F. A. K. van Benthem. *The Logic of Time: A Model-Theoretic Investigation into the Varieties of Temporal Ontology and Temporal Discourse*, Reidel, Dordrecht, 1983.

[Brazier *et al.*, 1998] F. M. T. Brazier, F. Cornelissen, R. Gustavsson, C. M. Jonker, O. Lindeberg, B. Polak and J. Treur. Compositional design and verification of a multi-agent system for one-to-many negotiation. In *Proceedings of the Third International Conference on Multi-Agent Systems, ICMAS'98*. pp. 49–56. IEEE Computer Society Press, 1998.

[Brazier *et al.*, 1997] F. M. T. Brazier, B. Dunin-Keplicz, N. R. Jennings and J. Treur. DESIRE: modelling multi-agent systems in a compositional formal framework. *International Journal of Cooperative Information Systems*, M. Huhns and M. Singh, eds. Special issue on Formal Methods in Cooperative Information Systems: Multi-Agent Systems, **6**, 67–94, 1997. Preliminary and shorter version in ICMAS'95.

[Brazier *et al.*, 2000] F. M. T. Brazier, C. M. Jonker and J. Treur. Compositional design of multi-agent systems: modelling dynamics and control. In this volume.

[Chang and Keisler, 1973] C. C. Chang and H. J. Keisler. *Model Theory*, North Holland, 1973.

[Engelfriet *et al.*, 1999] J. Engelfriet, C. M. Jonker and J. Treur. Compositional verification of multi-agent systems in temporal multi-epistemic logic. In this volume. Preliminary version in *Intelligent Agents V. Agents Theories, Architectures, and Languages*, J. P. Mueller, M. P. Singh and A. S. Rao, eds. Lecture Notes in Computer Science, vol. 1555, Springer Verlag, 1999.

[Hodges, 1993] W. Hodges. *Model Theory*, Cambridge University Press, 1993.

[Jonker and Treur, 1998] C. M. Jonker and J. Treur. Compositional verification of multi-agent Systems: a formal analysis of pro-activeness and reactiveness. In *Proceedings of the International Workshop on Compositionality, COMPOS'97*, W. P. de Roever, H. Langmaack and A. Pnueli, eds. pp. 350–380. Lecture Notes in Computer Science, vol. 1536, Springer Verlag, 1998.

[Jonker *et al.*, 1998] C. M. Jonker, J. Treur and W. de Vries. Compositional verification of agents in dynamic environments: a case study. In *Proc. of the KR98 Workshop on Verification and Validation of KBS*, F. van Harmelen, ed. 1998.

[Łukasiewicz and Madalinska-Bugaj, 2000] W. Łukaszewicz and E. Madalinska-Bugaj. Reasoning about action and change: actions with abnormal effects. In this volume.

[McCarthy and Hayes, 1969] J. McCarthy and P. J. Hayes. Some philosophical problems from the standpoint of artificial intelligence. *Machine Intelligence*, **4**, 463–502, 1969.

[Reiter, 1991] R. Reiter. The frame problem in the situation calculus: a simple solution (sometimes) and a completeness result for goal regression. In *Artificial Intelligence and Mathematical Theory of Computation: Papers in Honor of John McCarthy*, V. Lifschitz, ed. pp. 359–360. Academic Press, 1991.

[Reiter, 1993] R. Reiter. Proving properties of states in the situation calculus. *Artificial Intelligence*, **64**, 337–351, 1993.

[Sandewall, 1994] E. Sandewall. *Features and Fluents. The Representation of Knowledge about Dynamical Systems, Volume I*, Oxford University Press, 1994.

FRANCES BRAZIER, FRANK CORNELISSEN,
RUNE GUSTAVSSON, CATHOLIJN M. JONKER,
OLLE LINDEBERG, BIANCA POLAK AND JAN TREUR

COMPOSITIONAL VERIFICATION OF A MULTI-AGENT SYSTEM FOR ONE-TO-MANY NEGOTIATION

1 INTRODUCTION

When designing multi-agent systems, it is often difficult to guarantee that the specification of a system actually fulfils the needs, i.e., whether it satisfies the design requirements. Especially for critical applications, for example in real-time domains, there is a need to prove that the designed system has certain properties under certain conditions (assumptions). While developing a proof of such properties, the assumptions that define the bounds within which the system will function properly, are generated. For nontrivial examples, verification can be a very complex process, both in the conceptual and computational sense. For these reasons, a recent trend in the literature on verification is to exploit compositionality and abstraction to structure the process of verification; cf. [Abadi and Lamport, 1993], [Hooman, 1997], [Jonker and Treur, 1998].

The development of structured modelling frameworks and principled design methods tuned to the specific area of multi-agent systems is currently underway; e.g., [Brazier *et al.*, 1997], [Fisher and Wooldridge, 1997], [Kinny *et al.*, 1996]. Mature multi-agent system design methods should include a verification approach. For example, in [Fisher and Wooldridge, 1997] verification is addressed using a temporal belief logic. In the approach presented below, a compositional verification method for multi-agent systems (cf. [Jonker and Treur, 1998]) is used for formal analysis of a multi-agent system for one-to-many negotiation, in particular for load balancing of electricity use; see [Brazier *et al.*, 1998]. In short, the properties of the whole system are established by derivation from assumptions that themselves are properties of agents, which in turn may be derived from assumptions on sub-components of agents, and so on. The properties are formalised in terms of temporal semantics. The multi-agent system described and verified in this paper has been designed using the compositional design method for multi-agent systems DESIRE; cf. [Brazier *et al.*, 1997].

J.J.Ch. Meyer and J. Treur (eds.),
Handbook of Defeasible Reasoning and Uncertainty Management Systems, Vol. 7, 455–475.

2 COMPOSITIONAL VERIFICATION

The purpose of verification is to prove that, under a certain set of conditions (assumed properties), a system will adhere to a certain set of desired properties, for example the design requirements. In the compositional verification approach presented in this paper, this is done by a mathematical proof (i.e., a proof in the form mathematicians are accustomed to do) that the specification of the system together with the assumed properties implies the properties that it needs to fulfil.

2.1 The Compositional Verification Method

A compositional multi-agent system can be viewed at different levels of process abstraction. Viewed from the top level, denoted by L_0, the complete system is one component S; internal information and processes are hidden. At the next, lower level of abstraction, the system component S can be viewed as a composition of agents and the world. Each agent is composed of its sub-components, and so on. The compositional verification method takes this compositional structure into account. Verification of a composed component is done using:

- properties of the sub-components it embeds,
- the way in which the component is composed of its sub-components (the composition relation),
- environmental properties of the component (depending on the rest of the system, including the world)

Given the specification of the composition relation, the assumptions under which the component functions properly are the environmental properties and the properties to be proven for its sub-components. This implies that properties at different levels of process abstraction are involved in the verification process. The primitive components (those that are not composed of other components) can be verified using more traditional verification methods. Often the properties involved are not given at the start: to find them is one of the aims of the verification process.

The verification proofs that connect properties of one process abstraction level with properties of the other level are compositional in the following manner: any proof relating level i to level $i+1$ can be combined with any proof relating level $i-1$ to level i, as long as the same properties at level i are involved. This means, for example, that the whole compositional structure within a certain component can be replaced by a completely different design as long as the same properties of the component are achieved. After such a modification only the proof for the new component has to be provided. In this sense the verification method supports reuse of verification proofs. The compositional verification method can be formulated as follows:

A. Verifying one Level Against the Other
For each abstraction level the following procedure for verification is followed:

1. Determine which properties are of interest (for the higher level).
2. Determine which assumed properties (at the lower level) are needed to guarantee the properties of the higher level, and which environment properties.
3. Prove the properties of the higher level on the basis of these assumed properties, and the environment properties.

B. Verifying a Primitive Component
For primitive components, verification techniques can be used that are especially tuned to the type of component; both for primitive knowledge-based components and non-knowledge-based components (such as databases or optimisation algorithms) techniques (and tools) can be found in the literature.

C. The Overall Verification Process
To verify the entire system

1. Determine the properties that are desired for the whole system.
2. Apply **A** iteratively. In the iteration the desired properties of each abstraction level L_i are the assumed properties for the higher abstraction level L_{i-1}.
3. Verify the primitive components according to **B**.

Notes:

- The results of verification are two-fold:
  1. Properties at the different abstraction levels.
  2. The logical relations between the properties of adjacent abstraction levels.
- process and information hiding limits the complexity of the verification per abstraction level.

- a requirement to apply the compositional verification method described above is the availability of an explicit specification of how the system description at an abstraction level L_i is composed from the descriptions at the lower abstraction level L_{i+1}.
- in principle different procedures can be followed (e.g., top-down, bottom-up or mixed).

2.2 *Semantics behind Compositional Verification*

Verification is always relative to semantics of the system descriptions to be verified. For the compositional verification method, these semantics are based on compositional information states which evolve over time. In this subsection a brief overview of these assumed semantics is given.

An *information state* M of a component D is an assignment of truth values {true, false, unknown} to the set of ground atoms that play a role within D. The compositional structure of D is reflected in the structure of the information state. A more detailed formal definition can be found in [Brazier *et al.*, 1999]. The set of all possible information states of D is denoted by IS(D).

A *trace* $\mathcal{M}$ of a component D is a sequence of information states $(M_t)_{t\in\mathbf{N}}$ in IS(D). The set of all traces (i.e., $\mathsf{IS(D)}^{\mathbf{N}}$) is denoted by Traces(D). Given a trace $\mathcal{M}$ of component D, the information state of the input interface of component C at time point t of the component D is denoted by $\mathsf{state_D}(\mathcal{M}, t,$ input(C)), where C is either D or a sub-component of D. Analogously, $\mathsf{state_D}$ $(\mathcal{M}, t,$ output(C)) denotes the information state of the output interface of component C at time point t of the component D.

3 ONE-TO-MANY NEGOTIATION PROCESSES

In this section the application domain is briefly sketched, and the one-to-many negotiation process devised within this domain is presented.

3.1 *Load Balancing of Electricity Use*

The purpose of load management of electricity use is to smoothen peak load by managing a more appropriate distribution of the electricity use among consumers. A typical demand curve of electricity is depicted in Figure 1.

Flexible pricing schemes can be an effective means to influence consumer behaviour; cf. [Gustavsson, 1997]. The assumption behind the model presented in this paper is that, to acquire a more even distribution of electricity usage in time, consumer behaviour can be influenced by financial gain. One way to acquire a more even distribution of electricity over a period of time

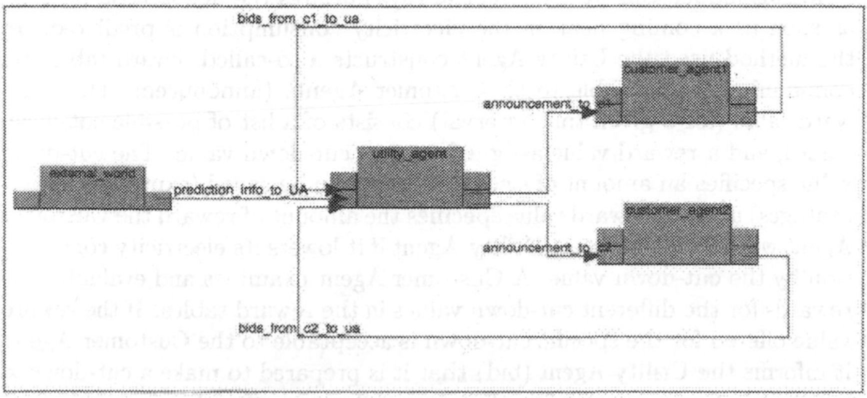

Figure 1. Demand curve with peak

is to influence consumer behaviour by offering flexible pricing schemes: by, for example, rewarding consumers if they co-operate in the efforts to lower the demand peak financial gain.

Consumers are autonomous in the process of negotiation: each individual consumer determines which price/risk he/she is willing to take and when. As consumers are all individuals with their own characteristics and needs (partially defined by the type of equipment they use within their homes), that vary over time, models of consumers used to design systems to support the consumer, need to be adaptive and flexible (cf. [Akkermans *et al.*, 1996]). Utility companies negotiate price in a one-to-many negotiation process with each and every individual separately, unaware of the specific models behind such systems for individuals. In the model discussed in this paper the negotiation process is modeled for one utility company and a number of consumers, each with their own respective agent to support them in the negotiation process: one Utility Agent and a number of Customer Agents.

3.2 Modelling the Negotiation Process

In [Rosenschein and Zlotkin, 1994], [Rosenschein and Zlotkin, 1994] a number of mechanisms for negotiation are described. A protocol with well-defined properties, called the *monotonic concession protocol*, is described:

during a negotiation process all proposed deals must be equally or more acceptable to the counter party than all previous deals proposed. The strength of this protocol is that the negotiation process always converges. The monotonic concession protocol has been applied to the load management problem, to obtain a model for the one-to-many negotiation process between one Utility Agent and a (in principle large) number of Customer Agents.

In this model, the Utility Agent always initiates the negotiation process, as soon as a coming peak in the electricity consumption is predicted. In the method used the Utility Agent constructs a so-called reward table and communicates this table to all Customer Agents (announcement). A reward table (for a given time interval) consists of a list of possible cut-down values, and a reward value assigned to each cut-down value. The cut-down value specifies an amount of electricity that can be saved (expressed in percentages) and the reward value specifies the amount of reward the Customer Agent will receive from the Utility Agent if it lowers its electricity consumption by the cut-down value. A Customer Agent examines and evaluates the rewards for the different cut-down values in the reward tables. If the reward value offered for the specific cut-down is acceptable to the Customer Agent, it informs the Utility Agent (bid) that it is prepared to make a cut-down x, which may be zero to express that no cut-down is offered.

As soon as the Customer Agents have responded to the announcement of a reward table, the Utility Agent predicts the new balance between consumption and production of electricity for the stated time interval. The Utility Agent is satisfied by the responses if a peak can be avoided if all Customer Agents implement their bids. If the Utility Agent is not satisfied by the responses communicated by the Customer Agents, it announces a new reward table (according to the monotonic concession protocol mentioned above) to the Customer Agents in which the reward values are at least as high, and for some cut-down values higher than in the former reward table (determined on the basis of, for example, the formulae described in Section 4.2 below). The Customer Agents react to this new announcement by responding with a new bid or the same bid again (in line with the rules of the monotonic concession protocol). This process continues until (1) the peak is satisfactorily low for the Utility Agent (at most the capacity of the utility company), or (2) the reward values in the new reward table have (almost) reached the maximum value the Utility Agent can offer. This value has been determined in advance. For more details on this negotiation method, see [Brazier *et al.*, 1998].

4 COMPOSITIONAL DESIGN OF MULTI-AGENT SYSTEMS

The example multi-agent system described in this paper has been developed using the compositional design method DESIRE for multi-agent systems

(DEsign and Specification of Interacting REasoning components); cf. [Brazier *et al.*, 1997]. In DESIRE, a design consists of knowledge of the following three types: process composition, knowledge composition, the relation between process composition and knowledge composition. These three types of knowledge are discussed in more detail below.

4.1 Process composition

Process composition identifies the relevant processes at different levels of (process) abstraction, and describes how a process can be defined in terms of lower level processes.

Identification of processes at different levels of abstraction

Processes can be described at different levels of abstraction; for example, the process of the multi-agent system as a whole, processes defined by individual agents and the external world, and processes defined by task-related components within individual agents. The identified processes are modeled as components. For each process the input and output information types are modeled. The identified levels of process abstraction are modeled as abstraction/specialisation relations between components: components may be composed of other components or they may be primitive. Primitive components may be either reasoning components (i.e., based on a knowledge base), or, components capable of performing tasks such as calculation, information retrieval, optimisation. These levels of process abstraction provide process hiding at each level.

Composition

The way in which processes at one level of abstraction are composed of processes at the adjacent lower abstraction level is called composition. This composition of processes is described by the possibilities for information exchange between processes (static view on the composition), and task control knowledge used to control processes and information exchange (dynamic view on the composition).

4.2 Knowledge composition

Knowledge composition identifies the knowledge structures at different levels of (knowledge) abstraction, and describes how a knowledge structure can be defined in terms of lower level knowledge structures. The knowledge abstraction levels may correspond to the process abstraction levels, but this is often not the case.

Identification of knowledge structures at different abstraction levels

The two main structures used as building blocks to model knowledge are: information types and knowledge bases. Knowledge structures can be identified and described at different levels of abstraction. At higher levels details can be hidden. An information type defines an ontology (lexicon, vocabulary) to describe objects or terms, their sorts, and the relations or functions that can be defined on these objects. Information types can logically be represented in order-sorted predicate logic. A knowledge base defines a part of the knowledge that is used in one or more of the processes. Knowledge is represented logically by formulae in order-sorted predicate logic, which can be normalised by a standard transformation into rules.

Composition of knowledge structures

Information types can be composed of more specific information types, following the principle of compositionality discussed above. Similarly, knowledge bases can be composed of more specific knowledge bases. The compositional structure is based on the different levels of knowledge abstraction distinguished, and results in information and knowledge hiding.

4.3 Relation between process composition and knowledge composition

Each process in a process composition uses knowledge structures. Which knowledge structures are used for which processes is defined by the relation between process composition and knowledge composition.

The semantics of the modeling language is based on temporal logic (cf., [Brazier *et al.*, 1999]). The development of multi-agent system is supported by graphical design tools within the DESIRE software environment. Translation to an operational system is straightforward; the software environment includes implementation generators with which formal design specifications can be translated into executable code of a prototype system.

5 COMPOSITIONAL DESIGN OF THE NEGOTIATION SYSTEM

The prototype Multi-Agent System has been fully specified and (automatically) implemented in the DESIRE software environment. The top level composition of the system consists of a Utility Agent, two Customer Agents, and an External World.

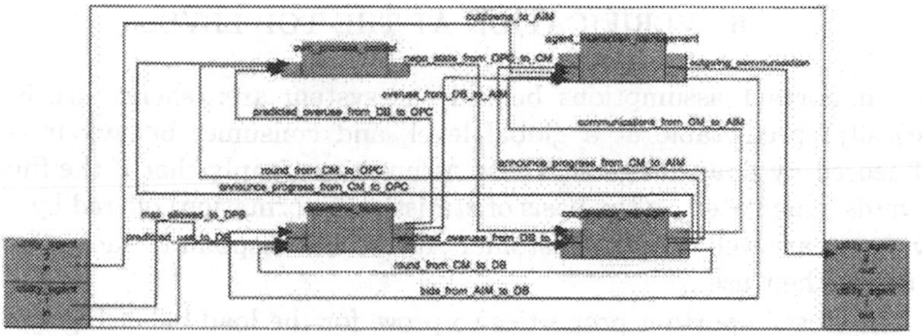

Figure 2. Process composition at the first level within the Utility Agent

5.1 Top Level Composition of the Utility Agent

The first level composition within the Utility Agent is depicted in Figure 1 (taken from the graphical design tool within the DESIRE software environment). This picture shows part of the graphical interface of the DESIRE software environment; in addition, interfaces to the agents have been implemented which are specific for this prototype (see [Brazier *et al.*, 1998]).

5.2 Knowledge used within the Utility Agent

In this prototype system the Utility Agent communicates the same announcements to all Customer Agents, in compliance with Swedish law. The predicted balance between the consumption and the production of electricity, is determined by the following formulae (here CA is a variable ranging over the set of Customer Agents):

predicted_use_with_cutdown(CA) =
        predicted_use(CA)
                if (1 - cutdown(CA)). allowed_use(CA) $\geq$ predicted_use(CA)
        (1 - cutdown(CA)). allowed_use(CA)
                otherwise

predicted_overuse	=	Σ_{CA} predicted_use_with_cutdown(CA) - normal_use
overuse	=	predicted_overuse/normal_use
new_reward	=	reward+beta.overuse.(1-reward/max_reward).reward

In the prototype system, the factor beta determines how steeply the reward values increase; in the current system it has a constant value. The reward value increases more when the predicted overuse is higher (in the beginning of the negotiation process) and less if the predicted overuse is lower. It never exceeds the maximal reward, due to the logistic factor (1 - reward/max_reward).

6 VERIFICATION AT THE TOP LEVEL

Two important assumptions behind the system are: energy use is (statistically) predictable at a global level, and consumer behaviour can be influenced by financial gain. These assumptions imply that if the financial rewards (calculated on the basis of statistical information) offered by a Utility Agent are well chosen, Customer Agents will respond to such offers and decrease their use.

The most important properties to prove for the load balancing system S as a whole are that

1. the negotiation process satisfies the monotonic concession protocol,
2. at some point in time the negotiation process will terminate, and
3. the agents make rational decisions during the negotiation process.

These properties are formally defined in Section 5.1. An important property for the Utility Agent, in particular, is that after the negotiation process the predicted overuse has decreased to such an extent that is at most the maximal overuse the utility company considers acceptable. To prove these properties several other properties of the participating agents (and the external world) are assumed. These properties of agents and the external world are defined in Section 5.2. Some of the proofs of properties are briefly presented in Section 5.3. Next, Section 6 shows how these assumed properties can be proven from properties assumed for the sub-components of the agents.

6.1 Properties of the System as a Whole

The properties defined at the level of the entire system are based on combinations of properties of the agents.

S1. Monotonicity of negotiation
The system *S* satisfies ***monotonicity of negotiation*** if the Utility Agent satisfies monotonicity of announcements and each Customer Agent satisfies monotonicity of bids. This is formally defined as the conjunction of the Utility Agent announce monotonicity property U7 and for each Customer Agent the bid monotonicity property C5 (see below).

S2. Termination of negotiation
The system S satisfies *termination of negotiation* (for a given time interval) if a time point exists after which no announcements or bids (referring to the given time interval) are generated by the agents. This is formally defined by: for all Customer Agents CA it holds

$$\begin{aligned} &\forall \mathcal{M} \in \mathsf{Traces(S)} \exists t \forall t' > t,\ \mathsf{CD, R, N} \\ &\quad \mathsf{states_S}(\mathcal{M}, t',\ \mathsf{output(UA)}) \not\models \mathsf{announcement(CD, R, N)} \quad \& \\ &\quad \mathsf{states_S}(\mathcal{M}, t',\ \mathsf{output(CA)}) \not\models \mathsf{cutdown(CD, N)} \end{aligned}$$

S3. Rationality of negotiation
The system S satisfies *rationality of negotiation* if the Utility Agent satisfies announcement rationality and each Customer Agent satisfies bid rationality. This is formally defined as the conjunction of the Utility Agent rationality property U9 and for each Customer Agent the Customer Agent rationality property C4 defined below.

S4. Required reward limitation
The system S satisfies *required reward limitation* if for each Customer Agent and each cut-down percentage, the required reward of the Customer Agent is at most the maximal reward that can be offered by the Utility Agent.

$$\forall \mathsf{CA}\ \forall\ \mathsf{CD}\quad rr_{\mathsf{CA}}(\mathsf{CD}) \leq mr_{\mathsf{UA}}(\mathsf{CD})$$

The above property is an assumption for the whole system, used in the proofs. In addition to these properties a global *successfulness property* for the whole negotiation process could be defined. However, as successfulness depends on the perspective of a specific agent, the choice has been made to define successfulness as a property of an agent (cf. property U1 below).

6.2 Properties of the Agents and the World

The properties of the Utility Agent, the Customer Agents, and the External World are defined in this section. Note that each of the properties is presented as a temporal statement either about all traces of the system S or about all traces of an agent. In the latter case the truth of the property does not depend on the environment of the agent. Section 5.3 discusses how the various properties are logically related.

Properties of the Utility Agent

U1. Successfulness of negotiation
The Utility Agent satisfies *successfulness of negotiation* if at some point in time t and for some negotiation round N the predicted overuse is less than or equal to the constant max_overuse.

$$\forall \mathcal{M} \in \text{Traces(S)} \exists t, \text{N} \exists \text{U} \leq \text{max\_overuse}$$
$$\text{state}_S(\mathcal{M}, t, \text{ output(UA)}) \models \text{predicted\_overuse(U, N)}$$

U2. Negotiation round generation effectiveness

The Utility Agent satisfies *negotiation round generation effectiveness* if the following holds: if and when predicted overuse is higher than the maximal overuse, a next negotiation round is initiated.

$$\begin{aligned}
&\forall \mathcal{M} \in \text{Traces(UA)} \forall t, \text{N, U, CD, R} \\
&\quad [\ \ \text{state}_{\text{UA}}(\mathcal{M}, t, \text{ output(UA)}) \models \text{round(N)} \\
&\quad\quad \& \ \ \text{state}_{\text{UA}}(\mathcal{M}, t, \text{ output(UA)}) \models \text{predicted_overuse(U, N)} \\
&\quad\quad \& \ \ \text{U} > \text{max_overuse} \\
&\quad\quad \& \ \ \text{state}_{\text{UA}}(\mathcal{M}, t, \text{ output(UA)}) \models \text{announcement(CD, R, N)} \\
&\quad\quad \& \ \ \text{R} < \text{mr}_{\text{UA}}(\text{CD})] \\
&\quad \Rightarrow \exists t' > t \text{state}_{\text{UA}}(\mathcal{M}, t', \text{ output(UA)}) \models \text{round(N + 1)}
\end{aligned}$$

U3. Negotiation round generation groundedness

The Utility Agent satisfies *negotiation round generation groundedness* if the following holds: if the predicted overuse is at most the maximal overuse, then no new negotiation round is initiated.

$$\begin{aligned}
&\forall \mathcal{M} \in \text{Traces(UA)} \forall t, \text{N, U} \\
&\quad \text{state}_{\text{UA}}(\mathcal{M}, t, \text{ output(UA)}) \models \text{predicted_overuse(U, N)} \ \& \ \text{U} \leq \text{max_overuse} \\
&\quad \Rightarrow \forall t', \text{N}' > \text{N} \quad \text{state}_{\text{UA}}(\mathcal{M}, t', \text{ output(UA)}) \not\models \text{round(N}')
\end{aligned}$$

U4. Announcement generation effectiveness

The Utility Agent satisfies *announcement generation effectiveness* if for each initiated negotiation round at least one announcement is generated:

$$\begin{aligned}
&\forall \mathcal{M} \in \text{Traces(UA)} \forall t, \text{N} \\
&\quad [\text{state}_{\text{UA}}(\mathcal{M}, t, \text{ output(UA)}) \models \text{ round(N)} \\
&\quad \Rightarrow \exists t' \geq t \forall \text{CD} \exists \text{R} \ \ \text{state}_{\text{UA}}(\mathcal{M}, t', \text{output(UA)}) \models \text{announcement(CD, R, N)}]
\end{aligned}$$

U5. Announcement uniqueness

The Utility Agent satisfies *announcement uniqueness* if for each initiated negotiation round at most one announcement is generated:

$$\begin{aligned}
&\forall \mathcal{M} \in \text{Traces(UA)} \forall t, t', \text{N} \ \forall \text{CD, R, R}' \\
&\quad \text{state}_{\text{UA}}(\mathcal{M}, t, \text{ output(UA)}) \models \text{ announcement(CD, R, N)} \\
&\quad \text{state}_{\text{UA}}(\mathcal{M}, t', \text{ output(UA)}) \models \text{ announcement(CD, R}', \text{N)} \\
&\quad \Rightarrow \text{R} = \text{R}'
\end{aligned}$$

U6. Announcement generation groundedness

The Utility Agent satisfies *announcement generation groundedness* if an announcement is only generated for initiated negotiation rounds:

$$\begin{aligned}
&\forall \mathcal{M} \in \text{Traces(UA)} \forall t, \text{N} \ \forall \text{CD, R} \\
&\quad \text{state}_{\text{UA}}(\mathcal{M}, t, \text{ output(UA)}) \models \text{ announcement(CD, R, N)} \\
&\quad \Rightarrow \exists t' \leq t \ \text{state}_{\text{UA}}(\mathcal{M}, t', \text{ output(UA)}) \models \text{round(N)}
\end{aligned}$$

U7. Monotonicity of announcement

The Utility Agent satisfies *monotonicity of announcement* if for each announcement and each cut-down percentage the offered reward is at least the reward for the same cut-down percentage offered in the previous announcements:

$$
\begin{aligned}
&\forall \mathcal{M} \in \mathsf{Traces(UA)} \forall t, t', \mathsf{N}, \mathsf{N}' \,\forall \mathsf{CD}, \mathsf{R}, \mathsf{R}' \\
&\quad \mathsf{state_{UA}}(\mathcal{M}, t,\ \mathsf{output(UA)}) \models\ \mathsf{announcement(CD, R, N)} \\
&\quad \&\ \ \mathsf{state_{UA}}(\mathcal{M}, t',\ \mathsf{output(UA)}) \models\ \mathsf{announcement(CD, R', N')} \\
&\quad \&\ \ \mathsf{N} \leq \mathsf{N}' \\
&\quad \Rightarrow\ \mathsf{R} \leq \mathsf{R}'
\end{aligned}
$$

U8. Progress in announcement

The Utility Agent satisfies *progress in announcement* if for at least one cut-down percentage the difference between the currently announced reward and the previously announced reward is at least the positive constant m (announce margin):

$$
\begin{aligned}
&\forall \mathcal{M} \in \mathsf{Traces(UA)} \forall t, t', \mathsf{N}\ \exists \mathsf{CD}\ \forall \mathsf{R}, \mathsf{R}' \\
&\quad \mathsf{state_{UA}}(\mathcal{M}, t,\ \mathsf{output(UA)}) \models\ \mathsf{announcement(CD, R, N)} \\
&\quad \&\ \ \mathsf{state_{UA}}(\mathcal{M}, t',\ \mathsf{output(UA)}) \models\ \mathsf{announcement(CD, R', N+1)} \\
&\quad \Rightarrow\ \ \mathsf{R + m} \leq \mathsf{R}'
\end{aligned}
$$

U9. Announcement rationality

The Utility Agent satisfies *announcement rationality* if no announced reward is higher than the maximal reward plus the announce margin:

$$
\begin{aligned}
&\forall \mathcal{M} \in \mathsf{Traces(UA)} \forall t, \mathsf{N}\ \forall \mathsf{CD}, \mathsf{R} \\
&\quad \mathsf{state_{UA}}(\mathcal{M}, t,\ \mathsf{output(UA)}) \models\ \mathsf{announcement(CD, R, N)} \\
&\quad \Rightarrow\ \ \mathsf{R} \leq \mathsf{mr_{UA}(CD)} + \mathsf{announce\_margin}
\end{aligned}
$$

U10. Finite termination of negotiation by UA

The Utility Agent satisfies *finite termination of negotiation* if a time point exists such that UA does not negotiate anymore after this time point:

$$
\begin{aligned}
&\forall \mathcal{M} \in \mathsf{Traces(S)} \exists t \forall t' > t, \mathsf{CD}, \mathsf{R}, \mathsf{N} \\
&\quad \mathsf{state_S}(\mathcal{M}, t',\ \mathsf{output(UA)}) \not\models\ \mathsf{announcement(CD, R, N)}
\end{aligned}
$$

Properties of each Customer Agent

C1. Bid generation effectiveness

A Customer Agent CA satisfies *bid generation effectiveness* if for each announced negotiation round at least one bid is generated (possibly a bid for reduction zero):

$$
\begin{aligned}
&\forall \mathcal{M} \in \mathsf{Traces(CA)} \forall t, \mathsf{N} \\
&\quad \mathsf{state_{CA}}(\mathcal{M}, t,\ \mathsf{output(CA)}) \models\ \mathsf{round(N)} \\
&\quad \Rightarrow \exists \mathsf{CD}, t' \geq t\ \mathsf{state_{CA}}(\mathcal{M}, t',\ \mathsf{output(CA)}) \models\ \mathsf{cutdown(CD, N)}
\end{aligned}
$$

C2. Bid uniqueness
A Customer Agent CA satisfies ***bid uniqueness*** if for each negotiation round at most one bid is generated:

$$\begin{aligned}&\forall \mathcal{M} \in \text{Traces(CA)} \forall t, t', \text{N}, \text{CD}, \text{CD}'\\ &\quad \text{state}_{\text{CA}}(\mathcal{M}, t, \text{ output(CA)}) \models \text{cutdown(CD, N)} \ \&\\ &\quad \text{state}_{\text{CA}}(\mathcal{M}, t', \text{ output(CA)}) \models \text{cutdown(CD}', \text{N)}\\ &\quad \Rightarrow \text{CD} = \text{CD}'\end{aligned}$$

C3. Bid generation groundedness
A Customer Agent CA satisfies ***bid generation groundedness*** if a bid is only generated once a negotiation round is announced:

$$\begin{aligned}&\forall \mathcal{M} \in \text{Traces(CA)} \forall t, \text{N}, \text{CD}\\ &\quad \text{state}_{\text{CA}}(\mathcal{M}, t, \text{ output(CA)}) \models \text{cutdown(CD, N)}\\ &\quad \Rightarrow \exists t' \leq t \ \text{state}_{\text{CA}}(\mathcal{M}, t', \text{ input(CA)}) \models \text{round(N)}\end{aligned}$$

C4. Bid rationality
A Customer Agent CA satisfies ***bid rationality*** if for each bid the required reward for the offered cut-down is at most the reward announced in the same round, and the offered cut-down is the highest with this property:

$$\begin{aligned}&\forall \mathcal{M} \in \text{Traces(CA)} \forall t, t', \text{N}, \text{CD}, \text{R}\\ &\quad [\text{state}_{\text{CA}}(\mathcal{M}, t, \text{ output(CA)}) \models \text{cutdown(CD, N)} \ \&\\ &\quad\ \ \text{state}_{\text{CA}}(\mathcal{M}, t', \text{ input(CA)}) \models \text{announcement(CD, R, N)}\\ &\quad \Rightarrow [rr_{\text{CA}}(\text{CD}) \leq \text{R} \ \&\\ &\quad [\forall \text{CD}', \text{R}', t''[\ \text{state}_{\text{CA}}(\mathcal{M}, t'', \text{ input(CA)}) \models \text{announcement(CD}', \text{R}', \text{N)}\\ &\quad \& \ \ rr_{\text{CA}}(\text{CD}') \leq R']\\ &\quad \Rightarrow \ \text{CD} \geq \text{CD}']]]\end{aligned}$$

C5. Monotonicity of bids
A Customer Agent CA satisfies ***monotonicity of bids*** if each bid is at least as high (a cut-down percentage) as the bids for the previous rounds.

$$\begin{aligned}&\forall \mathcal{M} \in \text{Traces(S)} \forall t, t', \text{N}, \text{N}' \ \forall \text{CD}, \text{CD}'\\ &\quad \text{state}_{\text{S}}(\mathcal{M}, t, \text{ output(CA)}) \models \text{cutdown(CD, N)} \ \&\\ &\quad \text{state}_{\text{S}}(\mathcal{M}, t', \text{ output(CA)}) \models \text{cutdown(CD}', \text{N}') \ \ \& \ \ \text{N} \leq \text{N}'\\ &\quad \Rightarrow \ \text{CD} \leq \text{CD}'\end{aligned}$$

C6. Finite termination of negotiation by CA
A Customer Agent CA satisfies ***finite termination of negotiation by*** CA if a time point exists such that CA does not negotiate anymore after this time point:

$$\begin{aligned}&\forall \mathcal{M} \in \text{Traces(S)} \exists t \forall t' > t, \text{CD}, \text{N}\\ &\quad \text{state}_{\text{S}}(\mathcal{M}, t', \text{ output(CA)}) \not\models \text{cutdown(CD, N)}\end{aligned}$$

A successfulness property of a Customer Agent could be defined on the basis of some balance between discomfort and financial gains.

Properties of the External World

The External World satisfies ***information provision effectiveness*** if it provides information about the predicted use of energy, the maximum energy level allocated to each Customer Agent, and the maximal overuse of the Utility Agent. The External World satisfies ***static world*** if the information provided by the external world does not change during a negotiation process.

6.3 Proving Properties

To structure proofs of properties, the compositional structure of the system is followed. For the level of the whole system, system properties are proved from agent properties, which are defined at one process abstraction level lower.

Proofs of the System Properties

Property S4 is an assumption on the system, which is used in the proofs of other properties. The other top level properties can be proven from the agent properties in a relatively simple manner. For example, by definition monotonicity of negotiation (S1) can be proven from the properties monotonicity of announcement (U7) and monotonicity of bids (C5) for all Customer Agents. Also S2 (termination) can be proven directly from U10 and C6, and S3 (rationality) immediately follows from U9 and C4.

Proofs of Agent Properties

Less trivial relationships can be found between agent properties. As an example, the termination property for the Utility Agent (U10) can be proven from the properties U1, U3, and U6. The termination property of a Customer Agent depends on the Utility Agent, since the Customer Agents are reactive: the proof of C6 makes use of C3, and the Utility Agent properties U1 and U3, and the assumption that the communication between UA and CA functions properly (CA should not receive round information that was not generated by UA). In the proofs of an agent property, also properties of sub-components of the agent can be used: the proof can be made at one process abstraction level lower. This will be discussed for the Utility Agent in Section 6.

7 VERIFICATION WITHIN THE UTILITY AGENT

To illustrate the next level in the compositional verification process, in this section it is discussed how properties of the Utility Agent can be related to properties of components within the Utility Agent. First some of the properties of the components Agent Interaction Management and Determine Balance are defined.

7.1 *Properties of Components within UA*

Properties are defined for the components Agent Interaction Management (AIM), Determine Balance (DB), Cooperation Management (CM), and Own Process Control (OPC) of the Utility Agent (see Figure 1).

Properties of AIM

The following two properties express that the component Agent Interaction Management (1) distributes the relevant information extracted from incoming communication, and (2) generates outgoing communication if required.

AIM1. Cut-down provision effectiveness
The component Agent Interaction Management satisfies ***cut-down provision effectiveness*** if AIM is effective in the analysis of incoming communication: the cut-down information received by AIM of the form received(cutdown_from (CD, CA, N)) is interpreted and translated into cut-down information required by other components of the form offered_bid(cutdown(CD, CA, N)) and made available in AIM's output interface:

$$\forall \mathcal{M} \in \mathsf{Traces(AIM)} \forall t, \mathsf{N}, \mathsf{CD}, \mathsf{CA}$$
$$\mathsf{state_S}(\mathcal{M}, t,\ \mathsf{input(AIM)}) \models \mathsf{received(cutdown\_from(CD, CA, N))}$$
$$\Rightarrow \exists t' > t\ \mathsf{state_S}(\mathcal{M}, t',\ \mathsf{output(AIM)}) \models \mathsf{offered\_bid(cutdown(CD, CA, N))}$$

AIM2. Communication generation effectiveness
The component Agent Interaction Management satisfies ***communication generation effectiveness*** if AIM is effective in generation of outgoing communication on the basis of the analysis of input information received from other components of the form next_communication (round(N)), next_communication (announcement(CD, R, N)) and made available in statements own_communication(round(N)), and own_communication(announcement(CD, R, N)):

$$\forall \mathcal{M} \in \mathsf{Traces(AIM)} \forall t, \mathsf{N}, \mathsf{CD}$$
$$\mathsf{state_{AIM}}(\mathcal{M}, t,\ \mathsf{input(AIM)}) \models \mathsf{next\_communication(X)}$$
$$\Rightarrow \exists t' > t\ \mathsf{state_{AIM}}(\mathcal{M}, t',\ \mathsf{output(AIM)}) \models \mathsf{own\_communication(X)}$$

Properties of Determine Balance

The following two properties express that the component Determine Balance calculates predictions in a reasonable manner.

DB1. Overuse prediction generation effectiveness
The component Determine Balance satisfies *overuse prediction generation effectiveness* if the predicted overuse is determined if and when normal capacity, predicted use and cut-downs are known.

$$\begin{aligned}
&\forall \mathcal{M} \in \text{Traces(DB)} \forall t, \text{N}, \text{C}\\
&\quad [\text{state}_{\text{DB}}(\mathcal{M}, t,\ \text{input(DB)}) \models\ \text{predicted_use(U)}\\
&\quad \&\ \ \text{state}_{\text{DB}}(\mathcal{M}, t,\ \text{input(DB)}) \models\ \text{normal_capacity(C)}\\
&\quad \&\ \ \forall \text{CA} \exists \text{CD}\ \ \text{state}_{\text{DB}}(\mathcal{M}, t,\ \text{input(DB)}) \models \text{cutdown_from(CD, CA, N)}\\
&\quad \&\ \ \text{state}_{\text{DB}}(\mathcal{M}, t,\ \text{input(DB)}) \models\ \text{round(N)}\\
&\quad \Rightarrow \exists \text{U}', t' > t\ \text{state}_{\text{DB}}(\mathcal{M}, t',\ \text{output(DB)}) \models \text{predicted_overuse(U}', \text{N)}
\end{aligned}$$

DB2. Overuse prediction monotonicity
The component Determine Balance satisfies *overuse prediction monotonicity* if the following holds: if based on received cut-downs CD_{CA} for each Customer Agent CA, a predicted overuse U is generated by DB, and based on received cut-downs CD'_{CA} for each Customer Agent CA, a predicted overuse U′ is generated by DB, then $\text{CD}_{\text{CA}} \leq \text{CD}'_{\text{CA}}$, for all CA implies $\text{U}' \leq \text{U}$:

$$\begin{aligned}
&\forall \mathcal{M} \in \text{Traces(DB)} \forall t, t', \text{N}, \text{N}', \text{C}, \text{U0}, \text{U}, \text{U}'\\
&\quad \text{state}_{\text{DB}}(\mathcal{M}, t,\ \text{input(DB)}) \models\ \text{predicted_use(U0)}\\
&\quad \&\forall\ \text{CA}\ [\ \ \text{state}_{\text{DB}}(\mathcal{M}, t,\ \text{input(DB)}) \models\ \text{cutdown_from}(\text{CD}_{\text{CA}}, \text{CA}, \text{N})\\
&\quad \&\ \ \text{state}_{\text{DB}}(\mathcal{M}, t',\ \text{input(DB)}) \models\ \text{cutdown_from}(\text{CD}'_{\text{CA}}, \text{CA}, \text{N}')\\
&\quad \&\ \ \text{CD}_{\text{CA}} \leq \text{CD}'_{\text{CA}}\ ]\\
&\quad \&\ \ \text{state}_{\text{DB}}(\mathcal{M}, t,\ \text{output(DB)}) \models\ \text{predicted_overuse(U, N)}\\
&\quad \&\ \ \text{state}_{\text{DB}}(\mathcal{M}, t',\ \text{output(DB)}) \models\ \text{predicted_overuse}(\text{U}', \text{N}')\\
&\quad \Rightarrow \text{U}' \leq \text{U}
\end{aligned}$$

Note that in this property the monotonicity is not meant over time, but for the functional relation between input and output of DB.

DB3. Overuse prediction decrease effectiveness
The component Determine Balance satisfies *overuse prediction decrease effectiveness* if the following holds: cut-down values exist such that, if the Utility Agent receives them, the predicted overuse will be at most the maximal overuse. Formally, a collection of numbers CD_{CA} for each Customer Agent CA exists such that:

$$\begin{aligned}
&\forall \mathcal{M} \in \text{Traces(DB)} \forall t, \text{N}\\
&\quad \forall \text{CA}\ \ \ \text{state}_{\text{DB}}(\mathcal{M}, t,\ \text{input(DB)}) \models\ \text{cutdown_from}(\text{CD}_{\text{CA}}, \text{CA}, \text{N})\\
&\qquad \Rightarrow \exists t' > t, \text{U} \leq\ \text{max_overuse}\\
&\qquad \text{state}_{\text{DB}}(\mathcal{M}, t',\ \text{output(DB)}) \models\ \text{predicted_overuse(U, N)}
\end{aligned}$$

Properties of Cooperation Management

Cooperation Management fulfills a number of properties, for example on properly generation announcements: announcement generation effectiveness, announcement uniqueness, and announcement generation groundedness. These are defined similarly to the corresponding properties of the Utility Agent. In this paper only the property that guarantees that new rounds are initiated is explicitly stated.

CM1. Round generation effectiveness
The component Determine Balance satisfies *round generation effectiveness* if CM determines the value of the next round and makes this information available to other components in its output interface:

$$\forall \mathcal{M} \in \text{Traces(CM)} \forall t, \text{N}$$
$$\text{state}_{\text{CM}}(\mathcal{M}, t, \ \text{input(CM)}) \models \ \text{round(N)}$$
$$\Rightarrow \exists t' > t \ \text{state}_{\text{CM}}(\mathcal{M}, t', \ \text{output(DB)}) \models \ \text{round(N + 1)}$$

Properties of own Process Control

One of the properties of the component Own Process Control guarantees that decisions about continuation of a negotiation process are made:

OPC1. New announce decision effectiveness If the predicted overuse is still more than the maximum overuse, then a new announcement is warranted.

$$\forall \mathcal{M} \in \text{Traces (OPC)} \ \forall t \ \text{N, U}$$
$$\text{state}_{\text{OPC}}(\mathcal{M}, t, \text{input(OPC)}) \models$$
$$\text{current\_negotiation\_state(predicted\_overuse(U, N))}$$
$$\& \ \text{state}_{\text{OPC}}(\mathcal{M}, t, \text{input(OPC)}) \models \text{current\_negotiation\_state(round(N))}$$
$$\& \ \text{U} > \text{max\_overuse}$$
$$\Rightarrow \exists t' > t, \text{state}_{\text{OPC}}(\mathcal{M}, t', \text{output(OPC)}) \models \text{new\_announce}$$

7.2 Proofs within the Utility Agent

To verify the UA property U2 (*negotiation round generation effectiveness*), a number of properties of sub-components are of importance, and also the interaction between the components through the information links (the arrows in Figure 1) should function properly. The following gives a brief sketch of the proof of the UA property negotiation round generation effectiveness.

The round number itself is determined by CM; to guarantee this, CM needs to satisfy the property of *round generation effectiveness* (CM1). This round value is transferred to the component AIM. The component AIM must fulfil the property of *communication generation effectiveness* (AIM2) to enable this value to be placed in the Utility Agent's output interface, once the relevant link has been activated. Activation of the link to the Utility Agent's output interface depends (via task control) on whether the component OPC derives the need for a new announcement. To guarantee this, the property *new announce decision effectiveness* (OPC1), is needed.

Based on the properties mentioned, the proof runs as follows. Whenever the component AIM has received all the cut-downs for the current round, the link bids from AIM to DB is activated (via task control). Because of the property BD1 (*overuse prediction generation effectiveness*), this component then derives the current predicted overuse (assuming predicted use, normal capacity and round are known). It can be assumed that the overuse for this round is above max_overuse (otherwise the conditions for U2 are not satisfied). The component OPC is then activated (by task control) and, given property OPC1 (*new announce decision effectiveness*) this component will derive the atom new_announce. Then Cooperation Management is activated and given property CM1, this component will derive a new round. Given property AIM2, this new round information will be available on the output interface of AIM; the link outgoing communications transfers the desired result: round(N+1) at the output of UA. This proves Utility Agent property U2.

The first level composition within the Utility Agent is depicted in Figure 8 (again taken from the graphical design tool within the DESIRE software environment).

8 DISCUSSION

To come to clearer understanding of strengths and weaknesses of a compositional approach to verification it is important to address real world problems where size and/or complexity are characteristic. The load balancing problem of electricity use, as addressed in this paper, belongs to the class of real world problems. This paper focuses on one-to-many negotiation between a Utility Agent and its Customer Agents, using a (monotonic) negotiation strategy based on announcing reward tables.

The compositional verification method used in this paper is part of the compositional design method for multi-agent systems DESIRE, based on compositionality of processes and knowledge at different levels of abstraction, but can also be useful to other compositional approaches. Two main advantages of a compositional approach to modeling are the transparent structure of the design and support for reuse of components and generic models. The compositional verification method extends these main advantages to (1) the complexity of the verification process is managed by compositionality, and (2) the proofs of properties of components that are reused can be reused.

The first advantage entails that both conceptually and computationally the complexity of the verification process can be handled by compositionality at different levels of abstraction. The second advantage entails: if a modified component satisfies the same properties as the one it replaces, the proof of the properties at the higher levels of abstraction can be reused to show that the new system has the same properties as the original system. This increases the value for a documented library of reusable generic and instantiated components.

Also due to the compositional nature of the verification method, a distributed approach to verification is facilitated: several persons can work on the verification of the same system at the same time. It is only necessary to know or to agree on the properties of these sub-components with interfaces in common.

A main difference in comparison to [Fisher and Wooldridge, 1997] is that our approach exploits compositionality. An advantage of their approach is that it uses a temporal belief logic. A first step to extend our approach a compositional variant of temporal logic can be found in [Engelfriet *et al.*, 1999]. A main difference to the work described in [Benjamins *et al.*, 1996] and [Fensel *et al.*, 1996] is that in our approach compositionality of the verification is addressed; in the work as referred only domain assumptions are taken into account, and no hierarchical relations between properties are defined.

A future continuation of this work will address both the embedding of verification proofs in a suitable proof system for temporal logic, and the development of tools for verification. At the moment only tools exist for the verification of primitive components; no tools for the verification of composed components exist yet. To support the handwork of verification it would be useful to have tools to assist in the creation of the proof. This could be done by formalising the proofs of a verification process in a suitable proof system.

Frances M.T. Brazier, Frank Cornelissen, Catholijn M. Jonker, Bianca Polak and Jan Treur
Vrije Universiteit Amsterdam, The Netherlands.

Rune Gustavsson and Olle Lindeberg
University of Karlskrona/Ronneby (HK/R), Sweden.

BIBLIOGRAPHY

[Abadi and Lamport, 1993] M. Abadi and L. Lamport. Composing Specifications. *ACM Transactions on Programming Languages and Systems*, **15**(1), 73–132, 1993.

[Akkermans *et al.*, 1996] H. Akkermans, F. Ygge and R. Gustavsson. HOMEBOTS: Intelligent Decentralized Services for Energy Management. In: *Proceedings of the Fourth International Symposium on the Management of Industrial and Corporate Knowledge, ISMICK'96*, 1996.

[Benjamins *et al.*, 1996] R. Benjamins, D. Fensel and R. Straatman. Assumptions of problem-solving methods and their role in knowledge engineering. In: W. Wahlster (ed.), *Proceedings of the 12th European Conference on AI, ECAI'96*, John Wiley and Sons, 408–412, 1996.

[Brazier *et al.*, 1998] F.M.T. Brazier, F. Cornelissen, R. Gustavsson, C.M. Jonker, O. Lindeberg, B. Polak and J. Treur. Agents Negotiating for Load Balancing of Electricity

Use. In: M.P. Papazoglou, M. Takizawa, B. Krämer and S. Chanson (eds.), *Proceedings of the 18th International Conference on Distributed Computing Systems, ICDCS'98*, IEEE Computer Society Press, 622–629, 1998.

[Brazier *et al.*, 1997] F.M.T. Brazier, B. Dunin-Keplicz, N.R. Jennings and J. Treur. Formal specification of Multi-Agent Systems: a real-world case. In: V. Lesser (ed.), Proc tasks. In: B.R. Gaines and M.A. Musen (eds.), *Proceedings of the First International Conference on Multi-Agent Systems, ICMAS'95*, MIT Press, Cambridge, MA, 25–32, 1995. Extended version in: *International Journal of Cooperative Information Systems*, M. Huhns and M. Singh, (eds.), special issue on *Formal Methods in Cooperative Information Systems: Multi-Agent Systems*, **6**, 67–94, 1997.

[Brazier *et al.*, 1999] F.M.T. Brazier, J. Treur, N.J.E. Wijngaards and M. Willems. Temporal semantics of compositional task models and problem solving methods. *Data and Knowledge Engineering*, **29**(1), 17–42, 1999.

[Engelfriet *et al.*, 1999] J. Engelfriet, C.M. Jonker and J. Treur. Compositional Verification of Multi-Agent Systems in Temporal Multi-Epistemic Logic. In: J.P. Müeller, M.P. Singh and A.S. Rao (eds.), *Proceedings of the Fifth International Workshop on Agent Theories, Architectures and Languages, ATAL'98*, pp. 177–194. Lecture Notes in AI, Vol. 1555, Springer Verlag, 1999.

[Fensel *et al.*, 1996] D. Fensel, A. Schonegge, R. Groenboom and B. Wielinga. Specification and verification of knowledge-based systems. In: B.R. Gaines and M.A. Musen (eds.), *Proceedings of the 10th Banff Knowledge Acquisition for Knowledge-based Systems workshop, KAW'96*, Calgary: SRDG Publications, Department of Computer Science, University of Calgary, 4/1–4/20, 1996.

[Fisher and Wooldridge, 1997] M. Fisher and M. Wooldridge. On the Formal Specification and Verification of Multi-Agent Systems. *International Journal of Cooperative Information Systems*, M. Huhns, M. Singh, (eds.), special issue on *Formal Methods in Cooperative Information Systems: Multi-Agent Systems*, **6**, 67–94, 1997.

[Gustavsson, 1997] R. Gustavsson. Requirements on Information Systems as Business Enablers. Invited paper. In *Proceedings of DA/DSM Europe DistribuTECH'97*, PennWell, 1997.

[Hooman, 1997] J. Hooman. Compositional Verification of a Distributed Real-Time Arbitration Protocol. *Real-Time Systems*, **6**, 173–206, 1997.

[Jonker and Treur, 1998] C.M. Jonker and J. Treur. Compositional Verification of Multi-Agent Systems: a Formal Analysis of Pro-activeness and Reactiveness. In: W.P. de Roever, H. Langmaack and A. Pnueli (eds.), *Proceedings of the International Workshop on Compositionality, COMPOS'97*. Lecture Notes in *Computer Science*, **1536**, Springer Verlag, 350–380, 1998.

[Kinny *et al.*, 1996] D. Kinny, M.P. Georgeff and A.S. Rao. A Methodology and Technique for Systems of BDI Agents. In: W. van der Velde, J.W. Perram (eds.), Agents Breaking Away, *Proceedings 7th European Workshop on Modelling Autonomous Agents in a Multi-Agent World, MAAMAW'96*, Lecture Notes in *AI*, **1038**, Springer Verlag, 56–71, 1996.

[Rosenschein and Zlotkin, 1994] J.S. Rosenschein and G. Zlotkin. *Rules of Encounter: Designing Conventions for Automated Negotiation among Computers*, MIT Press, 1994.

[Rosenschein and Zlotkin, 1994] J.S. Rosenschein and G. Zlotkin. Designing Conventions for Automated Negotiation. In: *AI Magazine*, **15**(3), 29–46, 1994.

INDEX